METAL/CYANIDE-CONTAINING WASTES

METAL/CYANIDE CONTAINING WASTES

Treatment Technologies

by

S.A.K. Palmer, M.A. Breton, T.J. Nunno, D.M. Sullivan, N.F. Surprenant

Alliance Technologies Corporation
Bedford, Massachusetts

NOYES DATA CORPORATION
Park Ridge, New Jersey, U.S.A.

Library of Congress Catalog Card Number: 88-17878
ISBN: 0-8155-1179-5
ISSN: 0090-516X
Printed in the United States

Published in the United States of America by
Noyes Data Corporation
Mill Road, Park Ridge, New Jersey 07656

10 9 8 7 6 5 4 3 2 1

Library of Congress Cataloging-in-Publication Data

Metal/cyanide-containing wastes.

(Pollution technology review, ISSN 0090-516X ; no. 158)
Includes bibliographies and index.
1. Metal wastes. 2. Cyanide wastes. 3. Factory and trade waste. I. Palmer, S.A.K. II. Title.
III. Series.
TD898.8.M47M47 1988 628.5'4 88-17878
ISBN 0-8155-1179-5

Foreword

Treatment options for metal- and/or cyanide-containing wastes are discussed in this book. These options include waste minimization, recycling, and treatment of waste streams. Emphasis has been placed on collection and interpretation of data for proven technologies. Other potentially viable technologies are also covered. Each technology is described in terms of actual performance in removing constituents of concern, their associated process residuals and emissions, and restrictive waste characteristics which impact their ability to treat effectively the metal/cyanide wastes under consideration. Performance data, costs, and capacity data are provided so that the reader may assess the applicability of a particular technology to a specific waste stream.

The Resource Conservation and Recovery Act (RCRA), as amended by the Hazardous and Solid Waste Amendments of 1984 (HSWA), prohibits the continued placement of RCRA-regulated hazardous wastes in or on the land, including placement in landfills, land treatment areas, waste piles, and surface impoundments (with certain exceptions for surface impoundments used for the treatment of hazardous wastes). The amendments call for banning the land disposal of hazardous wastes which have "free" liquid fractions containing metals and/or cyanides in excess of certain specified concentrations. Restrictions for disposal of hazardous wastes with metal/cyanide concentrations follow varying timetables depending on the presence and concentrations of other hazardous constituents.

The book provides a review of regulatory background, a summary of the current hazardous waste management data base regarding waste sources and characteristics as well as waste quantities, existing management practices, and EPA estimates of available waste treatment capacity. This is followed by information concerning waste minimization techniques and an evaluation of a wide range of treatment/recovery processes for metal- and cyanide-containing wastes.

These technologies are examined in terms of identifying process design and operating factors and waste characteristics which affect treatment of metal/cyanide wastes. Cost data presented will assist the user in evaluating and selecting options. Approaches to the selection of treatment and recovery options are also reviewed.

The information in the book is from *Technical Resource Document: Treatment Technologies for Metal/Cyanide-Containing Wastes,* prepared by S.A.K. Palmer, M.A. Breton, T.J. Nunno, D.M. Sullivan, and N.F. Surprenant of Alliance Technologies Corporation for the U.S. Environmental Protection Agency, December 1987.

The table of contents is organized in such a way as to serve as a subject index and provides easy access to the information contained in the book.

Advanced composition and production methods developed by Noyes Data Corporation are employed to bring this durably bound book to you in a minimum of time. Special techniques are used to close the gap between "manuscript" and "completed book." In order to keep the price of the book to a reasonable level, it has been partially reproduced by photo-offset directly from the original report and the cost saving passed on to the reader. Due to this method of publishing, certain portions of the book may be less legible than desired.

ACKNOWLEDGMENTS

The authors would like to thank Robert C. Thurnau, the Hazardous Waste Engineering Research Laboratory Work Assignment Manager, whose assistance and support was utilized throughout the program. The authors also extend thanks to other members of the HWERL staff for their assistance and to the many industrial representatives who provided design, operating, and performance data for the waste treatment technologies. Special thanks are also due to Mark Arienti, Michael Kravett, and Andrew Shayer of Alliance who participated in the preparation of this document.

NOTICE

The materials in this book were prepared as accounts of work sponsored by the U.S. Environmental Protection Agency. It has been subject to the Agency's peer and administrative review and has been approved for publication. On this basis the Publisher assumes no responsibility nor liability for errors or any consequences arising from the use of the information contained herein.

Mention of trade names or commercial products does not constitute endorsement or recommendation for use by the Agency or the Publisher. Final determination of the suitability of any information or product for use contemplated by any user, and the manner of that use, is the sole responsibility of the user. The book is intended for informational purposes only. The reader is warned that caution must always be exercised when dealing with hazardous chemicals, hazardous wastes such as metal/cyanide-containing wastes, or hazardous processes; and expert advice should be obtained before implementation is considered.

Contents and Subject Index

1. Introduction

This Technical Resource Document provides information that can be used by environmental regulatory agencies and others as a source of technical information for waste management options for hazardous liquid wastes containing heavy metals and/or cyanide compounds. These options include waste minimization, recycling, and treatment of waste streams. Emphasis has been placed on the collection and interpretation of performance data for proven technologies. These include:

Metals:

- precipitation
- coagulation/flocculation
- chemical reduction
- membrane separation technologies
- activated carbon adsorption
- ion exchange
- electrolytic recovery
- thermal recovery

Cyanides:

- alkaline chlorination
- ozonation
- biological treatment
- thermal destruction

These, and other potentially viable technologies, are described in terms of their actual performance in removing constituents of concern, their associated process residuals and emissions, and those restrictive waste characteristics which impact their ability to effectively treat the metal/cyanide wastes under consideration. Although emphasis is placed on performance data, cost and capacity data are also provided to assist the user of this document in assessing the applicability of technologies to specific wastes streams. References are cited throughout to identify additional sources of background information for the user.

This document provides a review of regulatory background (Section 2), a summary of the current hazardous waste management data base regarding waste sources and characteristics (Section 3) as well as waste quantities, existing management practices, and EPA estimates of available waste treatment capacity (Section 4). This is followed by information concerning waste minimization techniques (Section 5) and an evaluation of a wide range of treatment/recovery processes for metal (Sections 6 through 12) and cyanide (Sections 13 through 15) containing wastes. In order of their presentation, these include:

Metals Treatment/Recovery Technologies:

6 Membrane Separation Processes
7 Extraction
8 Adsorption
9 Electrolytic Processes
10 Chemical Treatment/Removal Processes
11 Biological Treatment/Removal
12 Thermal Treatment/Recovery of Metal Wastes

Cyanide Treatment/Destruction Technologies

13 Physical Removal Processes
14 Chemical Destruction
15 Miscellaneous Destruction Technologies

These technologies are examined with emphasis placed on identifying process design and operating factors and waste characteristics which affect treatment of metal/cyanide wastes. Cost data are also presented to assist the user in evaluating and selecting options. Approaches to the selection of treatment and recovery options are reviewed in the final section of this document (Section 16).

2. Background

Section 3004 of the Resource Conservation and Recovery Act (RCRA), as amended by the Hazardous and Solid Waste Amendments of 1984 (HSWA), prohibits the continued placement of RCRA-regulated hazardous wastes in or on the land, including placement in landfills, land treatment areas, waste piles, and surface impoundments (with certain exceptions for surface impoundments used for the treatment of hazardous wastes). The amendments specify dates by which these prohibitions are to take effect for specific hazardous wastes. After the effective date of a prohibition, wastes may only be land disposed if: (1)they comply with treatment standards promulgated by the Agency that minimize short-term and long-term threats arising from land disposal or (2)the Agency has approved a site-specific petition demonstrating, to a reasonable degree of certainty, that there will be no migration from the disposal unit for as long as the waste remains hazardous. In addition, the statute authorizes the Agency to extend the effective dates of prohibitions for up to 2 years nationwide if it is determined that there is insufficient alternative treatment, recovery or disposal capacity.

2.1 LAND DISPOSAL PROHIBITION FOR METAL/CYANIDES

Effective July 8, 1987, the amendments call for banning the land disposal of hazardous wastes which have "free" liquid fractions containing metals and/or cyanides in excess of the concentrations specified in Table 2.1.1.[1] Restrictions for disposal of hazardous wastes with lower metal/cyanide concentrations follow varying timetables depending on the presence and concentrations of other hazardous constituents.

TABLE 2.1.1. SCHEDULING FOR PROMULGATION OF REGULATIONS BANNING LAND DISPOSAL OF SPECIFIED HAZARDOUS WASTES

Waste category	Effective date*
Dioxin containing waste	11/8/86
Solvent containing hazardous wastes numbered F001, F002, F003, F004, F005	11/8/86
California List:	
Liquid hazardous wastes, including free liquids associated with any solid or sludge containing:	
- Free and complex cyanides at $\geq$1,000 mg/L - As $\geq$500 mg/L - Cd $\geq$100 mg/L - Cr^{6+} $\geq$500 mg/L - Pb $\geq$500 mg/L - Hg $\geq$20 mg/L - Ni $\geq$134 mg/L - Se $\geq$100 mg/L - Tl $\geq$130 mg/L	7/8/87
Liquid hazardous wastes with:	
- pH $\leq$2.0 - PCBs $\geq$50 ppm	7/8/87
Hazardous wastes containing halogenated organic compounds in total concentration $\geq$1,000 mg/kg	7/8/87
Other listed hazardous wastes (Sections 261.31 and 261.32), for which a determination of land disposal prohibition must be made:	
- One-third of wastes	8/8/88
- Two-thirds of wastes	6/8/89
- All wastes	5/8/90
Hazardous wastes identified on the basis of characteristics under Section 3001	5/8/90
Hazardous wastes identified or listed after enactment	Within 6 months

*Not including underground injection.

The Paint Filter Test Method (50 FR 18370, April 30, 1985) classifies a waste as containing free liquids if any liquid passes through a paint filter in a period of five minutes.[2] If wastes are determined to contain free liquids, use of the Toxicity Characteristic Leaching Procedure (TCLP) is required to determine if leachate from the waste contains metals or cyanides in excess of the restricted concentrations.[3] However, since determination of these characteristics is to be made at the point of disposal, these wastes may be rendered non-liquid (e.g., via solidification) and, thus, be exempted from the disposal restrictions. Conversely, dilution as a means of attaining sufficiently low concentration levels is expressly prohibited.

The specified metals concentrations apply to both individual constituents as well as the metal portion of any metal-containing compounds. Cyanides are defined as any substance that can be shown to exhibit the resonance structure associated with a carbon-nitrogen triple bond. Although not all such compounds exhibit the same level of toxicity as free cyanides, some complexes are capable of being converted to free cyanides under conditions which may exist in disposal environments. Hence, the EPA has proposed a broad interpretation of cyanides which are subject to the requirements of Section 3004(d)(2).[4] Total cyanide determination is to be made using Method 9010.[4]

At this writing, the EPA has yet to define treatment standards for heavy metal and cyanide containing wastes. In general, treatment standards can be defined in terms of either treatment technology (BDAT) requirements or concentration based thresholds. Although the agency has recognized several effective and readily available treatment technologies (e.g., alkaline chlorination of cyanides, chemical precipitation of metals, solidification), it did not have adequate waste characterization data to determine whether application of these techniques would be sufficient in all cases to result in sufficiently low residual concentrations. Similarly, the agency did not have data to suggest that leachate concentrations below those specified in Table 2.1.1 were necessary to protect human health and the environment. Thus, the EPA proposed to codify the levels stated in Table 2.1.1 and adopt these as the statutory concentration requirements.[4]

2.2 METAL/CYANIDE WASTE CODES AFFECTED BY THE DISPOSAL RESTRICTIONS

The metal/cyanide waste disposal restrictions apply to any liquid containing wastes which exceed the specified concentrations of total metals or cyanides in their leachate. Table 2.2.1 summarizes the EPA hazardous waste codes (40 CFR Part 261) which are expected to exceed the disposal restrictions for certain waste streams. These include streams which posses the characteristics of EP toxicity (D004 through D010), corrosivity (D002), or reactivity (D003), industrial process wastes (Fxxx, Kxxx codes), and off-specification or discarded commercial products, manufacturing intermediates, or spill residues (Uxxx, Pxxx codes).

TABLE 2.2.1. METAL/CYANIDE RCRA WASTES

Code	Waste description	Toxic constituent: Heavy metal	Toxic constituent: Cyanide
D002	Corrosive waste not listed in subpart D	Y	N
D003	Reactive solid waste not listed subpart D	N	Y
D004	Arsenic (5 mg/L)	Y	N
D005	Barium (100 mg/L)	Y	N
D006	Cadmium (1 mg/L)	Y	N
D007	Chromium (5 mg/L)	Y	N
D008	Lead (5 mg/L)	Y	N
D009	Mercury (0.2 mg/L)	Y	N
D010	Selenium (1 mg/L)	Y	N
D011	Silver (5 mg/L)	Y	N
F006	Wastewater sludges, electroplating operations	Y	Y
F007	Spent cyanide solutions, electroplating	Y	Y
F008	Plating sludges (bottoms), electroplating	Y	Y
F009	Spent stripping/cleaning solutions, electroplating	Y	Y
F010	Quench bath, metal heat treating operations	Y	Y
F011	Spent cyanide solutions, metal heat treating operations	Y	Y
F012	Quench wastewaters, metal heat treating operations	Y	Y
F019	Wastewater treatment sludges, aluminum coating	Y	Y
K002	WWT* sludge from pigment production	Y	N
K003	WWT sludge, molybdate orange pigment production	Y	N
K004	WWT sludge from zinc yellow pigment production	Y	N
K005	WWT sludge from chrome green pigment production	Y	N
K006	WWT sludge from chrome oxide green pigment production	Y	N
K007	WWT sludge from iron blue pigment production	Y	Y
K008	Oven residue from Cr oxide green pigment production	Y	N
K011	Bottom stream from acrylonitrile production	N	Y
K013	Column bottom stream, acrylonitrile production	N	Y
K014	Purification column bottoms, acrylonitrile mfg.	N	Y
K021	Antimony catalyst waste, fluoromethanes production	Y	N
K027	Distillation residue from TDI production	N	Y
K031	By-product salts, MSMA and cacodylic acid production	Y	N
K044	WWT sludge from explosives manufacture	Y	N
K046	WWT sludges, lead based initiating compounds	Y	N

(continued)

TABLE 2.2.1 (continued)

Code	Waste description	Toxic constituent: Heavy metal	Toxic constituent: Cyanide
K048	DAF float from the petroleum refining industry	Y	N
K049	Slop oil, emulsion solids from petroleum refining	Y	N
K050	Heat exchanger sludge from petroleum refining	Y	N
K051	API separator sludge from petroleum refining	Y	N
K052	Tank bottoms from petroleum refining	Y	N
K060	Ammonia lime sludge from coking operations	N	Y
K061	Dust/sludge, electric furnace steel production	Y	N
K062	Spent pickle liquor, steel finishing	Y	N
K069	Dust/sludge from secondary lead smelting	Y	N
K071	Brine purification muds, chlorine production	Y	N
K084	Ar WWT sludges, veterinary pharmaceuticals mfg.	Y	N
K086	Ink washes and sludges containing Cr and Pb	Y	N
K087	Decanter tank tar sludges from coking operations	N	Y
K100	Leaching solution from secondary lead smelting	Y	N
K101	Distillation residues (Ar), veterinary pharmaceuticals mfg.	Y	N
K102	Activated carbon residue (Ar), veterinary pharmaceuticals mfg.	Y	N
K106	Wastewater sludge from chlorine production	Y	N
K116	Organic condensate from TDI production	N	Y
P006	Aluminum phosphide	Y	N
P010	Arsenic acid	Y	N
P011	Arsenic (V) oxide	Y	N
P012	Arsenic trioxide	Y	N
P013	Barium cyanide	Y	Y
P015	Beryllium dust	Y	N
P021	Calcium cyanide	N	Y
P027	Propanenitrile, 3-chloro-	N	Y
P029	Copper cyanides	N	Y
P030	Cyanides (soluble cyanide salts), miscellaneous	N	Y
P031	Cyanogen	N	Y
P033	Cyanogen chloride	N	Y
P036	Dichlorophenylarsine	Y	N
P038	Diethylarsine	Y	N
P055	Ferric cyanide	N	Y
P063	Hydrogen cyanide	N	Y
P064	Methyl isocyanate	N	Y
P065	Mercury fulminate	Y	N
P069	2-Methylacetonitrile	N	Y

(continued)

TABLE 2.2.1 (continued)

Code	Waste description	Toxic constituent: Heavy metal	Toxic constituent: Cyanide
P073	Nickel tetracarbonyl	Y	N
P074	Nickel cyanide	Y	Y
P087	Osmium tetroxide	Y	N
P091	Phenyl dichloro arsine	Y	N
P092	Phenylmercuric acetate	Y	N
P098	Potassium cyanide	N	Y
P099	Potassium silver cyanide	Y	Y
P101	Ethyl cyanide	N	Y
P103	Selenourea	Y	N
P104	Silver cyanide	Y	Y
P106	Sodium cyanide	N	Y
P107	Strontium sulfide	Y	N
P110	Tetraethyl lead	Y	N
P113	Thallium(III) oxide	Y	N
P114	Thallium(I) selenite	Y	N
P115	Thallium(I) sulfate	Y	N
P119	Ammonium vanadate	Y	N
P120	Vanadium pentoxide	Y	N
P121	Zinc cyanide	N	Y
P122	Zinc phosphide, at concentrations 10%	Y	N
U003	Acetonitrile (I,T)	N	Y
U009	Acrylonitrile	N	Y
U032	Calcium chromate	Y	N
U136	Cacodylic acid	Y	N
U139	Ferric dextran	Y	N
U144	Lead acetate	Y	N
U145	Lead phosphate	Y	N
U146	Lead subacetate	Y	N
U149	Propanedinitrile	N	Y
U151	Mercury	Y	N
U152	Methacrylonitrile (I,T)	N	Y
U204	Selenium dioxide	Y	N
U205	Selenium disulfide (R,T)	Y	N
U214	Thallium(I) acetate	Y	N
U215	Thallium(I) carbonate	Y	N
U216	Thallium(I) chloride	Y	N
U217	Thallium(I) nitrate	Y	N
U223	Touene diisocyanate (R,T)	N	Y
U246	Bromine cyanide	N	Y
U249	Zinc phosphate, at concentrations 10%	Y	N

Y = Yes; N = No

*WWT = Wastewater Treatment

Source: Reference 5.

REFERENCES

1. Federal Register. Environmental Protection Agency Regulations on Land Disposal Restrictions. 40 CFR Part 268; 51 FR 19305. May 28, 1986.

2. Federal Register. Paint Filter Test. 50 FR 18370. April 30, 1985.

3. Federal Register. Toxicity Characteristic Leaching Procedure (TCLP). 51 FR 21648. June 30, 1986.

4. U.S. EPA Office of Solid Waste. Background Document for California List Wastes to Support 40 CFR Part 268, Land Disposal Restrictions. Proposed Rule: California List, Volumes, Characteristics, and Required Treatment Capacity. LDR 4 - Reference No. 15. U.S. EPA Public Docket, 401 M Street, S.W. Washington, D.C. November 1986.

5. Federal Register. Environmental Protection Agency Regulations for Identifying Hazardous Waste. 40 CFR Part 261. EPA Regulations for Identifying Hazardous Waste. 45 FR 33119. May 29, 1980.

3. Metal/Cyanide Waste Sources and Characteristics

3.1 BACKGROUND

Metals and cyanides are used in a broad range of industrial processes and products; from metal plating to paint additives, ore benefication to photographic films. Only some of these uses result in the generation of RCRA wastes, and only a fraction of these are sufficiently high in metal/cyanide concentration to be affected by the land disposal ban. Quantifying these wastes is a particularly difficult task, primarily due to the inadequacy of current data regarding waste stream characteristics.

Over the last 10 years, the EPA has devoted significant resources to determining the extent and nature of hazardous waste generation. During the late seventies and into the eighties, the EPA conducted studies of waste generation by major industrial categories (e.g., Effluent Guidelines Limitations and similar documents). These formed the basis for the original identification and quantification of hazardous wastes.

The subsequent promulgation of RCRA regulations in 1980 had a profound effect on the generation and management of hazardous wastes, thus rendering earlier data obsolete. The first national estimate of RCRA waste generation was obtained through a compilation of Part A applications. Unfortunately, these data were limited in usefulness because the agency did not have data on waste characteristics (e.g., hazardous constituent concentrations) with which to interpret the reported waste quantities. Recognizing these limitations, the EPA conducted a comprehensive national survey (see Section 4), designed to characterize RCRA wastes generation and management.[1]

Additional surveys were undertaken to supplement this national survey by improving the existing data on waste characteristics. Data collection efforts began with a compilation of information obtained through the earlier Effluent Guidelines studies. This was then updated with new, post-RCRA data to generate the Industries Studies Data Base (ISDB).[2] The ISDB represents a significant improvement over previous characterization efforts, primarily because it provides a precise correspondence between individual waste stream quantities, characteristics, and management practices.

Analyses of the ISDB are being performed by the EPA Office of Solid Waste, Waste Characterization Branch, to support development and implementation of the land disposal restrictions. However, these data are still in the process of being collected and few industries have thus far been adequately characterized. Specifically, data from the Chemical and Allied Products Industries Standard Industrial Classification (SIC) 28 has been compiled and processed and data on petroleum refining, inorganic chemical production, coke making, and wood preserving should be available in 1987.[3]

The ISDB data suggest that the majority of SIC 28 metal wastes are sufficiently dilute to be discharged under NPDES regulations.[4] Although the industries covered in the ISDB account for a majority of total hazardous waste generation, they represent only a small fraction of metal/cyanide wastes that may be affected by the land disposal ban.[5] Since existing data characterizing concentrated metal/cyanide waste generation are incomplete, a comprehensive compilation of waste characteristics was not considered to be appropriate for this report. Instead, a discussion is provided summarizing industrial uses of metals and cyanides (Section 3.2) and types of wastes produced by high volume generators (Section 3.3). Some waste characteristic data are also presented to give an indication of the variability which can be expected in metal/cyanide wastes.

3.2 INDUSTRIAL USES OF METALS AND CYANIDES

This section describes major industrial uses of California List metals and cyanides. Major uses are summarized in Table 3.2.1.[1,6] Except where otherwise indicated, information was taken from the Kirk-Othmer Encyclopedia of Chemical Technology.[7]

TABLE 3.2.1. METAL AND CYANIDE INDUSTRIAL USES

Chemical	Metal Finishing	Pharmaceuticals	Refining	Photographic Chemicals	Pesticides	Agricultural Chemicals	Plastics & Organic Chemicals Manufacture	Pesticides	Pigments	Wood Preservation	Explosives	Paint	Metal Parts Fabrication	Batteries	Printing	Alloys	Ceramics/Glass	Paper Manufacture	Electronics, Electrical Applications	Leather Tanning
Silver	X			X									X	X						
Cyanide	X	X	X	X	X	X	X	X												
Arsenic					X		X			X		X		X		X				
Cadmium	X						X		X	X		X		X						
Chromium VI	X								X	X			X		X	X				X
Lead	X						X		X		X	X	X	X	X	X				
Mercury		X	X				X				X	X							X	
Nickel	X		X		X		X		X					X		X	X	X		
Selenium					X		X		X			X				X	X		X	
Thallium							X		X							X			X	

3.2.1 Cyanides

Hydrogen cyanide is the single largest cyanide compound in use by industry. It is used as a chemical intermediate in the production of methyl methacrylate (60 percent), cyanuric chloride (15 percent), sodium cyanide (10 percent), chelators such as NTA and EDTA (10 percent), and other chemical compounds (5 percent). Sodium cyanide is used for electroplating of zinc, copper, brass, cadmium, and to a lesser extent, gold and silver. It is also used in the following applications: heat treating small metal parts when case hardening is required; cyanidation recovery of gold and silver and froth flotation of metals; as a chemical intermediate in production of dyes, optical brighteners, chelators, pharmaceuticals, and agricultural chemicals.

Miscellaneous compounds produced using HCN include cyanogen, potassium cyanide (used for silver plating), ferrocyanides, acrylates, lactic acid, pharmaceuticals, and specialty chemicals. Acrylonitrile used to be produced with HCN as a raw material but is now produced using Sohio's process or similar technology, generating HCN as a by-product.

Other significant volume cyanide compounds include cyanuric acid, calcium cyanamide, and toluene diisocyanate. Most cyanuric acid is chlorinated to produce various chemical intermediates which are used as ingredients in swimming pool disinfectants, cleansers, and bleaches. Calcium cyanamide is used for steel nitration, desulferization, pesticides, and as a chemical intermediate. Toluene diisocyanate is used in the production of flexible urethane foams, and to a lesser extent, in protective coatings and castable urethane elastomers.

3.2.2 Lead

Of the California List metals, lead is consumed in the largest quantities, with over half of domestic production resulting from secondary lead. Consumption in 1979 was 1.08 million metric tons, nearly two-thirds of which was consumed in the production of storage batteries. Storage batteries represent the largest single waste source from small quantity generators (90 percent of metal/cyanide wastes) and are largely recycled.[8]

Other significant uses include production of tetraethyl and tetramethyl lead anti-knock additives for gasoline (17 percent), pigments (7.7 percent), ammunition (4.9 percent), solder (3.7 percent), and brass, bronze, bearings, casting metal, sheet lead, and other uses. Cadmium and lead salts of long-chain organic acids are used as stabilizers for nearly all plastic materials, except those used for food processing.[9]

3.2.3 Chromium

Sodium dichromate is the primary raw material for the production of other chromium chemicals, both trivalent and hexavalent. These compounds and their uses include chromic acid (41 percent), chrome pigments (27 percent), leather tanning (14 percent), drilling fluids (8 percent), water treatment chemicals (5 percent), and others. Chromic acid is used for decorative plating, primarily over undercoats of copper and nickel for automobiles, plumbing, appliances, and other metal parts. However, relative to functional chrome plating, decorative use has declined in recent years. Functional chrome plating (60% of total chrome plating consumption) is used to build up worn surfaces and improve abrasion resistance on printing plates, machine parts, bearings and other steel part coatings.

Pigment production is the second highest use for sodium dichromate. These are primarily lead chromates including chrome oranges, yellow, and oxide green. These pigments are used in paints, inks, paper and floor coverings and are known for being highly stable, having good hiding power, and providing bright colors.

Chromic acid is also used in the production of chromated copper arsenate preservatives for wood products intended for interior use. Since this compound is water soluble, disposal options may require some form of solidification or encapsulation to ensure minimal leachate generation. Relative to other consumptive uses of chromic acid, this use has been increasing steadily.

Other industrial uses for chromium (e.g., leather tanning) primarily use compounds that contain chromium in its less toxic, trivalent oxidation state. In some cases, these wastes are exempt from RCRA provisions and, in all cases, are not affected by the land disposal ban which pertains solely to hexavalent chromium.

3.2.4 Nickel

The most important industrial use of nickel is as a hydrogenation catalyst, such as Raney nickel. Nickel catalysts are used for methanation, steam-hydrocarbon reforming, petrochemical hydrogenation, and fat hardening. The second largest application is as an electrolyte in nickel electroplating, often as an underplate for decorative chrome. Commonly plated products include automobile bumpers and trim, appliances, wire products, jewelry, and other consumer items. Nickel oxides are widely used in the ceramic industry, as plastic additives, and in the production of organic dyes and pigments.

3.2.5 Cadmium

U.S. consumption of cadmium was approximately 4,226 tons in 1977, over half of which was imported. The principal use of elemental cadmium is as an electroplated coating on fabricated steel and cast iron parts for corrosion protection. It is usually plated from a cyanide bath but can also be applied in powdered form through vacuum deposition and other means. Other significant uses are in pigment manufacture (e.g., cadmium sulfide and sulfoselenide pigments), and as the negative electrode in nickel-cadmium and silver-cadmium batteries. Cadmium pigments are inherently insoluble and often encapsulated in a plastic or glass matrix.[10] Therefore, leachate levels from pigment containing wastes would be minimal, and thus, would not result in restriction of these wastes from land disposal.

Cadmium is used to a lesser extent as heat and light stabilizers in poly vinyl chlorides, as a silver alloy for producing electrical contacts, with Group-VI elements in semiconductor applications and as catalysts in organic polymerization reactions.[11]

3.2.6 Arsenic

The major uses of arsenic which result in the generation of hazardous wastes are in the production of various chemical intermediates (63 percent) and wood preservatives (20 percent). The former include orthoarsenic acid, monosodium methylarsonate, disodium methlyarsonate, cacodylic acid, calcium arsenate, lead arsenate and sodium arsenate. Roughly 90 percent of the arsenic containing wood preservatives use chrome copper arsenate with the remainder accounted for by ammoniacal copper arsenite.[12]

3.2.7 Silver

Estimated consumption of silver in the U.S. was 3,726 tons in 1981. Silver is primarily consumed in the form of silver nitrite and silver halides for photographic materials (41 percent), and silver cyanide complexes, for electronic product contacts and conductors (22 percent). Other uses include production of brazing alloys and solders (7 percent), sterling ware (7 percent), silver oxide batteries (5 percent), jewelry, electroplated esthetic ware, catalysts, coins, and other products. Due to its expense, silver is widely recovered for reuse through processes such as ion exchange and electrolysis.

3.2.8 Mercury

Mercury consumption in the U.S. was approximately 1,853 metric tons in 1978, the bulk of which was imported. The primary use of mercury is in electrical applications (37 percent). These include batteries, fluorescent and high pressure mercury lamps, and wiring and switching devices. The second major use is in the electrolytic preparation of chlorine and caustic soda (23 percent), in which mercury is used as a flowing cathode and is largely recycled. However, mercury use in this application has declined since chlor-alkali production has shifted in favor of more cost-effective processing methods (see Section 5.3.8).[13]

Other uses of mercury include mildew proofing in paints (18 percent), industrial and control instruments (7 percent), and as catalysts in production of vinyl chloride monomer, urethane foam, pharmaceuticals, and other miscellaneous end products. Use in paints and pharmaceuticals has declined sharply due to growing awareness of mercury's toxicity and difficulties in handling and disposal.

3.2.9 Selenium

Fifty percent of selenium is consumed in the decolorization and production of colored glass and in the production of cadmium sulfide selenide pigments which are primarily used in plastics. Another 25 to 30 percent is consumed in xerography and 5 percent in metallurgical applications. The latter includes use in ferrous metals, and copper, lead, and nickel alloys. Selenium decreases porosity, and enhances impact resistance, machinability, and other material properties. The remainder (15-20 percent) consists of miscellaneous uses including chromium plating for corrosion protection, and production of catalysts (e.g., selenium dioxide in the synthesis of organic chemical and drug products), feed additives, anti-dandruff preparations and others.

Selenium exhibits fairly unique properties and thus has been difficult to replace in many uses. As semiconductors, selenium and its compounds posses asymmetric conduction. They also exhibit increasing electrical conductivity upon exposure to light and thus find use in photovoltaic cells and rectifiers. However, silicon and germanium have displaced selenium usage in rectifier applications. Attempts have also been reported to replace the selenium used in xerography with organoselenium compounds in a polymeric matrix and other semiconductor materials.

3.2.10 Thallium

Thallium is obtained as a flue-gas by-product during the production of sulfuric acid. Annual production is very small; roughly 5 to 15 metric tons per year. Thallium and mercury alloys have been substituted for mercury in equipment (e.g., switches and seals) used in the polar region and the stratosphere. Silver-thallium alloys are used in contact points and an alloy of thallium, tin and lead is used in the production of anodes. Thallium (III) derivatives have been used as oxidants in organic synthesis in research laboratories.

Overall, thallium has not enjoyed wide use in commercial applications because of its toxic nature and resulting environmental problems associated with its disposal.

3.3 WASTE SOURCES AND CHARACTERISTICS

Many of the industrial uses of metals and cyanides identified above result in the generation of off-spec products, spills, processing wastes, and hazardous by-products. Metal/cyanide waste sources and characteristics have been documented for individual industries (e.g., EPA's Effluent Limitations and Guidelines series), by many state environmental agencies (e.g., Illinois, California, Massachusetts, etc.), and on a national basis (e.g., 1981 RIA National Survey[1] and the 1983 Small Quantity Generator Survey[8]). These and other data sources have been used in this section to provide an overview of metal/cyanide waste sources and characteristics.

Table 3.3.1 identifies key industries that reported the generation of metal and cyanide RCRA wastes in the RIA survey. These are listed in decreasing order of number of facilities generating metal/cyanides. Unfortunately, the data did not permit a similar ranking by waste quantity. Nevertheless, it can be seen that metal and cyanide wastes are generated in large part by metal fabrication facilities, which perform forming, plating, polishing, and coating operations (SIC 33 and 34), and manufacturers of metallic parts and equipment (SIC 35 through 37). These facilities also have a high tendency to generate concentrated solutions (e.g., spent baths) which are likely to be affected by the land disposal ban. Companies within the Chemicals and Allied Products industries (SIC 28) generate spent metal catalysts and wastewaters containing metals and organometallics.

Small quantity generators (SQG), although numerous, generate only a small fraction of the nation's hazardous waste (see Section 4.1). However, these facilities may suffer significant adverse impacts as a result of the land disposal ban because they are generally less capable of implementing alternative waste management techniques. Table 3.3.2 summarizes data on SQG waste quantities and number of facilities generating metal/cyanide wastes for high volume facility categories.[8] As shown, vehicle maintenance facilities generate metal wastes (e.g., lead-acid batteries) in the largest numbers and volume. Other large categories of waste generators include printing and photography facilities, which primarily generate silver wastes, and metal manufacturing facilities which generate metal plating and cyanide streams.

TABLE 3.3.1. NUMBER OF METAL/CYANIDE WASTE GENERATORS BY SIC CODE

# Facilities (Dun's 1983)	SIC Description	SIC	Weighted # Facilities CN	Metals	Metals + CN
4,287	Plating and Polishing	3471	750	263	1,013
2,145	Paints and Allied Products	2851	20	212	232
2,902	Metal Coating and Allied Products	3479	46	149	195
4,151	Motor Vehicle Parts and Accessories	3714	62	107	169
2,183	Industrial Inorganic Chemicals	2819	3	131	134
876	Metals, Nonferrous, Secondary	3341	37	93	130
55,380	Fabricated Metal Products	3400	51	73	124
393	National Security	9711	16	84	100
966	Motors and Generators	3721	40	61	100
32,867	Misc. Manufacturing Industries	3900	23	68	91
384	Metal,Nonferrous, Rolling, Drawing	3356	30	60	90
609	Printing Ink	2893	0	89	89
1,229	Blast Furnaces, Steel Mills	3312	11	78	89
1,229	Foundries, Gray Iron	3321	0	82	82
2,614	Electric Services	4911	0	81	81
1,160	Industrial Organic Chemicals	2869	5	80	85
1,529	Plastics Material	2821	7	76	83
4,656	Radio & TV Communication Equipment	3662	30	49	79
5,392	Electronic Components	3679	35	42	77
1,040	Motor Vehicle Bodies	3711	41	34	75
3,432	Machine Tool Accessories	3545	40	0	40

Source: Reference No. 5.

TABLE 3.3.2. METAL/CYANIDE WASTE GENERATION FOR SMALL QUANTITY GENERATORS

	Chemical manufacturing		Wood preserving		Formulators		Photography		Vehicle maintenance		Metal manufacturing		Motor freight terminals		Printing industry		Other manufacturing	
	# of Gen.	Waste Qnt.	# of Gen.	Waste Qnt.	# of Gen.	Waste Qnt.	# of Gen.	Waste Qnt.	# of Gen.	Waste Qnt.	# of Gen.	Waste Qnt.	# of Gen.	Waste Qnt.	# of Gen.	Waste Qnt.	# of Gen.	Waste Qnt.
Cyanide wastes	3	<1			2	2					1,208	1,572			51	63		
Heavy metal dust	5	61			30	85												
Heavy metal solutions					30	52												
Heavy metal waste Materials	21	121															93	411
Ink sludges containing chromium or lead					32	71									51	56		
Paint wastes containing heavy metals									156	7								
Photographic wastes							2,302	7,885							2,578	6,071		
Spent plating wastes											1,034	4,528						
Arsenic wastes																	19	104
Solutions or sludges containing silver							2,496	7,749										
Used lead-acid batteries									77,608	302,576					23	53		
Waste inks containing flammable solvents or heavy metals															705	1,337		
Wastewater containing wood preservatives			107	688														
Wastewater sludges containing heavy metals											790	2,216						

Source: Adapted from Reference No. 8.

Table 3.3.3 shows the number of facilities and quantity of waste generated by small and very small waste quantity generators. The latter generate less than 100 kg/month and are therefore exempt from RCRA regulations and the land disposal ban provisions.

Waste characteristics, including the relative quantities of metal/cyanide constituents in land disposed RCRA wastes, are difficult to determine. However, available data provide a rough ranking and further suggest that the distribution between constituents differs between all RCRA wastes and those wastes which exceed restricted land disposal concentrations.

The RIA data did not specify metal contaminants or concentrations in Fxxx and Kxxx code wastes making it difficult to rank constituents in order of volume. However, an approximate ranking, in decreasing volume of combined waste, appears to be as follows: chromium, cyanides, lead, cadmium, and nickel, with lesser amounts of mercury and arsenic, and small quantities of selenium and thallium.[6]

This ranking is in general agreement with industrial usage patterns (Section 3.2), publically owned treatment works (POTWs) discharges, and other hazardous materials surveys. For example, a waste generation study for New England showed F006 (contains cyanide, chromium, nickel, cadmium) as the metal/cyanide waste generated in the highest volume followed by F007 (cyanide electroplating baths), D007 (chromium), D008 (lead), and D006 (cadmium).[14] Similarly, data describing raw waste loadings to the nations POTWs, in decreasing order, are estimated to be: cyanide (15 Mkg/yr), chromium (5 Mkg/yr), lead (4 Mkg/yr), nickel (3 Mkg/yr), cadmium (0.6 Mkg/yr), silver (0.4 Mkg/yr), and arsenic (0.1 Mkg/yr). Mercury and selenium are only discharged in small quantities (e.g., 20,000 kg/yr).[4]

Only one study, performed by the State of California, was identified which specifically quantified wastes that exceeded the land disposal concentration limits for metals and cyanides.[15,16] In decreasing order of waste volume, the constituents were ranked as follows: chromium (61 percent), nickel (26 percent), arsenic (23 percent), lead (11 percent), cyanide (6 percent), cadmium (3 percent), and small quantities of mercury. No thallium or selenium wastes were identified. Nearly 50 percent of the heavy metal wastes in the state had a pH of less than 2.0 and, therefore, were also categorized as RCRA corrosive wastes. However, this study may be limited somewhat since it only included wastes which were land disposed offsite.

TABLE 3.3.3. NUMBER OF SMALL QUANTITY GENERATORS AND WASTE QUANTITY GENERATED BY WASTE STREAM

	VSQGs: Generators of <100 kg of waste/month		SQGs: Generators of ≥100 kg to 1,000 kg of waste/month		Total	
	No. of generators	Waste quantity (MT/yr)	No. of generators	Waste quantity (MT/yr)	No. of generators	Waste quantity (MT/yr)
Arsenic wastes	21	7	19	104	40	111
Cyanide wastes	587	17	1,384	2,129	1,972	2,146
Heavy metal dust	48	10	40	163	88	173
Heavy metal solutions	15	6	30	52	45	58
Heavy metal waste materials	121	31	117	537	238	568
Ink sludges containing chromium or lead	1,093	90	83	127	1,176	217
Mercury wastes	19	1	0	0	19	1
Paint wastes containing heavy metals	381	12	156	7	537	19
Photographic wastes	21,287	4,408	4,949	14,023	26,236	18,431
Spent plating wastes	3,960	493	1,422	5,275	5,382	5,768
Solutions or sludges containing silver	4,482	938	2,648	7,981	7,130	8,919
Used lead-acid batteries	119,747	64,903	77,880	304,194	197,627	369,097
Waste inks containing flammable solvents or heavy metals	3,642	263	718	1,359	4,360	1,622
Wastewater containing wood preservatives	88	26	108	693	196	719
Wastewater sludges containing heavy metals	894	188	790	2,216	1,684	2,404
Total:		71,392		338,860		410,252

Source: Reference No. 8

The high proportion of arsenic waste in the California study resulted from a relatively high tendency for wood preserving facilities to ship these wastes offsite for treatment. Also, metal containing wood preservatives (i.e., chrome copper arsenate and ammoniacal copper arsenite) are soluble and, therefore, would be expected to result in high leachate concentrations.[12] Although the RCRA surveys identified above indicate that cyanides are disposed in high volume relative to heavy metals, the California study suggests that these rarely exceeded the 1,000 mg/L concentration limit which would prohibit them from being land disposed. Concentrated cyanide wastes tended to be spent processing baths since rinses and combined wastes typically had concentrations under 100 mg/L.[16]

Due to a lack of detailed data on metal/cyanide waste stream compositions, minimal effort was made to characterize these wastes beyond providing a relative ranking of constituents. The ISDB, which represents the most comprehensive and accurate source of RCRA waste characteristics, did not cover most of the key industries which generate these wastes. Other data sources are dated and/or restricted to specific industries or geographical locations. Compilation of these diverse data sources was both beyond the scope of this report and likely to result in an inaccurate representation of currently disposed wastes. Nevertheless, some composition data has been provided in the following discussion for metal/cyanide wastes which exceed disposal ban concentrations. It should be remembered that these data are not necessarily representative of their waste codes due to the wide variability in industrial wastes which is found in practice.

A summary of RCRA waste compositions, compiled in an earlier EPA study, is provided in Table 3.3.4.[5] The remainder of this section is devoted to a discussion of metal/cyanide waste generation and characteristics for several major industrial sectors. Unless otherwise indicated, this information was taken from a recent study performed for the EPA Office of Solid Waste to identify waste minimization practices employed by industry.[13]

3.3.1 Acrylonitrile Production

Acrylonitrile is produced by the Sohio ammoxidation process through the exothermic reaction of propylene, ammonia and air in a fluidized-bed catalytic

TABLE 3.3.4. METAL/CYANIDE WASTE STREAM COMPOSITION DATA (in ppm)

List of waste streams	CN	As	Cd	Cr	Pb	Hg	Ni	Se
F006	5.8[1] 131[2] 0.1[3]	6.24	1,320	39,730 25.4[5]	408	0.32	14,760	23
F007	14,547	--	--	--	--	--	--	--
F009	350,000	--	21.6	525.9	25.39	--	2,954	--
F010	9.23%	--	--	--	--	--	--	--
F011	12% 35%	-- --	-- --	-- --	-- --	-- --	-- --	-- --
F012	26,800	--	--	--	--	--	--	--
F019	597 539[3]	<0.5	0.62	11,361 2.2[5] 11,361[6]	<17	0.34	1,275	0.5
K002-K008	65	40	2,300	22,000	28,000	10	350	--
K011	7%[4]	--	--	--	--	--	--	--
K013	225[4]	--	--	--	--	--	--	--
K011,K013 K014 solids	25%[4] 5%	--	--	Trace	Trace	--	Trace	--
K031	--	<1%	4.6	26	--	--	--	--
K046	0.41	--	--	--	0.45%	--	0.1%	--
K048	<50	33	<50	3,003	1,250	0.27	4.9	4.0
K049	26	7.4	1.0	7,764	3,960	0.59	50	4.0
K050	1.7	10.6	3.4	4,185	1,200	1.9	116	27.2
K051	0.29	6.2	1.7	10,800	6,200	1.0	260	140
K052	0.001	2,940	6.3	973	5,800	5.0	314	6.95

(continued)

TABLE 3.3.4 (continued)

List of waste streams	CN	As	Cd	Cr	Pb	Hg	Ni	Se
K060	7,400	1,086	--	--	--	--	--	--
K061	--	750	17,900	2,190[1] 4,600[5] 20,600[6]	38,100	--	5,900	2,850
K062	4.6	5.8	0.43	12,400[1] 19[5] 6,690[6]	1,550	<0.005	10,450	<0.2
K069	--	--	900	150	12%	--	--	--
K071	--	17.0	6.4	10.0	81.0	1,000	153	--
K084	--	7% <0.35% <20%	--	--	--	--	--	--
K086	--	<0.05	4.3	200	900	<0.01	2.4	0.05
K101	--	15%	--	--	--	--	--	--
K102	--	14% 2.8%	--	--	--	--	--	--
K106	--	--	--	--	--	15%	--	--
K071 & K106	--	--	--	--	0.7%	--	--	--

Source: Adapted from Reference No. 5. See Reference for RCRA code descriptions or refer to Table 2.2.1.

[1]Total

[2]Complexed

[3]Free

[4]As HCN

[5]As Cr(VI)

[6]As Cr(III)

reactor at approximately 750 to 950 °F and 5 to 30 psig. The reaction product is quenched generating an absorber wastewater effluent and an acrylonitrile/acetonitrile product stream. These are separated and purified, generating product streams, heavy ends (K013, K014) and crude HCN. The HCN stream is typically incinerated. Acrylonitrile bottoms are usually combined with the aqueous effluent from the quench/adsorption section and then steam stripped. The resulting aqueous effluent (K011) is combined with the acetonitrile purification bottoms and typically disposed via underground injection. As shown previously in Table 3.3.1, these RCRA wastes may exceed land disposal limits for total cyanide concentration.[5]

3.3.2 Metal Surface Treatment

The metal surface treatment industry is dominated by small job shops and captive operations in larger manufacturing industries. Chemical surface treatment results in significant quantities of hazardous wastes from a wide variety of unit operations including electroplating (see Section 3.3.3), electroless plating, conversion coating, anodizing, cleaning, etching, and milling. All these chemical surface treatments essentially involve dipping metal objects in a bath to achieve the desired surface modifications. Primary sources of hazardous waste include: 1) wastewater treatment sludges from rinse bath treatment; 2) filter residues from in-line filtration of process baths (F010, F012); and 3) spent baths (F011). The latter, together with sludges and solids (e.g., contaminated bath filters) are frequently sent offsite for recycle or disposal.

Commonly used toxic metals and metal-containing reagents include chromium, cadmium, nickel, silver, vanadium, metal phosphates, sodium dichromate, sodium cyanate, and chromic acid. Common acids found in metal surface treatment wastes include hydrochloric, sulfuric, phosphoric, hydrofluoric, and boric acid. The primary alkalis in use are sodium hydroxide and sodium carbonate. Metals are also frequently discharged in streams that contain complexing agents including cyanides, amines, ammonia and its derivatives, EDTA, NTA, citrate, tartrate, oxalate, and gluconate. Finally, combined waste streams will also frequently contain smaller quantities of organics such as oils and grease, solvents (e.g., 1,1,1-trichloroethane, toluene, xylene), and bath additives (e.g., phenols, aldehydes, ketones, alcohols). A summary of metal finishing bath constituents is provided in Table 3.3.5 for baths which contain California List metals and/or cyanides.[17]

Nickel is widely applied to metal surfaces via electroless plating. These baths contain complexing agents to maintain ions in solution (e.g., lactic acid, dicarboxylates) as well as stabilizers, reducing agents and other additives. Highly complexed baths complicate subsequent metals precipitation in the plant's treatment unit and are therefore often segregated for batch treatment. Filtration is used to extend bath life and minimize dumping frequency but the eventual build-up of impurities and equipment fouling (plate-out of the metal ions) necessitate periodic dumping.

Chemical conversion coatings are applied to metal surfaces to inhibit corrosion or to improve surface adhesion properties. Chromic acid rinses are used as a rust-inhibiting treatment for phosphatized substrates. Chromate coatings are applied directly on zinc, cadmium, aluminum, magnesium, copper, brass, bronze, and silver to minimize rust and promote paint adhesion.[5] Sodium dichromate is used for passivation of stainless steels. Chromic acid is also used as a chemical etchant.

Cyanide baths are used to provide surface hardening of carbon and alloy steels. Baths normally contain from 30 to 97 percent sodium cyanide and are replenished to make up for losses through drag-out. Quench solutions (e.g., water or oil) following the bath are potentially hazardous as a result of the drag-out and are thus treated for cyanide destruction (e.g., chlorination for aqueous solutions, incineration for oils).

Depending on the basis material and previous processing steps, aqueous and solvent based metal parts cleaning wastes can contain significant concentrations of California List metals. Example wastes include those resulting from the acid pickling of specialty steels containing chromium and nickel.[1] Spent solvent-based degreasers, cleaners, developers, and paints can exceed metal concentration limits and become further concentrated during recovery operations. Concentrations of nickel, lead, and chromium in the 10,000 mg/L range have been widely reported for raw wastes.[18]

TABLE 3.3.5. METAL FINISHING BATHS CONTAINING CALIFORNIA LIST METALS AND/OR CYANIDES

Process	Bath type	Constituents
Electroplating	Brass & Bronze	Copper cyanide Zinc cyanide Sodium cyanide Sodium carbonate Ammonia Rochelle salt
	Cadmium Cyanide	Cadmium cyanide Cadmium oxide Sodium cyanide Sodium hydroxide
	Cadmium Fluoborate	Cadmium fluoborate Fluoboric acid Boric acid Ammonium fluoborate Licorice
	Copper Cyanide	Copper cyanide Sodium cyanide Sodium carbonate Sodium hydroxide Rochelle salt
	Fluoride Modified Copper Cyanide	Copper cyanide Potassium cyanide Potassium fluoride
	Chromium	Chromic acid Sulfuric acid
	Chromium with Fluoride Catalyst	Chromic acid Sulfate Fluoride
	Gold Cyanide	Metallic gold Potassium cyanide Sodium phosphate

(continued)

TABLE 3.3.5 (continued)

Process	Bath type	Constituents
Electroplating	Lead Fluoborate	Lead fluoborate Fluoboric acid Boric acid Gelatin or glue Hydroquinone
	Lead-Tin	Lead fluoborate Tin fluoborate Boric acid Fluoboric acid Glue Hydroquinone
	Nickel (Watts)	Nickel sulfate Nickel chloride Nickel fluoborate Boric acid Nickel sulfate Nickel chloride Nickel sulfamate Boric acid Phsophoric acid Phosphorous acid "Stress-reducing agents
	Nickel-Acid Fluoride	Hydrofluoric acid Nickel carbonate Citric acid Sodium lauryl sulfate (wetting agent)
	Black Nickel	Nickel ammonium sulfate Nickel sulfate Zinc sulfate Ammonium sulfate Sodium thioscyanate
	Silver	Silver cyanide Potassium cyanide or Sodium cyanide Potassium carbonate or Sodium carbonate Potassium hydroxide Potassium nitrate Carbon disulfide

(continued)

TABLE 3.3.5 (continued)

Process	Bath type	Constituents
Electroplating	Tin-Copper Alloy	Copper cyanide Potassium stannate Potassium cyanide Potassium hydroxide Rochelle salt
	Tin-Nickel Alloy	Stannous chloride Nickel chloride Ammonium fluoride Ammonium bifluoride Sodium fluoride Hydrochloric acid
	Tin-Zinc Alloy	Potassium stannate Zinc cyanide Potassium cyanide Potassium hydroxide
	Zinc Cyanide	Zinc oxide Sodium cyanide Sodium hydroxide Zinc cyanide
	Silver	Silver cyanide Sodium cyanide Sodium hydroxide Dimethylamine borane Thiourea
	Nickel	Nickel chloride Sodium glycollate Sodium hypophosphite or Nickel carbonate Hydrofluoric acid Citric acid Ammonium acid fluoride Sodium hypophosphate Ammonium hydroxide

(continued)

TABLE 3.3.5 (continued)

Process	Bath type	Constituents
Electroless Plating	Arsenic	Zinc sulfate Arsenic trioxide Sodium citrate Sodium cyanide Sodium hydroxide Ammonium hydroxide Sodium hypophosphite
	Chromium (acidic)	Chromic bromide Chromic chloride Potassium oxalate Sodium acetate Sodium hypophosphite
	Chromium (alkaline)	Chromic bromide Chromic iodide Sodium oxalate Sodium citrate Sodium hypophosphite
	Cobalt-Nickel	Cobalt chloride Nickel chloride Rochelle salt Ammonium chloride Sodium hypophosphite
	Gold	Potassium gold cyanide Ammonium chloride Sodium citrate Sodium hypophosphite
	Gold over Cu, Ni, Kovar	Potassium gold cyanide Citric acid Nonopotassium acid phthalate Tungstic acid Sodium hdyroxide N,N diethylglycine (Na salt)

(continued)

TABLE 3.3.5 (continued)

Process	Bath type	Constituents
Immersion Plating	Gold on Copper Alloys	Potassium gold cyanide Sodium cyanide Sodium carbonate
	Gold on Iron & Steel	Denatured alcohol Gold chloride
	Lead on Copper Alloys and on Zinc	Lead monoxide Sodium cyanide Sodium hydroxide
	Lead on Steel	Lead nitrate Sodium cyanide Sodium hyrdoxide
	Nickel on Aluminum	Nickel sulfate Ammonium chloride
	Nickel on Copper Alloys	Nickel sulfate Nickel ammonium sulfate Sodium thiosulfate
	Nickel on Steel	Nickel chloride Boric acid
	Nickel on Zinc	Nickel sulfate Sodium chloride Sodium carbonate
	Arsenic on Aluminum	White arsenic Sodium carbonate
	Arsenic on Copper Alloys	White arsenic Ferric chloride Muriatic acid
	Arsenic on Steel	White arsenic Muriatic acid
	Brass on Alluminum	Zinc oxide Sodium hydroxide Copper cyanide Sodium cyanide Lead carbonate

(continued)

TABLE 3.3.5 (continued)

Process	Bath type	Constituents
Immersion Plating	Cadmium on Aluminum	Cadmium sulfate Hydrofluoric acid
	Cadmium on Copper Alloys	Cadmium oxide Sodium hydroxide
	Copper on Aluminum	Copper sulfate Ammonia Potassium cyanide
	Silver on Copper Alloys	Silver cyanide Sodium cyanide
	Silver on Zinc	Silver cyanide Potassium
	Tin on Copper Alloys	Tin chloride Sodium cyanide Sodium hydroxide
Chemical Etching	Ammonium persulfate solutions:	Ammonium persulfate Mercuric chloride Sulfuric acid Ammonium chloride Sodium chloride Copper Base material
	Chromic-sulfuric acid solutions:	Chromic acid Sodium sulfate Sulfuric acid Copper Base material
Chemical Milling	For various metals:	Nitric acid Chromic acid Hydrochloric acid Base metal
Bright Dip	For Zinc and Cadmium:	Chromium acid Sulfuric acid
	For Silver:	Sodium cyanide Hydrogen peroxide

Source: Adapted from Reference No. 17.

Distillation bottoms from solvent reclaimers can be expected to increase substantially in volume as a result of the land disposal ban on solvent wastes. These bottoms may contain 25 percent or more organic solvent content, oil and grease, and high solids content in addition to lead, nickel, chromium and cadmium. These wastes are restricted to 2 ppm or less solvent content in their leachate and thus may require incineration to meet land disposal restrictions.[19]

3.3.3 Electroplating

The electroplating industry is dominated by small, captive shops concentrated in the mid-west, California, and New York. Cadmium, chromium, lead, nickel and silver are toxic heavy metals which are commonly electroplated, and cyanides are used as components of plating and cleaning baths, as shown previously in Table 3.3.5. Additional data provided in Table 3.3.6 show that spent baths are likely to contain cyanides in concentrations which will make them subject to the land disposal prohibition. However, plating rinses will generally contain well under the 1,000 mg/L concentration limit.[20]

Electroplating process wastes are summarized in Table 3.3.7.[13] Rinse water, contaminated with bath drag-out and residue from work pieces, accounts for the large majority of waste produced. Spent cleaning and plating baths are another significant waste source. These may be acidic or alkaline and may also contain toxic organics. Spent plating solutions are typically disposed only when impurities from drag-in accumulate to levels which impair product quality. These concentrated wastes are either recovered, shipped offsite for disposal, or bled into onsite treatment systems.

The largest volume hazardous waste which is ultimately land disposed is the toxic heavy metal sludge (F006) produced in onsite wastewater treatment processes.[1] Metals are most commonly precipitated as hydroxides or carbonates, with separate pretreatment for chromium reduction and cyanide oxidation.

TABLE 3.3.6. CONCENTRATIONS OF CYANIDE IN PLATING WASTEWATERS

Process	Range (mg/L)
Plating rinse	14-256 (56 average)
Bright dip	15-20
General (separate cyanide)	9-115 (72 average)
General (combined stream)	1-103 (28 average)
Alkaline cleaning bath	4,000-8,000
Plating bath	45,000-100,000
Plating bath:	
Brass	16,000-48,000
Bronze	40,000-50,000
Cadmium	20,000-67,000
Copper	15,000-52,000
Silver	12,000-60,000
Tin-zinc	40,000-50,000
Zinc	4,000-64,000

Source: JAPCA, May 1986.

TABLE 3.3.7. ELECTROPLATING PROCESS WASTES

Waste Description	Process Origin	Composition	RCRA Codes
Spent alkaline cleaning solution	Aqueous cleaning	$NAOH$, Na_2CO_3, Na_2SiO_3, Na_3PO_4, Na_2HPO_4 $(NaPO_3)_4$, cyanide, soils, EDTA+ Mg/Ca, saponified and/or emulsified grease	F009
Spent acid cleaning solutions	Acid pickling	HCl, H_2SO_4, HNO_3, H_2CrO_4, H_3PO_4, HBF_4, Me^+, oils, soils	- -
Degreaser sludges	Solvent cleaning	kerosene, naphtha, toluene, ketones, alcohols, ethers, halogenated hydrocarbons, oils, soils, water	F001, F002 F003, F005
Solvent recycle still bottoms	Solvent recycling	same as above solvents. May contain HCl from solvent decomposition	F001, F002 F003, F005
Spent plating solutions	Electroplating	see Table 6-1	F007
Filter sludges	Electroplating	Silica, silicides, carbides, ash, plating bath constituents	F008
Waste rinse water	Drag-out, equipment cleaning, spills	same as No. 1 and 2, but in lower concentrations	--
Wastewater treatment sludge	Wastewater treatment	Metal hydroxides, sulfides, carbonates	F006
Vent scrubber wastes	Vent scrubbing	similar to No. 7	- -
Ion exchange resin reagents	Demineralization of process water	brine, HCl, NaOH	- -

Source: Reference No. 13.

3.3.4 Printed Circuit Board Manufacture

Printed circuit board manufacturers are represented by both small captive firms and job shop, as well as large manufacturers. Roughly two-thirds of these facilities are located in California and the Northeastern states.[5] Toxic heavy metal and cyanide compounds of concern are used in electroplating baths (e.g., tin/lead, nickel, silver, cyanide) and in lesser quantities in etchants (e.g., chromic acid) and resist stripping solutions (e.g., sulfuric/dichromate). The most prevalent metal contaminant in wastewater discharges is copper which can be removed from spent baths containing California list constituents to permit their reuse. The primary metal/cyanide hazardous waste streams from printed circuit board manufacture are contaminated rinses, sludges from onsite wastewater treatment operations (F006) and spent cyanide-containing baths from electroplating (F007, F008) and stripping (F009) operations. Similar to electroplating facilities, spent baths are either regenerated, shipped offsite for recovery or land disposal, or bled into onsite wastewater treatment systems.

3.3.5 Inorganic Pigment Manufacture

The inorganic pigments manufacturing industry is dominated by a larger firms with the majority clustered in the industrial region between Illinois and New Jersey. Primary pigments which contain California list heavy metals include chrome yellow and orange, molybdate chrome orange, chrome oxide green, cadmium sulfide pigments, and white lead pigments.[5]

3.3.6 Petroleum Refining

Hazardous metal-containing wastes generated by the petroleum refinery industry include 1) spent conversion catalysts contaminated with nickel, vanadium and chromium; 2) leaded bottoms from crude oil storage tanks (K052); 3) heat exchanger cleaning waste (K050); 4) cooling water blowdown and its resulting treatment sludge, which are contaminated with chromium; and 5) wastewater treatment sludges which include API separator sludge (K051), dissolved air flotation waste (K048), and slop oil tank emulsion waste

(K049). Each of these treatment sludges contain lead, chromium and other contaminants in a waste matrix consisting of oil, water, and inert solids. Table 3.3.8 summarizes hazardous waste generation for the refining industry including projected total quantities of toxic metals.[21]

3.3.7 Wood Preserving

The wood preserving industry is one of the largest generators of arsenic containing wastes.[16] Unlike other wood preservatives, chromated copper arsenate consumption has experienced rapid growth (800 percent over the last 15 years), primarily for use in the preservation of lumber and plywood. Consumption in 1981 was estimated to be 46.4 million pounds, with another 1.2 million pounds of arsenic containing preservatives consumed in the form of ammoniacal copper arsenate.

The wood preservation process consists of batch soaking and steam/vacuum drying of wood in a retort followed by open-air curing and drying on drip pads. Steam condensate is commonly recycled in a closed loop with small amounts being bled off to avoid excessive accumulation of dissolved wood sugars. Liquid wastes are filtered in a "work" tank, the filtrate is reused and the sludge is land disposed. The latter consists primarily of wood fiber, dirt, silt and residual preservative. Runon, cooling water and other liquid wastes are treated in onsite wastewater systems, generally through chromium reduction and precipitation with lime, hydrogen sulfide, or sodium sulfide. Sludge generated from this treatment (K001) contains chromium, copper, arsenic, lead and other contaminants and is generally landfilled. Finally, smaller facilities still receive preservative in bags or drums, thus generating waste in the form of contaminated containers.

3.3.8 Miscellaneous Industries

Photographic films contain high concentrations of silver and are used in the photography and printing industries. California List constituents of concern in film processing wastes include silver and ferrocyanide. Ferricyanide bleaches, used in photo processing solutions, contain from 10 to

TABLE 3.3.8. TOTAL WASTE GENERATION IN PETROLEUM REFINERIES IN THE UNITED STATES IN 1981

Waste Stream	Total Waste (tons)	Oil (tons)	Arsenic (pounds)	Cadmium (pounds)	Chromium (pounds)	Copper (pounds)	Lead (pounds)	Mercury (pounds)	Nickel (pounds)	Selenium (pounds)	Silver (pounds)	Vanadium (pounds)	Zinc (pounds)	Cyanide (pounds)	Phenols (pounds)
1. Slop Oil Emulsion Solids	158,890	76,260	280	7	20,140	1,840	1,080	23	1,920	38	16	960	9,590	<1	4,770
2. Silt from Storm Water Runoff	39,090	1,570	300	12	19,690	1,570	2,960	17	4,730	94	28	3,840	12,790	134	590
3. Exchanger Bundle Cleaning Sludge	12,610	1,380	100	12	2,850	650	710	18	1,060	249	<1	230	1,780	43	340
4. API Separator Sludge	124,840	28,710	380	25	15,480	1,140	1,590	25	1,180	<1	27	600	18,210	<1	3,400
5. Products Tanks Bottoms (leaded and unleaded)	353,080	225,970	950	62	1,680	11,340	5,700	57	4,420	1,560	79	2,800	65,640	5,230	90,250
6. Crude Tank Bottoms	1,890	910	30	<1	30	100	30	<1	30	<1	<1	27	230	<1	60
7. DAF Skimming	919,270	119,500	200	<1	14,210	710	760	27	3	203	25	5	8,630	510	11,950
8. Kerosene Filter Clays	13,870	550	60	7	70	290	210	<1	1	248	13	900	650	<1	72
9. Lube Oil Filter Clays	184,110	40,500	190	203	6,210	1,150	340	51	2,970	294	133	8,840	15,480	44	1,180
10. Neutralized HF Alkylation Sludge	56,750	630	100	3	120	640	320	3	2,460	316	8	120	400	2,620	1,010
11. Spent Lime	987,360	3,160	80	3	1,770	3,050	3,050	32	2,010	8	400	40	12,040	2	4,150
12. Waste Biosludge	87,010	240	80	7	6,670	210	110	4	<1	<1	7	2	2,710	<1	780
13. Once Through Cooling Water Sludge	234,550	1,010	1,890	1	22,060	53,590	19,960	178	12,260	560	175	13,660	43,790	51	940
14. Cooling Tower Sludge	1,640	10	10	<1	450	40	30	<1	5	<1	<1	7	540	<1	12
15. FCC Catalyst Fines	25,850	50	50	<1	3,650	900	2,570	<1	12,400	<1	92	12,350	2,010	7	109
16. Coke Fines	110	<1	<1	<1	<1	<1	3	<1	130	<1	<1	100	3	<1	<1
Total Waste	3,200,910	500,460	4,710	342	115,100	77,230	39,440	434	45,580	3,570	1,000	44,480	192,493	8,640	119,590

Source: Reference No. 21.

100 g/L ferricyanide, and fixing baths contain from 1 to 10 g/L silver thiosulfate complex and ferrocyanide. Silver recovery is widely applied in this industry.[22]

Chromium salts are used in the production of organic dyes and pigments as is lead peroxide (catalyst). Lead and chrome account for nearly 100 million pounds of pigments per year which are consumed primarily in the manufacture of red lead primer and chrome yellow. The latter is used as traffic paint due to its bright color and low cost. Inks containing chromium based pigments are also used in the printing industry.

Vanadium and nickel are widely used as catalyst bases (e.g., vanadium tetrachloride, vanadium oxytrichloride, Raney nickel) in the synthetic rubber, chemical manufacturing, and organic dyes and pigments industries.

Non-ferrous metals industries generate large volume wastewaters containing heavy metals. However, the majority of these streams are sufficiently dilute so as to be below the restricted concentrations specified by the land disposal ban. Exceptions include the following: (1) spent sulfuric acid electrolyte from electrolytic refining of primary copper, which can contain from 500 to 12,000 mg/L arsenic and 2,000 to 20,000 mg/L nickel; (2) arsenic plant wash down water which can contain 310 mg/L arsenic; and (3) waste sulfuric acid containing lead from battery recovery operations.[23]

The steam-electric industry generates waste boiler cleaning solutions (e.g., ammoniated citric acid, EDTA, hydrochloric acid) which contain elevated nickel concentrations (e.g., 65 to 400 mg/L) in addition to high concentrations of iron and other metals.[24] The presence of complexers increases the difficulty of removing toxic metals from these solutions.

A small quantity of explosive hazardous wastes are generated by explosive/ordinance sites which contain heavy metals. These wastes, including lead azide and mercuric fulminate, are typically disposed through detonation under controlled conditions to prevent explosion hazards.[13]

Many wastes containing metals/cyanides can result from metal forming and production industries. For example, cyanide concentration in ferromanganese blast furnace scrubber water can be 4,000 mg/L or more.[25] Wastes containing metals/cyanides resulting from iron and steel production include electric furnace emission control sludge (K061) and rolling mill sludge. The latter consists of iron and iron oxide, 5 to 15 percent oil and grease, and trace metals including 250 ppm Ni and 300 ppm Cr.[26] Electric furnace emission control sludge is principally iron and silica oxides and lime with reported chromium levels of 1,300 ppm and nickel concentrations of 300 ppm. Similarly, analysis of combined sludges from electrolytic zinc manufacture shows an average lead concentration of 15,300 ppm.[26] However, these wastes often contain little liquid and leachate levels of toxic metals may not be high enough to exclude them from being land disposed under the new restrictions.

REFERENCES

1. Dietz, S., et al. Westat, Inc. Rockville, MD. National Survey of Hazardous Waste Generators and Treatment, Storage, and Disposal Facilities Regulated under RCRA in 1981. Prepared for U.S. EPA, Office of Solid Waste. EPA 530/SW-84-005. March 1984.

2. The Industries Studies Data Base. Maintained by Science Applications International Corporation for U.S. EPA Office of Solid Waste, Waste Characterization and Assessment Division. 1987.

3. Conversation between Martin Huppersart, Science Applications International Corporation and Marc Breton, Alliance Technologies Corporation. August 26, 1986.

4. U.S. EPA. Report to Congress on the Discharge of Hazardous Wastes to Publicly Owned Treatment Works. U.S. EPA Office of Water Regulations and Standards, Washington, D.C. EPA/530-SW-86-004. February 1986.

5. Versar, Inc., Springfield, VA. Technical Assessment of Treatment Alternatives for Wastes Containing Metals and/or Cyanides. Draft Report prepared for U.S. EPA Office of Solid Waste under Contract No. 68-03-3149. October 1984.

6. Peer Consultants, Inc., Rockville, MD. Background Document for Land Disposal Restrictions of Hazardous Waste Listed in Section 3004(d) of the Resource Conservation and Recovery Act (LDR4 - S0001). Performed for U.S. EPA Office of Solid Waste under Contract No. 68-03-3312. December 1986.

7. Kirk-Othmer Encyclopedia of Chemical Technology. John Wiley & Sons, New York, N.Y., 3rd Edition. 1978.

8. Ruder, E., et al., ABT Associates. National Small Quantity Generator Survey. Prepared for U.S. EPA, Washington, D.C. EPA/530/SW-85-004. February 1985.

9. Cole, C.A., and S.A. Kressin. Pennsylvania Stat University. Technology of Recovery, Recycle, and Reuse of Hazardous Waste in Plastics and Synthetics Industries in Pennsylvania. In: Toxic and Hazardous Wastes, Proceedings of the 16th Mid-Atlantic Industrial Waste Conference. Technomic Publishing Company. June 24-26, 1984.

10. Barbour, A.K. Cadmium - Environmental Aspects of Production and Use. UNEP Industry and Environment. ISSN 0378-9993. United Nation Environment Program. February-March 1982.

11. Congress of the United States. Office of Technology Assessment. Serious Reduction of Hazardous Waste: For Pollution Prevention and Industrial Efficiency. U.S. Government Printing Office, Washington, D.C. 1986.

12. Radian Corporation. Research Triangle Park, N.C. Preliminary Study of Sources of Inorganic Arsenic. Prepared for U.S. EPA. EPA-450/5-82-005. August 1982.

13. Versar, Inc., and Jacobs Engineering, Inc. Waste Minimization: Issues and options, Volume 2. Prepared for U.S. Office of Solid Waste. EPA 530/SW-86-042. 1986.

14. New England Congressional Institute. Hazardous Waste Generation in New England. Washington, D.C. February 1986.

15. Radimsky, J. et al. Recycling and/or Treatment Capacity for Hazardous Wastes Containing Dissolved Metals and Strong Acids. Alternative Technology and Policy Development Section, Toxic Substances Control Division, California Department of Health Services. October 1983.

16. Radimsky, J. et al. Recycling and/or Treatment Capacity for Hazardous Wastes Containing Cyanides. Alternative Technology and Policy Development Section, Toxic Substances Control Division, California Department of Health Services. March 1983.

17. U.S. EPA. Development Document for Effluent Guidelines and Standards for the Metal Finishing Point Source Category. U.S. EPA Effluent Guidelines Division. EPA/440/1-80-091a. 1980.

18. Cheng, S.C., et al. Monsanto Research Corporation. Alternative Treatment of Organic Solvents and Sludges for Metal Finishing Operations. EPA-600/2-83-094. Prepared for U.S. EPA IERL, Cincinnati, OH. September 1983.

19. Breton, M. et al., Alliance Technologies Corporation. Technical Resource Document: Treatment Technologies for Solvent Containing Wastes. Prepared for U.S. EPA HWERL under Contract No. 68-03-3243. August 1986.

20. Grosse, D.W. A Review of Alternative Treatment Processes for Metal Bearing Waste Streams. Journal of the Air Pollution Control Association. May 1986.

21. Bider, W.L., and R.G. Hunt, Franklin Associates. Industrial Resource Recovery Practices: Petroleum Refineries and Related Industries. Prepared for U.S. EPA Office of Solid Waste under Contract No. 68-01-6000. June 1982.

22. Campbell, M.E., and W.M. Glenn. Profit from Pollution Prevention: A Guide to Industrial Waste Reduction and Recycling. Pollution Probe Foundation, Toronto, Canada. 1982.

23. Coleman, R. T. et al., Radian Corporation. Sources and Treatment of Wastewater in the Nonferrous Metals Industry. Prepared for U.S. EPA IERL, Cincinnati, OH. EPA-600/2-80-074. April 1980.

24. U.S. EPA. Development Document for Effluent Limitations Guidelines and Standards, and Pretreatment Standards for the Steam Electric Point Source Category. U.S. EPA Effluent Guidelines Division, Washington, D.C. EPA-440/1-82-023. November 1982.

25. U.S. EPA. Development Document for Effluent Limitations Guidelines and Standards, and Pretreatment Standards for the Nonferrous Metals Forming and Iron and Steel, Copper Forming, Aluminum Metal Powder Production, and Powder Metallurgy Point Source Category. U.S. EPA Effluent Guidelines Division, Washington, D.C. EPA-440/1-84/019B. February 1984.

26. Seitter, L.E., and R.G. Hunt. Industrial Resource Recovery Practices: Metals Smelting and Refining (SIC 33). Prepared for U.S. EPA Office of Solid Waste under Contract No. 68-01-6000. January 1983.

4. Waste Quantity, Management Practices, and Treatment Capacity

This section summarizes waste management and available treatment capacity for metal/cyanide containing wastes. Metal/cyanide RCRA waste quantities and management practices are discussed in Section 4.1 and available treatment capacity is summarized in Section 4.2.

The bulk of this data was obtained from the Background Document for California List Wastes, prepared by the EPA Office of Solid Waste to support the proposed rule for the 40 CFR Part 268 land disposal restrictions.[1] This information, in turn, relied heavily upon the 1981 EPA Regulatory Impact Analysis (RIA) national survey of Treatment, Storage, and Disposal Facilities (TSDFs) and hazardous waste generators.[2] This represents the most comprehensive, available survey performed to date describing U.S. hazardous waste generation in terms of RCRA waste codes and management practices.[1]

Data on waste generation from small quantity generators was obtained from a 1984 survey performed by ABT Associates.[3] This data was combined with the EPA analysis to develop a more complete description of metal/cyanide waste generation and available capacity. While this survey was less prone to errors than the RIA Survey, this class of generators only represent a small fraction of total metal/cyanide waste generation. Thus, the combined quantities should still be regarded as approximations of existing industry conditions.

Recognizing these data limitations, the EPA has recently initiated an effort to more accurately characterize current waste management practices and available treatment capacity.[4] However, results of this 1986 survey were not yet available at the time of this writing.[1] As available, these will be incorporated into the final rule regarding California List waste disposal restrictions which will be published in the Federal Register prior to July 8, 1987.

4.1 WASTE QUANTITY AND MANAGEMENT PRACTICES

4.1.1 Data Sources

4.1.1.1 Large Quantity Generators-

As stated above, the EPA relied primarily on the results of the 1981 RIA Survey to generate its waste quantity estimates. The RIA Survey data has undergone repeated refinement since it was initially compiled in 1983. However, it is still subject to significant uncertainty due to the widely skewed distribution of the waste quantity data. For example, for each category of landfilled waste, the single largest waste stream accounts for over 50 percent of the total waste volume disposed. As another example, 0.5 percent of the individual waste streams accounted for 90 percent of the total metal/cyanide waste volume with the three largest waste streams alone accounting for 50 percent of the entire volume.[1] Since the survey only included a fraction of the waste streams managed in the country, excluding or including a disproportionate number of large volume streams may have greatly skewed the results.

These deficiencies become increasingly evident when the waste quantity data are compared to that of a second national survey performed in 1983 by the Congressional Budget Office[5] or to surveys conducted by individual states or regions (see References 6 and 7 for comparisons). For example, while both the RIA and CBO surveys projected an annual metal/cyanide waste generation figure of roughly 13 to 14 billion gallons, the RIA survey attributes 41 percent of this to cyanide wastes whereas the CBO survey figure is a modest 16 percent. In addition, the RIA survey results show 15 percent of the waste to be non-liquid whereas the CBO estimates that this quantity is approximately three times as high (45 percent).[1,5]

Other factors also contribute to data uncertainties. For example, in response to regulatory changes and waste disposal cost increases since 1981, waste generating and management industries have instituted corresponding changes in waste management practices. Significant regulatory changes include revised waste listings, changes in small quantity generator status, changes in reporting and other requirements for certain recycled wastes, the ban on free liquid disposal in landfills, and perhaps most significantly, the regulatory changes occurring under the 1984 Hazardous and Solid Waste Amendments (HSWA).

Details of the impacts of these changes on projections of current waste management practices and other shortcomings of the RIA Survey can be found in the references.[1,3] In general, the results presented in this section probably err on the conservative side; i.e., overestimate waste quantities. Thus, they should be interpreted with caution and viewed only as providing an approximate characterization of the current hazardous waste management industry.

4.1.1.2 Small Quantity Generators--

The small quantity generator (SQG) survey was conducted between January 1983 and October 1984 by ABT Associates for the U.S. EPA Office of Solid Waste.[3] This survey identified 378,000 generators of 598,000 metric tons/year of hazardous waste from which ABT projected a total population of 600,000 to 660,000 generators of 940,000 metric tons of waste. The survey was designed to categorize waste generation by waste type, industry source, and regulatory status of the generator. However, no data were collected on waste characteristics. Thus, as a worst case scenario, it can be assumed that all metal/cyanide wastes will exceed the land disposal ban concentration limits.

In the SQG survey, generators were categorized according to the total quantity of hazardous waste generated by the facility. As of 1981, when the RIA survey was conducted, generators of less than 1,000 kg/month were considered to be SQGs and therefore were not included in the national survey. However, the SQG exclusion limit has since been reduced to 100 kg/month. As a result, generators of waste exceeding this quantity are now subject to RCRA waste management regulations similar to those pertaining to large quantity generators. This regulatory change was effective September 22, 1986 for wastes shipped offsite and March 24, 1987 for wastes managed onsite.[8]

4.1.2 Metal/Cyanide Waste Management Summary

Data is provided in this section summarizing waste management practices by waste type and industrial source. Detailed data are presented separately for large and small quantity generators for wastes which are land disposed and potentially subject to the land disposal restrictions. Gross figures are provided for other management methods (i.e., incineration and deep well injection).

4.1.2.1 Large Quantity Generators--

Of the waste streams reported in the RIA data, only liquid-containing metal/cyanide wastes managed in land disposal units subject to regulation were considered for this analysis. Wastes which were disposed via underground injection were not included since they are subject to separate restrictions under the HSWAs. However, if by the statutory deadline (August 8. 1988), the EPA determines that this practice should be discontinued or subjected to similar concentration restrictions, potentially large additional quantities of metal/cyanide wastes may require alternative management practices (1,157 million gallons).[9] Similarly, unused underground injection capacity was not assumed to be a viable alternative waste management practice for the analysis presented in Section 4.2.

Wastes reported in the RIA Survey were identified by RCRA code and thus could be identified as containing metals or cyanides on the basis of the definitions of these wastes. Due to lack of data, all wastes known to contain at least one of these constituents were assumed to exceed the land disposal ban treatment standards. More than any other assumption, this would tend to overestimate the quantity of waste subject to the disposal restrictions. For example, the largest single volume stream reported was a DAF float waste discharged to a disposal impoundment by a small to medium size Wyoming-based refinery. This stream, which accounted for 28 percent of the metal/cyanide RCRA waste generated in the country, is clearly a dilute liquid. For example, in order to be characterized as a California List waste which would be prohibited from land disposal, it would have to have 1,000 times the estimated chromium concentration in all K048 refinery wastes generated in the country.[10]

Cyanide wastes were identified as any waste containing a carbon-nitrogen triple bond. Thus, free and complexed cyanides, cyanogens, cyanogen halides, cyanide salts, and organonitriles were included in this category. Although only some of the wastes reported as code D003 (exhibiting the characteristic of reactivity) actually contain cyanides, the entire D003 volume reported was included in the quantity estimates provided here.

Metal wastes included in this analysis include those containing any of the California List constituents; i.e., As, Cd, Cr, Pb, Hg, Ni, Se, and Tl.

Barium and silver were excluded since these are not subject to the land disposal restrictions. All wastes identified as containing chromium were conservatively assumed to contain hexavalent chromium.

Metal and cyanides waste quantities which are land disposed (excluding deep well injection) are presented by treatability group and current management method in Table 4.1.1. Treatability groups include wastes that could be treated via cyanide oxidation, chromium reduction, metals precipitation, or some combination of these processes.

As shown, 91 percent (10,360,106 gal/yr) of metal-containing wastes may include chromium as a constituent and approximately 16 percent (2,007,106 gal/yr) may contain cyanides. Of the latter, roughly half of the wastes also contain metals, and may require cyanide oxidation, chromium reduction and some form of metals removal; e.g., precipitation.

However, due to the methodology employed in the RIA survey, the EPA probably overstated the proportion of wastes which would require treatment for either chromium or cyanide as a result of the land ban. As discussed in Section 3.3, a California state survey of offsite disposal found that only 61 percent contained chromium in concentrations exceeding 500 mg/L and only 6 percent contained cyanides above 1,000 mg/L.[11]

The EPA lacked similar data to quantify constituent concentrations. However, it acknowledged that at a minimum, waste streams which were treated prior to disposal would be likely to contain cyanide concentrations at levels below 1,000 mg/L, thus rendering them exempt from the land disposal restrictions. If these wastes are eliminated from the totals presented in Table 4.1.1, quantities of California List wastes currently land disposed which may include cyanides are reduced by 48.5 percent to 1,034.5 million gallons/year.[1]

Despite this adjustment, the projected waste quantities are significantly overstated. The California study[11] concluded that the only significant source of cyanide wastes exceeding 1,000 mg/L was from spent process solutions. Contaminated rinse waters typically contained under 100 mg/L and thus would not be restricted. Of cyanide wastes received at offsite land disposal facilities, which are likely to be significantly more concentrated than wastes managed onsite, a minimum of two-thirds of liquids and sludges contained less than 1,000 mg/L and another 5 percent of cyanides were solids.

TABLE 4.1.1. METAL/CYANIDE WASTE QUANTITIES BY MANAGEMENT METHOD AND TREATABILITY GROUP[a]

Waste management practice	Cyanide wastes	Metal wastes: Without Cr^{6+}	Metal wastes: With Cr^{6+}	Metals/ cyanides	Chromium/ cyanides	Total metal/ cyanide waste
Surface impoundments:						
Treatment only	450.7	559.2	1664.3	0	624	3,298.2
Treatment and storage	213	263.0	524.3	0	163	1,163.3
Storage only	363	111.7	2513.7	0	17.3	3,005.7
Disposal	3.9	57.0	3708.3	0.6	76.9	3,846.7
Waste pile storage	1.6	14.2	149.0	1.3	19.2	185.3
Land application	<0.1	1.2	613.4	0	<0.1	614.6
Landfill	9.1	60.5	223.0	0	63.7	356.3
Total land disposal	1,041.3	1,066.8	9,396.0	1.9	964.1	12,470.1

Source: Adapted from Reference No. 1.

[a]1,000 gal/yr.

Thus, maximum liquid cyanide waste generation containing over 1,000 mg/L total cyanides is roughly 1.5 million gallons for the state of California.[11] Since California generates a disproportionate share of the country's metal/cyanide waste, projected annual national estimates should be well under 50 million gallons as compared to the EPA's estimate of 1,554 million.

Surface impoundments dominate the management practices used for metal/cyanide wastes in terms of both waste volume (91 percent) and fraction of total waste streams (48 percent). Waste codes handled in large volumes in impoundments include D007, F006, K048, D008, and wastewater from acrylonitrile production. Nearly 85 percent of the waste streams which are disposed via land application are oil refinery wastes (K048 through K052) although one chromium containing stream (D007) accounted for 85 percent of the volume. Overall, the dominant waste codes in each category are as follows:

Chromium/cyanides:	F006;
Chromium wastes:	D007 with lesser quantities of K048 through K051 and K061 and K062; and metal containing wastes: lead (D008, K052, and others).

4.1.2.2 Small Quantity Generators (SQG)--

The SQG survey identified 410,252 MT/yr of metal/cyanide waste generation.[3] This data has been summarized previously in Table 3.3.3 by generator size and waste type. Roughly 83 percent of this waste originates from facilities which generate between 100 and 1,000 kg/month and are therefore subject to RCRA waste management requirements similar to those of large quantity generators. The largest waste category is used lead-acid batteries which account for approximately 90 percent of SQG wastes. Other significant volume categories include photographic, plating, and silver containing wastes. The majority of metal/cyanide containing SQG wastes are generated by vehicle maintenance shops, metal manufacturers, printers, and equipment repair facilities.

For the most part, waste management practices employed by SQGs were not detailed by waste category in the summary document.[3] One exception was used lead-acid batteries, which accounts for the majority of metal/cyanide waste generation. For generators subject to the land disposal restrictions, 98 percent of these wastes are managed offsite with 89 percent shipped to recycling facilities. The remainder is disposed in solid waste landfills (2 percent), reclaimed onsite (1.5 percent) or handled in an unknown manner. The photography industry also has a high tendency to recycle wastes (roughly one-third). Finally, approximately one-fourth of wood preserving wastes, many of which contain arsenic, are reused or reclaimed. Other industrial waste management processes were summarized by industry, but could not be segregated by waste stream using the data presented in the reference.[3]

As a result of recycling, total waste quantity which is ultimately land disposed is significantly less than that generated. A rough estimate of the quantity potentially affected by the land disposal ban is 60,000 MT/yr. Assuming this waste has a specific gravity equal to that of water results in an estimate of less than 16 million gallons, or less than one-tenth of a percent of metal/cyanide wastes land disposed by large quantity generators.

4.1.3 Metal/Cyanide Waste Quantities Affected by the Ban

After calculating waste quantities currently land disposed, the EPA determined the quantities of metal/cyanide wastes which would be potentially affected by the ban.[1] This quantity excludes solids and all wastes managed in storage surface impoundments or landfills. Solids were excluded because the land disposal ban pertains only to wastes with free liquid fractions. Wastes stored in impoundments were not included since this does not constitute land disposal. Furthermore, at least a fraction of these wastes must be subsequently managed in other treatment or disposal processes and thus would be double counted. Finally, since free liquids have been prohibited from being landfilled since May 1985, these wastes were also not expected to be affected by the ban.

Of the 12,470 million metric tons of California List metal/cyanides reportedly generated in the RIA survey, the EPA estimates that 73 percent will potentially be affected by the land disposal ban. These data are summarized below:

Waste group	Waste quantity (10^6 gal/yr)	Waste quantity affected by ban (10^6 gal/yr)
Cyanide waste	2,007.3	1,554.4
Metal waste with Cr+6	10,360.1	7,542.0
Metal waste, no Cr+6	1,068.7	896.3
Total metal/cyanide	12,470.1	9,108.1

Note that there is overlap among waste categories so that these totals are not truly additive. Also note that if treated cyanide streams were assumed to be below the 1,000 ppm disposal limit, projected waste quantity and quantity affected by the ban would be reduced to 1,035 and 651 million gal/yr, respectfully.

Wastes generated by SQGs which will not be subject to the land disposal ban include those which are currently recycled (e.g., most lead-acid batteries and silver containing solutions), non-liquids, and wastes currently disposed via landfilling. As discussed above, the resulting quantity is expected to be approximately 16 million gallons; i.e., only a small fraction of large quantity generator wastes affected by the land disposal ban. Thus, ignoring these generators does not significantly affect the available capacity determination.

4.2. AVAILABLE TREATMENT CAPACITY DETERMINATION

In addition to data on waste generation and management practices, the EPA analysis prepared for the Background Document also included a determination of available treatment capacity, as required under the 1984 Hazardous and Solid Waste Amendments. Lack of available capacity to meet treatment objectives is to be the basis for granting extensions to the statutory deadline, of up to 2 years, for affected waste types.

Available capacity can originate from the following sources:

- Commercial facilities;
- Private capacity which can be used to process additional wastes generated by the owner;

- Private capacity which may become available to other generators; and
- Planned capacity which will be available prior to implementation of the ban.

Of these four sources, the EPA was only able to quantify commercially available capacity. Data regarding available capacity at private firms and planned capacity were not available, and thus could not be included in the analysis. As discussed below, the agency estimated that quantifiable, (i.e., commercial) available capacity falls far short of satisfying the demand for wastes requiring alternative treatment. However, the EPA made a qualitative determination that sufficient planned capacity would make up the difference between capacity demand and supply. The agency anticipated that firms would be able to install additional treatment units, as necessary, between the time the disposal ban rules were codified (December 11, 1986)[12] and the effective date of the land disposal ban. Thus, all metal and cyanide containing wastes will be subject to the disposal ban effective July 8, 1987.

4.2.1 Alternative Capacity Needs Assessment

The EPA made certain assumptions in order to project available commercial capacity. SQGs were excluded from consideration due to their almost negligible contribution to total waste generation. In addition, demand for available capacity resulting from wastes generated through CERCLA remedial and removal actions, or RCRA corrective actions, were not included. These were anticipated to contain less than the statutory concentration limits for metals and cyanides and, therefore, would not be subject to the disposal ban regulations.

Impoundments categorized as disposal units will require alternative treatment capacity whereas units used only for temporary storage are not regulated under the 1984 HSWAs. The majority of impoundments currently used for treatment will probably be available for use following enactment of the disposal restrictions. However, these units must meet design and operating criteria such as double liners, leachate collection systems, and ground water monitoring systems.[13] The EPA did not have adequate data to determine which wastes are currently treated in exempt impoundments. Therefore, it conservatively assumed that all such wastes would require alternative capacity on an interim basis while facilities complied with the appropriate impoundment standards.

Another requirement for impoundments is that they must be dredged on a periodic basis. Thus, the quantity of impoundment sludges requiring disposal will increase to an unknown extent. Since EPA included the entire impoundment waste volume in its capacity assessment, it did not include additional capacity for these wastes. However, it should be noted that sludges will impose increased requirements for treatments such as sludge consolidation and solidification.

Finally, the EPA recognized that the current ban on landfilling liquid wastes, imposed after the RIA data was collected, would have already resulted in adoption of alternative management methods. The EPA assumed that these wastes are currently being solidified or chemically stabilized prior to landfilling and, therefore, do not require additional alternative capacity.

Based on the above assumptions, EPA estimated that 72.8 million gallons of cyanide waste, 286 million gallons of waste containing hexavalent chromium, and 60.5 million gallons of other metal containing wastes which are currently land disposed will not be affected by the land disposal restrictions. When these results are combined with the alternative capacity demand figures presented in the previous section, the total additional annual treatment capacity required becomes:

Waste type	Waste quantity (10^6 gal/yr)
Cyanide waste	1,554
Metal waste, with Cr+6	7,542
Metal waste, no Cr+6	896
Total metal/cyanides	9,927

Note again that these categories are not mutually exclusive, thus the total is somewhat less than the summation of the individual quantities. Of this total, all but 8 percent is currently managed in surface impoundments.

EPA further characterized capacity requirements as being either an interim or a long-term need, as shown in Table 4.2.1. Interim requirements included all wastes requiring long-term disposal capacity, as well as those which are currently treated in surface impoundments. Impoundment capacity is considered interim because many of these will require upgrading to meet RCRA operating and design specifications. Interim requirements account for 93 percent of cyanide capacity requirements, 39 percent for chromium containing wastes, and 92 percent for other California List metal wastes.

In the above figures, EPA assumed that all wastes exceeded the concentration limits which would make them subject to the disposal ban. However, certain waste streams which were treated prior to disposal would be likely to contain cyanide concentrations at levels below 1,000 mg/L, thus rendering the waste exempt from land disposal restrictions. If these wastes are eliminated from the totals presented above, cyanides requiring alternative treatment capacity are reduced by 58 percent to 651 million gal/yr. Of this, 644 million gallons requires interim capacity and 7 million gallons requires long-term management alternatives.[1]

4.2.2 Available Capacity Determination

EPA did not establish Best Developed Available Technology (BDAT) for metal/cyanide wastes. However, for the purposes of determining available treatment capacity, it assumed that all additional treatment (e.g., alkaline chlorination, chromium reduction, metals precipitation, and solidification) would be performed in tanks. As discussed above, the EPA conservatively assumed that only existing (as of 1981) commercial capacity would be certain of being available for general use. Thus, availability of unused commercial TSDF tank capacity, obtained from the RIA survey, was used as the determinant of available treatment capacity for metal/cyanide wastes. Data were presented such that tank capacity could be categorized according to function (e.g., neutralization, precipitation, reduction, oxidation, and others) and extrapolated to provide national estimates of available capacity.

TABLE 4.2.1. TOTAL VOLUME OF CALIFORNIA LIST WASTES REQUIRING ALTERNATIVE TREATMENT CAPACITY (MILLION GALLONS PER YEAR)

Waste group	Volumes requiring treatment capacity		
	long-term	interim	total
Acidic corrosive waste (pH ≤ 2)	357.5	6,387	6,745
Cyanide wastes	103.4	1,451	1,554.4
Metal wastes:			
Without chromium (+6)	74.3	822	896.3
With chromium (+6)	4,566	2,976	7,542
Halogenated organic wastes	398.8	52.4	451.2
Mixed RCRA/PCB wastes	7.0	0	7.0
Total	5,507	11,688.4	17,195.4

Source: Reference No. 1.

EPA recognized that competition for this capacity would result from treatment requirements for other wastes being banned from land disposal (see Table 2.1.1). Thus, with a similar methodology as that described in Section 4.1, waste quantities and current management methods were determined for other California List wastes (e.g., liquid acids, PCB's, and halogenated organics), solvents, and dioxins. Resulting capacity needs have been previously summarized in Table 4.2.1. Waste quantities were then compared to BDAT capacities to determine whether extensions to the land disposal prohibitions were warranted.

BDAT for solids and sludges which contain solvents, halogenated organics, dioxins, and PCBs was determined to be incineration. However, as a result of the solvent and dioxin ban effective November 8, 1986, incineration capacity is expected to be fully utilized, especially the relatively few available units which are capable of handling sludges and solids. In addition, bringing new units on-line is expected to take approximately 2 years due to the complex problems of siting, testing and permitting these units. These factors caused EPA to grant a 2-year variance for the management of these wastes.[14] A similar variance was granted for treatment of solvent containing wastewaters since capacity requirements (2,481,106 gal/yr) exceeded unused commercial capacity (2,103,106 gal/yr) which was further determined to be available at only four facilities nationwide.[14]

These variances extend the effective date of the solvent and dioxin bans (excluding organic solvent liquids) to November 8, 1988, thus leaving some of this capacity available for the management of organometallic and organocyanide sludges and wastewaters. It is evident from the above discussion that wastes requiring incineration, particularly those which cannot be handled in liquid injection incinerators, will face the most highly restricted supply of disposal units and rapidly rising disposal costs. However, this volume should be relatively small since organic liquids/sludges account for less than 10 percent of the total cyanide wastes and less than 1 percent of the total metal wastes generated. In addition, nearly 40 percent of the organic cyanides generated are already disposed via incineration.[15]

Total available tank capacity for the management of aqueous metal/cyanide wastewaters was estimated on the basis of the anticipated type of treatment required; i.e., cyanide oxidation, chromium reduction, and metals precipitation. RIA data specified the type of treatment performed in tanks and/or the type of waste handled. From this data, the EPA determined available capacity at commercial facilities. As shown in Table 4.2.2, a comparison of available commercial treatment capacity with demand shows a large shortfall. For each waste category, available capacity represents less than five percent of capacity required to manage metal/cyanide wastes which will require alternative treatment.

Taken at face value, the EPA analysis suggests that over the short-term, the disposal restrictions could cause severe capacity shortfalls. In addition, the analysis did not take into account discrepancies between geographical demand and supply of available capacity which could exacerbate shortages on a local level, possibly requiring transport of wastes over large distances. However, despite the inability of commercial capacity to handle the volume of waste potentially affected by the ban, the conclusion that sufficient capacity will be available is probably sound. This results from two considerations: 1) the EPA analysis was overly conservative, essentially assuming a worst case scenario; and 2) the vast majority of metal/cyanide wastes should be amenable to uncomplicated, well-established treatment techniques which can be brought on-line in a minimal time period.

The realities of the hazardous waste industry which could not be quantified and, therefore, were not included in the EPA's analysis, resulted in an overly conservative estimate of excess available capacity. As described above, waste quantities requiring alternative capacity were probably overestimated and available capacity was underestimated. For example, onsite private treatment capacity was not included in this analysis. In all likelihood, this unused volume dwarfs available offsite capacity (onsite management accounts for 96 percent of total RCRA waste management).[2]

Secondly, demand for capacity and waste generation patterns can be expected to change significantly in response to the new regulations. Increased disposal costs will offset increases in demand through adoption of lower cost waste management alternatives; e.g., source reduction or onsite treatment. For example, as discussed above, a very small percentage of the waste streams (0.5 percent) account for a large fraction (90 percent) of the total waste volume. The economics of waste management suggest that at best,

TABLE 4.2.2. COMPARISON OF AVAILABLE METAL/CYANIDE TREATMENT CAPACITY WITH CAPACITY NEEDS

	Annual waste quantity (10^6 gal/yr)		
Metal/cyanide waste type	Capacity requirement	Commercially available capacity	Percentage of waste requiring additional capacity
Cyanide Oxidation	1,554	65	95.8
Chromium Reduction	7,542	35	99.5
Metals Precipitation	8,438	165	98.0

Source: Adapted from Reference No. 1.

only residuals from these huge wastewater streams are likely to compete for offsite capacity. Construction or upgrading of existing onsite facilities will prove to be much more cost-effective, primarily as a result of significant savings in transportation costs.

Although the EPA lacks available information to quantify these effects, it based its final determination that sufficient capacity would be available prior to the deadline on two assumptions: (1) applicable treatment technologies are widely available, technically uncomplicated, and require minimal installation time; and (2) sufficient advance notice was provided by the agency to enable waste generators and TSDFs to install the required processing equipment.

Liquids have been banned from disposal in landfills since 1984. Since that time, the agency believes that significant capacity for solidification of inorganic sludges has been implemented at TSDFs.[12] The disposal restrictions apply only to liquid wastes, permitting disposal of containerized wastes rendered non-liquid through addition of adsorbants.[12] Thus, chemical stabilization is only required for wastes disposed in bulk. Cyanide oxidation, chromium reduction, and metals precipitation have also been widely applied for liquid wastes. Therefore, treatment of these wastes is not likely to present insurmountable technical difficulties. Residuals generated from such treatment will consist primarily of inorganic sludge which will be amenable to solidification.

Time constraints also do not warrant an extension of the effective disposal ban date. California List constituent concentration limits were first published in 1984 when the HSWA were passed. EPA took further steps to codify these levels in December, 1986.[12] Although EPA has reserved the right to more strictly interpret the Congressional mandate to impose more stringent restrictions on constituent concentrations, it has thus far elected to maintain them at the originally proposed levels. Thus, the industrial community has had significant time to meet these treatment standards through installation or upgrading of treatment systems, or implementation of source reduction programs.

However, the ban is likely to impose hardship on some sectors of the economy. In particular, small quantity generators that cannot afford to install treatment equipment will incur additional storage costs and will be forced to utilize increasingly expensive offsite disposal services. In addition, all firms generating metal/cyanide wastes with high organic content will find available disposal options to be limited. Incineration facilities will be operating at peak capacities and therefore will be reluctant to accept wastes which create handling or residual disposal problems unless they are compensated by charging premium prices. Currently, incineration is used for less than 4 million gallons of metal/cyanide wastes annually.[9] Other waste management alternatives, such as chemical fixation and stabilization, have not yet proven to be effective on wastes with high organic content.[6]

Metal types which may be more difficult to treat include liquid arsenic wastes, and wastes containing thallium or selenium. The literature does not contain much information on the ability of commercially available processes for treating these wastes. However, these wastes can be rendered non-liquids or be granted case-by-case extensions on the ban effective date if it is determined that no treatment technology capacity is available. The number of such exemptions is expected to be small, particularly since available data suggest that little selenium and thallium waste generation occurs.[1,11]

REFERENCES

1. U.S. EPA Office of Solid Waste. Background Document for California List Wastes to Support 40 CFR Part 268, Land Disposal Restrictions. Proposed Rule: California List Waste Volumes, Characteristics, and Required Treatment Capacity. LDR 4 - Reference No. 15. U.S. EPA Public Dockett, 401 M Street, S.W. Washington, D.C. November 1986.

2. Deitz, S. et al., Westat, Inc., Rockville, MD. National Survey of Hazardous Waste Generators and Treatment, Storage, Disposal Facilities Regulated under RCRA in 1981. Prepared for U.S. EPA Office of Solid Waste. April 1984.

3. Ruder, E. et al., ABT Associates. National Small Quantity Generator Survey. Prepared for U.S. EPA, Washington, D.C. EPA/530/SW-85/004. February 1985.

4. Conversation between Jim Craig, U.S. EPA Economic Analysis Branch and Marc Breton, Alliance Technologies Corporation, February 3, 1986.

5. U.S. Congressional Budget Office. Hazardous Waste Management - Recent Changes and Policy Alternatives. CBO Congress of the United States. May 1985.

6. Breton, M. et al., Alliance Technologies Corporation. Technical Resource Document: Treatment Technologies for Solvent Containing Wastes. Prepared for U.S. EPA Hazardous Waste and Environmental Research Laboratory, Cincinnati, OH. Contract No. 68-03-3243. August 1986.

7. Wilk, L. et al., Alliance Technologies Corporation. Technical Resource Document: Treatment Technologies for Corrosive Wastes. Prepared for U.S. EPA Hazardous Waste and Environmental Research Laboratory, Cincinnati, OH. Contract No. 68-03-3243. October 1986.

8. Federal Register, Volume 51, 51FR10146. 40 CFR Part 261. March 24, 1986.

9. U.S. EPA. Assessment of Incineration as a Treatment Method for Liquid Organic Hazardous Wastes. Background Report III: Assessment of the Commercial Hazardous Waste Incineration Market. U.S. EPA Office of Policy, Planning, and Evaluation. Washington, D.C. March 1985.

10. Bider, W.L., and R.G. Hunt, Franklin Associates, Ltd. Industrial Resource Recovery Practices: Petroleum Refineries and Related Industries (SIC 29). Prepared for U.S. EPA Office of Solid Waste. June 15, 1982. Draft Final.

11. Radimsky, J. et al. Recycling and/or Treatment Capacity for Hazardous Wastes Containing Dissolved Metals and Strong Acids. Alternative Technology and Policy Development Section, Toxic Substances Control Division, California Department of Health Services. October 1983.

12. Federal Register, 51 FR 44714; EPA Proposal to Codify Rules Limiting Land Disposal of Untreated Hazardous Waste on RCRA California List. December 11, 1986.

13. Federal Register. Resource Conservation and Recovery Act. Section 3005(j)(11)(A and B)

14. Federal Register, Volume 51, No. 216. 40 CFR Part 260. Hazardous Waste Management System; Land Disposal Restrictions; Final Rule. November 7, 1986.

15. Versar, Inc., Springfield, VA. Technical Assessment of Treatment Alternatives for Wastes Containing Metals and/or Cyanides. Performed for U.S. EPA Office of Solid Waste under Contract No. 68-03-3149. October 31, 1984.

5. Waste Minimization Processes and Practices

Waste minimization consists of two distinct aspects of hazardous waste management: source reduction and recycling/reuse. Source reduction refers to preventive measures taken to reduce the volume or toxicity of hazardous waste generated at a facility. Recycling/reuse refers to procedures and processes aimed at the recovery of generated waste or its direct reuse. The two approaches will be described separately in this section, using examples to illustrate the potential of these activities for the control of hazardous metal/cyanide waste. This will be followed by a summary of available waste minimization practices and options for each of the major waste producing industrial categories identified in Section 3.3. As will become apparent, both source reduction and recycling/reuse are practices that often are carried out concurrently by a facility as management implements multifaceted programs to achieve waste minimization.

5.1 SOURCE REDUCTION

Source reduction is defined as any onsite activity which reduces the volume and/or hazard of waste generated at a facility. Source reduction represents a preventive approach to hazardous waste management since the reduction of waste volume or hazard reduces problems associated with waste handling, treatment, disposal, or liability. Source reduction practices may impact all aspects of waste generating industrial processes, from raw material procurement, to equipment requirements, to product characteristics. A primary motivation for plants to implement source reduction practices is the potential economic benefit they may accrue. These economic benefits increase as restrictions on waste management practices become more stringent.

Source reduction involves a wide variety of practices, some of which may be applicable at virtually any plant generating metal/cyanide wastes. Because the potential application of these practices is so diverse, there are little documented data which indicate the significance of waste source reduction on nationwide industrial waste generation patterns. The EPA and State environmental agencies believe that some form of source reduction is applicable to most industrial plants generating hazardous wastes and will result in a significant reduction in waste generation as more companies implement waste minimization programs.[1]

Waste source reduction practices vary widely from plant to plant, reflecting the variability of industrial processes and waste characteristics. However, in general, source reduction practices may be classified as follows:

- raw material alteration;
- product reformulation;
- process redesign/modernization; and
- improved operating practices.

These options are summarized in Figure 5.1.1. A description of each type of practice is presented below.

5.1.1 Raw Material Alteration

Raw material alteration can take the form of purification of existing raw materials or substitution of a feedstock, catalyst, or other material involved in production for another. The substitute is either less hazardous or results in lower hazardous waste generation but must continue to satisfy end-product specifications. The ideal raw material substitution would be the replacement of a hazardous material with a nonhazardous material, without compromising product quality. However, case studies described in the literature indicate that frequently either product quality is affected or some alteration in process equipment is required. An example of raw material alteration is the use of deionized water in plating baths and rinses to extend bath life and improve product quality. An example of raw material substitution is use of

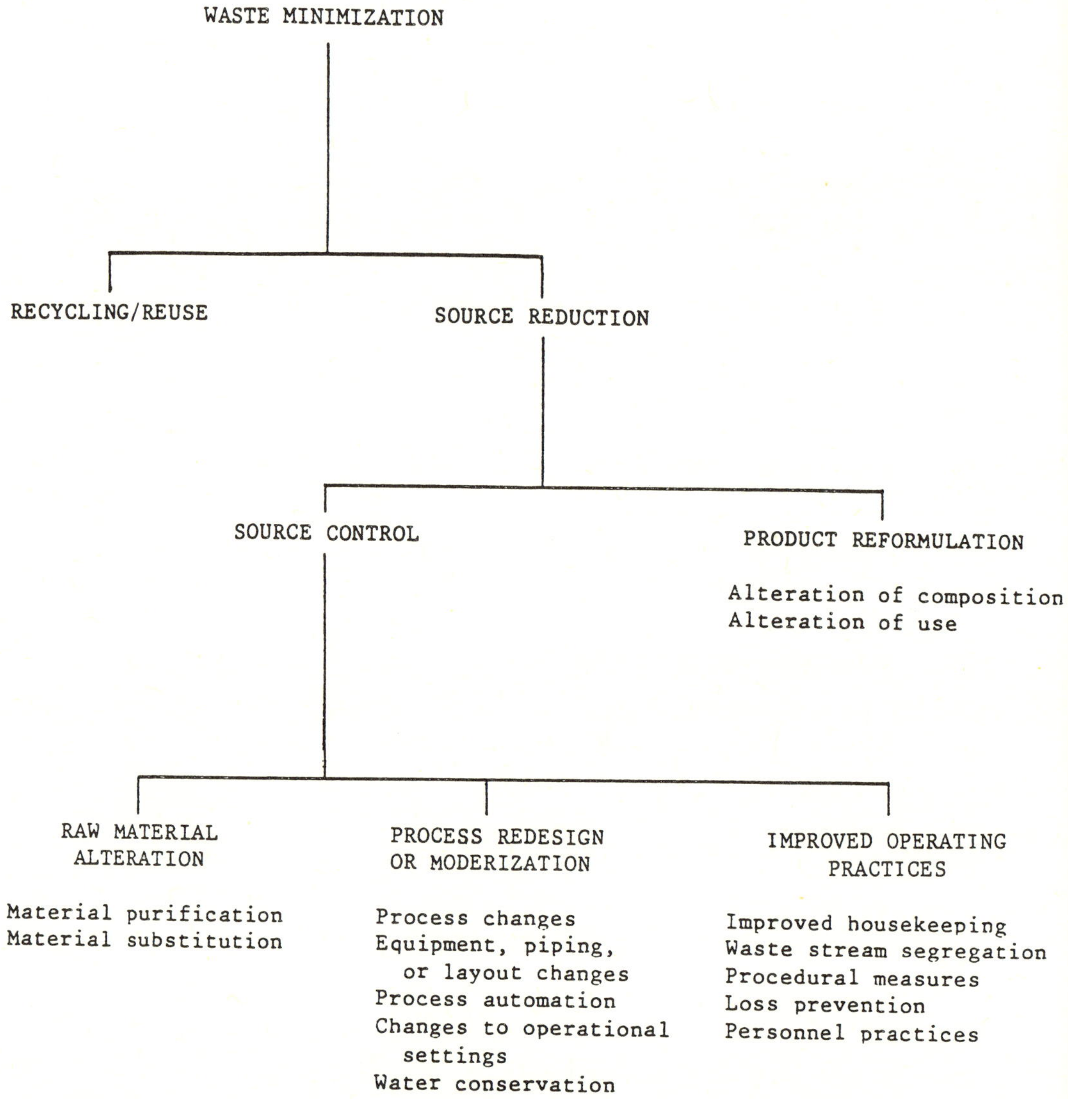

Figure 5.1.1 Source Reduction Options.

zinc in place of cadmium for certain plating operations (e.g., acidic environments or applications which do not require exceptionally thin coats).[2]

5.1.2 Product Reformulation

Another method that is employed to reduce the volume or toxicity of wastes produced by a plant is to alter product specifications. Product reformulation is considered to be relatively common in industry, particularly among manufacturers of specialty chemicals.[3] However, competition from imported goods or client restrictions on product specifications (e.g., those imposed on manufacturers of military components) can severely restrict the ability of manufacturers to pursue this method of source reduction.

Product reformulation can be accomplished by either altering the composition of the product or altering its end use to permit more flexibility in its manufacture. An example of product alteration is the current trend in manufacturing automobile bumpers out of urethane elastomers instead of chromium coated metal parts. Another example is the reconfiguration of printed circuit boards to permit surface mounting of components, reductions in overall board size, or use of injection-molded thermoplastics which can eliminate some plating steps and reduce waste generation from surface cleaning.[4]

5.1.3 Process Redesign/Equipment Modification

Process redesign includes 1) the alteration of the existing process design to include new unit operations; 2) the implementation of new technologies to replace older operations; or 3) changes in operating conditions employed in processing. Process redesign can, therefore, vary widely in terms of the effect upon production, product quality, and operating expenses. Many processes which utilize metals and cyanides were designed in an era when pollution control was not a priority or when energy and raw material costs were low. Thus, many equipment redesign efforts have been undertaken to meet increasingly stringent environmental protection standards or to address relative changes in input costs.[5]

Equipment modification or modernization appears to be a prevalent method for achieving source reduction, despite the potentially high initial costs involved. New or better equipment may achieve the goals of source reduction in three ways. First, it may allow for the elimination of a hazardous material by performing mechanically an equivalent operation to a chemical process. As an example, mechanical cleaning or stripping of metal-based coatings may replace chemical methods which generate a high volume liquid waste. Second, new equipment may allow for the replacement of a hazardous material by a less hazardous one. For example, installation of high efficiency metal precleaning equipment can permit substitution of cyanide based plating baths with pyrophosphate copper solutions in certain applications. Third, new or better process equipment may simply provide better environmental control. An example of this would be the installation of counter-flow and stagnant rinses to minimize water discharge and improve recovery opportunities by concentrating waste rinse solutions.

Improving process controls is considered a particularly important aspect of equipment modification. Process controls may be less costly and more technically feasible to implement than replacement or modification of large-scale equipment. Process controls include manual, automatic, and computer-controlled systems. An example of the use of improved process controls to reduce waste generation is the increased usage of computerized controls for paint formulation which minimizes the potential for generating off-specification products and excess formulations which may otherwise be disposed. In the printed circuit board industry, process control techniques are commonly employed to minimize drag-out by maximizing drip time and to regulate the flow of make-up water to minimize unnecessary dilution.

The manner in which a process is operated may also be changed to effect waste reduction. This may be accomplished through the use of different temperatures, concentrations, or flow rates, by reducing the frequency of process startups or shutdowns, or by changing maintenance schedules. For example, reduction in plating bath metal concentration reduces water requirements for rinsing due to reduced drag-out.

5.1.4 Improved Operating Practices

Operating practices which can result in source reduction include improved housekeeping, waste stream segregation, and changes in procedural methods and personnel practices. Improved housekeeping practices are the most commonly employed and often the most cost-effective method to achieve source reduction.[6] These practices include optimizing equipment cleaning and maintenance, shutting down ancillary equipment when not in use, replacing gaskets, tightening valves, and other measures. Another manner in which source reduction can be achieved is through increased management attention to pollution control and waste generation. For example, many companies offer employee incentive programs for identifying cost-cutting measures, some of which involve source reduction of wastes.[7]

Waste segregation entails special storage or handling procedures to avoid the mixing of different waste streams. The segregation of wastes allows for certain streams to be treated, recovered, reused, or disposed of in a more environmentally and perhaps economically sound manner. Segregation is particularly desirable in eliminating the mixing of toxic waste streams with nontoxic streams, which otherwise results in a larger volume of waste requiring treatment. Waste segregation most often will require implementation of new equipment to collect the separated streams. The technical and economic feasibility of waste segregation, therefore, may be somewhat limited. Waste segregation is widely practiced between wastes which are amenable to different forms of treatment. For example, commonly segregated streams include cyanide, chromium, and other heavy metal wastes; bath dumps and high volume, dilute rinse waters; and highly complexed versus noncomplexed metal waste streams.

5.2 RECYCLING

According to the EPA, "recycling" is defined as practices in which wastes are either reclaimed or reused.[7] A reclaimed waste is one which is processed or treated through some means to purify it for subsequent reuse, or to recover specific constituents for reuse. Reused wastes are those which serve directly as feedstocks, without any treatment. Recycling of wastes may be done by either the original generator or other firms, although data indicate that the vast majority of recycling is performed onsite.[8]

This section summarizes the available technologies for spent metal/cyanide waste recycling. More detail on performance data, costs, and applicable waste types for specific technologies can be found in the corresponding chapters of this document.

5.2.1 Recycling Practices

Previous researchers have divided metal recycling technologies into the following unit operation categories: agglomeration, metal concentration, metal reduction, and metal substitution.[9] Agglomeration includes any process which gathers small particles into larger particles, where the smaller particles can still be identified. Waste dusts and vapors from particulate and vapor recovery equipment can be recovered through agglomeration technologies including low temperature bonding, hot briquetting, direct reduction, and green balling. These are used to create feedstocks which are high in metal content and of a physical size which facilitates material and process handling. One case study reported a one year payback of agglomeration equipment and handle filters used to collect and recycle metal dusts from secondary metal smelters.[10]

Metal concentration techniques include various membrane separation (reverse osmosis, liquid membranes, Donnan Dialysis, coupled transport), precipitation, extraction/leaching, adsorption (activated carbon, resin adsorption, ion exchange), thermal (calcination, evaporation, crystallization, smelting), biological, and flotation processes. Metal reduction techniques include electrolytic recovery, sodium borohydride, and thermal processes. Characteristics and limitations of these technologies are summarized in Table 5.2.1[11-17] and covered in detail in Sections 6.0 through 12.0.

Due to their low raw material price and relative ease of treatment, cyanides are not commonly recycled except in situations where the solution contains valuable metal or other components. For example, waste cyanide solutions in gold benefication and some cyanide plating baths are processed to recover the metal values which then permits reuse of the cyanide solution. As an example of the latter, the Department of Defense has patented a recovery process for plating baths which contain a sodium carbonate weight ratio in

TABLE 5.2.1. SUMMARY OF RECYCLING TECHNOLOGIES FOR METALS-BEARING WASTE STREAMS

Type of process	Description	Applications	Limitations of use
Metal concentration processes			
Hydrometallurgical processing (leaching)	Metals can be leached out of solids and sludges by extended contact with specific acids.	Extraction of metals from hazardous sludges, brine muds.	Concentration of desirable metals must be reasonably high (over 5,000 ppm) to make leaching attractive. Moderate cost of acids used is an economic constraint imposing lower limits on contents of waste to be handled.
Solvent extraction	Selective solvents used to extract and concentrate metal cations from aqueous solutions.	Economically feasible for recovery of vanadium pentoxide. Evaporation of the amine solvent leads to recovery of reasonably pure ammonium vanadate.	Solvent losses can be a problem with volatile or soluble solvents. High cost is not feasible for many metal-bearing wastes unless offset by metal recovery value.
Ion exchange	Ion exchange resins are produced which will selectively remove certain metal ions but permit others to pass when wastewater is pumped through the packed bed.	Same as above. Good for sulfuric acid anodizing, phosphoric acid bright dip; and chromic acid anodizing and chromating bath rinse recovery.	Expected life of resins is a concern in that frequent resin replacement will make the process more costly. Poisoning of resin with nonremovable impurities is also a major concern. For many applications the process is costly. Not capable of generating highly concentrated streams; excess regenerate required which becomes waste.
Precipitation	Metals dissolved in wastewater are precipitated out of solution by reacting them to form insoluble compounds.	Frequently used.	Recovered sludges need further processing to recover metal values such as dewatering, leaching and recovery of metals.
Chemical reduction	Addition of reducing agents to waste solution containing toxic metals causes precipitation of elemental silver and mercury, or the reduction of Cr^{+6} to Cr^{+3}.	Recovery of silver or mercury in useable form from wastes. Converts hazardous Cr^{+6} to nonhazardous Cr^{+3}.	Useful only for wastes containing easily reducible toxic constituents.
Crystallization	Solid metal compounds removed from solution by cooling it to lower solubility of metal salts.	Used only in limitedcases in which concentrated recoverable material is saleable; e.g., copper sulfate crystals from hydrogen peroxide solutions; sodium aluminate from caustic etch in aluminum finishing; ferrous sulfate from pickling liquors.	Practiced only for concentrated solutions (i.e., above 20 percent concentration).

(continued)

TABLE 5.2.1 (continued)

Type of process	Description	Applications	Limitations of use
Calcination	Consists of reacting metal-bearing sludges at high temperatures to drive off water and other volatiles, incinerate residual organics, and oxidize remaining inorganic compounds including metals.	Converts waste to oxide that is easily handled as feedstock by a smelter. Used only in limited cases.	Not applicable to wastes containing arsenic or selenium, which form volatiles oxides.
Evaporation	Concentration for recovery by evaporation.	Allows for recovery of concentrated solutions. Used for chromic acid etch and chromium plating solutions.	Energy costs place lower limits on concentrations to which technology is applicable. Cation exchange may be required to remove metal impurities.
Membrane separation	Solids larger than pore openings in the filter media are removed. The openings must be smaller to achieve metal separations than those used for organic separations. Driving force is pressure (reverse osmosis), liquid head (ultrafiltration). Liquid membranes make use of differential solubility and diffusion coefficients.	Allows for recovery of dilute solutions. Reverse osmosis widely used for nickel plating rinses. Less energy intensive than evaporation.	Membrane materials must be selected based on their ability to withstand degradation by the waste; chromic acid and high pH cyanide baths have been particularly difficult streams to treat with this operation. Cannot typically achieve desired level of concentration for return to plating bath. May be supplemented by evaporation. Feed filtration essential to minimize fouling.
Adsorption	Similar to ion exchange in selectively removing materials when wastewater is passed through a column of adsorptive media. Various natural materials including redwood bark and sphagnum moss are in commercial use for removal of various metals.	Removes metals from wastewaters, e.g., mercury removal in chloroalkali plants. Not frequently used due to higher costs.	Recovery of metals from adsorbents such as high surface area clay or silica is difficult. Not frequently used due to higher costs.
Retorting	Process used to recover mercury from sludges; waste is heated in an oxidizing environment. Mercury is recovered by condensation.	Recovery of mercury. If retorting is done properly, residue may be nonhazardous.	Energy-intensive operation. Value of recovered mercury may be insufficient to cover costs unless wastes with high mercury content are processed.
Foam flotation	Involves air flotation of foams after addition of polyelectrolyte and adjusting pH. Relatively new process - no commercial installations to date.	Effectively removes copper, zinc, chromium, and lead. Rarely used due to higher costs.	Raw material to process must be ore-like. Many waste types unacceptable as feeds. Rarely used due to higher costs.

(continued)

TABLE 5.2.1 (continued)

Type of process	Description	Applications	Limitations of use
Metal reduction and recovery			
Electrolytic recovery	Current passed through electrodes immersed in the metal solution. Metal ions migrate to the electrode where they give up an electron and are plated out. A variation, electrodialysis, was a membrane across alternating anion and cation exchangers.	Recovery of precious metals. Highly effective on copper pickling and milling solutions including sulfuric acid, cupric chloride and ammonium chloride solutions. Electrodialysis can recover ions from dilute solutions and generate a concentrated solution with low impurities.	Process becomes inefficient when handling dilute solutions (concentrations below 100 mg/L). Plastic substrates reduce value of reclaimed material due to undesirability of plastic in smelter.
Sodium borohydride	Addition of sodium borohydride to neutral or alkaline solutions of metals will result in precipitation of the metallic powders out of solution.	Recovery of mercury from chloroalkali production, recovery of metals from mixed metal finishing wastes.	Process limited to recovering more noble metals, i.e., precious metals, nickel, cobalt, copper, and mercury. Process limited to salts for which metals are easily formed by reduction and to neutral or alkaline solutions. Used in limited cases due to higher operating costs.
Reduction in furnaces	Sludge is mixed with coke or other reducing agent and heated.	Metal refining.	High cost limits this process to metal refining.
Other reducing processes	Copper can be removed from electroless solutions in metallic form by addition of formaldehyde and raising the pH. Copper will plate onto steel in acidic copper baths.	Recovery of material in metallic form.	Metal salt must be easily reducible. This limits process to precious metals, nickel, cobalt, copper, and mercury. Value of there covered material must justify cost of using the process. Used only in limited cases due to higher costs.
Agglomeration			
Low temperature bonding	Waste stream mixed with a binder; briquettes or pellets pressed out, which are then used as feedstock in metals operations (steelmaking, iron).	Allows for reuse of collected particulate materials.	Briquettes prepared by this method may not have desired integrity at elevated temperatures. Use of waste by metals procedure is probably preferable to any onsite use of such a process.
Hot briquetting	Feed material heated between 1600°F and 1800°F in fluidized bed, then pressed into briquettes.	Same as above.	Applicable only to solids with low vapor pressure at briquetting temperature. Process is not widely used.
Direct reduction	The process mixes, pelletizes, and preheats the waste stream on a grate and reduces the pellets on a rotary kiln by making use of the carbon in the pellets as the reductant.	Some oxide/hydroxide wastes from plating operations, if kept segregated by metals, could be a useful feedstock for a smelter using such a process to convert ore to metal.	Useful only with easily reducible substances (i.e., some metal oxides). Recovered metal must justify cost. Process is used as part of smelting industry to reduce ores to metals. Shipment of waste to smelter in lieu of onsite processing is probably preferred.

Source: Adapted from References 11 through 17.

excess of 6 to 1 relative to cyanide. The technique involves freeze crystallization of sodium carbonate crystals on a cold surface immersed in the bath.[11]

Table 5.2.2 presents a summary of metal/cyanide RCRA waste recycling activities as reported by waste generators and treatment facilities in the 1981 EPA National Survey.[18] Wastes which were not specifically identified as containing heavy metals, such as corrosives, spent halogenated solvents, and other wastes, have not been included in the table. In addition, since this study was not specifically designed to measure recycling (some recycled wastes are not considered RCRA wastes) and since recycling activities have increased substantially since 1981, these figures are probably underestimated. As shown in Table 5.2.2, wastes containing metals are recycled in the highest volume whereas wastes which only contain cyanides are not frequently recovered (2.9 percent of total recovery). With the exception of spent pickle liquors, high volume waste recycling is typically conducted onsite (77 percent, 94 percent if k062 is excluded).[18]

Offsite recycling activities include the recovery of scrap metals for re-refining as well as other applications. Recyclers may pay generators as much as 50 percent of the current market price for that metal for easily recovered wastes (e.g., metals recovered via electrowinning) or, conversely, charge prices comparable to disposal costs for dilute or highly complexed solutions.[15,18] Pure isolated sludges of tin, nickel, cadmium, copper, and zinc have excellent potential for being sold as byproducts.[15]

Examples of metal wastes recycled offsite are:[18]

- Recovery of zinc from steel mill flue dust for production of zinc and zinc salts;
- recovery of vanadium from spent sulfuric acid catalysts;
- reuse of copper, zinc, and nickel solutions as raw materials in chemical manufacture;
- recovery of copper, boron, manganese, zinc, and magnesium trace metals for fertilizer manufacture;
- recovery of concentrated metal hydroxides from sludges for production of metal salts;

TABLE 5.2.2. METAL/CYANIDE RCRA WASTES RECYCLED DURING 1981

Description of waste stream	WESTAT waste code	RCRA waste code	Volume recycled onsite (mg/l)	(%)	Volume recycled offsite (mg/L)	(%)	Total volume recycled (mg/L)
Metal-bearing wastes							
Chromium		D007	470	(99)	0.3	(0.1)	470
Lead		D008	32.0	(66)	17	(34)	49
Slop oil emulsion solids (petroleum refining)		K049	40	(98)	0.8	(2)	40
Dissolved air flotation float (petroleum refining)		K048	35	(97)	0.9	(3)	36
Emission control dust/sludge from primary production of steel in electric furnaces		K061	11	(38)	18	(62)	29
Emission control dust/sludge from secondary lead smelting		K069	5.6	(56)	4.5	(44)	10
Mixture of barium, cadmium, chromium, lead, and mercury	X039	mixture of D005,D006, D007,D008, D009	9.5		NR		9.5
API separator sludge from petroleum refining; hexavalent chromium and lead		K051	7.2	(96)	0.3	(4)	
Ignitable solid waste		D001	4.3	(90)	0.5	(10)	4.8
Washes and sludges from ink formulation		K086	<0.1	(<1.0)	2.4	(99)	2.4
Spent pickle liquor (steel finishing operations)		K062	28	(9.8)	260	(90.2)	290
Sulfuric acid, thallium salt (1)		P115	NR		1.6		1.6
Corrosive characteristic waste containing lead	X052	mixture of D002,D008	0.4		NR	0.4	0.4
Metal/Cyanide-bearing wastes							
Wastewater treatment sludges from electroplating operations		F006	430	(96)	19	(4)	449
Reactive characteristic waste		D003	18	(98)	0.3	(2)	18
Plating bath sludges from electroplating		F008	0.1	(24)	0.2	(76)	0.3
Cyanide-bearing wastes							
Spent plating bath solutions from electroplating operations		F007	3.3	(72)	1.3	(28)	4.6
Spent stripping and cleaning bath from electroplating		F009	0.6	(33)	1.1	(67)	1.7
Ammonia still lime sludge from coking		K060	1.3		NR		1.3
Sodium cyanide		P106	NR		0.5		0.5
Still bottoms from final purification of acrylonitrile		K012	0.2		NR		0.2
Cyanides		P030	<0.1	(80)	<0.1	(20)	<0.1

Source: Reference 18. Adapted from 1981 National Surveys of RCRA Treatment, Storage, and Disposal Facilities. U.S. EPA.

- precious metals recovery (e.g., silver from photographic paper, film, and spent developing solutions); and
- recovery of cobalt, molybdenum, nickel, and vanadium from petroleum refining hydrotreating catalysts.

Certain metal/cyanide wastes are unlikely to exhibit significant potential for recovery. These typically contain contaminants which are of low economic value, difficult to separate, or are unwanted since they are originally intended to be removed from the process. For these wastes, source reduction alternatives may be the most viable means to achieve waste minimization. A list of Fxxx and Kxxx waste codes which fall into this category is provided in Table 5.2.3.

5.2.2 Selection of Recycling Alternative

Economic considerations play a major role in determining the recyclability of a hazardous waste. The primary economic considerations are the capital and operating costs of the recycle system, residual disposal costs and value of recovered products. The economic benefits of recycling a wastestream or mixture of waste streams are dependent upon the physical and chemical characteristics of the waste stream and the quantity of waste to be recycled.

Physical and chemical characteristics of a waste determine the technical constraints of the treatment process. In general, physical form and corrosivity determine whether or not a process can be used whereas chemical characteristics affect ease of separation and selection of optimal processing conditions. The types of constituents in a waste will determine the extent of competing reactions, reagent requirements, processing efficiency, chemical reactivity, and/or ease of separation. In particular, the presence of multiple metal species, chelators, organics, and suspended solids present practical limits on the effective application of recycling technologies. Pretreatment of waste streams through the use of physical separation processes (e.g., filtering, decanting, settling, skimming) is common in recycling applications. Similarly, post treatment processes (e.g., addition of depleted chemical constituents) may be required to bring recycled materials up to process specifications.

TABLE 5.2.3. Fxxx AND Kxxx METAL/CYANIDE WASTES UNLIKELY TO BE RECYCLED IN SIGNIFICANT VOLUMES

EPA waste code	Waste	Reason for limited or no recycling
F007, F008 and F009	Spent cyanide plating solutions	CN content is usually destroyed rather than attempting recycle. Some recovery reported (e.g., zinc or copper cyanides).
F010, F011 and F012	Spent cyanides containing metal treating solutions	No metals of value to recover.
K002 - K005	Treatment sludges from chrome pigments production	Contain both trivalent chromium hydroxide and varying amounts of heavy metal chromate salts which are not easily reducible or separable.
K007	Sludges from iron blue production	These contain iron blue (iron ferrocyanide) in addition to other insoluble iron compounds. The ferrocyanide is not easily destructible.
K011	Bottoms from acrylonitrile production	Waste are higher molecular weight cyanides: not useful in a production process. Only option for recycling is burning for fuel value.
K013	Bottoms from acetonitrile	Same as above.
K014	Purification wastes from acetonitrile	Same as above.
K027	Residues from toluene diisocyanate production	Polymeric isocyanates useful only for fuel.
K031	Wastes from arseno-pesticides	Contains unwanted organoarsenates.
K044 to K046	Explosive wastes	Safety considerations limit reuse. Controlled detonation preferred.
K084, K101 and K102	Pharmaceutical wastes	Unwanted arsenic-containing byproducts limit reuse.

Source: Adapted from Reference 18.

The quantity of waste to be recycled is also a significant factor in the selection of an appropriate recycling technology. Waste quantity will determine the size of equipment, volume of raw materials to be used in recycling (e.g., carbon for carbon adsorption), pollution control equipment needs, and disposal requirements. Certain technologies may be preferable for small quantity processing but, in general, the larger the quantity of waste to be recycled, the more economically attractive recycling becomes. Disposal costs and value of recovered product become increasingly important as waste volume increases. If recovered products are used onsite, their value is reflected in the reduced demand for virgin raw materials. If sold for offsite use, their value is dependent on the market price of virgin materials and the degree of purity.

Another factor to be considered in selecting a recycling technology is whether the operation should be conducted onsite, or at an offsite facility such as a commercial recycler. In addition to costs, the choice between onsite and offsite recovery is dependent on many factors including availability of equipment, personnel and markets, facility size, technical capability of personnel, and use of recovered product. Transportation cost must also be considered for offsite recycling. The cost for transportation is a function of the distance from the generating facility to the recycling facility, the volume of the waste being transported, and the transportation method used. Smaller quantity generators of spent lead acid batteries and many firms in the Primary Metals Industry (SIC 33) have found it more economical to ship wastes offsite for recycling.[18]

In addition to the economic factors discussed above, the size of a facility and its technical expertise may also influence the decision to recycle metal/cyanide wastes onsite or offsite. Large facilities usually have the advantage of a strong technical staff to manage onsite recovery. However, onsite recovery has been found to be a competitive option to offsite recovery for both small and large generators. For example, generators of low volumes of wastes can significantly reduce their handling and transportation cost by participating in cooperative storage arrangements with other small quantity generators of similar wastes. The success of such a program, however, may depend on the similarity and chemical compatibility of the waste streams.

There are currently several cooperative metal recovery facilities in operation. Typically, they provide waste collection vehicles which make pickups at member waste generating facilities. Examples of cooperative ventures include the Metropolitan Recovery Corporation in Minneapolis which is an organization comprised of twenty printed circuit board and metal finishing facilities. It handles all wastes generated by its members and provides each with ion exchange canisters for metal recovery. Another cooperative venture is an ion exchange treatment facility, run by Tricil, which was sponsored by an association of thirty generator facilities located in Cleveland. A third venture of 100 facilities located around New York city has resulted in the formation of the Metal Finishers Foundation. This organization expects to establish a centralized metal recovery facility by December, 1987.[18]

Ultimate selection of a recycling technology and offsite versus onsite operation will be highly site specific. A selection methodology has been presented in Section 16 outlining economic and other considerations. Details on costs and capabilities of specific recycling technologies can be found in Sections 6 through 12.

5.2.3 Reuse of Metal/Cyanide Wastes

Certain metal hazardous wastes may be directly used for a different purpose in another process. The principal use of metals recovered from hazardous waste is onsite recycling as a feedstock.[18] Examples include direct reuse of plating rinse waters as rinses for compatible etchants or metal cleaning operations, mill scale recycled to steel mills, and lead oxide recycled for tetra-ethyl lead manufacturer. Another waste with potential reuse is hydroxide sludges containing chromium which could be solubilized through reaction with sulfuric acid to recover chromium sulfate for leather tanning. Reused wastes have the advantage of much lower costs relative to use of virgin materials, either through reduction in raw material purchase costs and/or through reduced costs for waste management. In general, reusable wastes are producea by large manufacturing operations, or those which require high purity, and are consumed by smaller facilities, often batch processors, which do not necessarily require high levels of purity in feedstocks.

Three primary factors should be considered when evaluating reuse as a potential waste management option. First, the ability to reuse a waste depends upon its chemical composition and effect of the various waste contaminants on the reuse process. For example, reuse of recovered plating baths may yield unacceptable plating quality if there is excessive buildup of carbonate (e.g., formed through the anodic oxidation of cyanide) or undesirable organics (e.g., formed through the breakdown of brighteners, wetting agents, and other compounds).[14] Second, the economic value of the reused waste must justify the expense incurred in changing a process to accommodate it. Third, the availability and consistency of the waste must be considered. A processor using a secondary material must be sure that the material will be available to satisfy his demand and that it will be of consistent quality to ensure minimal process upsets.

Waste Exchanges--

Reuse of wastes may be accomplished either by the generator itself, or through sales to a different processor or intermediary; i.e., waste exchange. An example of a direct transfer of metal containing waste which results in further waste minimization is the transfer of spent pickle liquor from Andrews Wire Company (South Carolina) to Diamond Shamrock. Andrews generates approximately 1.5 million gallons of waste pickle liquor containing 10 to 15 percent ferrous chloride and 5 to 10 percent HCl. Diamond accepts the waste without a fee provided that the acid content exceeds 5 percent. The waste is used to reduce hexavalent chromium compounds to trivalent chromium hydroxide in the plant's waste treatment system.[11]

Marketing of wastes for reuse is often facilitated through use of waste exchanges. Waste exchanges are institutions which serve as brokers of wastes or clearing houses for information on wastes available for reuse. In some waste exchanges, potential buyers of wastes are brought into contact with generators, while other waste exchanges accept or purchase wastes from a generator for sales to other users. Waste exchanges are considered by EPA to be of great potential value in future waste management since generators are often unfamiliar with characteristics of wastes generated by other industries.[19]

A wide variety of wastes have been recycled via the waste exchange system. A listing of metal hazardous wastes types available through waste exchanges is presented in Table 5.2.4. Table 5.2.5 provides details on waste quantities and estimated values that were recycled by three waste exchanges for which data were available. Of all hazardous waste types handled by these organizations, metal wastes are among those which have been sought most highly due to their versatile reuse potential.[8] In contrast, few examples of cyanide waste exchanges have been identified.[8,9]

In general, the "exchangeability" of a waste is enhanced by higher concentration and purity, quantity, availability, and higher offsetting disposal costs. Some of the limitations to waste exchangeability are the high costs and other difficulties associated with transportation and handling, costs of purification or pretreatment required and in certain cases, the effect on process or product confidentiality. In general, waste exchange involves transfer of either: (1)products from large, continuous processors to small, batch processors; (2)of manufacturing products from basic chemical manufacture to chemical formulators; or (3)products from high purity processors such as pharmaceutical manufacturers, to low purity processors such as paint manufacturers.[7]

Waste exchanges are operated by both private firms and public organizations. Several waste exchanges are listed below:[7]

- California Waste Exchange (California);
- Canadian Waste Materials Exchange (Ontario);
- Chemical Recycle Information Program (Texas);
- Colorado Waste Exchange (Colorado);
- Georgia Waste Exchange (Georgia);
- Great Lakes Regional Waste Exchange (Michigan);
- Industrial Materials Exchange Service (Illinois);
- Industrial Waste Information Exchange (New Jersey);
- Inter-Mountain Waste Exchange (Utah);
- Louisville Area Waste Exchange (Kentucky);
- Midwest Industrial Waste Exchange (Missouri);
- Montana Industrial Waste Exchange (Montana);

TABLE 5.2.4. TYPES OF WASTES LISTED BY WASTE EXCHANGES

Wastes available	Wastes wanted
Metals	
Zinc hydroxide filter cake	Alumina, aluminum, and aluminum sludge
Chrome drag-out solution	Nickel
Metal-plating sludge	Tungsten carbide
Electrodeless nickel bath	Copper solutions
Copper filter cake	Tin residue
Magnesium sludge	Precious metals
Aluminum oxide slag	Zirconia and zirconium compounds
Slag (60-70% Fe; 6% Cr; 3% Ni; 1% Si)	Residues, grindings, spent catalysts, sludges, and waste byproducts containing nonferrous and precious metals
Zinc cyanide	
Zinc-containing dust from baghouses and scrubbers	
Pickle liquors ($FeCl_2$, or $FeSO_4$)	
Chromic acid	
Cyanides/Reactives	
Sodium cyanide solution	No listings found
Cyanides; sodium potassium, or metal cyanide	
Cyanide solution from cyanide recovery process	
Zinc cyanide	

Source: Adapted from Reference 9.

TABLE 5.2.5. SUMMARY OF METAL/CYANIDE WASTES RECYCLED VIA THREE MAJOR WASTE EXCHANGES

Type of wastes	Waste exchange	Time period	Quantity (tons)	Distance hauled (miles)	Estimated value ($)
Copper sulfate crystals	--	--	3	300	840
Potassium cyanide (e)	--	--	0.11	125	150
Metals and metal sludges	IME[a]	1985	56.9	--	24,000 1,000 (d)
Metals	PWE[c]	1983 - 1984	6.9	--	15,000 (f) 186,020 (g)
Metal/metal solutions	NEIWE[b]	1983	--	--	--
Copper oxide	NEIWE	1983	10.0	1,025	5,000
Copper oxide	NEIWE	1983	5.0	25	2,400
Copper sulfate	NEIWE	1983	41.7	50	37,905 (d)
Nickel sludge	NEIWE	1983	80.0	200	5,000
Metal/metal sludges					
Copper sulfate solution (e)	NEIWE	6/81 - 12/81	7	400	1,000
Metal/metal sludges					
Copper sulfate solution (e)	NEIWE	2/82 - 2/83	Unknown	150	400

[a]IME = Industrial Material Exchange.
[b]NEIWE = Northeast Industrial Waste Exchange.
[c]PWE = Piedomont Waste Exchange.
[d]Unit cost estimate obtained from Chemical Marketing Reporter, May 28, 9184 issue.
[e]One-time only transaction.
[f]Savings/earnings.
[g]Average Replacement Value (Aggregate).
Source: Reference 9

- Northeast Industrial Waste Exchange (New York);
- Piedmont Waste Exchange (North Carolina);
- Southern Waste Information Exchange (Florida);
- Techrad (Oklahoma);
- Tennessee Waste Exchange (Tennessee);
- Virginia Waste Exchange (Virginia);
- Western Waste Exchange (Arizona); and
- World Association for Safe Transfer and Exchange (Connecticut).

The following is a list of the private material exchanges currently in business:[7]

- Zero Waste Systems, Inc. (California);
- ICM Chemical Corporation (Florida);
- Environmental Clearinghouse Organization - ECHO (Illinois);
- American Chemical Exchange - ACE (Illinois);
- Peck Environmental Laboratory, Inc. (Maine);
- New England Materials Exchange (New Hampshire);
- Alkem, Inc. (New Jersey);
- Enkarn Research Corporation (New York);
- Ohio Resource Exchange - ORE (Ohio); and
- Union Carbide Corporation (in-house operation only, West Virginia).

5.3 EXAMPLES OF WASTE MINIMIZATION PRACTICES

There is a growing incentive for companies to undertake waste minimization programs as a consequence of increasing waste disposal costs and liability. Besides protecting human health and the environment by drastically lowering the amount of waste generated, waste minimization programs can, in many cases, provide substantial economic benefits. The following is a summary of waste minimization practices employed by various industry categories. Much of this information was obtained through the results of a 1986 study commissioned by EPA to provide the U.S. Congress with information on the status of current waste minimization efforts in the country.[11] The industrial waste generator categories discussed below parallel those identified in Section 3.3 as being high volume metal/cyanide waste generators.

5.3.1 Acrylonitrile Production

The most significant means of minimizing waste generation can be accomplished by improving product yields thereby reducing the formation of heavy metal impurities. This could be accomplished through improvements in catalyst development and gas-catalyst contact in the ammoxidation reactor or through staged addition of NH_3. Segregation of acrylonitrile and acetonitrile purification bottoms from the quench-absorption aqueous effluent would permit incineration of the concentrated streams and reduce the toxicity of the wastewater.[11]

5.3.2 Metal Finishing

Waste minimization efforts in the metal finishing industry consist primarily of methods to minimize consumption of rinse water, extend bath life, recover baths and rinses, or to use some form of raw material substitution. Rinse water accounts for roughly 90 percent of raw waste generation in the industry.[4] Methods to achieve waste reduction which are unique to metal surface finishing are discussed below whereas waste minimization efforts which are similar to those employed in electroplating operations (e.g., reduction of drag-out, use of counter-flow rinses, etc.) are discussed in the following section.

The use of air-dried, no-rinse chromate conversion coatings for steel, galvanized steel, and aluminum in the coil coating industry has been reported by the EPA.[11] The literature also documents successful implementation of chromic acid recovery through both ion exchange and evaporation, and nickel recovery from rinses via electrodialysis, each involving subsequent recycling to the bath and reuse of rinse water.[11]

Raw material substitution can effectively reduce quantities of contaminated rinses. Cyaniding can be replaced by gas phase carbonitriding which eliminates the need for the rinse step.[11] Chromic acid rinses following zinc-based phosphating have been replaced by nonchrome rinses, although some loss in effectiveness has been observed.[11]

Generation of spent baths can be reduced by various methods aimed at extending bath life or removing contaminants. Filtering bath solutions is widely practiced to remove insoluble metallic salts.[20] These otherwise precipitate onto heating/cooling process equipment, thereby reducing energy efficiency, or precipitate onto metal parts resulting in impaired product quality.[21] Soluble salts can also lower bath activity and have been removed from electroless nickel baths by crystallization with subsequent filtration. The U.S. Bureau of Mines has experimented with chromic acid etchant recovery through use of an electrolytic diaphragm cell. Trivalent chromium is oxidized and reused along with the simultaneous recovery of copper.[22]

Contamination of baths can also be reduced by taking precautionary measures, such as thorough rinsing, to reduce drag-in. Also, rack maintenance (e.g., application of fluorocarbon coatings) is effective in preventing contaminant build-up resulting from dissolution of rack metals.

Primary bath treatment methods resulting in recovery include electrolytic recovery, ion exchange, crystallization and evaporation. Evaporation has been used successfully to recover plating solutions, chromic acid, nitric/hydrofluoric acid pickling liquors, and metal cyanide baths.[23] An example of recovery of metal finishing wastes is recovery of electroless nickel plating sodium phosphate salts by using ion exchange resins activated with hypophosphorous acid.[22] Liquid membranes have been used by Bend Research Inc. to recover contaminated dichromate rinses and baths.[24]

Raw material substitution has been applied to eliminate or reduce the amount of hazardous waste generated by metal finishing processes. As stated previously, cyaniding baths can be replaced by gas phase carbonitriding which utilizes ammonia gas instead of cyanide to provide nascent nitrogen. However, this is less economical for solutions which are used to treat many small batches requiring different cycle times and high heating rates.[11] Polysiloxanes, substitutes for cyanide-based stress relievers in electroless copper plating, are currently marketed by General Electric.[11] Ferric chloride or ammonium persulfate solutions can be substituted for chromic-sulfuric etchants and strippers if it is compatible with the basis metal. Peroxide-based secondary pickle solutions have successfully replaced chromic acid pickling liquor at a wire manufacturer resulting in improved product quality and economic savings due to recovery of the resulting pure copper oxide sludge.[25] As another example, at least 5 companies currently offer trivalent chrome systems for conversion coating applications which currently use hexavalent chrome.[11] Other substitutions include electroless copper for electroless nickel, plating of zinc instead of nickel, and varying substitutes for cadmium and silver depending on the application.[11]

Improved operating practices can also contribute to bath life. More frequent monitoring of bath activity and temperature can result in timely correction of deviations thereby improving both product quality and bath life.[11]

Several processing alternatives may provide potential substitutes for waste generating metal plating operations. However, these methods are either in the developmental stage or otherwise have not yet been widely applied in the U.S. These include vacuum evaporation methods for coating nickel, aluminum, and other metals; ion plating of chromium and cadmium; and chemical vapor deposition. Similarly, ion beam processing may provide an alternative to case-hardening treatments.[11]

5.3.3 Electroplating

Plating bath life can be extended by taking methods to reduce plating bath contamination. Examples of practices that will extend bath life include: use of purer anodes; improved rinsing, rack design, and extended drip time to reduce drag-in; use of deionized water to compensate for evaporative loss; and use of treatment techniques to selectively remove contaminants. Examples of treatment methods include filtering to remove suspended solids, use of carbon adsorption or chemical oxidation to remove organic breakdown products, and freezing to effect carbonate precipitation from cyanide solutions.[21]

Another plating bath waste minimization option is substitution of hazardous plating bath materials with nonhazardous compounds. For example, cyanide solutions have been effectively replaced by cyanide-free zinc solutions and pyrophosphate copper plating solutions. However, more stringent precleaning of the metal substrate is required to ensure high quality plating. Cadmium-based plating baths can be replaced with zinc graphite plating, titanium dioxide vapor deposition, and aluminum ion vapor deposition.[11]

Cadmium can also be replaced by zinc except in applications in alkaline environments or when the plate must be exceptionally thin.[2] Hexavalent chrome can be replaced by the less toxic trivalent chrome in certain applications, thus eliminating the need for a separate chrome reduction treatment process. In addition, significant sludge volume reductions (70 percent) have been reported due to the elimination of excess sulfate ions that are introduced during reduction.[11] Much of chromium plating is also used strictly for decorative purposes and is thus being replaced by other coating operations (e.g., painting of automobile bumpers).[26]

Waste rinse water generation can be minimized through drag-out reduction, by optimizing design and configuration of the rinse system, or through recovery of contaminated water. These can be accomplished through a variety of ways including:[11]

- Use of counter-current, multiple rinses;
- use of drip tanks, drain boards or stagnant rinses with recovery apparatus immediately following baths;
- use of rinse water for plating bath make-up;
- lowering bath surface tension (e.g., use of nonionic wetting agents) and viscosity (high temperature, changes in chemical composition);
- lowering concentrations of toxic chemicals;
- reshaping work pieces and rack layouts to improve drainage;
- increasing drip time; and
- use of methods to increase rinse efficiency such as agitation of immersion rinses or use of spray or fog rinses.

The EPA has reported that acceptable chromium plating can be achieved with CrO_3 concentrations as low as 25 to 50 g/l versus traditional concentration levels of 250 g/l.[27] Since drag-out is directly proportional to concentration (more so if viscosity effects are included), rinse contamination can therefore be reduced by up to a factor of ten. One author estimated that the use of wetting agents to decrease surface tension can reduce drag-out by as much as 50 percent.[27]

Automatic process controls enable drip time to be maximized by changing it to correspond to variations in process throughput or changes in work pieces. Use of counter-flow rinse tanks arranged in series are capable of achieving theoretical reductions in water requirements up to 90 percent.[27] Spray or fog rinsing is widely used and most effectively applied on rack-mounted, simple shaped parts with high area of exposed surface. Finally, rinse waters are frequently reused in the same process through recovery (e.g., electrolytic recovery from stagnant rinses) or reused in another process operation when its contaminants will not adversely affect the subsequent processing step (e.g., in nickel plating, the same rinse can be used following alkali cleaning, acid dip and nickel plating tanks). Cost evaluations of rinsing options have been presented in detail in the literature.[27]

Sludges generated from onsite wastewater treatment can be reduced by modifying treatment operations. Waste segregation, use of more effective precipitation agents, and sludge dewatering are the major categories of waste reduction options. Segregation of wastes containing highly complexed solutions permits specialized batch treatment and thus optimizes reagent requirements and subsequent sludge generation. Segregation of streams containing different metals can result in waste products that are more amenable to recovery or reuse. For example, nickel hydroxide sludges have reportedly been reused as plating bath make-up, as have chromium bath scrubber wastes.[11]

Primary recovery techniques employed by the industry include evaporation, electrolytic recovery, reverse osmosis, ion exchange, and electrodialysis. In addition, several plating bath suppliers (e.g., MacDermid, Harshaw, CP Chemical) reprocess spent baths for their customers.

Evaporation is simple and reliable but also energy intensive and nonselective; i.e., it concentrates impurities as well as metal components. Thus, to be economically attractive, evaporation is often utilized in conjunction with other reduction techniques such as counter-current rinsing, to concentrate solutions, and deionization of rinse water to reduce build-up of calcium and magnesium salts in the recovered concentrate. Multiple effect and vapor recompression evaporators can recover 90 to 99 percent of heavy metals and are currently used most commonly on chromium baths.[28]

Relative to evaporation, membrane technologies are, in general, more selective and lower in operating costs but are also more complex processes. Reverse osmosis has been most effectively applied in the recovery of nickel rinses and has also been used for cadmium recovery.[29] Its use is restricted to dilute, prefiltered solutions with moderate pH levels to ensure sufficient membrane life.[7] Ion exchange is more versatile in its application, currently applied in nickel, chromium, cyanide, silver and other metal-containing rinse solutions.[30] However, it requires a high level of process control and maintenance. Electrolytic recovery is highly effective for recovering wastes from concentrated rinses and has found more recent application in direct recovery from process and treatment baths (e.g., cyanide destruct tank).[11] Electrodialysis has also found application for rinse recovery to remove silver, cadmium and other metals.[11] As with reverse osmosis, membrane stability and fouling potential restrict its application.

A national survey of electroplaters and metal finishing facilities conducted in 1983 identifies evaporation, ion exchange, and reverse osmosis as the three most widely applied recovery technologies.[31] Use of these methods for recovery of specific solutions is summarized in Table 5.3.1. Technologies which have potential for plating solution recovery, based on pilot testing, industrial application, or theoretical considerations are summarized in Table 5.3.2.[6,31] The EPA has estimated that the majority (48 percent) of the electroplating industry heavy metal discharges are accounted for by nickel and chromium. Since these are also the most expensive of the metals which are discharged in large volumes, they account for an even higher fraction of the the value of lost minerals in the industry (83 percent).[32] Recovery costs for these metals is expected to be significantly offset by the recovery value of the metals.[32]

5.3.4 Printed Circuit Boards

Waste reduction methods are similar to those previously discussed for electroplaters and metal finishing facilities. Those that are specific to printed circuit board manufacturers which result in the reduction of California list wastes are discussed below. These include direct substitution or recovery of these wastes as well as reduction of other metal containing wastes which contribute to wastewater sludge generation (F006).

Chromic acid used for desmearing has been successfully replaced by concentrated sulfuric acid and, more recently, potassium permanganate. The advantages of the latter are: 1) it does not introduce chromium into the waste effluent; 2) it is not hydroscopic like sulfuric acid, therefore bath lives are extended; and 3) recent developments by Morton Thiokol have resulted in production of a proprietary additive which will reoxidize the permanganate [11]

TABLE 5.3.1. APPLICATION OF LEADING RECOVERY TECHNIQUES FOR ELECTROPLATING AND METAL FINISHING

Application	Units in operation*		
	Evaporation	Ion exchange	Reverse osmosis
Chromium plating	158	50	-
Nickel plating	63	38	106
Copper plating	19	-	3
Zinc plating	7	-	3
Cadmium plating	68	-	-
Silver/gold plating	13	20	-
Brass/bronze plating	10	-	-
Other cyanide plating	6	-	-
Mixed plating wastes	-	11	6
Chromic acid etching	6	-	-
Other	16	2	1

Source: Reference 31.

*According to a survey of the U.S. Electroplating and Metal Finishing industries cited in the reference.

TABLE 5.3.2. POTENTIAL METAL FINISHING BATH RECYCLING PROCESSES

Metal finishing baths commonly used	Electrolytic Recovery	Evaporation	Electrodialysis	Ultrafiltration	Ion Exchange	Reverse Osmosis
Plating - Hard and Decorative		X			X	X
Nickel	X	X	X		X	X
Nickel Iron	X		X			
Copper Cyanide	X	X				X
Copper Acid	X					X
Copper Pyrophosphate	X					
Tin, Acid	X		X			
Tin, Alkaline	X		X			
Tin Fluoborate	X	X	X			
Zinc Cyanide	X					X
Zinc, Acid	X	X				X
Tin/Lead, Fluoborate	X		X			X
Cadmium Cyanide	X	X	X			X
Gold Cyanide, Alkaline	X		X		X	
Gold Cyanide, Acid	X		X		X	
Silver Cyanide	X	X			X	X
Electroless Baths: Copper	X				X	
Electroless Baths: Nickel	X				X	
Pickling: Sulfate Copper	X					
Pickling: H_2O_2/H_2SO_4 Copper	X					
Pickling: HNO_3 Copper	X					
Cleaning: Alkaline Cleaners				X		
Cleaning: Acid Cleaners				X		

Source: Reference 5 and 31.

thereby increasing bath life and reducing sludge generation in the etch tank. However, this solution is relatively expensive and can spontaneously combust if it is exposed to air and allowed to dry.

Plating baths are commonly replenished and treated to enable reuse. For example, firms commonly remove organic breakdown products (e.g., from stabilizers and brighteners) from copper, nickel and solder plating baths by oxidation (e.g., potassium permanganate) followed by carbon adsorption and filtration (e.g., diatomaceous earth).[21]

A large number of printed circuit board manufacturers have switched from panel plating to pattern plating. Since the latter only involves electroplating board holes and circuitry, its use reduces the amount of noncircuit copper which must be subsequently etched away.[21] This, in turn, reduces the amount of etching waste generated and discharged to onsite treatment processes. Other processing techniques which can reduce or eliminate the generation of etching wastes include using dry plasma etching techniques (e.g., using reactive gaseous radicals, using nonreactive ion bombardment), using additive or semi-additive instead of subtractive board manufacturing, using less toxic etchants (e.g., ammonium persulfate, peroxide-sulfuric acid which are widely used in place of chromic acid), or using in-line recovery methods to extend etchant life (e.g., liquid membrane copper recovery).[11] Peroxide-sulfuric use in etchants, only recently adopted by industry, has the advantage of not introducing additional chelators into the plant's discharge stream. It is also easily regenerated through crystallization which results in the precipitation of copper sulfate crystals. These can be easily removed from the etch tank and have potential reuse applications.[11]

5.3.5 Inorganic Pigments Manufacture

Cadmium and other metal dusts collected in air pollution control equipment have reportedly been recycled for use in low grade paint. (Versar 1980) Substitutes for red lead primer and chrome yellow (used in traffic paint) have been identified but generally do not result in comparable performance, cost, or color characteristics.[11]

Waste reduction at pigment production facilities has also been achieved through modifications in conventional wastewater treatment systems.[33] Conventional treatment consists of chromate reduction followed by filtration and landfilling of the collected solids. A modified process, employed by at least two facilities, consists of the following: (1)use of improved filtration systems to minimize wastewater particulate content; (2)addition of soluble barium salts to precipitate barium chromate which can then be used to produce a light yellow pigment; and (3)pH adjustment to alkaline conditions to precipitate lead and zinc (e.g., as hydroxides or carbonates) which can then be recycled to the process as feedstock salts.[33]

5.3.6 Petroleum Refining

Since the generation of hazardous waste from petroleum refining is a direct result of the attempt to remove existing impurities from the crude feed, waste minimization in the industry is primarily accomplished by sludge consolidation. This includes maximization of slop oil recovery and separation of water and oil from other, nonrecyclable waste products. For facilities with cokers, much of the API, DAF, and slop oil sludges can be converted to coke, according to industry representatives.[11] In addition, the current trend away from production of leaded gasoline will reduce wastes generated from tetra-ethyl lead production and leaded gasoline storage (i.e., K052)

Leaded tank bottoms can also be reduced by agitation of the tanks which effectively transfers solids downstream, eventually ending up in either asphalt or coke byproducts. Methods have also been developed to recover this sludge by dissolving it in a heated, low viscosity distillate with the resulting liquid sent to slop oil recovery systems.[11] Other processes currently in use by refineries include the Victor extraction process, which uses steam and air to separate residual oil trapped in the sludge; physical sludge consolidation processes such as vacuum filtration; thermal, chemical, or ultrasonic emulsion breaking; solvent extraction (e.g., B.E.S.T. process from Resource Conservation Co.); and electroacoustic dewatering.[11] Leaded tank bottoms have reportedly been treated by calcination to recover lead oxide. Tank bottoms are reacted at high temperatures to drive off water and other volatiles, incinerating residual organics, and oxidizing the lead.[18,34]

Many of the above processes are applicable for recovery of oil from API separator sludge and DAF float. Other methods include installation of floating roofs which was found to reduce the oxidation of oil and the resulting formation of heavy waste material in API separators. Conversion from induced air to pressurized air in DAF units has resulted in the generation of less than one half the float volume for the same degree of solids removal.[11]

Methods to achieve waste minimization for other refinery waste streams of concern include 1)hydrotreating catalytic cracking feed to remove metal contaminants, thereby extending catalyst life; 2)substitution of chromium corrosion inhibitors in cooling water with organic chelating agents, nonoxidizing biocides, and other proprietary compounds;[11] and 3)recovery of Raney nickel catalysts through roasting, leaching of aluminate, and preparation of nickel carbonate.[35]

5.3.7 Wood Preserving

For chromium/arsenic preservatives received in drums or bags, closed systems are available which can minimize residual levels of metal contaminants in the containers. Alternatively, plastic liners or reusable drums can be used. Sludge from the work tank can be minimized by careful operating practices that ensure minimal amounts of dirt, silt, and loose wood fiber entering the retort before treatment. Most facilities have installed drip pads and spill basins to collect excess preservative which drips from the wood pieces after treatment.[11] Other measures to reduce the amount of water contaminated and thus requiring further treatment includes covering processing areas, increasing drip time, and diverting run-on. Use of nonchromate cooling water treatment chemicals would reduce the amount of this compound in the plant's combined treatment sludge.[11]

5.3.8 Chloralkali Industry

The membrane cell process for the production of sodium hydroxide has begun to replace the more costly mercury cell process. Since its introduction in 1980, six plants have opened in the U.S. and, in response, mercury cell plants have been closing.[11] The membrane cell process eliminates the generation of mercury containing hazardous waste. In fact, DuPont claims that its most recently built plant will completely eliminate the production of hazardous waste.[11] Over the long term, conversion or closing of non-competitive mercury cell facilities is likely since the membrane process not only results in less pollution control costs but also requires approximately 30 percent less energy per unit of production.[4]

For plants which continue to use the mercury cell process, waste minimization options are available. Retorting has been used in the chloralkali industry to remove mercury from mercury-bearing sludges and solid wastes. The waste is heated in an oxidizing environment forming mercury gas which is collected by condensation.[11] Alternatively, wastes are pretreated through hydrometallurgical processes. One facility leaches contaminated muds with sulfuric acid to concentrate mercury and convert the bulk of the solids to nonhazardous gypsum.[11] Although the latter is capable of recovering over 99 percent of the mercury contained in the sludge, its capital cost is several million dollars and thus appropriate only to large volume facilities.[10] Solvent extraction has also been suggested for stripping mercury from effluent wastewaters (see Section 7.2).

5.3.9 Other Industries

Raw material substitution and waste recovery technologies have been the predominant means of waste minimization in industries which use silver-containing photographic films.[29] Several companies, including Napp Systems, are marketing silver free films for lithography.[29] To a large extent, used and spoiled film is already sent to professional recyclers for silver recovery. Wastewaters containing silver are economically recovered using technologies such as metallic replacement, chemical precipitation, electrolytic recovery, reverse osmosis, and ion exchange.[11]

Printing inks may contain heavy metals such as chromium. Contaminated solutions can be recycled onsite or shipped to ink manufacturers who reuse these materials in the formation of black newspaper ink.[29]

Several examples of catalyst recovery by the inorganic chemicals industry have been documented. Solvent extraction has been used to recover vanadium pentoxide from spent sulfuric acid catalysts using a high molecular weight amine. The amine solvent is subsequently evaporated leaving a reasonably pure ammonium vanadate which is available for reuse.[11] Another example is fluidization and precipitation of spent nickel catalysts used by inorganic chemical manufacturers. A nickel salt is formed by dissolving the catalyst in a mineral acid. This is reacted with soda ash to precipitate nickel carbonate, which is then collected and reacted with sulfuric acid to form a nickel sulfate solution. Sodium sulfide is added to precipitate iron salts and the resulting solution is purified through filtration and evaporation. A similar process is employed by manufacturers of plating chemicals for recovery of nickel plating solutions.[11]

Cadmium is used as a stabilizer for polyvinyl chloride. In this application, it can be replaced by organotin compounds which are more efficient but also more expensive.[2]

5.4 WASTE MINIMIZATION SUMMARY

Regulatory trends appear to be moving towards the promotion of waste minimization. The EPA has recently proposed requirements that generators certify institution of hazardous waste reduction programs.[36] Generators would be required to reduce the volume or toxicity of hazardous wastes to a degree determined by the generator to be economically practicable. Three states currently have established source reduction/pollution prevention programs: North Carolina, Minnesota, and Massachusetts. In addition, Tennessee has established a "pilot program", and Kentucky, California, Maryland, and Washington have programs currently in development. These programs vary but, in general, include information exchange, technical assistance, and economic incentives to companies to encourage development of their programs.[20]

Table 5.4.1 presents a summary of several documented cases of waste reduction involving metal/cyanide hazardous wastes.[9,37] Additional case studies can be found in the appropriate sections of this document pertaining to specific recovery technologies. Although some of the data in Table 5.4.1 are incomplete, this compilation clearly demonstrated the potential economic benefits which can be achieved through implementation of waste minimization technologies. In particular, since disposal costs have increased sharply in recent years, payback periods indicated in the table can be interpreted as being conservative estimates.

TABLE 5.4.1. COMPILATION OF INDUSTRIAL WASTE MINIMIZATION

Company and location	SIC Code	Product	Waste minimization method description	Waste Reduction Percent	Waste Reduction Quantity	Year (*)	Capital investment ($1,000)	Annual cost savings ($1,000)	Payback period
Climax Molybdenum Co., Leadville, Colorado	1061	Raw molybdenum copper, zinc, iron, manganese	Installation of interceptor canals to pass runoff water through ion exchange unit. Other separation of heavy metals through coagulative electrolytic processing.	93.2 (Cu) 99.9 (Fe) 99.9 (Mn) 93.4 (Zn) 90.3 (Mb) 96.4 (Cn)					
Florida Steel Corp., Charlotte, N.C.	3312	Steel	Resale of collected high-zinc furnace dust instead of land filling.				No change	129.6	Immediate
Waupaca Foundry, Wauoaca, Wisconsin	3321	Grey and compacted graphite iron casting	Separation of wastewater to reduce amount requiring treatment to 25 percent.			1980		20.995	3.5 yrs
Stanadyne, Inc., Sanford, N.C.	3432	Plumbing products	Combination of equipment installations and revised operating procedures to reduce sludge volume and cyanide concentrations from plating operations.	46.0		1982			
Elkhart Products Division, Inc., Elkhart, Indiana	3471	Pipe fitting fabrication	Process redesign for waste reduction in combination with electrolytic recovery of copper.		182,000 gal/yr	1979	60	120	0.5 yr
Pioneer Metal Finishing, Inc., Franklinville, N.J.	3471	Electroplating job shop	Replacement of single-pass continuous treatment system with closed-loop batch treatment systems.	50.0 (sludge)	40,000 gal/yr (sludge)		210	52.460	3 yrs
Deche and Co., Moline, Illinois	3520 3530	Farm and construction equipment	Formation of a hazardous waste task force. Comprehensive treatment facility to reclaim and detoxify selected wastes.		330,000 gal/yr	1980	1,900	155.750	2.5 yrs
Emerson Electric Co., Special Products Division, Murphy, N.C.	35 36	Metal finishing Stationery manufacture	Equipment installation for waste reduction/elimination plus a chemical waste management program and an incentive program for cost reduction or product ideas.	100.0 solvent waste oil	68 lb/dy (paint solids) 90 lb/dy (plating acid, oil, caustic) 720 lb/mo (solvent)		874	1,800	1.1-5 yrs
GTE Sylvania, Chicago, Illinois	3661	Electronic telephone switching equipment	Installation of closed-loop treatment system and electrolytic copper recovery system.		55 gal/wk (CuOH sludge)	a1976		6	
Data General Corporation, Clayton, N.C.	3573 3679	Printed circuit boards	Sale of untreated wastes, process changes to obtain marketable copper sludge.	100.0 (process waste-water)	400 ton/yr (landfill waste)	1981	50	180	1.5 month
Ian Heath Co. Birmingham, England	-	Silver plating of giftware	Installed 3 Chemeles recovery units on nickel plating and pyrophosphate copper rinses.	-	5-10 kg, Au, 14 kg Ni, and 5 kg Cu per week	1979	43	-	-
Artistic Plating Co. Milwaukee, WI	-	Precious metals plating on plumbing accessories	Gold recovery through electrodialysis and ion exchange.	—	—	1982	24.4	-	9 months

(continued)

TABLE 5.4.1 (continued)

Company and location	SIC Code	Product	Waste minimization method description	Waste Reduction Percent	Waste Reduction Quantity	Year (*)	Capital investment ($1,000)	Annual cost savings ($1,000)	Payback period
Digital Equipment Corp., Tempe, AZ	3679	Printed wiring boards	Electrolytic recovery of Cu in treatment system, use of waste from other industries.			a1976			
Modine Manufacturing, Trenton, MO	3714	Metal radiators	Ion exchange and electrolytic equipment for recovery of copper.	100.0			27	22	14 months
Carolina Power and Light Co., New Hill, N.C.	4911	Electric power	Sale of fly ash and bottom ash for reuse.	30.0 (max)		1979			
Duke Power Co.	4911	Electric power	Sale of fly ash and bottom ash, waste stream segregation, revised equipment operation establishment of an ongoing waste minimization program.	60.0 (tritium and boron release)	345,000 ton/yr (ash)	1979	1,365	6,184	1 yr
PCA International Inc., Matthews, N.C.	7395	Mass portrait photography	Silver recovery through electrolysis and developer regeneration through ion exchange.		115,000 troy ounce silver 2,919 gal/dy				1 yr
E.I. duPont de Nemours & Co., Inc. Wilmington, Del.	28 13 29	Fibers Industrial and consumer products Polymer products Agricultural and industrial chemicals Biomedical products Coal Petroleum product	Process change in ADW manufacture. Marketing of landfilled waste, recycling of off-spec product. Filtration of paint sludge, incineration of liquid. Pretreatment of waste aluminum oxide for sale to recycler. Process change to reduce incinerator ash. Sale of waste ferric chloride instead of ocean dumping. Process modification to reduce load to treatment plant.	50.0 (water) 100.0 90.0 20.0		1980			
Exxon Chemical Americas, Houston, Texas	28 2221 2298 3079	Olefins, aromatics, polyolefins, elastomers, solvents, specialties, oil/fuel additives	Aluminum hydroxide removal from sludge for reclamation.	60.0		1982			
Olin Corporation, Standford, CN	26 28 34	Chemicals Brass and stainless strip and mill products	Replacement of cyanides, solvent-based paints, raw material substitutions.			1983			
California Electroplating, Los Angeles, Calif.	3471	Plating	Installation of four counterflow rinses after each plating stage, plus spray nozzles to wash drag-out back into tanks.						
Vasland Metal Services, Inc., Cincinnati, OH	-	Multipurpose plant	Installation of an Electrochemical Reactor for Cd recovery.	99.4		1980	8.5 (annual)	17.5	
Pontiac Division of General Motors, Pontiac, MI	-	Plating of bumpers							
Gillette Razor Boston, MA	-	Barrel Plating of Ni	Installation of Reverse Osmosis on Ni dragout tanks.	85		1979	8.5		

(continued)

TABLE 5.4.1 (continued)

Company and location	SIC Code	Product	Waste minimization method description	Waste Reduction Percent	Waste Reduction Quantity	Year (*)	Capital investment ($1,000)	Annual cost savings ($1,000)	Payback period
Allied Metal Finishing Baltimore, MD	3471	Plating of parts	Installation of H.S.A. electrochemical reactor.	31	Cd 1.22 ppm in effluent	1979		17	1.5 yrs
General Plating, Detroit, Michigan	3471	Metal plating	Installation of tising film evaporator unit.		350 lb/dy chromic acid				100
Ford Motor Co., Sabine, Michigan	3471	Metal plating	Installation of evaporator recovery units.						
Ford Motor Co., Sabine, Michigan	3471	Automotive parts plating	Installation of three evaporator recovery units.						
Advance Plating Co., Cleveland Ohio	3471	Automotive parts plating	Attachment of Innova Chrome Napper ion transfer system to allow closed-loop recovery of chromiu.	80.0-90.0 (Cr) 99.0 (water)					
Reliable Plating Works, Milwaukee. Wisconsin	3471	Plating of napkins, paper towels and toilet tissue dispensers	Attachment of Innova Chrome Napper ion transfer system to automatic hoist line.	80.0-90.0 (Cr) 99.0 (water)	39.4 kg/wk (H2CrO4) 92 Kl/dy	1980	15	-	Amortization by Cr and H_2O savings alone
Colorcraft, Rockford, Illinois	7395	Photo finishing	Installation of production prototype developer recycling unit (electrodialysis).	80.0					50
Deluxe Motion Picture Labora-tories, Hollywood, California	7819	Film processing	Introduction of ion exchange technology to purify waste washwater for reuse and recover silver.	62.0 (water) 90.0 (Ag)					
PCA International Matthews, N.C.	7395	Photo finishing	Recycle bleaches. Recover silver from washwater. Recover heavy metals from wastewater. Recycle pre-bath, final bath, paper color developer. Total water recycle with 3-stage evaporation.	100.0 (Ag) 100.0 (water)					
Phelps Dodge Corp., Hidalgo, New Mexico	33	Copper	Replacement of copper smelting using a reverberatory furnace to the Outokumpu flash smelting process to facilitate sulfur recovery and energy saving.						
Kennecott, Garfield, Utah	33	Copper	Use of Noranda continuous process for copper smelting to facilitate sulfur and energy saving.						
Ubis, Inc., Crankton, R.I.	3471	Electroplating job shop	Replacement of counter-current flow rinsing by the Providence method, which removes the majority of contaminating dragout in small volume before using a flowing rinse to reduce wastewater volume and to a low for recovery of plating solution.	94.0 (waste-water)					
Phillips Plating Co. Phillips, WI	3471	Cr Plating	Installation of closed-loop rising film evaporator to concentrate Cr plating bath drag-out for recycle.		1.8 kg Cr/hr	1979	-	-	-
Sommer Metalcraft Corp. Crawfordsville, IN	-	Paints, plates, steel products	Installation of an evaporator to recover Cr wastes.		180 kg/day	1980	-	100 based on Cr consumption alone	-
Allied Finishing, Grand Rapids, MI	3471	Zinc castings and sheet metal stampings	Installation of Cr recovery evaporator.	84% Cr consumption, 15-20% sludge generation		1979	-	-	2-2.5 yrs.

*a:after; b:before.

Source: Reference Nos. 9 and 37.

A survey of 610 hazardous waste generators in Massachusetts was conducted in 1985 to identify current and planned source reduction efforts.[6] Of these facilities, 238 were identified which practiced source reduction activities for metal containing wastes, 107 for cyanides, and 7 for petroleum refining wastes. Current source reduction activities and percent reduction in volume achieved are summarized in Table 5.4.2. Predominant methods employed include waste segregation, process modification, improved housekeeping, precipitation, improved rinsing, and chemical detoxification for cyanides.

This section was not intended to represent a complete survey of waste reduction practices available to generators of metal/cyanide wastes. The limited scope of this survey only permitted a broad overview of available methods to be presented, supplemented with specific examples for high volume waste sources. A comprehensive literature survey of waste reduction practices is being undertaken by the EPA Office of Solid Waste.[38] The survey data will be compiled in the form of a computerized data base and is intended to provide technical assistance for both states and private companies. This should be available by 1988 or early 1989 and is expected to represent a significant improvement over current compilations of waste minimization data.[39] Other useful sources of information which are currently available include several surveys which provide lists of articles, by industry, on waste reduction practices.[10,40,41]

TABLE 5.4.2. SOURCE REDUCTION ACTIVITIES PRACTICED BY RCRA WASTE GENERATORS IN MASSACHUSETTS

Source reduction technique	Aqueous metals		Cyanides		Petroleum wastes	
	Facilities using method[a] (%)	Waste reduction[b] (%)	Facilities using method[a] (%)	Waste reduction[b] (%)	Facilities using method[a] (%)	Waste reduction[b] (%)
Waste segregation	9.7	46	12.1	50	28.6	19
Process modification	11.3	39	13.1	50	14.3	5
Better housekeeping	8.8	42	9.3	30	14.3	5
Waste recycling	5.9	62	1.9	92	14.3	100
Raw material substitution	1.3	22	5.6	58	-	-
Waste rinse	5.0	43	1.9	55	14.3	100
Neutralization	5.9	65	3.7	54	-	-
Filtration	6.3	61	3.7	50	-	-
Distillation	0.4	25	-	-	14.3	100
Product reformulation	2.1	64	0.9	10	-	-
Precipitation	9.7	47	7.5	56	-	-
Improved rinsing	9.7	33	15.0	33	-	-
Chemical detoxification	2.1	60	12.1	61	-	-
Sedimentation	4.6	53	2.8	22	-	-
Clarification	5.9	93	3.7	74	-	-
Evaporation	2.1	71	1.9	82	-	-
Carbon adsorption	0.4	10	-	-	-	-
Ion exchange	3.8	47	1.9	57	-	-
Flotation	1.3	90	-	-	-	-
Electrodialysis	1.7	66	1.9	19	-	-
Other	2.1	22	0.9	15	-	-
Total source reduction	238[c]	51	107[c]	48	7[c]	50

[a]Percentage of facilities practicing some form of source reduction.

[b]Percentage of waste generation prior to implementation of source reduction technique.

[c]Total number of facilities practicing source reduction.

Source: Adapted from Reference No. 6.

REFERENCES

1. Minnesota Waste Management Board. Hazardous Waste Management Report. 1983.

2. Congress of the United States. Office of Technology Assessment. Serious Reduction of Hazardous Waste: For Pollution Prevention and Industrial Efficiency. U.S. Government Printing Office, Washington, D.C. 1986.

3. Carner, P. Telecon with M. Kravett, Alliance Technologies Corporation. W.R.'Grace and Company. Lexington,'MA. June'1986.

4. Versar, Inc. Technical Assessment of Treatment Alternatives for Wastes Containing Metals and/or Cyanides. Springfield, VA. Performed for U.S. EPA Office of Solid Waste under Contract No. 68-03-3149. October, 1984.

5. Breton, M. et al. Alliance Technologies Corporation. Technical Resource Document: Treatment Alternatives for Solvent Containing Wastes. Prepared for U.S. EPA HWERL, Cincinnati, OH under Contract No. 68-03-3243. August, 1986.

6. Roeck, D.R., et al. GCA Technology Division, Inc. Hazardous Waste Generation and Source Reduction in Massachusetts. Bedford,'MA. Contract No. 84-198, MA Dept. of Env. Mgt., Bureau of Solid Waste Disposal, June, 1985 (Draft).

7. Wilk, L. et al. Alliance Technologies Corporation. Technical Resource Document: Treatment Alternatives for Corrosive Wastes. Prepared for U.S. EPA HWERL, Cincinnati, OH under Contract No. 68-03-3243. Sept., 1986.

8. Versar, Inc. National Profiles for Recycling: A Preliminary Assessment. (Draft) Prepared for U.S. EPA Waste Treatment Branch under Contract No. 68-01-7053 July, 1985.

9. Versar, Inc. and Jacobs Engineering Corporation. Waste Minimization Issues and Options. Volume 3. U.S.EPA Office of Solid Waste. EPA 530/SW-86-043. October, 1986.

10. Overcash, Michael R. Techniques for Industrial Pollution Prevention. A Compendium for Hazardous and Non-Hazardous Waste Minimization. Lewis Publishers. 1983.

11. Versar, Inc. and Jacobs Engineering Corporation. Waste Minimization Issues and Options. Volume 2. U.S.EPA Office of Solid Waste. EPA 530/SW-86-042. October, 1986.

12. Methods Described to Minimize and Manage Electroplating Sludges. The Hazardous Waste Consultant. July/August, 1985.

13. Steward, F. A. and W. J. McLay. Waste Minimization: Part 1: Introduction, Alternate Recovery Technologies. Metal Finishing, August, 1985, pg. 23.

14. Steward, F. A. and W. J. McLay. Waste Minimization: Part 2: Concentrate Return Methods. Metal Finishing, September, 1985, pg. 55.

15. Steward, F. A. and W. J. McLay. Waste Minimization: Part 3: Non-Return Recovery Methods. Metal Finishing, October, 1985, pg 63.

16. Steward, F. A. and W. J. McLay. Waste Minimization: Part 4: Recovery and Regeneration of Process Baths. Metal Finishing, November, 1985, pg 69.

17. Steward, F. A. and W. J. McLay. Waste Minimization: Part 5: Recycle of Treated Wastewater. Metal Finishing, December, 1985, pg 47.

18. Versar, Inc. and Jacobs Engineering Corporation. Waste Minimization Issues and Options. Volume 1. U.S.EPA Office of Solid Waste. EPA 530/SW-86-043. October, 1986.

19. Tucker, S.P., and G.A. Carson, NIOSH. Deactivation of Hazardous Chemical Wastes. Cincinnati,'OH. Environmental Sci. Technol., 9(3):215-220. 1985.

20. U.S. EPA. Development Document for Effluent Limitations Guidelines and Standards for the Metal Finishing Point Source Category. Effluent Guidelines Division, Washington, D.C. EPA 440/1-82-091b. August, 1982.

21. Nunno, Thomas et al. Alliance Technologies Inc. Waste Minimization in the Printed Circuit Board Industry - Case Studies. Prepared for U.S. EPA HWERL under Contract No. 68-03-3243. 1986.

22. Basta, Nicholas et al. Total Metals Recycle is Metal Finisher's Goal. Chemical Engineering, August 8, 1983. pg 16.

23. Cushnie, G. C. Centec Corporation, Reston, VA. Navy Electroplating Pollution Control Technology Assessment Manual. Prepared for the Naval Civil Engineering Laboratory, Port Hueneme, CA. NCEL-CR-84.019. February, 1984.

24. Martin, M., Bend Research Corporation. Bend, OR. Telephone conversation with Lisa Wilk, Alliance Technologies Corp., Sept. 24, 1986.

25. Steward, F. A., ERC/LANCY Company. Process Changes to Reduce the Production of Industrial Sludges. Presented at the 1981 Annual Mtg. of the AIChE, New Orleans, LA. Nov. 8-12, 1981.

26. Kirk Othmer Encyclopedia of Chemical Technology. John Wiley and Sons, New York, NY. Third Edition. 1978.

27. U.S. EPA. Control and Treatment Technology for the Metal Finishing Industry: In-Plant Changes (Summary Report). U.S. EPA IERL Cincinnati, OH. EPA 625/8-82-008. January, 1982.

28. Clark, R. Massachusetts Bureau of Solid Waste Disposal. Massachusetts Hazardous Waste Source Reduction Conference Proceedings, MA Dept. of Environmental Management. October, 1983.

29. Campbell, M.E. and W.M. Glenn, Pollution Probe Foundation. Profit From Pollution Prevention: A Guide to Industrial Waste Reduction and Recycling. Toronto, Canada. 1982.

30. Kohl, J. and B. Triplett. Managing and Minimizing Hazardous Waste Metal Sludges: North Carolina Case Studies, Services, and Regulations. North Carolina University, Raleigh Industrial Extension Service. 1984.

31. Noll, K. E., et al. Illinois Institute of Technology. Recovery, Reuse, and Recycle of Industrial Waste. U.S. EPA Office of Research and Development. EPA 600/2-83-1. November, 1983.

32. Grosse, Douglas W. Treatment Technologies for Hazardous Wastes: Part 4 - Metal Bearing Hazardous Waste Streams. JAPCA Vol 36, No. 5. May, 1986.

33. Rissmann, E.L., et al. Waste Minimization Audits: How they can Lead to Reductions in Hazardous Waste Generation. DuPont Waste Minimization Symposium. Wilmington, Del. October 16, 1986.

34. Stoddard, S. K. Alternatives to the Land Disposal of Hazardous Wastes: An Assessment for California. Toxic Waste Assessment Group, Governor's Office of Appropriate Technology, State of California. 1981.

35. Tsuen-Ni, Lung, et al. Chemical Reclaiming of Nickel Sulfate from Nickel-Bearing Wastes. Conservation and Recycling. Vol. 6, No. 1/2. 1983.

36. U.S. Federal Register. 51 FR/10177. March 24, 1986.

37. Laughlin, R.G.W. et al. Ontario Research Foundation. Metal Finishing Industry Technical Manual: Waste Abatement, Reuse, Recycle, and Reduction Opportunities in Industry. Prepared for Environment Canada. Jan., 1984.

38. RCRA Performance Management Standards for Waste Minimization Said Not Needed Now. Environment Reporter. Current Developments. Nov. 7, 1986.

39. Elaine Eby, U.S. EPA OSW. Telephone conversation with T. Nunno, Alliance Technologies Corporation. March 9, 1987.

40. Hunt, Gary and Roger Schecter. Pollution Prevention Pays: Bibliography by Industrial Category. North Carolina Department of Natural Resources and Community Development. January, 1986.

41. Mathews, J.E. Industrial Reuse and Recycle of Wastewaters, Literature Review. U.S. EPA RSKERL, Ada, OK. EPA 600/2-80-183. December, 1980.

6. Membrane Separation Technologies for Metal Removal

6.1 PROCESS DESCRIPTION

The membrane processes considered here include commercially proven technologies such as ultrafiltration, reverse osmosis, and electrodialysis. Some discussion is also provided for other membrane technologies such as Donnan dialysis and coupled transport whose applicability for the treatment of hazardous waste streams has not yet been commercially demonstrated. Reverse osmosis and electrodialysis are used to recover plating compounds from rinsewater and to permit possible reuse of rinse waters as plating bath make-up. Ultrafiltration alone is of little value for these applications, but is used in combination with chemical treatment to physically contain metal sludges. It is also used as a pretreatment for other processes, such as reverse osmosis, which are subject to fouling and plugging due to the presence of particulates or high dissolved solids levels of certain salts in the feed. Although the waste treatment/recycling applications are not extensive, these processes have found growing acceptance in applications such as desalination of seawater and brackish waters, and as unit operations in the food and pharmaceutical industries.

According to Cheryan[1], the world-wide market for membranes, less than $10 million in 1960, reached $400 to $600 million/year in 1986. More than 30 manufacturers of membranes are identified by Cheryan in his ultrafiltration handbook. This handbook and several other books on membrane technology[2-5] are recommended for those concerned with the theory, development, and applications of membrane technology to individual process and waste streams.

The primary function of a membrane is to allow preferential containment and transport of certain components present within waste streams. Membranes can be classified in a number of ways in accordance with factors such as their origin, chemical composition, structure (e.g., pore size and asymmetry of pore structure), and mechanism of membrane action; e.g., adsorptive vs. diffusive, ion exchange, osmotic, or nonselective (inert) membrane.[1] Figure 6.1.1, taken from Cheryan, provides a classification of various separation processes based on particle or molecular size and the primary factor affecting the separation process. As shown in the Figure, membrane processes such as ultrafiltration, reverse osmosis, and electrodialysis permit separation of dissolved molecules down to the ionic range in size, provided the appropriate membrane is used.[1]

The distinction between membrane processes such as ultrafiltration and reverse osmosis is somewhat arbitrary and has evolved with usage and convention. Table 6.1.1 shows some characteristics of several membrane processes, including osmosis and dialysis, two processes with no apparent utility for hazardous waste treatment. They have been included for reference and completeness, along with microfiltration, a process similar to ultrafiltration that is often used as a pretreatment to remove suspended solids that may interfere with the operation of molecular separation processes.

Another useful classification system is the normal operating concentration range of membrane and other technologies. Figure 6.1.2 provides this information for a number of processes, including reverse osmosis and electrodialysis. These operating concentration ranges are based on a number of factors which include the increase of osmostic pressure with increasing concentration to levels that exceed membrane capacities, selectivity of the membrane process, flux (defined for processes such as reverse osmosis as the volume flow rate per unit area and pressure), and cost.

The flux obtainable with reverse osmosis, ultrafiltration, and other more conventional filtration media is shown in Figure 6.1.3. As shown in the Figure, ultrafiltration flow rates range from roughly 0.1 to 10 gal/ft^2/day/psi. Ultrafiltration systems are typically operated at pressures ranging from 10 to 100 psig, resulting in flow rates that are still several orders of magnitude below conventional filtration processes, but with size retention of the order of 10 to 200 angstroms (0.001 to 0.02μm), as opposed to one or more microns for conventional filters. These size values are arbitrary and it is customary to refer to a "molecular weight cutoff" when attempting to classify ultrafiltration membranes. Ultrafiltration is generally considered suitable for separation of molecules ranging from about 1,000 to 1,000,000 in molecular weight. Thus ultrafiltration, as designated

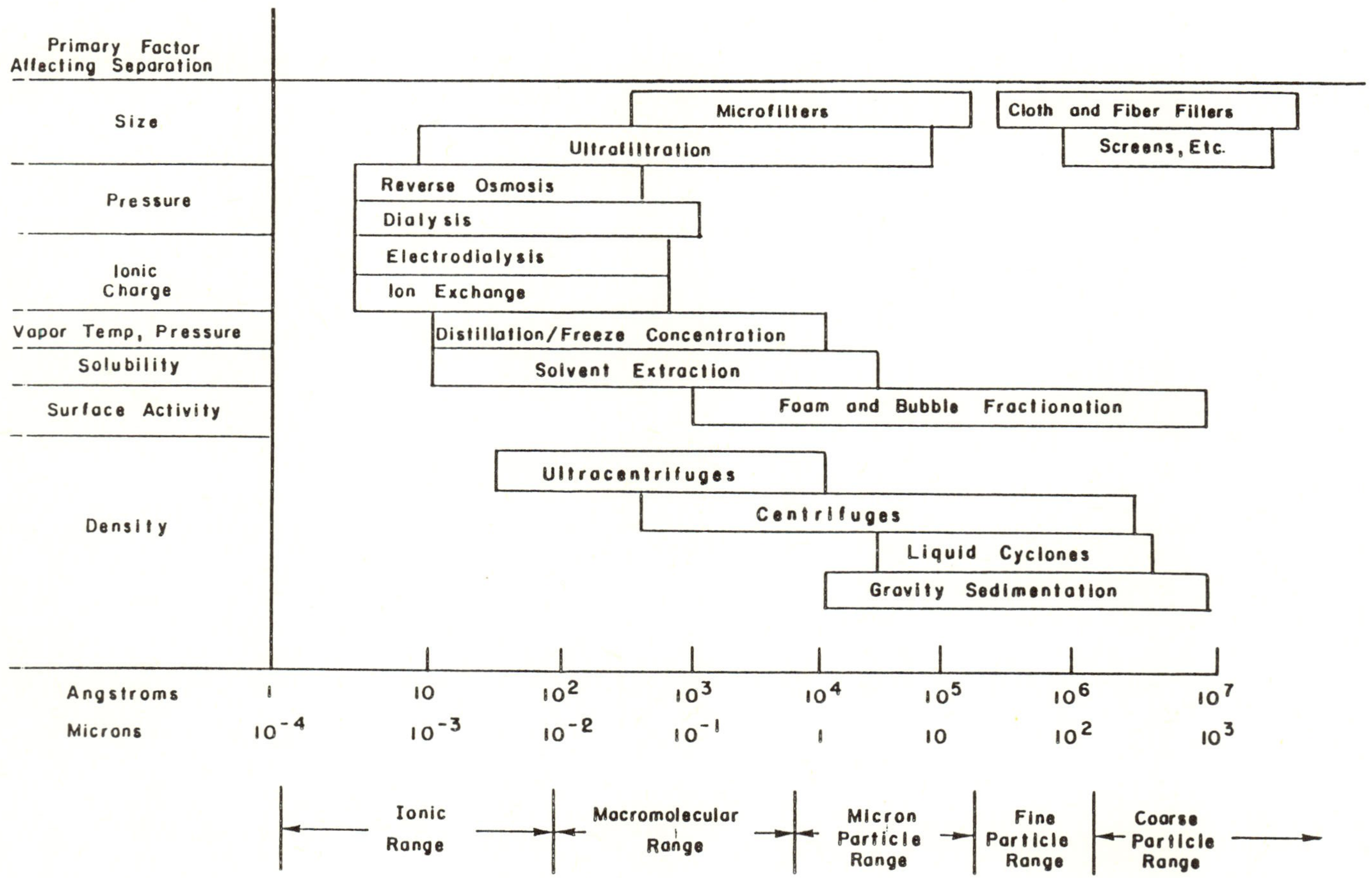

Figure 6.1.1. Useful range of separation processes, showing the range of particle or molecular size covered by each process and the primary factor governing each separation process.

Source: Reference 1.

TABLE 6.1.1. MEMBRANE SEPARATION PROCESSES

Process	Principal driving force	Function of membrane	Permeate	Retentate
Ultrafiltration	Pressure	Discriminates on the basis of molecular size, shape, and flexibility	Water and small molecules	Large molecules
Reverse osmosis	Pressure	Selective transport of water from concentrated solution	Water	Solvent
Electrodialysis	Electromotive Force	Selective ion transport	Water and ionic solutes	Nonionic solutes
Donnan dialysis	Concentration	Ion transport with charge equalization	Metal ions	Hydrogen ions from transfer fluid
Coupled transport	Concentration	Ion transport through complexing agent in membrane	Specific metal ions	Other solute ions
Dialysis	Concentration	Selective solute transport	Water and small molecules	Large molecules
Osmosis	Chemical potential	Selective transport of water into more concentrated solution	Water	Solutes
Microfiltration	Pressure	Removal of particulates	Water and dissolved solutes	Suspended particles

Source: References 1 and 6.

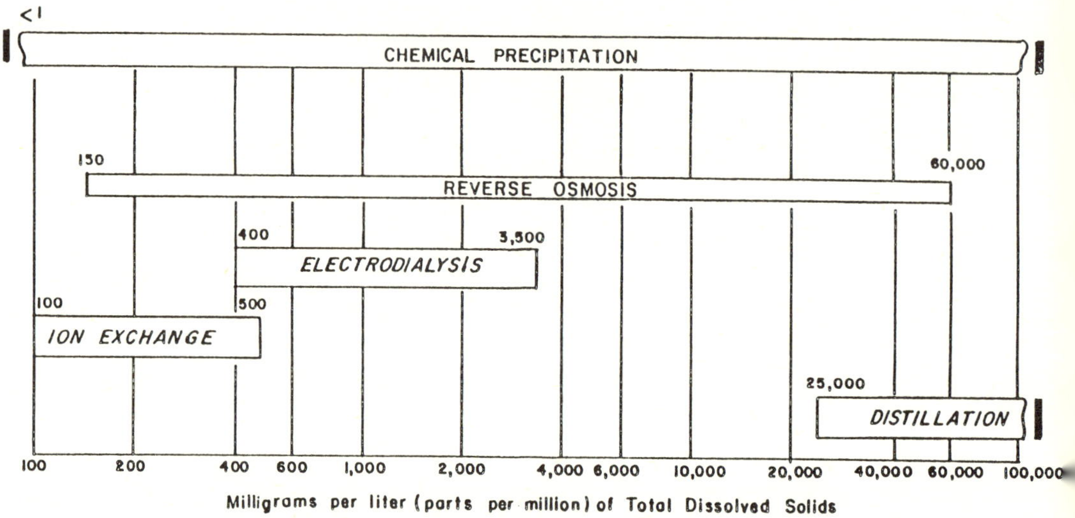

Figure 6.1.2. Normal operating concentration range of separation technologies. Source: References 7 and 8.

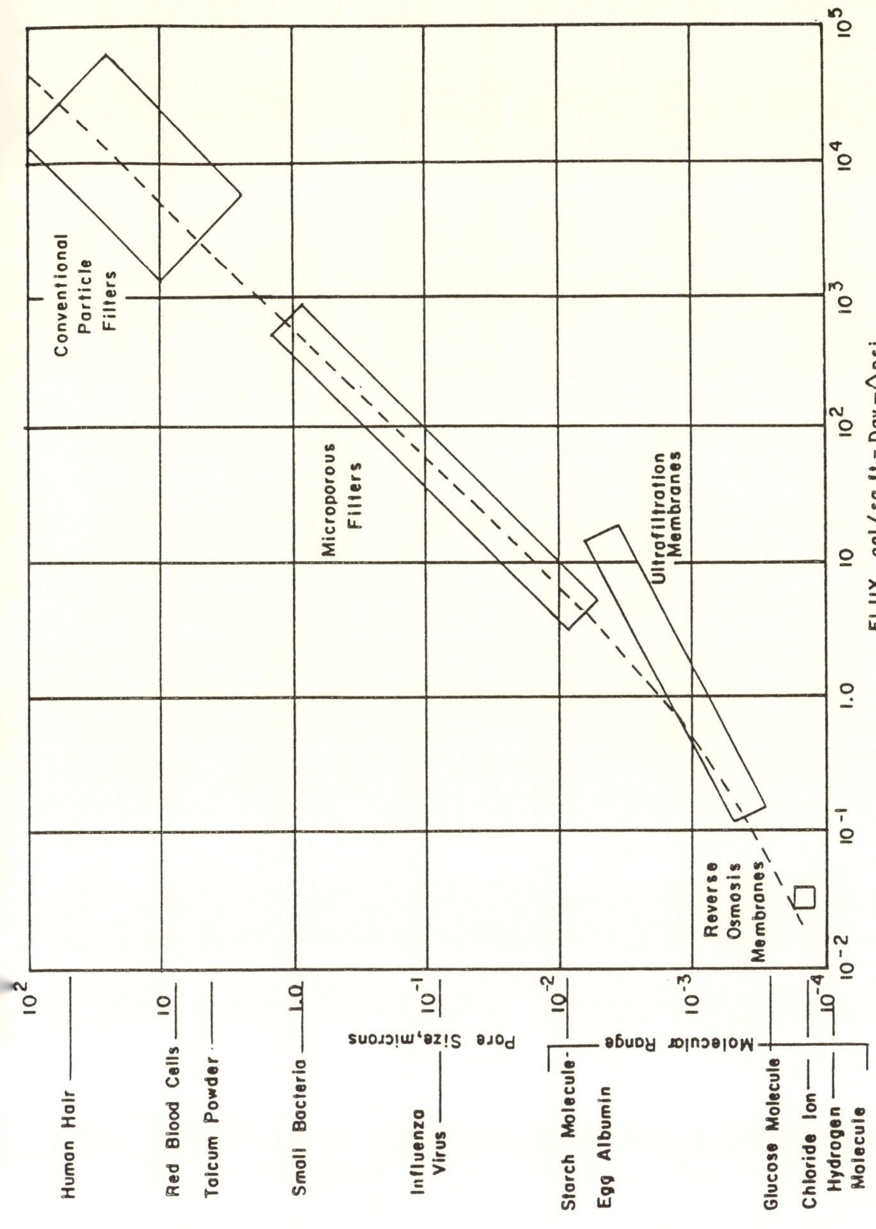

Figure 6.1.3. Pore size vs. flow rate for separation media.

Source: Reference 7, page 90.

by this retention characteristic, is not sufficient by itself to remove dissolved ionic species such as metal ions (and cyanides). Selective ion retention by ultrafiltration membranes, such as that which occurs in reverse osmosis, may take place, but the effect is slight and not significant in terms of effective separation. However, both ultrafiltration and microfiltration are capable of effectively collecting colloidal metal suspensions following precipitation of dissolved metal ions; and they also find application as a pretreatment for reverse osmosis and other processes to protect membranes from clogging.

Reverse osmosis, sometimes called hyperfiltration, may be used to concentrate dilute solutions of many inorganic species, including dissolved metal ions and many organic solvents. Reverse osmosis systems are available from many manufacturers for the treatment of metal-bearing waste streams. Ideally, they permit only the transfer of water, selectively retaining all other dissolved species within the waste stream. Operating pressures used for reverse osmosis are high; of the order of 300 to 1,500 psig,[1] in order to overcome the osmotic pressure of the solute and to provide adequate flux. As shown in Figure 6.1.2, normal operating concentration ranges vary from very low values to as high as 60,000 mg/L. As the concentration of the solute increases during reverse osmosis, additional pressure is required to maintain the water permeation rate. At some point in the separation process, further transfer of water through pressure increases will become impractical because of membrane and equipment limitations. Other processes (e.g., evaporation) will be needed to achieve higher concentration levels, if necessary.[9]

Electrodialysis processes use an electrical potential gradient and special synthetic membranes, usually ion exchange type resins, to produce an enriched stream and a depleted stream. Cation and anion exchange membranes are arranged alternatively to form compartments in a stack maintained between two electrodes. Upon application of an electrical field, the ions entering the compartments within the stack migrate in opposite directions. Depending upon the selectivity of the membrane, the ion will either pass through the first membrane it encounters or be held within its original compartment. Thus, salt solutions are concentrated or diluted in alternate compartments.

Other membrane processes that have been identified in the literature as potentially suitable for the recovery of metal ions from aqueous solutions include Donnan dialysis and coupled transport. Donnan dialysis operates on the principle that two solutions separated by a membrane will remain electrically neutral.[10] Thus, metal ions in a wastewater compartment will interchange with the hydrogen ions in an acidic solution contained in another compartment that is separated by a cation exchange membrane from the wastewater. In coupled transport, a process similar to liquid ion exchange, a porous membrane, containing a liquid complexing agent within its pores, is used to affect separation. The metal ions in the wastewater compartment combine with the complexing agent and migrate through the membrane to the second compartment. Here the complex is broken, releasing the metal ions to solution, with the regenerated complexing agent in turn becoming available for further reaction/interaction with the metal ions in the wastewater.

Despite some success in the laboratory, both Donnan dialysis and coupled transport have not been commercially applied, largely because existing membranes have short-life expectancy.[10] Accordingly, these processes are not discussed at the same level of detail provided for the more advanced, commercialized membrane technologies. Before proceeding with discussions of these technologies, the following subsection will discuss briefly the types of membranes and commercial designs available for membrane separations.

6.2 MEMBRANE STRUCTURE AND SYSTEM DESIGN

Membrane structures can be classified according to their ultrastructure as either microporous or asymmetric. The latter are also referred to as "skinned" membranes. Microporous membranes are designed to retain all particles above a certain size. However, particles that are approximately the same size as the pores may enter into the pores and plug them. Microporous structures with pore sizes in the ultrafiltration range (10 to 200 angstroms) generally have not been very successful. The few designs that are commercially available have low flux and are subject to rapid plugging.[1]

The development of the asymmetrical membrane by Loeb and Sourirajan in 1960 marked the beginning of modern membrane technology. These membranes are characterized by a thin "skin" on one surface, usually 0.1 to 0.5 μm in thickness, while the main body of the membrane supporting the skin is of the order of 40 to 200 μm in thickness and highly porous. The combination of a thin skin, supported by a highly porous substrate, results in high flux with good selectivity.

Asymmetric membranes rarely get plugged in the fashion that microporous structures do, although they are subject to flux lowering phenomena such as fouling and concentration polarization. These factors are controlled by pretreatment, system design, and operating conditions.[11] Cleaning cycles are also used, but there is a danger that the powerful cleaning agents required can damage or attack the membrane. For example, the cellulose acetate membrane used for many reverse osmosis applications has limited pH, temperature, and chlorine tolerance. Thus, cleaning to correct fouling can be a problem. Second generation membranes are available which minimize difficulties associated with cleaning. New membranes, such as those being developed from ceramic materials, may virtually eliminate problems such as irreversible fouling.

The following discussions identify the types of membranes and system designs available for specific membrane technologies. Since several variations are generally available, the user should contact the manufacturers of such equipment to identify the most appropriate system. Lists of manufacturers can be found in several references[1,2,12] and in McGraw Hill's Chemical Engineering Equipment Buyers' Guide.

6.2.1 Ultrafiltration/Microfiltration Systems

Ultrafiltration membranes are generally not defined by their pore sizes, which range from 10 to 200 angstroms and higher, but by the size or equivalent molecular weight of particles excluded. Although the size cutoff is arbitrary, one definition by Lonsdale[13] is that ultrafiltration membranes retain species in the 300 to 300,000 molecular weight range. Because ultrafiltration deals with the separation of larger molecules, it is not suitable for the separation of dissolved metal ions. However, it does find use as a pretreatment method or as a means of removing chemically precipitated metallic species. Microfilters with a pore size of greater than 0.1 μm are also used to effectively collect precipitates.[1]

Ultrafiltration/microfiltration membranes are made from a wider selection of polymers than are reverse osmosis membranes. Cellulose acetate and polyamide were the earliest of the commercial membranes. In addition to these, several other polymeric materials are available. These are comprised of thin skin composite membranes formed on the surface of a porous support polymer, usually a polysulfone. However, for specific applications, the composite structure can be tailored from other materials to enhance chemical and biological resistance and improve other properties such as selectivity.

Users of ultrafiltration membrane technology have their choice of four basic equipment designs: 1) tubular with inner diameters greater than 10 microns; 2) hollow fibers with inner diameters less than about 1.3 microns; 3) plate type units; and 4) spiral-wound modules.[1] The tubular module is the simplest design. However, because of its small surface area per module, it is only used in specialized applications. The membrane is either inserted into a porous tube or is cast in place. The feed is pumped through the tube and the permeate passes radially through the membrane and porous tube out through an exit line. The concentrate or reject stream exits from the downstream side of the tube. Although the tubular modules do have low surface area, they are easy to clean and are less susceptible to plugging by suspended solids than are other membrane types.

Hollow fiber membranes (acrylic copolymer) used in ultrafiltration employ a membrane skin on the inside of the hollow fiber. Each hollow fiber has a fairly uniform bore with available sizes ranging from 8 to 49 mil (0.19 to 1.25 mm) in diameter with a cross-sectional thickness of about 200 μm. Bundles of fibers are normally sealed in a shell and tube arrangement, although other configurations are possible.[12]

The permeate passes through the membrane and fiber wall and is collected on the outside of the fibers. The concentrate passes out the opposite end of the fiber bore. Hollow fibers have a fairly low pressure rating; thus, flow rates (flux) will also be low. A major advantage is the ease of cleaning achieved through backflushing, due to the self-supporting nature of the fibers.

The plate and frame-type units represent an early design and consist of membrane covered support plates stacked in a frame arrangement. Permeate exits via the support plate and concentrates leave the module opposite the feed end. Internal flow within the module may be arranged in a combination of parallel and series flow patterns by using section plates. The stacking arrangement is usually horizontal for ultrafiltration.

The spiral-wound design is compact, relatively inexpensible, and provides larger surface areas per unit volume of equipment. The spiral-wound modules consist of membrane and spacer materials that are wrapped around a perforated center tube which collects the permeate. Figure 6.2.1 shows a schematic of a spiral-wound structure along with a cross-section which illustrates the flow path of the permeate. Although economic treatment of larger volumes is possible with this design, it is more apt to plug than other designs. Also, it cannot be cleaned mechanically.

The characteristics of the feed will play a major role in determining which of the membrane materials and designs should be selected. Feeds containing larger suspended particles are best processed in large diameter tubular units. Other factors, such as ease of cleaning, pressure losses, degree of concentration, and other considerations, all contribute to the overall utility and cost of a system. Pilot plant studies, a service offerred by many manufacturers, should be undertaken before proceeding with final system selection.

6.2.2 Reverse Osmosis Systems

Three types of membranes are commercially available for reverse osmosis: cellulous acetate, aromatic polyamides, and thin film composits. Cellulose acetate membranes have high flux and high salt/metal rejection properties and are relatively easy to manufacture. Among the disadvantages of these membranes are: (1) a fairly narrow temperature range (maximum recommended temperature of 30°C); (2) a rather narrow pH range (preferably pH 3-6; (3) poor resistance to chlorine; (4) a tendency to "creep", bringing about a gradual loss of membrane properties (notably flux); and (5) susceptibility to microbial attack.

The aromatic polyamides (aramids), commercialized by DuPont in 1970, have an asymetric structure similar to cellulose acetate. They are not susceptible to biological attack, resist hydrolysis, and can be operated over a wider range of pH (pH 3 to 11) and slightly higher temperature (40°C) than cellulose acetate membranes. However, they are readily degraded by low levels (0.2 ppm) of free chlorine, a major drawback for some applications.

Thin film composites are formed by depositing a film of a polymeric material on a porous support structure, usually a polysulfone. An advantage of these thin film composite membranes is their ability to withstand more severe environments. However, not all polymers can be fabricated into

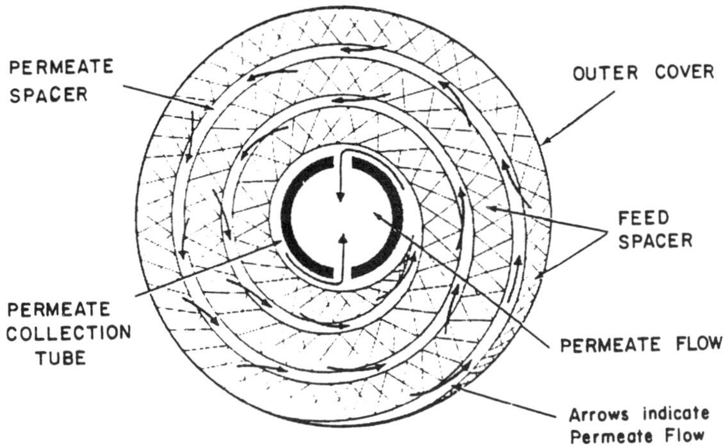

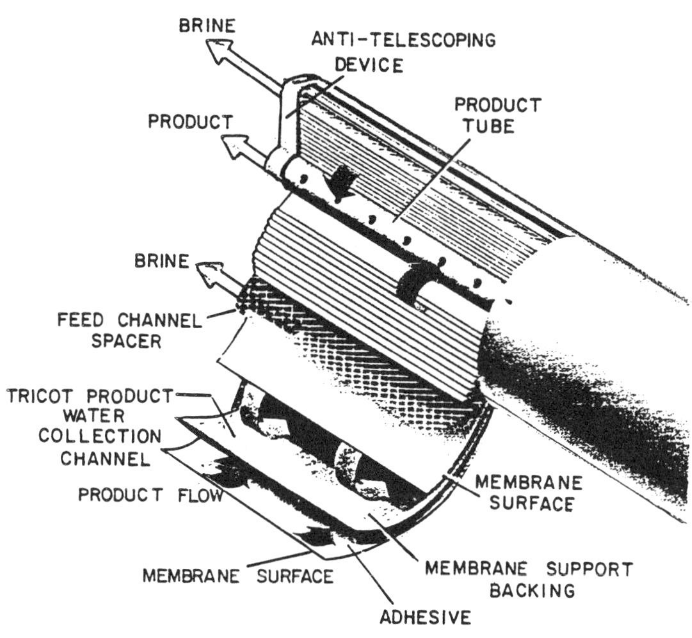

Figure 6.2.1. Spiral worm cartridge schematic and cross section showing flow or permeate (feed flow perpendicular).

structures that are suitable for reverse osmosis applications. Polysulfones, for example, cannot withstand high (e.g., above 100 psig) pressures, nor can they be fabricated with pore sizes with less than a 500 to 1,000 molecular weight cutoff. Thus, the polysulfones cannot be used directly as asymmetric membranes for reverse osmosis. Howerver, they find use in ultrafiltration applications and as backings for certain reverse osmosis barriers such as polyethylene imine/toluene disocyanates. Membrane development remains a major focus of membrane technology research.

The equipment used to conduct reverse osmosis separations is similar to that used for ultrafiltration. Tubular, spiral-wound, and hollow fiber systems are all, but the hollow fiber reverse osmosis system does differ from that used for ultrafiltration. This membrane is made from an aromatic polyamide, with an inside diameter of about 42 μm and an outside diameter of about 85 μm. The fiber has an asymetric structure. However, unlike the ultrafiltration hollow fibers, the skin is located on the outside of the fiber, necessitating the employment of a different system configuration to separate the permeate (which flows through the bore of the hollow fibers) and the concentrated feed. Up to 4.5 million of the fibers can be assembled into a bundle for use in reverse osmosis equipment. This system provides the highest membrane area per unit volume for any reverse osmosis system.[9]

6.2.3 Electrodialysis Systems

Electrodialysis is based on the migration of ions through sets of alternate cation and anion exchange selective membranes that permit the passage of positive and negative ions, respectively.[2] The selective membranes should possess the following characteristics:

- low electrical resistance;
- good selective qualities;
- good mechanical properties;
- good structural stability; and
- high chemical stability.

Since it is difficult to optimize these properties, only a few companies produce electrodialysis membranes commercially. Two general types of membranes are available, heterogeneous and homogeneous. Heterogeneous membranes are manufactured by mixing a commercial ion exchange resin (50 to 70 percent) with a binder polymer such as polyvinyl chloride. The ion exchange resins are usually crosslinked copolymers of styrene and divinylbenzene. Cation or anion exchange groups are introduced into the copolymer by sulfonation and chloromethylation/amination with a triamine. Plastic mesh or cloth is used as a support for the ion exchange/binder mixture that constitutes the membrane.

Homogenous membranes consist of a continuous homogeneous film onto which an active group is introduced. The membranes can be reinforced or nonreinforced. Some of the properties of some commercially available membranes are shown in Table 6.2.1, as provided in E. Korngold's chapter on electrodialysis in Reference 2. A more extensive description of these and other specialty membranes is provided in other references.[1,3-5]

In an electrodialysis stack, cation and anion exchange membranes are alternated between two electrodes. As shown in Figure 6.2.2, during electrodialysis, one cell will contain a concentrated solution and the other will contain a dilute solution. In industrial units, several hundred cell pairs can be assembled between two electrodes. The system can be used for desalination, separation of nonelectrolytes from electrolytes, and electrolyte concentration. Power consumption is directly proportional to ion concentration and operating costs are more favorable for low feed concentration. Also, because ion rejection varies from about 45 to 55 percent per pass (as opposed to up to 99 percent for reverse osmosis), a number of passes must be used if a low concentration of dissolved solids is required in the dilute stream.

The application of reverse osmosis for seawater conversion is much more advanced than electrodialysis technology. Because of the high energy consumption required at seawater concentration levels (approximately 32,000 ppm), electrodialysis is not economically attractive, although programs to develop high temperature operation and new membranes may improve the economics considerably. Cost data from the early 1980s indicate that electrodialysis becomes competitive for desalting at levels of roughly 2,000 ppm total dissolved solids.

TABLE 6.2.1. PROPERTIES OF COMMERCIALLY PRODUCED MEMBRANES

Manufacturer	Name of membranes	Membrane	Thickness (mm)	Capacity (meq/gm)	Electrical resistance (Ω cm² in 0.1 N NaCl)	Reinforcement
Ionac Chemical Co. New Jersey	Ionac	MC-3142	0.15	1.06	9.1	Yes
		MC-3470	0.35	1.05	10.5	Yes
		MA-3148	0.17	0.93	10.1	Yes
		MA-3475	0.40	1.13	23	Yes
		IM-12	0.13	—	4	Yes
American Machine and Foundry Connecticut	A.M.F.	C-60	0.30	1.5	6	No
		A-60	0.30	1.6	5	No
Ionics Inc. Massachusetts	Nepton	CR61 AZL 183	0.60	2.7	9	Yes
		AR 111 BZL 183	0.60	1.8	14	Yes
Asahi Glass Co. Ltd. Tokyo, Japan	Selemion	CMV	0.15	1.4	6.1	Yes
		AMV	0.14		4.0	Yes
Tokuyama Soda Ltd. Tokyo, Japan	Neosepta	CL 25 T	0.16	1.8–2.0	3.5	Yes
		AV 4 T	0.15	1.5–2.0	4.0	Yes
Asahi Chemical Industry Co. Ltd. Tokyo, Japan	A.C.I. or Acipex	DK 1	0.23	2.6	6.5	Yes
		DA 1	0.21	1.5	4.5	Yes
Ben-Gurion University of the Negev, Research & Development Authority Beersheva, Israel	Neginst	NEGINST-HD	0.35	0.8	12	Yes
		NEGINST-HD	0.35	0.8	10	Yes
		NEGINST-HC	0.2	1.6	6	No
		NEGINST-HC	0.2	1.7	8	No

Source: Reference 2.

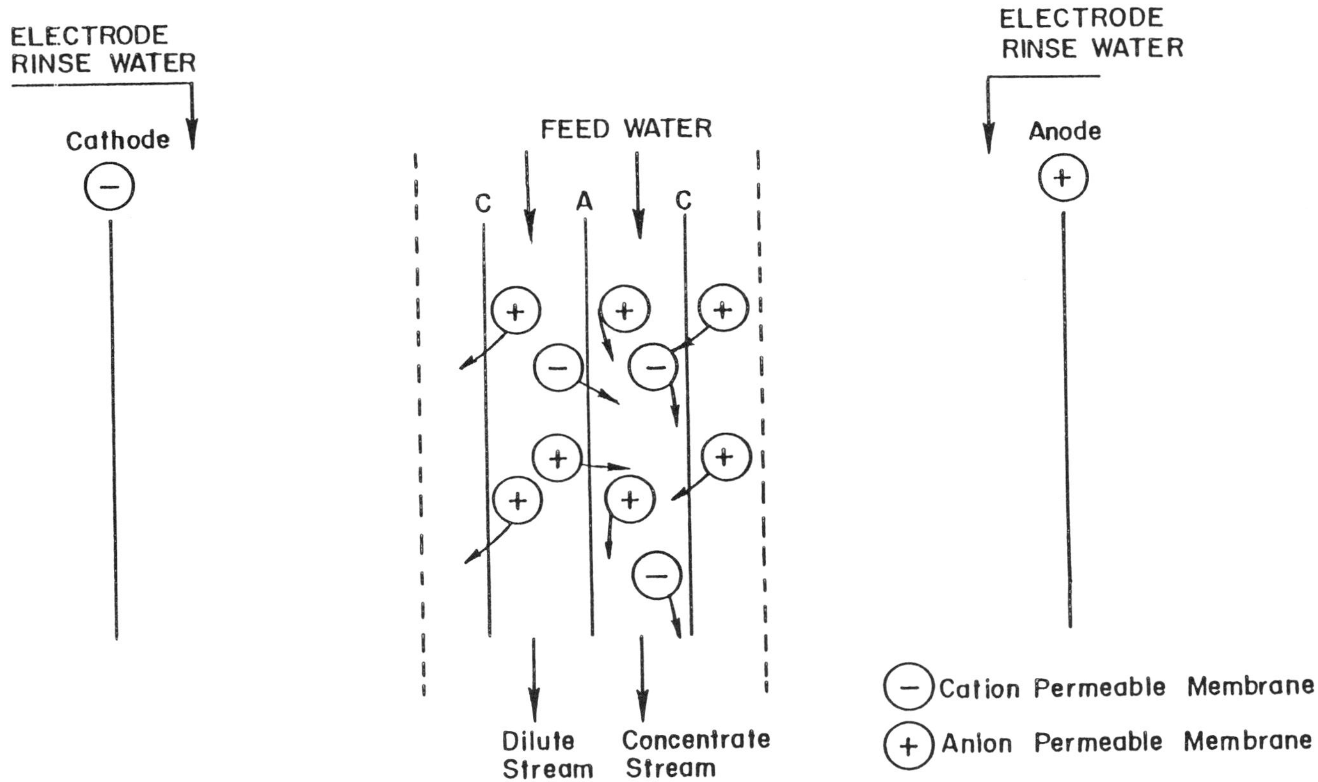

Figure 6.2.2. Diagrammatic representation of electrodialysis.

Source: Reference 6.

6.2.4 Other Membrane Systems

As noted previously, Donnan dialysis and coupled transport are two membrane technologies that appear to have some application for the treatment of metal-containing aqueous wastes. However, problems of membrane stability have limited their development. DuPont presently markets "Nafion", a perfluorosulfonic acid membrane that is being evaluated as a Donnan dialysis membrane for removal of nickel in electroplating wash water. Anion exchange membranes for the removal of copper, cadmium, and zinc cyanide complexes are also being evaluated. Reference 14 reports studies using quarternized polyvinyl pyridine and polyvinyl benzylchloride films grafted on a polyethylene base. Ion transport rates were reportedly proportional to ion exchange capacity.

The coupled transport process relies on a liquid, water immiscible, organic complexing agent held within the pores of a microporous membrane. The metal ion introduced with the feed solution to one side of a cell reacts with the liquid. It is then transported through the membrane to the product solution where it is released and the organic transport medium is regenerated. Both Donnan dialysis and the coupled transport processes operate without the need for electric current or high hydraulic pressures as required for electrodialysis and reverse osmosis, respectively. The only energy required is that needed to pump the feed and stripping solution through the cells.

6.3 ULTRAFILTRATION/MICROFILTRATION FOR TREATMENT OF METAL WASTES

As a result of the high molecular weight cutoff (approximately 1,000) of ultrafiltration membranes, they cannot be applied directly to recover metals present as dissolved solids in, for example, electroplating rinsewaters. Ultrafiltration has been used commercially to recover or treat electrophoretic paints, oil in water emulsions, proteins from the dairy industry, and rinsewaters from alkaline metal cleaning baths. However, there are no known applications for the recovery of metals from aqueous waste streams, with the exception of its use as a pretreatment method or as a means of recovering precipitated materials. Consequently, the following discussions will address membrane systems used to separate precipitated metals from waste streams.

The following discussion relies heavily on material provided by Memtek Corporation, a supplier of this technology.[15-17] Strictly speaking, the Memtek technology is microfiltration, rather than ultrafiltration, since a membrane with pore sizes of the order of 0.1 m is used. Although the nature of the membrane(s) is proprietary, it reportedly is inert and can withstand any solution pH.

6.3.1 Process Description

Advanced membrane processing of wastewater, as described by Memtek, utilizes a system where insoluble precipitated contaminants are separated from solution through chemical pretreatment followed by the use of cross-flow tubular membranes to contain the precipitate. These membranes have a nominal pore size of 0.1 microns which allows for complete rejection of all particulate or suspended solids larger than this size. The mechanism requires the solution to be pumped under low pressure and turbulent flow conditions down the center of a membrane module. The pressure exerted on the solution forces clean solution through the membrane. The solids rejected at the membrane surface are carried by the flowing liquid back to the beginning of the system. This design allows solutions of up to 10 percent solids to be filtered while producing particulate free effluent and concentrating the solution to a higher percent solids (see Figure 6.3.1).

The system differs from conventional membrane systems by incorporating chemical pretreatment to render metal ions insoluble. Since all precipitated solids and turbidity are retained by the membrane, the effluent quality is related to the residual soluble ions. When applied to heavy metal wastes with appropriate pretreatment chemistry, including co-precipitation effects, the toxic metal content of the effluent can be extremely low.

The basic components of the system are a chemical reaction section, concentration tank, process pump, membranes, and an integrated cleaning system. These systems utilize low pressure pumping (40 psi) and high turbulent flow (15 fps) through the membrane to affect good filtration rates. Typical design flow rates for these membrane systems used in industrial wastewater applications is 200 to 400 gfd (gal/ft^2 of membrane surface area per day). Filtration capacity is provided by installing the required number of modules in parallel to produce the design effluent.

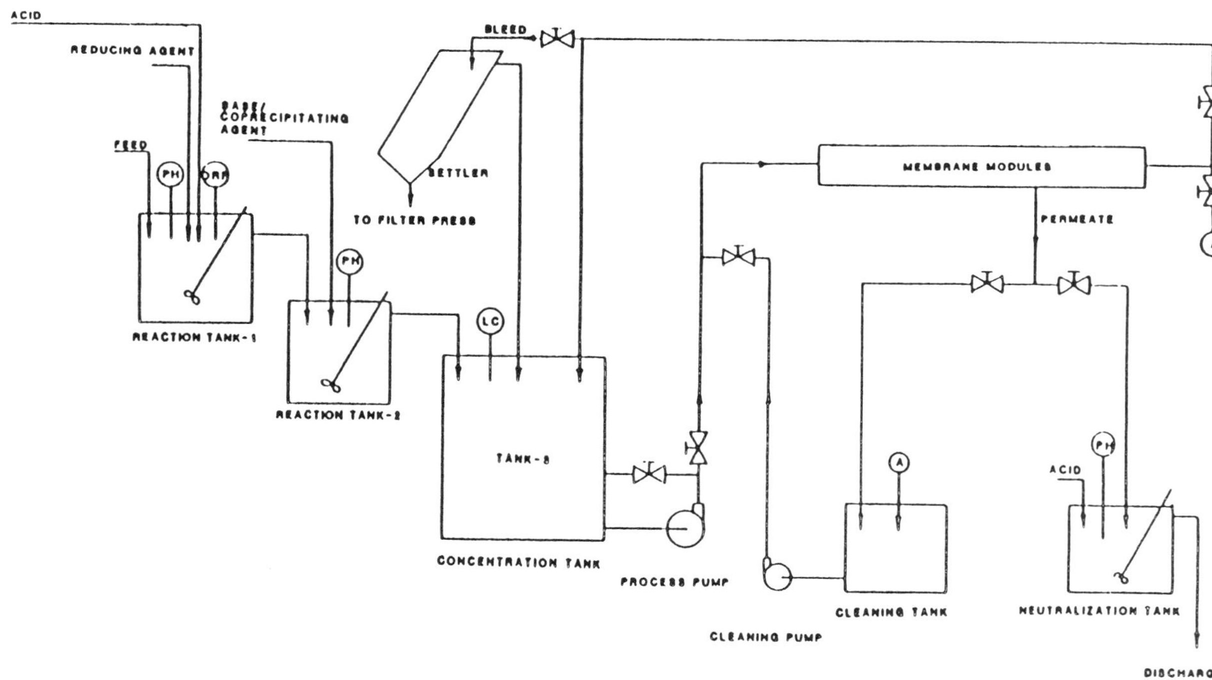

Figure 6.3.1. A typical simplified flow diagram of wastewater treatment for computer related industries.

Source: References 15 and 17.

The pretreatment required to attain effluent specifications is specific to the particular application. For example, a different pretreatment program is required for chelated compounds, hexavalent chrome, cyanide, and hydroxide reactions. The goal of all chemical pretreatment is simply to convert dissolved ions to precipitated compounds so that they can be effectively removed to produce the required effluent concentrations. Using variants of standard pretreatment chemistry, some systems have been designed to produce an effluent in the parts per billion range. However, the presence of oil and grease can cause premature fouling of the membrane. If present, additional pretreatment will be required to remove these constituents.

Each membrane section is provided with an integrated cleaning system to restore the membrane performance when fouling occurs. This procedure uses only chemical cleaning processes to restore flow rate to the design level. The chemical cleaning procedure is designed to complete the cleaning in a short time period, typically less than 2 hours, applied once each week.

During treatment, the solids content of the feed increases in the membrane modules to form a concentrated slurry with generally 2 to 5 percent solids. To increase this concentration, a portion of the slurry is removed from the system and routed to a settling tank or filter press where further sludge consolidation occurs; e.g., 5 to 20 percent or 30 to 40 percent solids, respectively.

6.3.2 Process Performance

Table 6.3.1 summarizes operational performance for a range of typical applications for ultrafiltration. The table shows the ability of the membrane to process high and low concentration wastes to extremely low effluent levels.

6.3.3 Costs

Costs for the various sizes of standard advanced membrane filtration units are presented in Table 6.3.2. The cost estimates include treatability study, engineering, all pretreatment equipment, control panel, piping, membranes, pumps, installation and start-up. These costs are based on typical pretreatment equipment, but do not include costs for pretreatment equipment such as cyanide destruction. Operation and maintenance costs include

TABLE 6.3.1. TYPICAL SYSTEM PERFORMANCE

Contaminants	Feed (mg/L)	Effluent (mg/L)
Aluminum	10-1000	0.5
Arsenic	1- 43	0.05
Cadmium	25- 115	0.05
Chromium	3- 275	0.1
Copper	1-1525	0.1
Cyanide	5- 300	0.1
Fluoride	18-5000	1.0
Gallium	4- 20	0.5
Germanium	20- 110	0.5
Gold	1- 12	0.15
Iron	2-1500	0.02
Lead	2- 25	0.05
Manganese	1- 10	0.02
Mercury	3- 30	0.02
Nickel	4- 300	0.02
Radium*	1- 10	0.6
Rhodium	20- 500	0.1
Silver	10- 200	0.1
Tin	20- 75	0.1
Uranium	1- 15	0.001
Zinc	2- 400	0.1
BOD	50-5000	**
COD	20-3500	**
Suspended solids	--	Non-detectable

*Concentration given in picocuries/liter.

** 95 percent removal.

Source: Reference 15.

TABLE 6.3.2. ESTIMATED CAPITAL AND OPERATING COSTS OF STANDARD ADVANCED MEMBRANE FILTRATION SYSTEM

Flow rate (gpm)	Dimensions	Capital cost ($)	Typical[a] O&M cost ($/yr)
20	6'W x 16'L x 11'H	90,000	8,500
45	6'W x 24'L x 11'H	130,000	15,400
75	12'W x 24'L x 11'H	170,000	19,100
100	12'W x 45'L x 11'H	225,000	25,500
150	18'W x 55'L x 11'H	330,000	37,400
500	40'W x 60'L x 11'H	900,000	113,300

[a]Depends greatly on influent waste characteristics.

Source: Reference 15, 1987 costs.

electricity, routine periodic repairs, and routine cleaning. They do not include costs for pretreatment chemicals, labor, post-treatments, or membrane replacement. As shown in Table 6.3.3, these costs are highly site-specific and can be appreciable.

6.3.4 Overall Status of Process

Availability--

Over 100 full-scale industrial systems, ranging in size from 10 to 400 gpm, have been installed by Memtek at facilities that include printed circuit board manufacturing, electroplating, battery manufacturing, and photographic processing. Each application will require treatability studies to optimize and integrate the chemical pretreatment and membrane systems.

Application--

This system can be applied to metal-containing aqueous waste streams provided solubility limits associated with the chemical precipitation step are within required limits. In most instances this does not appear to be a problem. Membrane fouling may also impose limitations, although Memtek reports that cleaning will restore flux rates.[16]

Environmental Impact--

Assuming that concentration levels found in the permeate are below regulatory limits, the principal environmental impact will result from the sludge generated by the process. Dewatering, followed by solidification, encapsulation, or some other treatment method, will be required.

Advantages and Limitations--

Chemical precipitation followed by microfiltration appears to be an effective means of reducing the heavy metal contaminant levels of aqueous waste streams. Treatment processes can effectively reduce contaminants levels down to 100 ppb or lower. The principal limitation results from the hazardous sludge generated which must be treated before it can be land disposed. The cost of treatment and disposal will depend upon the contaminant and its concentration and could be appreciable.

TABLE 6.3.3. ESTIMATED OPERATING COST FOR MEMBRANE FILTRATION SYSTEMS TREATING METAL-CONTAINING WASTEWATERS[a]

Waste description	System size (gpm)	Metal concentration in feed (ppm)	Costs ($/1,000 gallons of feed)			
			Chemicals	Electricity	Solids disposal	Total
General rinse water[b]	--	17	0.05	0.60	0.20	0.85
Chelated wastewater	40	480-665	9.10	0.80	5.20	15.10
Wastewater of chelates, non-chelates, and spent concentrates	100	27-70	0.80	0.70	0.83	2.33
Same as above	60	56-105	4.00	0.70	0.40	5.10
Same as above	35	46-85	3.40	0.75	0.60	4.75

[a]Wastes from printed circuit board producers containing primarily copper with some lead and nickel.

[b]Also contains ~2 ppm chrome.

Source: Reference 15, 1987 costs.

6.4 REVERSE OSMOSIS

6.4.1 Process Description

Reverse osmosis (RO) is a treatment technique used to remove dissolved organic and inorganic materials, and to control amounts of soluble metals, TDS, and TOC in wastewater streams. The technology has been applied in the metal finishing industry to recover plating chemicals from rinsewater, such that both plating chemicals and rinsewaters can be reused.

RO involves passing wastewater through a semipermeable membrane at a pressure greater than the osmotic pressure caused by the dissolved materials in the solvent. Thus, the osmotic flow, defined as the flow from a concentrated solution to a dilute solution, is reversed due to the increase in pressure applied to the system. The process is schematically presented in Figure 6.4.1.

To obtain reasonable water fluxes (approximately 10 gal/ft^2 day), the feed solution must be pressurized well above the equilibrium osmotic pressure. The expression for osmostic pressure can be written as:

$$\pi = CRT$$

where C is the volume concentration, R has the same value as the universal gas constant, and T is the absolute temperature. In practice, reverse osmosis systems are operated from about 4 to 20 times the equilibrium osmotic pressure, with pressures of 1,000 psi or greater not uncommon.[2] As shown in Figure 6.4.2, osmotic pressure increases with solute mass fraction and decreases with molecular weight. As the concentration of the solute increases during reverse osmosis, additional pressure must be applied to maintain flux. The detrimental effect of increasing concentration is further complicated by concentration polarization, a term which refers to accumulation of solute at the surface of the membrane resulting in a further increase in osmotic pressure. Proper design will minimize the polarization effects, e.g., through the use of turbulence within the feed stream.

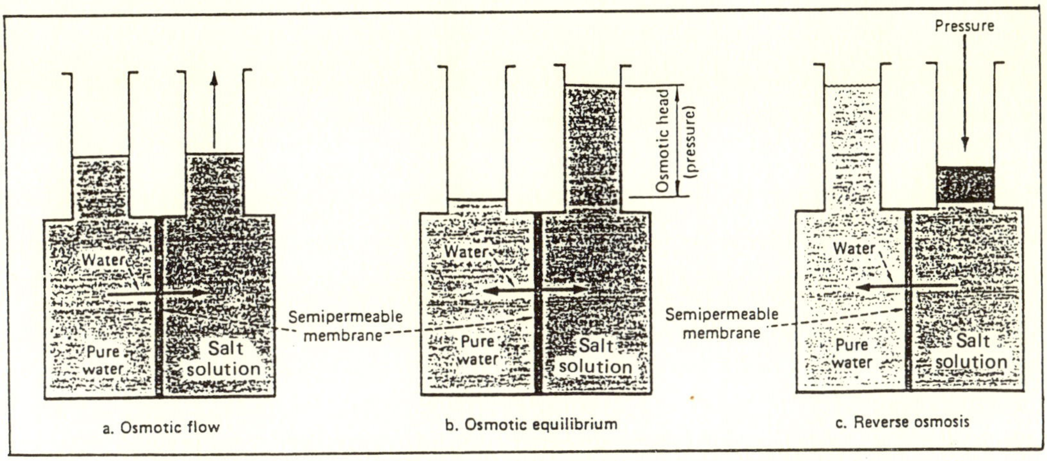

Figure 6.4.1. Principles of normal and reverse osmosis.

Source: Reference 9.

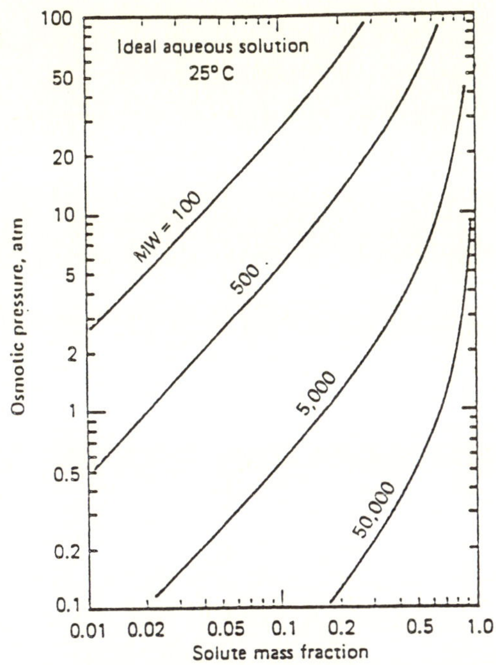

Figure 6.4.2. Osmotic pressure as a function of mass fraction and molecular weight.

Source: Reference 11.

The operation of a reverse osmosis system is affected primarily by the feed characteristics, operating pressure, and membrane type. These factors will affect the flux and percent rejection which, in turn, define system size requirements and effluent quality, respectively.

Flux determines the system size for a given waste flow rate; i.e., higher flux permits the use of smaller systems. Flux is the volume flow of permeate per unit membrane area. It is proportional to the effective pressure driving force, according to the following relationship:[18]

$$J = K\ (\Delta P - \Delta\pi)$$

where J is the flux, K is the membrane constant, $\Delta P$is the applied pressure across the membrane, and $\Delta\pi$ is the osmotic pressure across the membrane.

Since osmotic pressure is approximately proportional to molar feed concentration, flux increases with increasing operating pressure and decreases with increasing feed concentration. Thus, chemicals which form high-molecular weight complexes will have higher flux for a given weight percent in solution. More concentrated solutions can be achieved by utilizing a large effective driving pressure. Increases in temperature of the waste feed will also increase the flux by lowering viscosity. However, although increased operating temperatures will improve the performance of the system in the short-term, the lifetime of the membrane will be shortened.

Percent rejection is defined as follows:[18]

$$\%\ \text{Rejection} = \frac{\text{(feed concentration) - permeate concentration}}{\text{(feed concentration)}} \times 100\%$$

Higher percent rejections will result in better quality (higher purity) of the permeate and concentrated streams. Percent rejection is primarily affected by the membrane type, although rejection will decrease with increasing feed concentration.[18]

The application of reverse osmosis to the treatment of metal-containing wastes is often limited by the pH range in which the membrane can operate. Table 6.4.1 shows the characteristics of some commercially available

TABLE 6.4.1. COMMERCIALLY AVAILABLE MEMBRANE MATERIALS

Type	Description	Source	Allowable pH range	Typical operating pressure kg/cm^2(psig)	Flux rate $L/m^2/day$ ($gal/ft^2/day$) @ 77°F & 400 psi	Typical flux per module L/day(gal/day)	Module replacement costs ($)
Hollow fiber	Hollow fine fiber asymetric membranes or aromatic polyamide	E.I. DuPont Wilmington, DE	4-11	29.2-58.4	73(1.8)(9)	3785(1000)	750
Spiral wound cellulose acetate	Flat sheet composite membrane of cellulose acetate with mesh spacers, rolled into cartridge	Osmonics, Inc. Hopkins, MN Fluid System Div. of UOP, Inc. Dow Chemical USA Midland, MI	2.5-7	29.2-58.4 (400-800)	1140(28)	5680(1500)	350
RC-100	Flat sheet composite membrane of polyether/amide on polysulfone, rolled into cartridge	Fluid Systems Div. of UOP, Inc. San Diego, CA	1-12	29.2-58.4 (400-800)	530(13)	3785(1000)	1000

Source: Reference 19.

membranes. The cellulose acetates have a very small pH range, and thus cannot be used for recovery of cyanide plating baths where pH is well above 7. The polyether/amide on polysolfone material appears to result in the membrane that is least affected by pH.

Many semipermeable membranes can be fabricated either in the form of a sheet or tube, which is then assembled into modules. Figure 6.4.3 shows the three basic module designs, which include:

- Tubular--A porous tubular support with the membrane case in place or inserted into the tube. Feed is pumped through the tube, concentrate is removed downstream, and the permeate passes through the membrane/porous support composite.

- Spiral Wound--Large porous sheet(s) wound around a central permeate collector tube. Feed is passed over one side of the sheet and the permeate is withdrawn from the other.

- Hollow Fiber--Thousands of fine hollow fiber membranes (40 to 80 μm diameter) arranged in a bundle around a central porous tube. Feed enters the tube, passes over the outside of the fibers, and is removed as concentrate. Water permeates to the inside of the fibers and is collected at one end of the unit.

Reverse osmosis systems typically consist of a number of modules connected in series or parallel, or a combination of both arrangements. In a series arrangement, the reject stream from one module is fed directly to another module, such that greater product concentration is achieved. Alternatively, the reject stream may be recycled to the feed stream of the same unit. Series treatment may be limited in some cases by the ability of the membrane to withstand concentrated contaminants. The system capacity can be increased through the use of a parallel arrangement of modules; however, product quality will not be enhanced. Schematic flow diagrams of two series systems are shown in Figures 6.4.4 and 6.4.5.

To ensure a minimum permissible reject flow rate per module, and thus provide adequate turbulence, each successive stage contains a smaller number of modules than the preceding stage. The system shown in Figure 6.4.4 is designed for 87.5 percent water recovery. While the degree of rejection is dependant on the particular ion under consideration, Figure 6.4.5 shows how higher purity water can be obtained by feeding the product to a second stage.

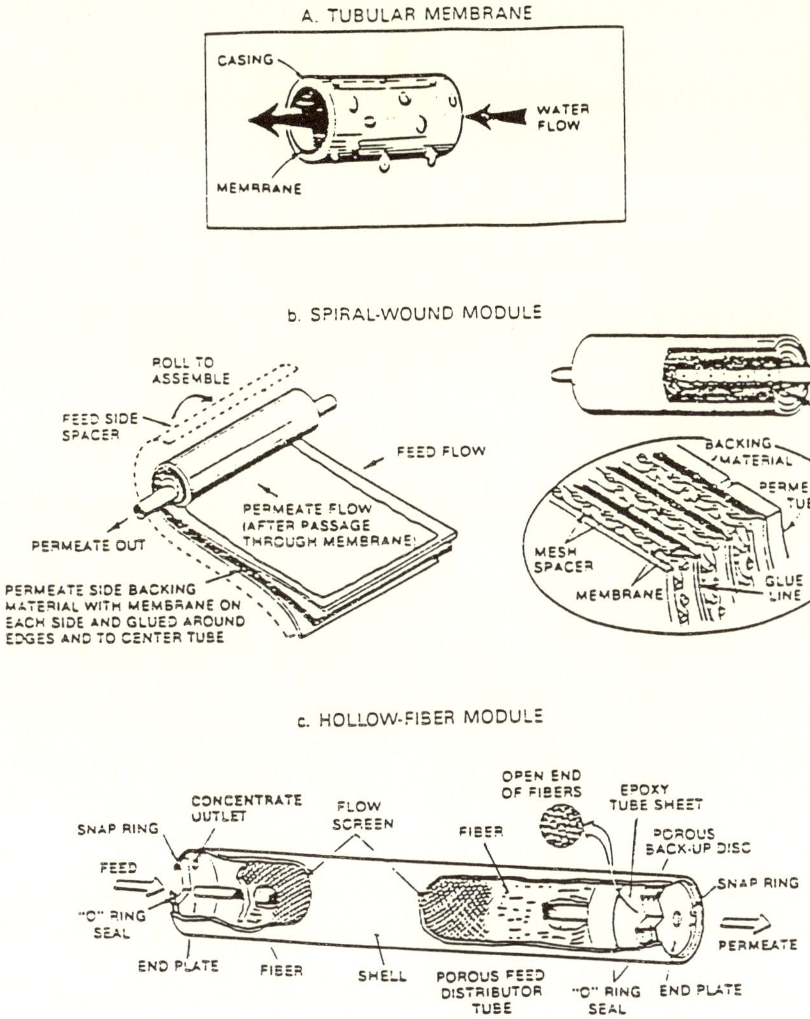

Figure 6.4.3. Reverse osmosis membrane module configurations.

Source: Reference 20.

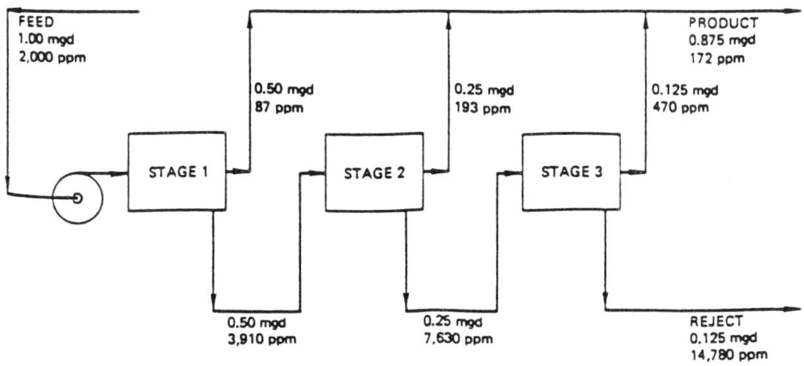

Figure 6.4.4. Schematic of a three-stage RO plant.

Source: Reference 2.

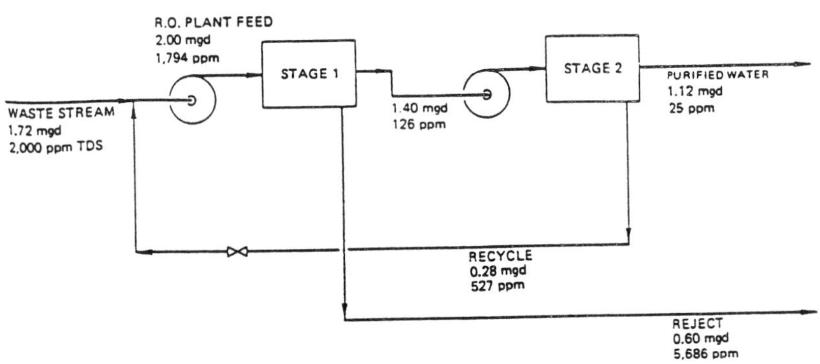

Figure 6.4.5. RO system to increase product purity.

Source: Reference 2.

Although they are able to operate at higher pressures, tubular modules are not applicable for most industrial applications because of large floor space requirements and high capital costs. Comparatively, hollow fiber and spiral wound modules have lower, and similar, capital costs. Hollow fiber modules require less floor space, but spiral wound modules are not as susceptible to plugging by suspended solids.[19] For best performance, the feed to any of these systems should be treated to remove gross amounts of solids and to prevent fouling by precipitation or biological growth.

Pretreatment Requirements--

Colloidal and organic matter can clog the membrane surface, thus reducing the available surface area for permeate flow. Also, low-solubility salts will precipitate on the membrane during the concentration process, similarly reducing membrane efficiency. Pretreatment techniques such as pH adjustment, activated carbon adsorption, chemical precipitation or filtration (approximately 5 m) may be required to ensure extended service life. Operating costs for membrane systems are a direct function of the concentration of the impurity to be removed, due in part to increased maintenance and membrane replacement costs.

Multi-charged cations and anions are effectively removed from the wastewater by reverse osmosis. However, most low molecular weight, dissolved organics are, at best, only partially removed. Their presence could present operational difficulties and may require expensive pretreatment for their removal. The use of reverse osmosis for recovery/reuse of process wastes is also currently somewhat limited because many membranes are attacked by solutions with a high oxidation potential (e.g., chromic acid) or excessive pH levels. However, future development of membranes which are able to withstand harsher environments is expected.[21]

Post-Treatment--

Reverse osmosis applied to plating bath wastes is usually supplemented with an evaporation system in order to adequately concentrate constituents for reuse.[21] The amount of feed concentration permitted in a unit is limited by the membrane characteristics. Reverse osmosis units can concentrate most divalent metals (nickel, copper, cadmium, zinc, etc.) from rinsewaters to a 10 to 20 percent solution.[21] Further concentration must be achieved through the use of a small evaporator. Evaporators are especially necessary for plating and other systems operated at ambient temperatures where atmospheric evaporative losses are minimal.[21]

6.4.2 Process Performance

Reverse osmosis applications in recycling many electroplating bath wastes are somewhat limited due to membrane degradation in the extreme pH regions. However, research has been conducted to recover both acidic and basic plating rinsewaters, which has led to the development of more chemically resistant membranes.

The Walden Division of Abcor, Inc. (Wilmington, MA) conducted a number of studies of reverse osmosis systems for recovery of plating rinsewaters. Initially, studies were conducted to test the applicability of membrane types to various plating rinses. Test samples were prepared by diluting actual plating bath solutions with de-ionized water.[18] Bath properties (total dissolved solids, pH) and test solution properties (concentration, pH) for the wastes tested is presented in Table 6.4.2. The percent TDS rejection values shown in the table were averages for tests at the low end of the concentration levels. Percent reduction values for the metals of interest were generally slightly higher. Overall, membrane performance was affected by feed constituent concentrations, operating pressure, operating temperature, flow rate, and pH. Flux and rejection data were not affected by changes in pH, but extreme pH values were found to decrease membrane life.

Additional tests were performed to evaluate the effects of feed concentrations on flux and percent-rejection. A summary of the operating parameters and results for some of these tests is presented in Table 6.4.3 for the chromic acid rinse. As shown, both flux and rejection decrease somewhat with increasing feed concentration. These results are typical for all of the wastes studied. While rejection and flux results were satisfactory for the chromic acid test runs, hydrolysis and degradation of the membranes occurred.

TABLE 6.4.2. SUMMARY OF REVERSE OSMOSIS EXPERIMENTS

Plating baths	Source of bath	Properties of bath		Properties of test solutions		Average % TDS rejection by module type		
		% TDS	pH	Concentration range (% TDS)	pH range	A	B	C
Chromic Acid (Neutralized)	Whyco Chromium Co.	27.5 (37.1)	--	0.3-4.5	4.5-6.1	98	94	97
Chromic Acid (Unneutralized)	Whyco Chromium Co.	27.5	0.53	0.4-9.0	1.2-1.9	90	94	97
Copper Pyrophosphate	Honeywell (M&T)	31.9	8.8	0.2-11.4	6.8-8.5	96	97	97
Nickel Sulfamate	Honeywell (Harstan)	31.0	4.2	0.5-12.0	4.9-6.1	91	93	91
Nickel Fluoborate	Hampden Colors & Chemicals	25.7	3.5	0.9-5.8	3.4-6.1	64	--	91
Zinc Chloride	General Electric (Conversion Chemical)	19.8	4.5	0.2-4.2	5.3-6.1	92	90	89
Cadmium Cyanide	American Electroplating Co.	26.3	13.1	0.3-3.1	11.5-12.5	95	--	--
Zinc Cyanide	American Electroplating Co.	11.4	13.9	0.5-2.4	12.3-13.3	95	--	--
Copper Cyanide	American Electroplating Co.	37.0	13.3	0.6-3.7	11.8-12.5	98	--	--
Rochelle Copper Cyanide	Whyco Chromium Co.	12.7	11.2	0.13-3.2	9.8-10.6	99	--	--

A - DuPont B-9 permeator, polyamide hollow-fiber membrane.
B - T.J. Engineering 97H32 spiral-wound module; cellulose acetate membrane.
C - Abcor TM 5-14 module, tubular configuration; cellulose acetate membrance.

Source: Adapted from Reference 18.

TABLE 6.4.3. ANALYTICAL RESULTS FOR REVERSE OSMOSIS TREATMENT OF SPENT CHROMIC ACID PLATING RINSE

Membrane module[a]	Waste feed		Operating conditions				% - Rejection	
	%-TDS	% of bath	Pressure (psig)	Temp. (°C)	pH of feed	Flux[b]	Basis: TDS	Basis: Cr^{+6}
Hollow fiber			400			2.59	84	97
Spiral	0.40	1.5	600	29	1.9	15.3	97	96
Tubular			800			10.0	99	98
Hollow fiber			400			1.97	95	87
Spiral	1.83	6.7	600	29	1.2	13.2	94	86
Tubular			800			8.58	97	91
Hollow fiber			400			1.20	90	91
Spiral	4.11	15	600	29	1.2	10.6	92	92
Tubular			800			7.31	95	7
Hollow fiber			400			leak	leak	leak
Spiral	9.43	34	600	28	0.9	leak	leak	leak
Tubular			800			6.60	94	97

[a]Three commercially-available membrane modules were tested:

DuPont B-9 hollow-fiber module (polyamide membrane);
T.J. Engineering 97H32 spiral-wound module (cellulose aretate membrane); and
Abcor, Inc., TM5-14 tubular module (cellulose acetate membrane).

[b]Gallon/minute/single DuPont B-9 permeator size 0440-035.
Gallon/day/ft^2 for spiral-wound and tubular modules.

Source: Reference 18.

Other available data, widely scattered throughout the literature, indicate that reverse osmosis is a proven technology for treating electroplating wastewater. Systems are being used commercially to recover brass, hexavalent chromium, copper, nickel, and zinc from metal finishing solutions. While the ultimate goal is zero discharge, evaporators may be needed to concentrate the solution to the required bath strength. Rinses such as Watts nickel, bright nickel, and nickel sulfamate, all can be treated in a zero discharge system; however, duplex nickel cannot.[22] Some typical membrane rejection values for cations and anions are shown in Table 6.4.4. They are similar to values reported by other sources.[2,10,22,24]

6.4.3 Cost of Treatment

Capital costs for reverse osmosis systems vary with operating parameters, membrane type, modular design, and waste feed characteristics. According to Reference 10, the capital and annual operating costs for a typical reverse osmosis system used in the electroplating industry were $20,000 and $5,000, respectively. Due to savings associated with recovery of plating chemicals, wastewater treatment, and sludge disposal, the payback period was 4.3 years. The above values represent 1979 dollars, but present day costs do not appear to have increased appreciably.

Capital costs are primarily a function of the membrane surface area needed to provide the necessary flux. The packaged reverse osmosis units available from manufacturers contain a fixed number of membrane modules along with auxiliaries such as a feed pump, prefilters, and other equipment needed for pretreatment. The packaged membrane modules can be readily replaced or expanded if the need arises. Installation costs are minimal since the units are normally skid mounted and require only utility connections.

The capital cost of a system is approximately the same for spiral-wound or hollow-fiber membrane units; tubular units are more expensive, but may be required if fouling is a problem for a particular waste. Costs are also very similar for most of the commercially available reverse osmosis membranes; i.e., cellulose acetate, polyamide, and polyether/amide. Figure 6.4.6 shows the relationship between equipment costs and membrane surface areas for a

TABLE 6.4.4. MEMBRANE REJECTIONS

Name	Percent rejection	Maximum concentration percent
Cations		
Sodium	94-96	3-4
Calcium	96-98	*
Magnesium	96-98	*
Potassium	94-96	3-4
Iron	98-99	*
Manganese	98-99	*
Aluminum	99*	5-10
Ammonium	88-95	3-4
Copper	96-99	8-10
Nickel	97-99	10-12
Strontium	96-99	--
Hardness	96-98	*
Cadmium	95-98	8-10
Silver	94-96	*
Anions		
Chloride	94-95	3-4
Bicarbonate	95-96	5-8
Sulfate	99*	8-12
Nitrate	93-96	3-4
Fluoride	94-96	3-4
Silicate	95-97	--
Phosphate	99*	10-14
Bromide	94-96	3-4
Borate	35-70**	--
Chromate	90-98	8-12
Cyanide	90-95**	4-12
Sulfite	98-99	8-12
Thiosulfate	99*	10-14
Ferrocyanide	99*	8-14

*Must watch for precipitation, other ion controls maximum concentration.

**Dependent on pH.

Source: Reference 23.

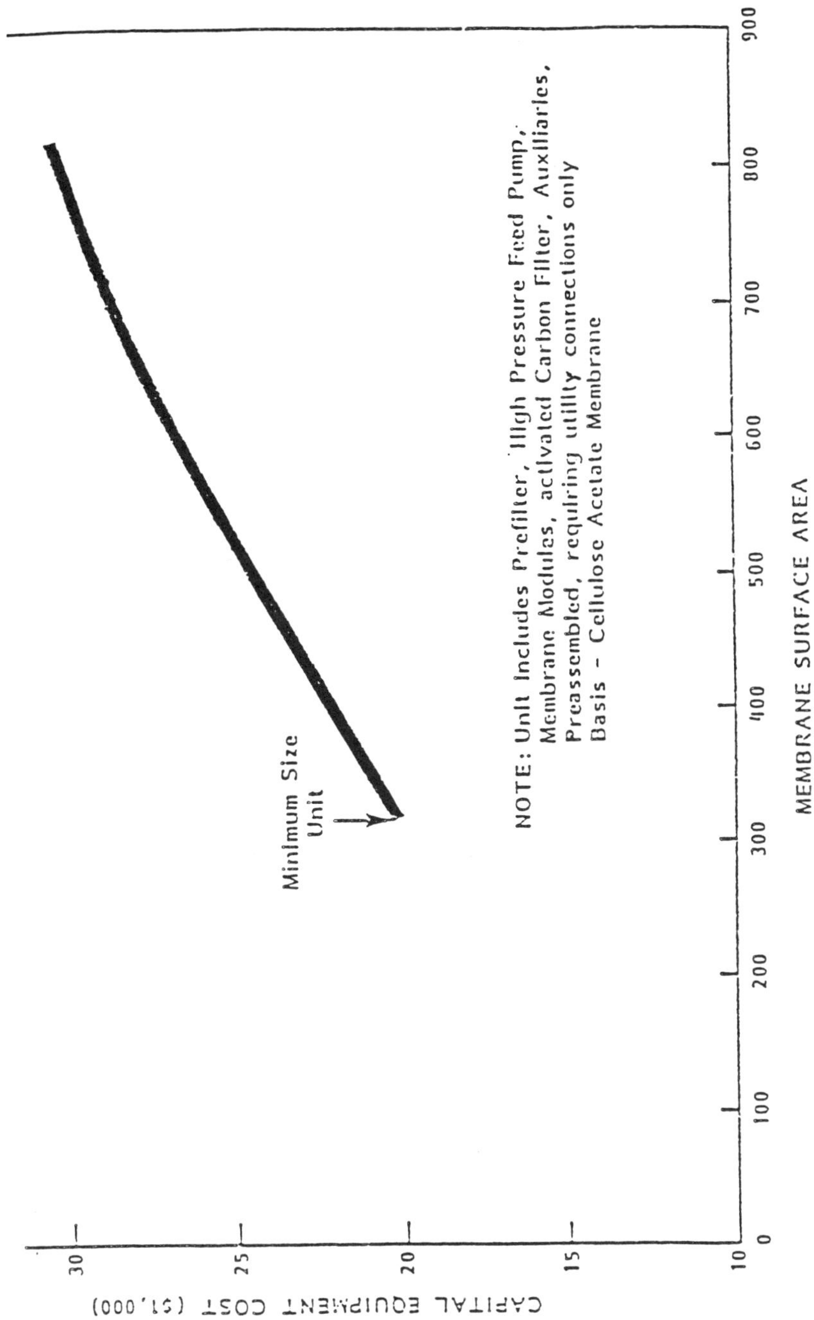

Figure 6.4.6. Reverse osmosis system capital cost vs. membrane surface area.

Source: Reference 25.

spiral-wound, cellulose acetate membrane system. The costs are in 1983 dollars, but are not appreciably different today if the Chemical Plant Index is used as an indicator (a value of 316.9 in 1983 versus 319.7 in April 1987).

The low operating cost of a reverse osmosis system is one of the most attractive features of the technology. The only utility needed for the system is electricity and the feed pumps generally draw less than one kilowatt of power. However, additional costs may be incurred for membrane replacement and feed pretreatment. Also, additional expense may be required for an evaporator when reverse osmosis alone is not capable of providing a output that is concentrated enough for direct reuse as an electroplating solution. In the case of a zinc cyanide system, a cost for an evaporator of $40,000 was estimated to supplement a $25,000 expenditure for a reverse osmosis system. A $10,000 annual savings was insufficient to offset the $12,000 operating costs for the combined reverse osmosis/evaporator system.

However, where reverse osmosis can be used to produce a satisfactory plating solution, chemical recovery benefits can be appreciable. Figure 6.4.7 shows the savings possible from the recovery of nickel salts from a Watts nickel plating line. A detailed analysis of the economics of a reverse osmosis installation for drag-out recovery is presented in Table 6.4.5. The values are given in 1983 dollars; as noted, values are not appreciably different from present-day values if the Chemical Plant Index is used as a cost indicator.

In summary, the cost-effectiveness of reverse osmosis is dependent upon the following factors: production rate, type and concentration of rinsewater constituents, water supply, wastewater disposal costs, and useful lifetime of membranes.[10] As more chemically resistant membranes are developed, reverse osmosis systems will have more cost-effective applications for metal containing waters. Also, with the implementation of the land disposal ban and the resulting rise in sludge disposal costs, reverse osmosis will become a more cost-effective alternative to conventional neutralization practices.

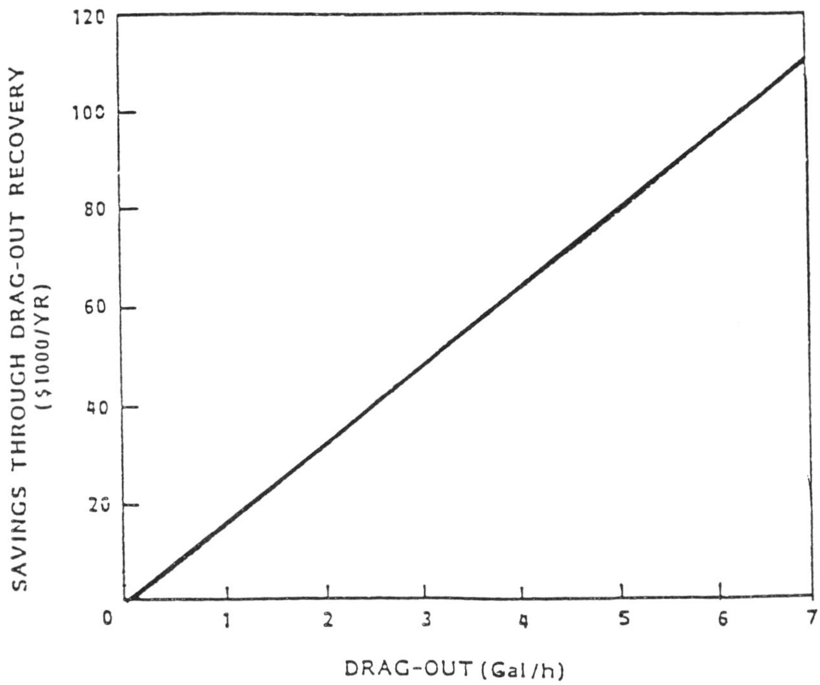

Basis: 4000 hours per year Operation on Watts Nickel Plating line Rinse Tank
Assume 90% Overall Drag-Out Recovery; Salvinos Based on Chemical Recovery Plus Reduced Pollution Control Costs.

Bath Composition--			
$NiSO_4$	44 oz/gal	@	$1.30/lb
$NiCl_2$	6 oz/gal	@	$1.75/lb
H_3BO_3	5 oz/gal	@	$0.45/lb

Figure 6.4.7. Annual savings from nickel plating drag-out recovery.

Source: Reference 25.

TABLE 6.4.5. ECONOMICS OF REVERSE OSMOSIS SYSTEM FOR NICKEL SALT RECOVERY, OPERATING 4,000 h/yr

Item	Amount
Installed cost, 550-ft^2 unit ($):	
Equipment:	
RO system including 25 um filter, pump less ten membrane units	17,000
Activated carbon filter	2,000
Auxiliaries, piping, and miscellaneous	3,000
Subtotal:	22,000
Installation, labor and material	3,000
Total installed cost	25,000
Annual operating cost ($/yr):	
Labor and maintenance at $10/hr	1,600
General plant overhead	1,000
Raw materials:	
Module replacement, 2-year life (10 x $350/module) x 0.5 yr	1,800
Resin for carbon filter	500
Prefilter element (25 um)	700
Electricity costs ($0.45/kwh)	1,100
Total operating cost:	6,700
Annual fixed costs ($/yr):	
Depreciation, 10 percent of investment	2,500
Taxes and insurance, 2 percent of investment	500
Total fixed costs:	3,000
Total cost of operation:	9,700
Annual savings ($/yr):	
Plating chemicals:	
4 lb/hr nickel-salt at $1/lb	16,000
1.5 oz/hr brightener at $0.10/oz	600
Water and sewer charges: saving 270 gal/hr at $0.80/1,000 gal	900
Total gross annual savings:	17,500
Net savings = annual savings - (operating cost + fixed cost) ($/yr)	7,800
Net savings after taxes, 45% tax rate 7,800 x 0.55 + 2,500[a] ($/yr)	6,800
Average ROI = net savings after taxes/total installed investment x 100 (%)	27
Cash flow from investment = net savings after taxes + depreciation ($/yr)	9,300
Payback period = total installed investment/cash flow (yr)	2.7

[a]10% investment tax credit = $2,500 (or 0.10 x 25,000).

Source: Reference 25.

6.4.4 Overall Status of Reverse Osmosis

Availability--

Reverse osmosis technology is available from a fairly large number of membrane and system equipment manufacturers. Many of these firms are listed in references previously cited[1,2] and in the Chemical Engineering Equipment Buyers' Guide which is published annually by McGraw Hill.

Application--

Reverse osmosis appears to have found widespread acceptance in the electroplating industry for the recovery of metals in rinsewater. A list of current reverse osmosis installations in the electroplating industry, adapted from Reference 24, is shown in Table 6.4.6. This list is expected to increase as new membranes are developed to meet the demands of the harsh electroplating environment.

Environmental Impacts--

The reverse osmosis technology, as applied in the electroplating industry, is a recovery technology capable of achieving zero discharge in the certain applications. There should be no detrimental environmental impacts associated with this technology provided the reject stream can be recycled.

Advantages and Limitations--

When properly applied, reverse osmosis systems should achieve economic benefits associated with chemical recovery and the elimination of the expense of hazardous waste disposal. The disadvantages of reverse osmosis are associated largely with the limited lifetimes of membranes in some applications, resulting in cost penalties for membrane replacement and pretreatment. Most suppliers will favor conducting treatability studies to ensure successful application of their systems to a specific waste stream.

TABLE 6.4.6. CURRENT RO INSTALLATIONS IN THE ELECTROPLATING INDUSTRY

Type of bath	Type of membrane and configuration	No. of installations/ zero discharge
Bright nickel Nickel sulfamate Watts nickel	Cellulose acetate Spiral wound	150/yes
Copper sulfate	Polyamide Cellulose triacetate Thin-film composite Hollow-fiber Spiral wound	12/no
Zinc sulfate	Thin-film composite Spiral wound	1/90% recovery
Brass cyanide	Polyamide Cellulose triacetate Hollow-fiber	5/90% recovery
Copper cyanide	Polyamide Hollow-fiber	2/90% recovery
Hexavalent chromium	Thin-film composite Spiral wound	Under investigation

Source: Reference 24.

6.5 ELECTRODIALYSIS

6.5.1 Process Description

Electrodialysis is one of the more recent technologies applied to the recovery of plating chemicals from rinse solution. Electrodialysis uses an electric field as the driving force to remove charged ionic species from a feed stream. Anion and cation exchange membranes allow anions and cations, respectively, to pass from the feed stream to a concentrated ionic solution.

As noted previously, several types of cation and anion exchange membranes are available commercially. Some properties of commercially produced membranes were shown previously in Table 6.2.1. As with reverse osmosis systems, electrodialysis systems are available as packaged units equipped with electrical components, pumps, motors, pretreatment features, recycle, temperature control, cleaning, and other features. These can be arranged in parallel or series as required by the application and its process streams. Properly designed and operated, electrodialysis units have proven to be effective and reliable.[25]

By packaging several cell pairs of membranes (typically 50 to 300 cell pairs[20]) between electrodes and manifolding the streams, a concentrated stream and a depleted stream, from which 45 to 55 percent of the ions have been removed, are generated.[2] Further ion reduction of the depleted stream can be accomplished in additional stages. However, electrodialysis cannot process highly deionized water because of the poor electrical conductivity of such waters. A flow sheet of a three stage electrodialysis system used for desalting is shown in Figure 6.5.1. As noted in Reference 2, the feed and recycle pumps operate at pressures of about 50 psig, a pressure sufficient to supply as many as six stages without intermediate pumps. Excessive feed pressures must be avoided to prevent leakage. However, the flow rate in any stage must be sufficient to create adequate turbulence to keep concentration polarization below scaling limits.

Electrodialysis removes dissolved matter from water, leaving nonionized material (such as many organics, suspended matter, silica, etc.) present in the ion depleted water. This can cause problems if, for example, a build-up

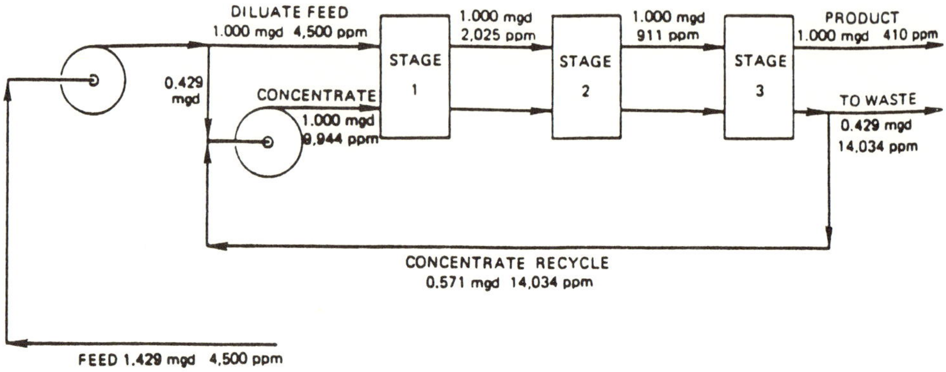

Figure 6.5.1. Schematic of a three-stage ED plant.

Source: Reference 2.

of organic compounds in the purified stream is undesirable. Also, all ionic species are nonselectively recovered. Pretreatment (for example, to reduce concentrations of hardness components and other dissolved impurities or organics) and maintenance requirements must consider these possibilities and their implications.

A potential problem with all applications is the possibility of reaching excessive current densities because of the high concentration of ions at the membrane interfaces. Possible consequences of this are the precipitation of metals such as calcium and magnesium and the electrolysis of water to hydrogen and hydroxide ions. Undesirable effects leading to membrane fouling and local overheating of membranes can result.

Pretreatment or system design features can avoid problems resulting from electrolysis. For example, introducing turbulence or reducing the total ionic content of the concentrate stream have been successful in reducing fouling and electrolysis. To avoid fouling tendencies, almost all manufacturers recommend periodic reversal of the applied voltage while simultaneously re-routing the feed and concentrate.

There are no fundamental limits, other than solubility, on the maximum concentration level obtained in the concentrate. However, power consumption is directly proportional to the ion content of the feed. This contrasts with reverse osmosis, in which separation costs are less strongly influenced by concentration. Consequently, electrodialysis operating costs are favorable for low feed ion concentration and become less so as concentration increases. In addition, electrodialysis, according to Reference 21, is generally used to produce a concentrated solution, such that evaporation units are not required. Where a valuable concentrate is being provided, salts may be concentrated to 20 percent or more, significantly beyond that feasible for reverse osmosis systems.[12]

6.5.2 Process Performance

As noted in Reference 25, there are now more than 100 applications of electrodialysis to process rinsewaters from electroplating processes. At least three vendors are currently manufacturing systems for treatment of wastes from gold, chromium, silver, and zinc cyanide plating operations and from nickel plating operations. Other plating baths treated successfully by electrodialysis include tin and tin lead fluoroborate, and trivalent chromium baths. Application to hexavalent chromium plating is questionable because of the potential for degrading presently available membranes.[25] Another application that has been successfully demonstrated involves the recovery of chromic acid and sulfuric acid from spent brass etchants. This particular electrodialysis system, used for acid recovery, was developed at the Bureau of Mines[26,27] and is now available from Scientific Control, Inc. in Chicago, Illinois. The system is applicable to wastes containing copper as the primary contaminant.[20] A more detailed description of this process and other electrodialysis processes can be found in References 10, 20, and 25, and in other primary references cited.

6.5.3 Cost of Treatment

Typical costs for electrodialysis systems to treat plating rinsewaters range from $30,000 to $45,000, depending on the application.[10] Capital costs for the "Chrome Napper" system available from Innova Technology, Inc. in Clearwater, Florida, range from $9,900 to $30,000, including installation and power supply.[28] This has been successfully used for the recovery of chromic acid from electroplating wastewaters. Systems are sized according to bath temperature, dragout concentrations, number of rinse tanks, concentration of the bath, and the volume of spent solution to be treated per unit time.

Scientific Control, Inc. sells electrolytic electrodialysis units to recover chromic/sulfuric acid brass etchants. Unit sizes are based on the amount of copper the system is capable of removing per unit of time. Available unit sizes range from 0.05 to 0.5 lb copper removal per hour. Capital equipment costs (1986 dollars) for these units range from $24,000 to $80,000. These costs do not include installation which would include a hoist, plumbing, and a ventilation/exhaust system. Additional costs for the exhaust system could range from $5,000 to $15,000, depending on the size required. Operating and maintenance costs are relatively low. Membranes will need to be replaced approximately every 9 months, depending on usage, at a replacement cost of approximately 10 to 15 percent of the original equipment costs. Additional maintenance costs will include approximately $10/month for replacement of filter cartridges (a pre-filter system is incorporated into the unit). The estimated payback period for the system is approximately 2 years, based on savings in treatment and disposal costs.[29]

6.5.4 Overall Status of Process

Availability--

Over 100 electrodialysis systems are now employed commercially for the recovery of metals from electroplating rinsewaters.[25] At least three manufacturers (Scientific Control, Inc., Chicago, IL; Innova Technology, Clearwater, FL; and Ionics Inc., Watertown, MA) are currently manufacturing electrodialysis equipment for this application. Other membrane-oriented equipment suppliers are listed in the references[1-5,12,20], as well as in the Chemical Engineering Equipment Buyers' Guide.

Application--

The principal area of application of electrodialysis appears to be the recovery of metals from electroplating bath rinsewaters. Electrodialysis and reverse osmosis are competitive processes for these applications. Electrodialysis would appear to have the advantage when concentration levels are low (operating costs are low), or when recovery values justify the expense of achieving concentration levels higher than those possible with reverse osmosis.

Environmental Impacts--

Because electrodialysis is a recovery process, the environmental impacts are limited to those resulting from pretreatment and post-treatment. Pretreatment operations generating wastes include filtration to remove solids, oil, and grease, chemical precipitation to remove scaling components, and ion exchange to remove organics that could lead to biological fouling or electroplating difficulties. Post-treatment requirements are minimal, but could involve treatment or purification of process streams suffering from a gradual accumulation of contaminants in a near zero discharge system.

Advantages and Limitations--

A significant advantage of electrodialysis over reverse osmosis is its ability to concentrate solutions up to their solubility limit, thus avoiding the need for auxiliary equipment such as evaporators. Further advantages include the following, as noted in Reference 2:

- The units operate continuously (ion exchange without regeneration);
- The only utility required for operation is a DC power source;
- The units are compact; and
- Operating cost is low; electrical power consumption averages \$0.25/hour.

Disadvantages of the process vary, depending on the application. All ionic species are nonselectively recovered, including ionic bath impurities. Conversely, organic brighteners, wetting agents, and other nonionized compounds will accumulate in the dilute stream, limiting its reuse potential.

A potential problem with any application is the possibility of exceeding the maximum voltage set by the solution conductivity at the membrane boundary layer. The consequence of this condition is electrolysis of water to hydrogen and hydroxide ions and the possible resulting precipitation of metal hydroxides which will foul the membranes.

6.6 OTHER MEMBRANE PROCESSES

Previous reference has been made to other membrane technologies, notably Donnan dialysis and coupled transport. Although these processes offer potential advantages over other technologies, including the commercially available membrane technologies, they have yet to achieve commercial status. Recent discussions with representatives of firms involved in the study and development of these processes have indicated that no additional work is in progress.[30,31] However, the Bend Research Corporation, a developer of a coupled transport process, is actively seeking licensing arrangements for a process they feel is viable and demonstrated.[31]

REFERENCES

1. M. Cheryan. Ultrafiltration Handbook. Technomic Publishing Company, Inc., Lancaster, PA. 1986.

2. G. Belfort. Synthetic Membrane Processes. Academic Press Inc., Orlando, FL. 1984.

3. R.E. Kesting. Synthetic Polymeric Membranes. 2nd Edition, John Wiley & Sons, Inc., New York, NY. 1985.

4. D.R. Lloyd. Material Science of Synthetic Membranes. Americal Chemical Society, Washington, D.C. 1985.

5. Sourirajan, S., and T. Matsuura. Reverse Osmosis/Ultrafiltration Process Principles. National Research Council of Canada, Ottawa, Canada. 1985.

6. J.D. Birkett. Dialysis/Electrodialysis. In: Unit Operations for Treatment of Hazardous Industrial Wastes. Noyes Data Corporation, Park Ridge, NJ. 1978.

7. J.B. Berkowitz. Unit Operations for Treatment of Hazardous Industrial Wastes. Noyes Data Corporation, Park Ridge, NJ. 1978.

8. E.I. DuPont DeNemours & Company. Permaset Permeators. Waste Treatment by Reverse Osmosis. October 1984.

9. L.E. Applegate. Membrane Separation Processes. Chemical Engineering. June 11, 1984.

10. T.E. Higgins. Industrial Processes to Reduce Generation of Hazardous Waste at DOD Facilities, Phase II Report. July 1985.

11. C.H. Gooding. Reverse Osmosis and Ultrafiltration Solute Separation Problems. Chemical Engineering. January 7, 1985.

12. The Hazardous Waste Consultant. May-June 1985.

13. H.K. Lonsdale. The Growth of Membrane Technology. Journal of Membrane Science, 10:81. 1982.

14. H.F. Hamil. Removal of Toxic Metals in Electroplating Wash Water by a Donnan Dialysis Process. U.S. EPA-600/2-82-098. December 1982.

15. C.J. Fournier. Memtek Corporation, Billerica, MA. Unpublished Material and Bulletins supplied to Alliance. April 1987.

16. Memtek Corporation, Bulletin No. PB-01. 1986.

17. T.V. Tram. Advanced Membrane Filtration Process Treats Industrial Wastewater Efficiently. Chemical Engineering Progress. March 1985.

18. Donnelly, R.G., Goldsmith, R.L., McNulty, K.J., and M. Tan. Reverse Osmosis Treatment of Electroplating Wastes. Plating. May 1974.

19. Crampton, P., and R. Wilmoth. Reverse Osmosis in the Metal Finishing Industry. Metal Finishing. March 1982.

20. Wilk, L. et al. Technical Resource Document - Treatment Technologies for Corrosive-Containing Wastes. Prepared for U.S. EPA, HWERL, Cincinnati, OH. October 1986.

21. Higgins, T.E., CH2MHILL. Industrial Processes to Reduce Generation of Hazardous Waste at DOD Facilities - Phase 2 Report, Evaluation of 18 Case Studies. Prepared for the DOD Environmental Leadership Project and the U.S. Army Corps of Engineers. July 1985.

22. Finkenbiner, K., and P. Cartwright. Reverse Osmosis Technology for Toxic Heavy Metals Treatment. Presented at HAZPRO '85, Baltimore, MD. May 12, 1985.

23. J.D. Birkett. Reverse Osmosis. In: Unit Operations for Treatment of Hazardous Industrial Wastes, Noyes Data Corporation, Park Ridge, NJ. 1978.

24. P.S. Cartwright. An Update on Reverse Osmosis for Metal Finishing. Plating and Surface Finishing. April 1984.

25. G.C. Cushnie, Centec Corporation, Reston, VA. Navy Electroplating Pollution Control Technology Assessment Manual. Final Report prepared for the Naval Facilities Engineering Command under Contract No. F08635-81-C-0258. NCEL-CR-84-019. February 1984.

26. Soboroff, D.M., Troyer, J.D., and A.A. Cochran. Regeneration and Recycling of Waste Chromic Acid-Sulfuric Acid Etchants. Bureau of Mines Report of Investigations No. 8377. 1979.

27. McDonald, H.O., and L.C. George. Recovery of Chromium From Surface-Finishing Wastes. Bureau of Mines Report of Investigations No. 8760. 1983.

28. D. Pouli, Innova Technology, Clearwater, FL. Telephone conversation with Lisa Wilk, Alliance Technologies Corporation. August 26, 1986.

29. S. Gary, Scientific Control, Inc., Chicago, IL. Telephone conversation with Lisa Wilk, Alliance Technologies Corporation. August 29, 1986.

30. J.B. Hsu, Southwest Research Institute, San Antonio, TX. Telephone conversation with Alliance Technologies Corporation. April 1987.

31. D. Friesen. Bend Research Corporation, Bend, OR. Telephone conversation with Alliance Technologies Corporation. April 1987.

7. Extraction for Metal Removal

7.1 BACKGROUND

Liquid-liquid extraction involves the separation of a component from a waste solution by transfer to a second liquid. The extractant is immiscible in the waste, but exhibits a preferential affinity for the constituent. Although not a widely applied treatment technology, liquid extraction has potential for removal of many toxic constituents from wastewaters. Liquid extraction is particularly attractive in cases where the solutes are present at high enough concentration levels to provide recovery value or when other treatment methods are less effective.

In the mining industry, the solvent extraction of metal salts from aqueous solutions has acquired commercial importance, particularly for the recovery of copper, nickel, cobalt, uranium, vanadium, and other metals from aqueous effluents. However, the application of solvent extraction to treat waste effluents remains undeveloped. As a unit operation, extraction lags in terms of the amount of research which has been conducted and the availability of quantitative design methods. Performance is highly waste and site-specific due to competing reactions, desired selectivity in multi-component wastes, and operating limitations of pH, temperature, and processing time. The impact of these variables can only be accurately determined through laboratory scale testing, using various extractants or combinations of extractant, to determine distribution isotherms and reaction kinetics. Thus, much of the technological development of extraction has been carried out by manufacturers of specialized equipment and has not been disseminated through open literature.

The simplest extraction system is comprised of three components: 1) the solute, or material to be extracted; 2) the carrier, or the nonsolute portion of the feed mixture to be separated; and (3) the solvent, which is immiscible with the carrier phase. Discussions of extraction require distinctions to be made between the light and heavy phases, the dispersed and continuous phases, and the raffinate and extract phases. The terminal streams from an extractor are the extract and the raffinate. This is shown in Figure 7.1.1 for the case of countercurrent extraction.

As a recovery process, the use of solvent extraction involves the following steps:

1. Extraction--Constituents are transferred from aqueous phase to organic phase using an organic solvent as an extractant.
2. Back-Extraction/Stripping--The constituent to be recovered is transferred from the organic phase to a concentrated aqueous phase.

Solvents used for the extraction of metals include three basic types: cation or acidic extractants; anion exchangers; and solvating agents, as shown in Table 7.1.1.[1] Metal cations react with the cation or acidic extractant, typically an organic acid, to form neutral complexes that are preferentially dissolved by the organic phase. The following equation describes a cation exchange:

$$M^{n+} + nRH \rightleftharpoons MR_n + nH^+$$

As shown, hydrogen ions are exchanged for the metal cation in proportion to its valence. Thus, Fe^{+3} is preferentially extracted by acid extractants in the presence of divalent ions such as Cu^{+2} or Ni^{+2}. The degree of extraction of the metal will also increase with the pH of the aqueous phase since, at low pH, the extractant cannot release its hydrogen ion in exchange for the cation.

As shown in Table 7.1.1, anion exchangers in solvent extraction are generally protonated forms of primary, secondary, and tertiary high-molecular weight amines and quaternary compounds. The extraction of metal complexes proceeds primarily by either ion exchange or an addition reaction as follows:[2]

$$R_4N^+X^- + MY^{n+1} \rightleftharpoons R_4N^+MY_{n+1} + X^-$$

or

$$R_4N^+X^- + MX_n \rightleftharpoons R_4N^+MX_{n+1}$$

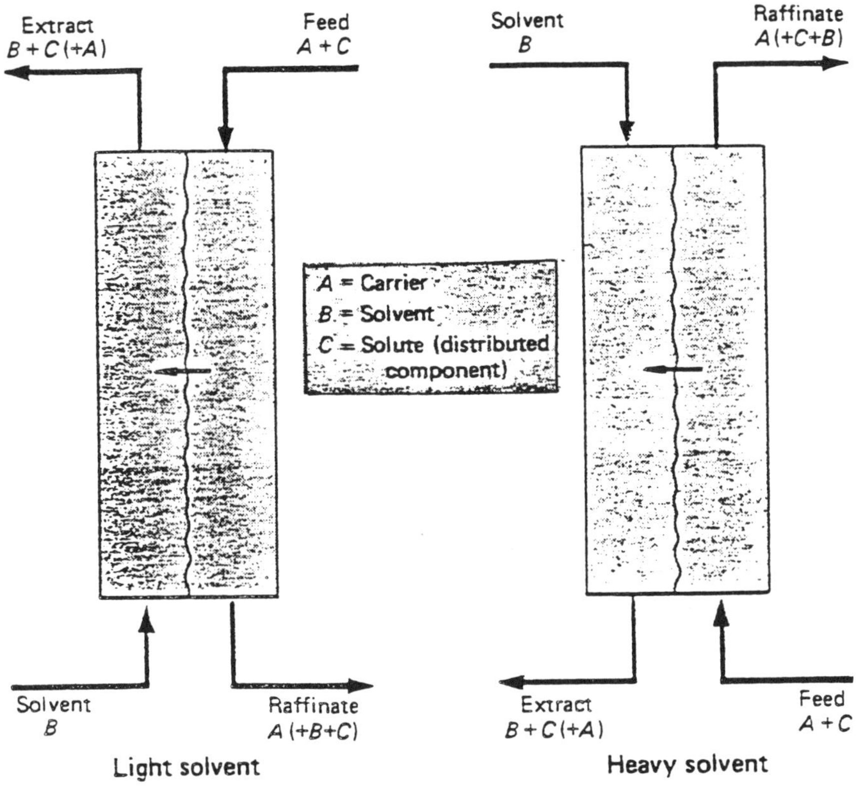

Figure 7.1.1. Simple system illustrating extraction technology.

Source: Reference 1.

TABLE 7.1.1. CLASSIFICATION OF EXTRACTION REAGENTS

Extractant type	Extraction mechanism	Example extractants
Cation or acid extractants	Extraction by compound formation	Organic acids, such as carboxylic, sulfonic, phosphoric, phosphonic, phosphinic acids; and acidic chelating agents.
Anion exchangers	Extraction by ion-pair formation	Polyphenylmetalloid type, polyalkylsulfonium type, polyalkylammonium type, and salts of high-molecular-weight aliphatic amines.
Solvating agents	Extraction by solvation	Carbon-, sulfur-, or phosphorus-bonded oxygen-bearing extractants; alkylsulfides; etc.

Source: Reference 2.

In any reaction, both extraction mechanisms occur. However, the controlling mechanism is determined by the free concentration of X^- in the aqueous phase and the dominant species present in the aqueous phase (i.e., $MX^{(n-1)+}$, $MX_2^{(n-2)}$, ..., MX_n, MX_{n+1}^-, etc.). Thus, a knowledge of individual system chemistry is of prime importance in determining reaction kinetics and optimal system design.

Extraction by solvation requires the transfer of a formally neutral species from the aqueous to the organic phase. This occurs through solvation of the metal ion of a neutral salt species or, in the case of formation of a complex acid species, solvation of the proton:[2]

$$MX_n + yS \rightleftharpoons MX_nS_y$$

or

$$HMX_{n+1} + xS \rightleftharpoons (HS_x)^+(MX_{n+1})^-$$

In acidic and solvation extraction of metal complexes, the extraction agent will replace primary and/or secondary waters of hydration, thus rendering the complex soluble in the organic phase.

The degree of extraction of a metal by a solvating extractant depends on a number of factors, including: 1) the nature and concentration of the anionic coordinating ligand X which, in turn, influences the type of metal complex formed; 2) the degree of hydration of these aqueous metal complexes; and 3) the relative strength of the water-metal and extractant-metal bonds. These factors determine the nature of the competition between water and the extractant for the solvation sites. Ideally, the metal ion will be completely stripped of its hydration layer.[2]

Tables 7.1.2 through 7.1.4 summarize the structures and properties of frequently used cation and acidic extractants, anion exchangers, and solvating exchangers.

7.2 PROCESS DESCRIPTION

Solvent extraction can be used for the recovery of concentrated solutions or for treatment of wastewater streams prior to discharge. The former is used for the treatment of spent process solutions such as used pickling and plating baths, and process bleed streams. The purpose of recovery in this case is to

TABLE 7.1.2. STRUCTURE AND PROPERTIES OF ACIDIC EXTRACTANTS

Name	Formula	Molecular Weight of Active Extractant	As Received Extractant		
			Active Extractant, wt. %	Specific Gravity	Flash Point, °C
Versatic 10	$R_1-C(R_2)(R_3)-COOH$ $R_1 + R_2 + R_3 = C_8$	175	99.6	0.91	129
Di-2-ethylhexyl-phosphoric acid	$(C_4H_9CH(C_2H_5)CH_2O)_2POOH$	322	100	0.98	
Octylphenyl-phosphoric acid	$ROPO(OH)_2 + (RO)_2POOH$ $R = (CH_3)_3CCH_2C(CH_3)_2C_6H_5$	–	–	–	
SYNEX 1051	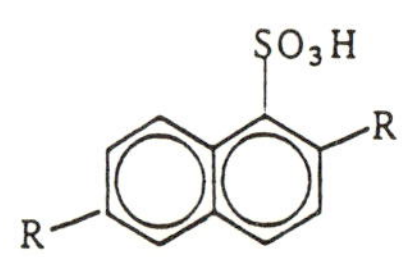 $R = C_9H_{19}$	458	50	0.92	

Source: Reference 2.

TABLE 7.1.3. STRUCTURE AND PROPERTIES OF ANION EXCHANGERS

Name	Formula	Molecular Weight of Active Extractant	As Received Extractant: Active Extractant, wt. %	As Received Extractant: Specific Gravity	As Received Extractant: Flash Point, °C
	Primary Amines (RNH_2)				
Primene JMT	$R = (CH_3)_3C(CH_2C(CH_3)_2)_4$—	269-325	100	0.84	–
	Secondary Amines (R_2NH)				
LA-2	$R = C_{12}-C_{13}$	351-393	100	0.83	180
Adogen 283	$R = C_{13}$	385	92	0.83	–
	Tertiary Amines (R_3N)				
Adogen 364	$R = C_8-C_{10}$	~380	96.2	0.81	–
Alamine 336	$R = C_8-C_{10}$	~392	95	0.81	168
Adogen 368	$R = C_8-C_{12}$	–	95.8	0.82	–
Hostarex A327	$R = C_8-C_{12}$	~395	–	0.81	203
Adogen 381	R = isooctyl	353	95.1	0.82	–
Alamine 308	R = isooctyl	–	–	–	–
Hostarex A324	R = isooctyl	~363	–	0.81	166
Adogen 382	R = isodecyl	437	95.1	0.82	–
	Quaternary Ammonium Compounds ($R_3N(CH_3)^+Cl^-$)				
Adogen 464	$R = C_8-C_{10}$	~431	90	0.84	–
Aliquat 336	$R = C_8-C_{10}$	~442	>88	0.88	132

Source: Reference 2.

TABLE 7.1.4. STRUCTURE AND PROPERTIES OF SOLVATING EXTRACTANTS

Name	Formula	Molecular Weight	Specific Gravity	Flash Point, °C	Viscosity (25°C), cP
Carbon-Oxygen-Bonded Donors					
Ethers (R_1OR_1) or ($R_2OCH_2CH_2OR_2$)					
Diisopropyl ether	$R_1 = (CH_3)_2CH$	102	0.726	-25	0.38
Dibutylcellosolve	$R_2 = C_4H_9$	174	0.837	–	1.34
Alcohols (ROH)					
n-Butanol	$R = C_4H_9$	74	0.81	32	2.46
n-Pentanol	$R = C_5H_{11}$	88	0.82	33	3.31
Ketones (R_1COR_2)					
Methyl isobutyl ketone	$R_1 = CH_3$, $R_2 = (CH_3)_2CHCH_2$	100	0.804	14	0.55
Phosphorus-Oxygen-Bonded Donors $R_1R_2R_3PO$					
Phosphoric acid esters					
Tri-*n*-butylphosphate	$R_1 = R_2 = R_3 = C_4H_9O$	266	0.97	193	3.56
Phosphonic acid esters					
Hostarex PO212	$R_1 = R_2 = C_4H_9O$, $R_3 = C_4H_9$	250	0.94	–	5
Hostarex PO224	$R_1 = R_2 = C_8H_{17}O$, $R_3 = C_8H_{17}$	418	0.91	–	16
Phosphine oxides					
Trioctylphosphine oxide	$R_1 = R_2 = R_3 = C_8H_{17}$	386		(solid)	
Sulfur-Containing Extractants					
Sulfides (RSR)					
Dihexyl sulfide	$R = C_6H_{13}$	202	–	–	–

Source: Reference 2.

purify or recycle the solution by removing or reducing the concentration of undesirable impurities. Process costs must be balanced against the value of recovered solution and, in some cases, the recovery value of the impurity. For example, built-up of iron and copper in pickling and etching operations, respectively, can be partly or entirely extracted. The regenerated solution can be reused after makeup and the recovered metal sold to secondary metal smelters.

Solvent extraction can also be used to treat low concentration liquid effluents. Typical applications often involve large volume flows, such as rinse waters used to remove dragout from pickling and plating baths. Mine waters, wet scrubber solutions, and drainage waters from dumps are also examples of dilute effluents which might utilize extraction.

System designs and configurations are highly waste and site specific. Configurations can include multiple stage extraction, use of more than one solvent for sequential extraction to obtain higher purity separations, and pretreatment (e.g., precipitation, filtration) and post-treatment (e.g., carbon adsorption to remove organics from the raffinate) options. These considerations are discussed below.

In commercial applications, the extractant is typically used at a 10 to 40 percent active level in a non-toxic, inexpensive solvent such as kerosene.[3] Depending on the application, this solution can be used in volumes equal to that of the waste stream. It is contacted countercurrently with the waste feed, usually at a slightly elevated temperature, to improve exchange kinetics and improve phase separation.[4,5] Initial solvent selection will be based on its extraction efficiency and selectivity as determined through laboratory testing. However, solvent purchase price, anticipated loss (e.g., resulting from incomplete phase separation), and breakdown will determine the overall economic viability of the extractant. Breakdown can occur as a result of build-up of organic additives which are present in the wastewater. These will be carried over into the organic phase and will eventually hinder phase separation. Alternatively, breakdown can occur due to long-term incompatibility with regenerants.

Solvent selection and processing conditions will determine overall system efficiency. Reaction kinetics can be enhanced by increasing the concentration of the extractant, raising the waste pH, increasing the relative volume of extractant to feed, and improving mixing.[6] However, since each of these impairs phase separation, an optimal balance must be determined. Phase separation can also be enhanced by using solvent modifiers (e.g., decyl alcohol, tributyl phosphate), optimal solvents (e.g., paraffinic solvents give better separation than kerosene or high aromatics), and reagent combinations. As an example of the latter, secondary reagents used with sulfonic acids can improve separation rates by 2 or 3 times over sulfonic acids alone.[6]

Choice of regenerant is dependant on the speed and efficiency of the separation.[4] A typical regenerant will contain 10 to 15 percent acid at an organic/aqueous volume ratio of 15 to 1.[6] Volume ratios of the original waste feed and the final metal-laden regenerant depend on the overall ease and required completeness of separation. Increases in metal concentrations of 20 to 30 times the feed level are not uncommon.[5]

The most prevalent type of equipment used in commercial applications is a mixer-settler which consists of a high intensity mixer and a large baffled settling chamber. Most applications use several units in series to provide high volumetric throughput. Equipment and maintenance costs are low, but expenses can escalate rapidly if significant solvent make-up or extensive post-treatment are required. Post-treatment equipment may include electrolytic or evaporative recovery units for the concentrated metal stream and carbon adsorption systems for organic removal from the raffinate. Other, more advanced, equipment includes reciprocating plate column contactors, centrifugal contactors, and electrostatic coalescers. Beaker tests have shown up to 90 percent reductions in phase separation time with the latter.[6]

A large number of processes using solvent extraction have been proposed for the treatment of liquid waste effluents. Some of these have included pilot or laboratory studies, as summarized in Section 7.4. However, very few proposed applications have actually been carried through to commercial operation. Those which have been identified are summarized below.

7.2.1 Recovery of Zinc from Pickling Liquor

Pickling in the galvanizing industry is commonly performed with hydrochloric acid. The spent solution contains about 1 to 2 percent free HCl and an iron content of 100 to 130 g/L. The solution will also contain 20 to 120 g/L of zinc, as well as smaller concentrations of California List metals such as chromium and nickel.[2] The high zinc content prohibits conventional treatment of the liquor via thermal decomposition to iron oxide and hydrochloric acid.

The Metsep process was designed to separate zinc from the iron chloride solution by continuous resin ion exchange.[2,8] Anionic zinc-chloro complexes are absorbed on a strong-base, ion-exchange resin. The resin is eluted with water to yield zinc chloride which is converted to a sulfate medium by solvent extraction with Di-2-ethylhexylphosphoric acid (D2EHPA) and stripped with sulfuric acid. The product is a zinc sulfate solution suitable for electrowinning (an electrolytic deposition process used for the recovery of metals from solution) and subsequent re-sale (e.g., sale to an electrolytic zinc refinery). The solvent extraction raffinate is used for hydrochloric acid production.

The MeS Process, an alternative to the Metsep process, has been developed for the recovery of zinc from pickle liquors. It uses a solvent extraction circuit for the initial separation of zinc from iron in the pickle liquor. Zinc is preferentially extracted, as a zinc chloride complex, with Tri-butylphosphate (TBP). Iron extraction is minimal, since it is primarily in the ferrous state. The preferential extraction of zinc over iron is somewhat less with TBP than with optimal amine extractants, but this is balanced by operational advantages such as higher loading.[9]

Zinc is stripped from the organic solution with water or dilute sulfuric acid. The zinc chloride strip solution is mixed with sulfuric acid mother liquor in a boiler, thus evaporating hydrochloric acid and crystallizing zinc sulfate. The zinc sulfate is separated by centrifugation. By adjusting the conditions in the boiler, chloride-free zinc sulfate suitable for electrowinning can be produced. The HCl and iron oxide by-products are also recovered. HCl is returned to the process and iron oxide is available as a saleable product although exact purities were not specified.

The MeS process was developed by MX-Processor AB in Sweden and has been piloted in Holland at a galvanizing plant with encouraging results.[9] The raffinate produced contained less than 100 ppm zinc. Extractant residues were removed by activated carbon adsorption.

7.2.2 Recovery of Zinc-Cyanide Plating Baths

Zinc electroplating is carried out from alkaline zinc cyanide solutions which generate contaminated rinsewater requiring treatment for cyanide destruction. Zinc cyanide can be efficiently extracted from alkaline solutions by quaternary amines. The Union Carbide Corporation[10,11] has developed a process based on the simultaneous extraction of both zinc and cyanide. The decontaminated raffinate was recycled as fresh rinse water. The amine extractant was regenerated by stripping with sodium hydroxide and recovered zinc and cyanide were recycled as plating bath make-up.

A typical composition of the contaminated rinse water was 40 ppm cyanide and 23 ppm zinc. Solvent extraction reduced these values to 0.4 ppm cyanide and 0.07 ppm zinc. Active carbon treatment further reduced these levels and, at the same time, reduced entrained and dissolved amine to 0.1 ppm. A ratio of feed to strip solution of 162:1 produced a strip solution containing 3 to 4 g/L zinc. The same procedure has also been demonstrated for cadmium cyanide plating rinse waters, however, it is not known whether the process has been operated commercially.

7.2.3 Recovery of Copper Plating and Etchant Baths

In general, the higher purchase price of copper relative to zinc and the simplicity of electrolytically recovering it from extractants makes its recovery comparatively more favorable. The hydrometallurgical copper industry has widely applied selective extractants that have a high affinity for copper in weakly acidic and ammoniacal solutions, while simultaneously rejecting ferric iron. Another application includes the recovery of copper containing etchants as described below.

Etching of copper, with the use of an ammoniacal solution, is a common procedure in the manufacture of printed circuit boards for the electronics industry. Spent ammoniacal etching solution contains free ammonia, one or more ammoniacal salts, copper, and oxidants. Maximum etching efficiency is obtained when the ammoniacal solution contains 110 to 130 g/L of copper. It gradually diminishes as the copper concentration approaches 150 to 170 g/L.[12,13] Thus, to keep etching efficiency constant and optimal, the etching solution must be continuously regenerated or replaced with fresh solution.

A process patented by the Criterion Corporation[14] (see Figure 7.2.1) completely removes copper from spent etchants to produce a fresh product that, after makeup, can be re-sold to printed circuit board producers. The process uses an (LIX64N) solvent to selectively extract copper and chloride ions; process of this type has been operated in the United Kingdom by Proteus Reclamation Ltd., recovering 300 kg/day of copper with Acorga P5100 used as extractant.

A similar process which has been used for onsite etchant recovery is the Mercer Process (see Figure 7.2.2).[10] This process withdraws etchant directly from the etching line and recirculates it through a solvent extraction circuit, maintaining the copper concentration in the etchant within the optimal range of 110 to 130 g/L. Treatment of the rinsewater obtained from rinsing the circuit boards after etching is also integrated into the process.

In the first extraction stage the etchant is mixed with an organic solution containing LIX54 in kerosene. The initial copper concentration in the etchant, approximately 130 g/L, is reduced to 90 g/L. The regenerated etchant is returned to the etching line after careful removal of entrained organic solvent. The copper in the rinsewater is extracted in a second extraction stage. At the same time, any entrained etchant from the first extraction stage is washed out. Copper is stripped from the organic solvent with barren copper electrolyte in the stripping stage. The solvent is reused and copper metal is produced by electrowinning on titanium cathodes.

The Mercer process is in successful operation in two prototype installations and commercial units are now marketed by P.R. Processutveckling AB, Sweden. Etchant makeup has reportedly been reduced by 95 percent and no negative influence on product quality has been encountered.[10]

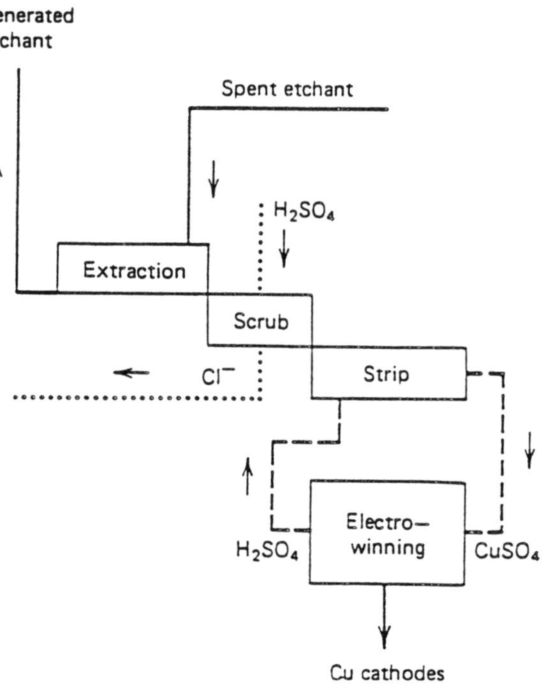

Figure 7.2.1. Process for recovery of copper from spent ammoniacal chloride etchant.

Source: Reference 14.

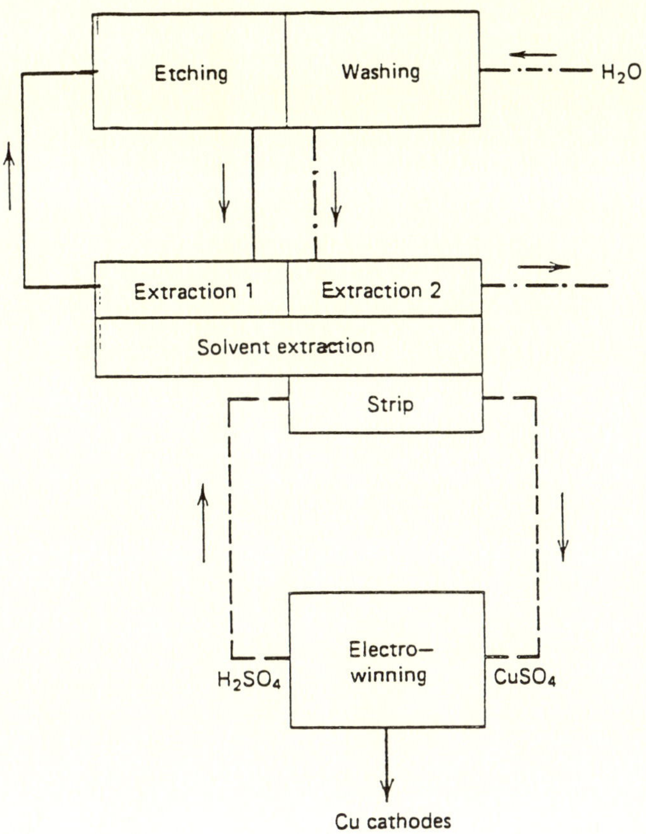

Figure 7.2.2. The MECER process for on-line regeneration and copper recovery from ammoniacal etchant.

Source: Reference 10.

7.2.4 Recovery of Nickel Plating Baths

Recovery of nickel from plating baths and rinsewaters is another feasible application of solvent extraction. One process proposed by Flett and Pearson[15] is based on extraction of nickel with D2EHPA. The solvent is used in its sodium form to avoid pH changes during extraction, as follows:

$$2NaD2EHP_{(org)} + Ni^{2+} \rightleftharpoons Ni(D2EHP)^{2}_{(org)} + 2Na^{+}$$

A typical feed solution may contain 1 to 2 g/L nickel. Laboratory tests showed that nickel could be effectively removed in two extraction steps with the resulting raffinate containing 4 mg/L.[2] By loading the solvent with nickel, the transfer of sodium to the strip solution was minimized. Nickel was stripped from the solvent with dilute sulfuric acid and recovered from the strip solution by electrowinning.

7.2.5 Recovery of Chromium Plating Baths

During the plating of chromium, a buildup of impurities such as Fe(III), Cr(III), Ni, Cu, and Zn gradually takes place, making the bath unusable for plating. Rinses containing similar contaminants in more dilute concentration are aso generated. Two alternative solvent extraction procedures are available for recovery of these wastes: extraction of Cr(VI), or extraction of impurities.

Chromate and dichromate ion extraction from acid solutions by use of TBP or an amine extractant have been well documented. Cuer et al.[16] have investigated the applicability of the extractants Alamine 336, LA-2, and TBP. They report excellent stability for TBP and have designed a process for the recovery of 99.5 percent of chromium (VI) from combined industrial effluents. The particular cases described in the reference refer to recovery of waste liquors originating from the production of chromium anhydride (CrO_3) and recovery from processes using this product (e.g., chromium plating and metal treatment). The extracted chromic acid is recovered as sodium chromate by stripping with a sodium hydroxide solution. Concentrations of 200 g/L of

Cr(VI) in the strip solution are reached. A similar process is reported in a Japanese patent, which includes the use of a chromic acid wash of the organic solvent to remove extracted impurities such as iron and chloride.[17]

The approach taken by MX Processor AB, Sweden[2] to recover spent chromium plating baths is to extract the impurities, thus regenerating the solution for reuse. A flow sheet of this process is outlined in Figure 7.2.3. The extraction is carried out by use of a mixture of HDNNS and TBP. Since the acidity of the plating bath limits the extraction efficiency, dilution of the bath with water significantly improves the operation. Water balance is partly maintained by the natural evaporation during plating. Small amounts of chromic acid may be extracted, but can be selectively scrubbed with water which can be used for dilution of the plating bath before extraction. Five molar H_2SO_4 or HCl is used for stripping, giving a metal concentration in the strip solution of more than 60 g/L.

7.2.6 Removal of Mercury from Chlor-Alkali Effluents

The mercury associated with brine effluents from chlor-alkali plants must be reduced from approximately 10 ppm to the parts-per-billion level prior to discharge. A solvent extraction method has been proposed by Gronier, which is based on the rapid extraction of mercury from chloride medium by high-molecular-weight tertiary and quaternery amines.[18] With a mercury contamination of 10 ppm in the effluent, a concentration factor of 2,500 is claimed to be obtainable, leaving a strip solution containing 25 g/L mercury. The amount of mercury that is extractable in a particular case depends, to some extent, on the pH of the brine. With tertiary amines, better than 99 percent is achieved at pH 3, but there is a decrease at higher pH. The most significant problem with this system is the loss of organic material to the brine effluent.

A closely related process, based on amine solvent extraction, has been patented by Chapman and Caban.[19] It is not known whether any commercial applications have been implemented.

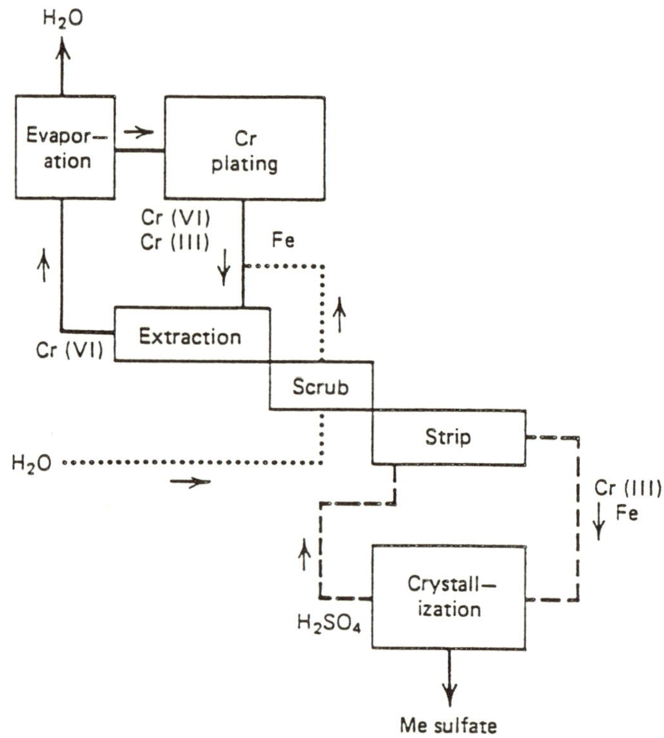

Figure 7.2.3. Process for removal of impurities and regeneration of chromium plating bath.

Source: Reference 2.

7.2.7 Miscellaneous Applications

Research on the recovery of noble metals, primarily from plating solutions, has been reported. Gold and silver are extracted from cyanide solutions with quaternary amines.[20,21] Gold can be stripped from the organic solvent with alkaline potassium cyanide solutions.

Rothmann et al.[22] have proposed the use of an amine extractant for recovery of chromium and vanadium in effluents from the processing of steelmaking slags. However, difficulties were reported with the precipitation of silica, which interfered with phase separation.

7.3 PRETREATMENT AND POST-TREATMENT REQUIREMENTS

Pretreatment requirements for liquid-liquid extraction are not nearly as stringent as those required for other physical separation techniques; e.g., carbon adsorption or membrane systems. Since the equipment used in extraction is not very susceptible to fouling or plugging, suspended solids removal is less critical. However, optimal operation will frequently require removal of organics which will otherwise dissolve in the extractant and interfere with subsequent operations; e.g., phase separation. Pretreatment operations can also be applied to prepare the waste for optimal separation efficiency. Pretreatments include pH adjustment, dilution, flow equalization, and temperature increases to optimize reaction and phase separation time.

Post-treatment will be required for the raffinate to remove residual metals/cyanides and solvent which has either dissolved in or been entrained in the coalescer effluent. Treatment options typically consist of adsorption or possibly biological treatment for organic destruction. Concentrated metal solutions from the regeneration step will be removed from the acidic media by electrowinning if recovery is economically viable. Alternatively, if concentrations are too dilute, they may be increased using evaporation or a membrane technology which is capable of operating in an acidic environment. Finally, if economic recovery is not viable, removal through precipitation may be most feasible.

7.4 PERFORMANCE DATA FOR LIQUID-LIQUID EXTRACTION

Performance data for solvent extraction is limited. Much of the development of processes and equipment has been carried out by manufacturers of customized equipment and is, therefore, considered to be proprietary. In other cases, pilot studies often do not supply enough information to indicate whether the corresponding commercial-scale process might be feasible. Data which were identified in this study are presented below.

7.4.1 Copper and Nickel Extraction from Metal Finishing Sludge

A pilot-scale study, conducted by the Department of Metallurgical Engineering of the Indian Institute of Technology, explored the use of LIX64N (10 percent by volume) to extract copper and nickel from metal finishing wastewater and sludge.[23] Samples were generated by redissolving metals from sludge with sulfuric acid to produce a solution containing 4.45 g/l copper, 0.16 g/L nickel and 2.5 g/L zinc. In addition, a synthetic sample containing 4.3 g/L copper, 7.9 g/L nickel, and 2.5 g/L zinc was made by dissolving metal sulfates in water. The extractant used was a 10 percent solution of LIX64N in kerosene and the stripper consisted of 4 N sulfuric acid. Extractions were carried out in a 500 mL separating funnel by manual shaking.

Table 7.4.1 illustrates the effect of pH on the extraction of copper with LIX64N from the synthetic solution. Multi-stage extractions were subsequently conducted at a pH of 2.0 as shown in Table 7.4.2. Tables 7.4.3 and 7.4.4 show similar diagrams for nickel. Finally, Table 7.4.5 is a summary of results for experiments conducted on the sludge leach solution. The data suggest that this process may be economical since the solvent is fully recovered.

7.4.2 Metal Extraction from Metal Finishing Wastewater

A study conducted for the EPA by Curtis W. McDonald, Texas Southern University, explored the use of high-molecular-weight amines for the removal of toxic metals from metal finishing wastewater such as cadmium, chromium, copper, nickel, and zinc.[24] The researchers used a 25 percent Alamine 336 solution diluted in xylene, with an extractant/water ratio of 1 to 100. The high ratio is desirable in concentrating the metal and in avoiding emulsions.

TABLE 7.4.1. EFFECT OF pH ON SINGLE-STAGE EXTRACTION OF COPPER USING 10 PERCENT VOL/VOL LIX64N

pH	Copper concentration in strip solution (g/L)	Distribution coefficient (D)	% Extraction
1.3	0.89	0.26	20.07
1.8	1.78	0.70	41.05
2.0	2.28	1.12	53.20
2.2	2.22	1.06	51.80
2.5	2.04	0.89	47.00

Composition of leach solution: Cu 4.3 g/L, Ni 7.9 g/L, and Zn 2.5 g/L; vol. of aq. phase 100 mL; vol. of organic phase 100 mL.
Source: Reference 23.

TABLE 7.4.2. MULTISTAGE CO-CURRENT EXTRACTIONS OF COPPER BY LIX64N

No. of stages employed	% Extraction calculated	% Extraction actual
3	89.5	73.6
4	95.0	80.6
5	97.6	93.3

Feed composition: Cu 4.3 g/L, Ni 7.9 g/L, and Zn 2.5 g/L; vol. of aq. phase 100 mL; vol. of org. phase 100 mL; pH 2.0.
Source: Reference 23.

TABLE 7.4.3. EFFECT OF pH ON SINGLE-STAGE EXTRACTION OF NICKEL

pH	Concentration of nickel in strip solution	Distribution coefficient (D)	% Extraction
8.0	4.4	1.26	55.7
8.2	5.3	2.02	67.0
8.5	5.9	2.96	74.8
9.0	4.3	1.18	54.3
9.5	4.0	1.01	50.4

Composition of feed solution: Cu 0.3 g/L, Ni 7.9 g/L, and Zn 2.5 g/L; vol. of aq. phase 100 mL; vol. of organic phase 100 mL.
Source: Reference 23.

TABLE 7.4.4. MULTISTAGE CO-CURRENT EXTRACTION OF NICKEL

No. of stages employed	% Extraction calculated	% Extraction actual
2	95.0	94.41
3	98.4	97.04

Feed composition: 0.3 g/L copper, Ni 7.9 g/L, and Zn 2.5 g/L; pH 8.5, vol. of aq. phase 100 mL; vol. of org. phase 100 mL.
Source: Reference 23.

TABLE 7.4.5. MULTISTAGE CO-CURRENT EXTRACTIONS OF Cu AND Ni CONTAINED IN THE LEACH SOLUTION OBTAINED FROM THE LEACHING OF HYDROXIDE SLUDGE

Metal	pH	No. of stages employed	% Extraction actual
Copper	2.0	4	93.97
Nickel	8.5	2	96.25

Original feed composition: 0.16 g/L nickel; 4.4 g/L copper, and 2.4 g/L zinc; vol. of aq. phase 100 mL; vol. of org. phase 100 mL.
Source: Reference 23.

Chromium extraction was found to be affected by the quantity of hexavalent versus trivalent chromium, since the latter is not as easily extracted. However, based on the effect of chloride concentration (HCl) on extraction, as shown in Table 7.4.6, selective extraction appears to have potential application. Results of a simultaneous extraction of the same three metals is shown in Table 7.4.7. A dose of 33 mL of concentrated HCl was mixed with 1-liter of the wastewater prior to extraction. For both of the experiments, no appreciable amount of copper or nickel was extracted. Stripping of the loaded extract was performed with 4.0 M NaOH with more than 99 percent of the metals being stripped. As previously indicated in Table 7.4.3, metals extraction is pH dependent and the addition of HCl or NaOH reagents affect extraction efficiency. The solvent was reused 15 times with no loss of efficiency. Reagent loss was estimated by an increase in TOC content of 50 ppm in the aqueous phase/extraction.

7.4.3. Lab Scale Study Using Sequential Extractions

A lab scale study was performed by Clevenger and Novak on a simulated regenerate waste from an electroplating ion exchange unit.[4] Four chelating compounds dissolved in chloroform were studied for recovery of Fe, Zn, Cu, Ni, and Cr. Results are shown in Figures 7.4.1 through 7.4.4 for single-stage extractions. It was shown that high metal removal efficiencies could be achieved with pronounced selectivity for copper at low pH for two of the chelates. Using these selectivity data, the researchers experimented with various schemes to identify optimal sequential extractions.

Although nearly complete removal and high selectivity could be achieved with sequential extractions, the investigators realized that recovery of both the metal and the chelator would be necessary for the process to be economically viable. Metals were efficiently extracted with 2.4 M HCl and 0.75 M HNO_3 solutions. However, the chelators could not successfully be reused due to significant loss of extraction capabilities following acid recovery. Since the chelators alone are generally more expensive than the recovery value of the metals, this process would not be economically viable on a commercial scale.

TABLE 7.4.6. SELECTIVE EXTRACTION OF CHROMIUM, CADMIUM, AND ZINC

	Chromium		Cadmium		Zinc	
	Mean % extd.	Std. dev.	Mean % extd.	Std. dev.	Mean % extd.	Std. dev.
First extraction 0.002M chloride	88.6	4.0	0.0	0.0	0.0	0.0
Second extraction 0.03M chloride	0.65	1.09	94.1	1.65	8.1	2.1
Third extraction 0.4M chloride	0.55	0.87	4.8	1.3	80.9	3.5
Total metal extracted	89.8		98.9		89.0	

Composition of wastewaters: Cr - 10.0 to 56.8 ppm
Cd - 4.1 to 5.9 ppm
Zn - 5.2 to 9.2 ppm
Cu - 0.3 to 0.5 ppm
Ni - 0.4 to 0.5 ppm

Source: Reference 24.

TABLE 7.4.7. SIMULTANEOUS EXTRACTION OF CHROMIUM, CADMIUM, AND ZINC

	Chromium		Cadmium		Zinc	
	Mean % extd.	Std. dev.	Mean % extd.	Std. dev.	Mean % extd.	Std. dev.
First extraction	90.6	1.2	98.0	0.11	83.3	0.11
Second extraction	0.0	0.0	1.4	0.86	15.4	1.8
Total metal extracted	90.6		99.4		98.7	

Composition of wastewaters: Cr - 8.0 to 9.0 ppm
Cd - 3.7 to 4.0 ppm
Zn - 4.8 to 5.2 ppm
Cu - 0.3 to 0.5 ppm
Ni - 0.4 to 0.5 ppm

Source: Reference 24.

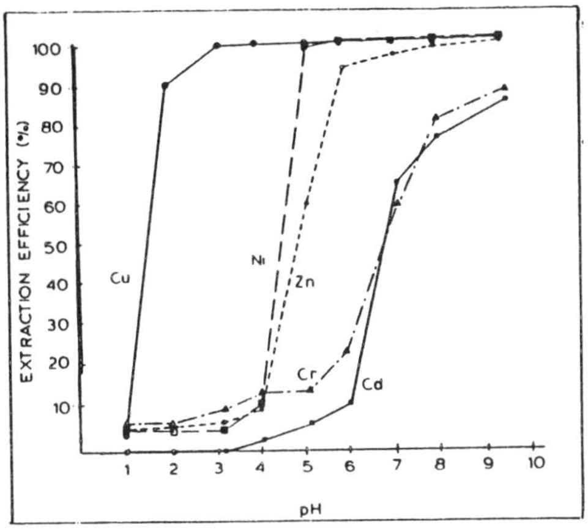

Figure 7.4.1. Metal extraction efficiency as a function of pH using 0.1-M thenoyltrifluoracetone and chloroform.

Source: Reference 24.

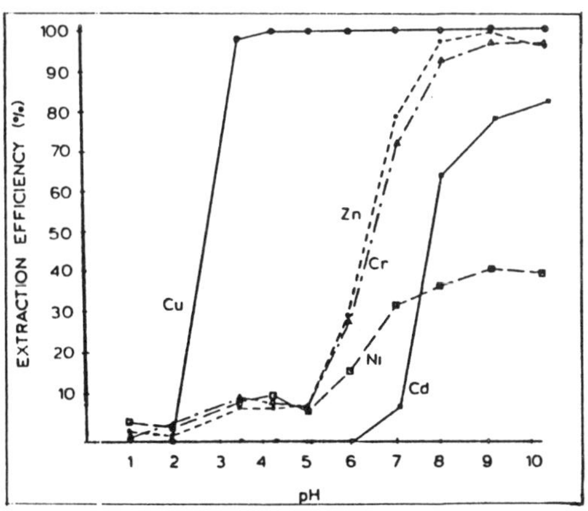

Figure 7.4.2. Metal extraction efficiency as a function of pH using 0.1-M acetylacetone and chloroform.

Source: Reference 24.

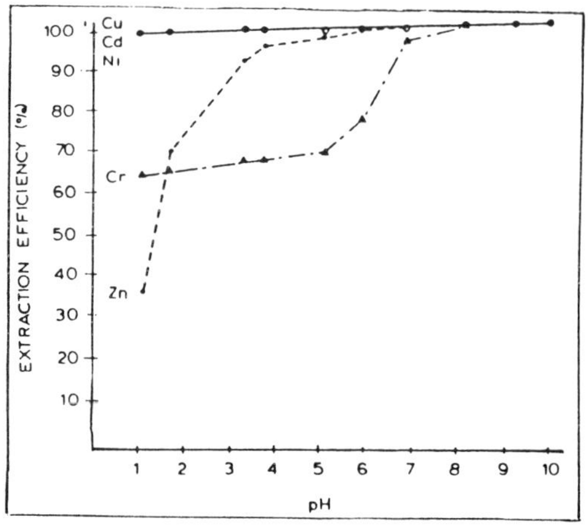

Figure 7.4.3. Metal extraction efficiency as a function of pH using 0.03-M sodium diethyidithio-carbamate and chloroform.

Source: Reference 24.

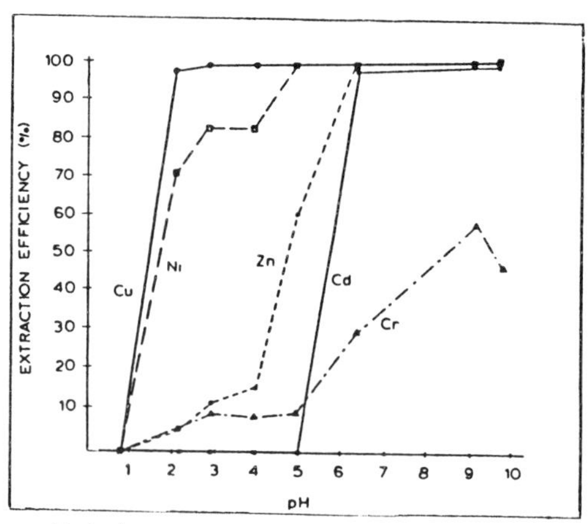

Figure 7.4.4. Metal extraction efficiency as a function of pH using 0.1-M 8-hydroxyquinoline and chloroform.

Source: Reference 24.

7.4.4. Metal Recovery From Scrap

An example of nickel, colbalt, and iron recovery from metal scrap (e.g., lathe turnings, mill shavings) was provided in the literature.[5] The process involved five processing steps:

1. Pyrometallurgical treatment to convert Mo and W into their carbides;
2. Electrolytic dissolution of Fe, Co, and Ni followed by partial stripping to concentrate the $CaCl_2$ electrolyte;
3. Separation of Fe, Co, and Ni by extraction with a high molecular weight amine;
4. Stripping the Co/Fe organic extract with the weakly acidic condensate from step (3); and
5. Cathodic depsoition of Co and Ni in separate half-cells. The stripped electrolyte then goes to step (2).

The process produces Ni/Fe and Co/Fe mixtures, which reportedly does not significantly affect the market value of the nickel or cobalt. The process, depicted schematically in Figure 7.4.5, does not generate any liquid discharge.

7.5 STATUS AND COSTS OF EXTRACTION

As stated previously, extraction is not a widely applied technology for the treatment of metal/cyanide wastes. The design and effectiveness of an extraction system will be highly specific to the waste type, constituent concentrations, and waste quantity. The difficulty in identifying an appropriate system and the relatively complicated nature of the process itself has undoubtably hindered its acceptance, particularly among smaller waste generators. Instead, its widest application has been in larger firms which have installed custom-designed systems through extensive support from the equipment and reagent suppliers.

Extraction will probably only be used in situations where recovery of valuable constituents or, recovery of baths via removal of contaminants, cannot be achieved through more conventional means. Due to the dependency of design variables on site-specific factors, generalizations on equipment

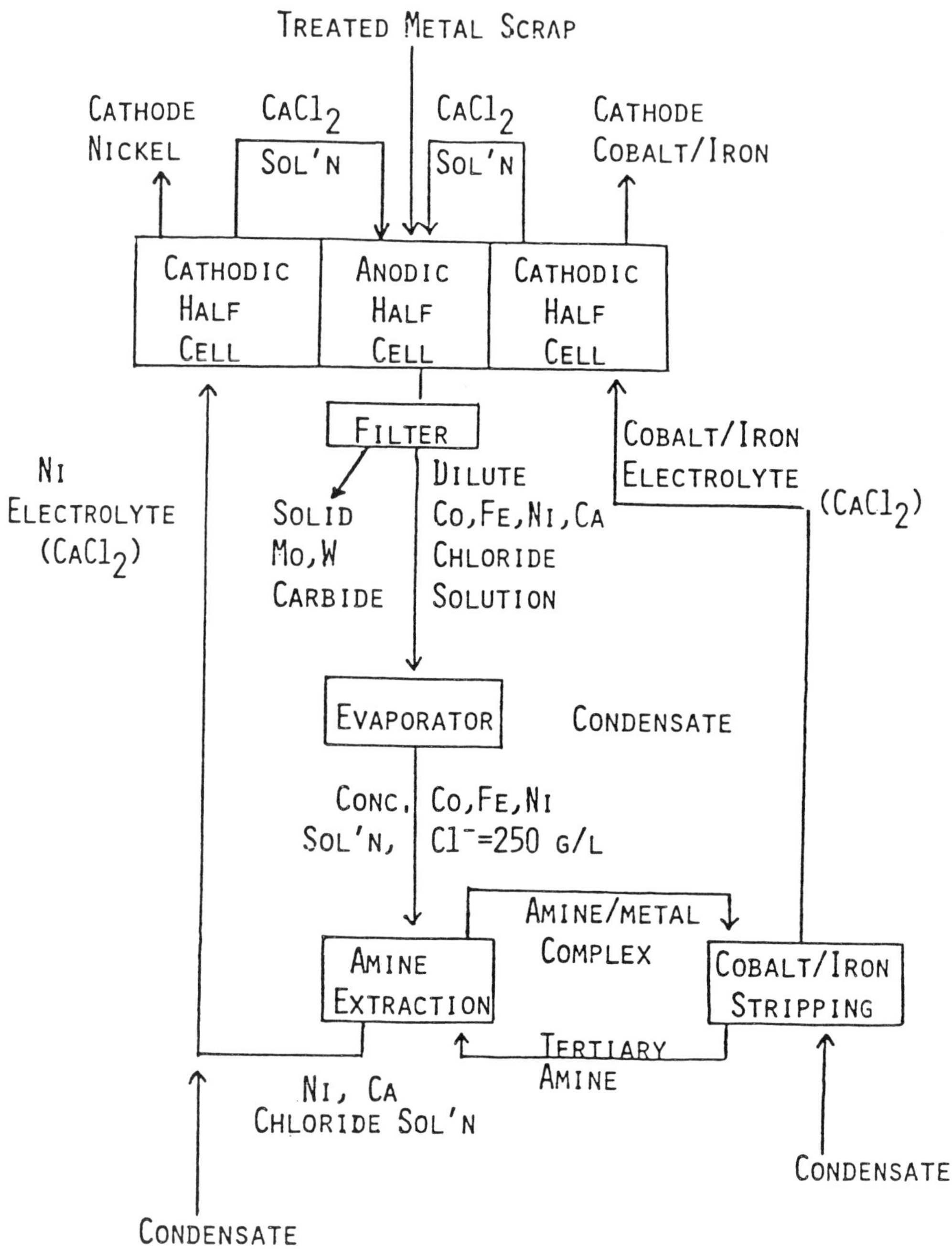

Figure 7.4.5. Schematic of metal recovery from scrap.
Source: Reference 5.

selection cannot be made. Similarly, the limited cost data presented in the literature cannot be generalized, since it is highly dependant on reagent loss and recovery value of the metals or baths. For example, solvent reagents cost roughly $1.50/pound, which can represent a significant operating cost if appreciable losses occur.[25]

REFERENCES

1. Humphrey, J.L., Rocha, J.A., and J.R. Fair. The Essentials of Extraction. Chemical Engineering. pp. 76-95. September 17, 1984.

2. Lo, T.C., Malcolm, H.I., and C. Hanson. Handbook of Solvent Extraction. John Wiley & Sons. pp. 53-89. 1983.

3. King Industries Product Bulletin SY2-378. August 18, 1978.

4. Clevenger, T.E., and J.T. Novak. Recovery of Metals from Electroplating Wastes Using Liquid-Liquid Extraction. Journal of the Water Pollution Control Federation, 55(7):984-989. July 1983.

5. L.V. Gallacher. Liquid Ion Exchange in Metal Recovery and Recycling. American Electroplaters Society Conference. April 14, 1980.

6. King Industries. Using Syvex Liquid Ion-Exchange Reagents. Product Bulletin.

7. Haines, A.K., Tunley, T.H., TeRiele, W.A.M., Cloete, F.L.D., and T.D. Sampson. J. South Afr. Inst. Min. Met. 74, 149. 1973.

8. Tunley, T.H., Kohler, P., and T.D. Sampson. J. South Afr. Inst. Min. Met. 77,423. 1976. R. Wood. Process Eng. 6, 6. 1977.

9. H. Reinhardt. Some Hydrometallurgical Processes for the Reclamation of Metal Waste. Paper presented at IWTU Conference, Waterloo. 1978.

10. Moore, F.L., and W.S. Groemier. Plating Surface Finishing, 26. August 1976.

11. F.L. Moore. Separation Science 19(4)489. 1975.

12. W. Hunter. The Use of Solvent Extraction for Purification of Silver Nitrate Electrolyte. Paper presented at the 2nd. International Conference of Precious Metals, New York, NY. May 1978.

13. W. Hunter. Electrolytic Refining. U.S. Patent 3,975,244. 1976.

14. Hamby, W.D., and M.D. Slade. Process for Regenerating and for Recovering Metallic Copper from Chloride-Containing Etching Solutions. U.S. Pattent 4,083,758. 1978.

15. Flett, D.S., and D. Pearson. Chem. Ind. 639. 1975.

16. Cuer, J.P., Stukens, W., and N. Texier. Proceedings of International Solvent Extraction Conference (ISEC), Lyons 1974, Vol. 2, Society of Chemical Industry, London. p. 1185. 1974.

17. Nishimura, S., and M. Watanabe. Japan Kokai 76, 90,999. 1976.

18. Reinhardt, H., and H.D. Ottertun. Swedish Patent Application 76-13686-0.

19. W.S. Gronier. Application of Modern Solvent Extraction Techniques to the Removal of Trace Quantities of Toxic Substances from Industrial Effluents. Oak Ridge National Laboratories, TN. Report ORNL-TM-4209. 1973.

20. Chapman, T.W., and R. Caban. Extraction of Mercuric Chloride from Dilute Solution and Recovery. U.S. Patent 3,899,570. 1975.

21. Ivanoviskii, M.D., Meretukov, M.A., Potekhin, V.D., and L.S. Strizhko. Izvest. Vyssh. Ucheb. Zadev. Tsvetnye Metally 17 (2), 36. 1974.

22. Rothmann, H., Bauer, G., Stuhr, A., and H.J. Retelsdorf. Metall. (Berlin) 30 (8), 737. 1976.

23. Phule, P.P., Dixit, S., and R. Mallikarjunan. The Recovery of Principal Metal Values from Metal Finishing Hydroxide Sludges. Indian Journal of Technology. Vol. 23, 462-464. December 1985.

24. C.W. McDonald. Removal of Toxic Metals from Metal Finishing Wastewater by Solvent Extraction. Industrial Environmental Research Laboratory, Texas State University. 1977.

25. King Industries. Price Schedule. Effective October 1, 1986.

8. Adsorption for Metal Removal

8.1 CARBON ADSORPTION

Adsorption involves the interphase accumulation or concentration of substances at an interface. The process can occur between any two phases, such as liquid-liquid or liquid-solid interfaces. The material being concentrated or adsorbed is the adsorbate, and the adsorbing phase is termed the adsorbent.[1]

Activated carbon adsorption involves separation of a substance from one phase, typically an aqueous solution, and the concentration of the substance at the surface of an activated carbon adsorbate. Activated carbon is most widely used for the removal of organic contaminants and is most effective when the organic solutes have a high molecular weight and low water solubility, polarity, and degree of ionization.[2] However, studies in the field of metallurgy have indicated that carbon adsorption of many metallic compounds can be successfully achieved and has found commercial application for certain aqueous waste streams.[3,4] However, adsorption efficiency varies considerably between different compounds.

Activated carbon is available as a powder (PAC) or in the form of granules (GAC). GAC is most commonly used because its larger size is most amenable to handling in conventional contacting and regenerating equipment.[5] However, despite handling and regeneration problems, PAC is preferred in some treatment schemes; e.g., when used in combination with biological treatment.[6,7] Both types of carbon have effective surface areas far in excess of their nominal external surface areas. Surface areas are on the order of 500 to 1,500 square meters per gram, resulting primarily from a network of internal pores 20 to 100 angstroms in diameter. Porosities can be as large as 80 percent.[8] The characteristics of the micropore structure are largely dependent on the activation process, which is a controlled sequence of dehydration, carbonization, and oxidation of raw materials including coal, wood, peat, shell, bone, and petroleum based residues.

The equilibrium capacity of an activated carbon for a contaminant is a function of the effective carbon surface area and the surface binding process. Adsorption equilibria are governed by two types of interactions: solute-adsorbent, which describes the carbon's affinity for the solute (contaminant), and solute-solvent, which involves the solubility of the solute in the liquid media. In general, an inverse relationship between the extent of adsorption of a substance from a solvent and its solubility in that solvent can be anticipated.[9] Overall, the relative affinity of a solute for either phase will be determined by the lyophobic (i.e., solvent-disliking) characteristics of the solute and the affinity of the solute for the adsorbate.

Activated carbon adsorption of inorganic compounds is more complex and compound specific than adsorption of organic compounds, primarily due to the charged nature of inorganic species in aqueous media. The important physical-chemical properties of activated carbon selected for inorganic electrolyte adsorption are: specific surface area, pore structure, and surface chemistry of the adsorbent.[6,10]

Specific surface area may be defined as that portion of the total surface area that is available for adsorption. Specific surface area is proportional to adsorption which, in turn, is dependent on pore size and pore size distribution. The pore size can range from less than 20 angstroms and size distribution is dependent on the source materials and the activation process employed.[10] Available surface area is nonpolar in nature, but depending upon the activation process, active sites can be formed yielding a slightly polar surface.

In addition to specific area, the adsorption capacity of activated carbon is primarily influenced by surface chemistry; e.g., the formation of carbon-oxygen complexes at the carbon surface and the anionic adsorption capacity.[6,11] Formation of surface functional groups is dependent on the activation process and carbon source.[11] Two broad categories of activated carbon can be identified based on activation temperature and atmospheric conditions:[12] L-type, which tends to adsorb bases, and H-type, which tends to adsorb acids.

L-type carbons are prepared by exposure to oxygen at 300°C to 400°C or by solution oxidation. H-type carbons are prepared by outgassing at 800°C to 1000°C followed by cooling in an inert atmosphere and exposure to oxygen at room temperature.[12] Typical surface oxide functional groups formed by these methods include:

- carboxyl,
- phenolic hydroxyl,
- lactone and quinone,
- carboxylic acid,
- anhydrides, and
- cyclic peroxides.

The surface oxide groups have significant effects on the adsorption capacity of activated carbon. Electrokinetic studies have shown that H-type activated carbon exhibits a positive surface potential whereas L-type carbons exhibit a negative surface potential. This is due partly to the high pH which developes when H-type carbon is brought in contact with water and the low pH which occurs when the surface functional groups of L-type carbon are hydrated.[10,11] The surface charge characteristics of both carbon types can be readily modified by the introduction of a strong base or acid to the system.[10]

In general, the following system parameters have the greatest influence on metals removal by activated carbon:

- pH,
- metal concentration,
- activated carbon dose,
- ionic strength,
- temperature, and
- presence of ligands.

The dominant solution parameter controlling adsorption of inorganic chemicals is pH.[13] As mentioned above, pH has a controlling influence on the surface charge characteristics of the adsorbent. The distribution of metal ions in solution is also a function of pH, with lower pH favoring solvation of metal ions.[14] The overall pH effect on adsoption of metal ions can be attributed to electrostatic attraction, which is a function of the charge of both the solid adsorbent and the adsorbate.

Studies have shown that the removal efficiency of inorganics from a waste stream by activated carbon increases with concentration of either the adsorbant or the solute. For example, researchers demonstrated that Cd^{+2}, Cr^{+6}, Cr^{+3}, and CN^{-} all show improved removal efficiencies at higher initial concentration throughout the pH range tested.[6,15] Similarly, by increasing the activated carbon/Cd ratio by 100 times, a threefold increase in removal efficiency was realized.[15]

Adsorption of Cd(II) was demonstrated to decrease with increasing ionic strength of the solvent.[14] This suggests that the extent of adsorption is sensitive to changes in concentration of supporting electrolyte, indicating that electrostatic interaction may be a significant component of adsorption in plating and other metal containing solutions.

Since adsorption is an exothermic process, adsorption efficiency might be expected to improve with decreasing temperature. However, it has been demonstrated that adsorption of Cd^{+2} increases with temperature.[14] This is unexpected thermodynamically and suggests that some extrinsic process which responds to temperature increase is at work.

Complexation of metal ions by inorganic and organic ligands can dramatically increase or decrease adsorption compared to a ligand-free system.[13] Studies have demonstrated that mercury and cadmium show a significant increase in adsorption efficiency using chelating agents such as APDC, TETA, NTA, and EDTA.[13,15]

System pH also influences adsorption of metal-ligand complexes. The effect of pH on cadmium removal was determined in a system using EDTA, tartate, citrate and TRIEN as complexing ligands.[11] Adsorption increased with pH for 3 out of 4 ligands tested. The other, EDTA, showed improvements with initial increases in pH up to pH 6, but adsorption capacity fell off as higher pH was approached.

The relative feasibility of using different adsorbents is usually determined in the laboratory by developing adsorption isotherms. The adsorption isotherm is a functional expression for the variation of adsorption with concentration of adsorbate in bulk solution at constant temperature.[1] Typically, the amount of adsorbed material per unit weight of adsorbent increases non-linearly with increasing concentration. It must be noted that adsorption isotherms can vary widely for different carbons, and isotherm data cannot be used interchangeably.

The two most common isotherm expressions used are the Freundlich Equation and the Langmuir Equation. The Freundlich equation in an empirical expression but is often useful as a means for data description.[1] The Langmuir model, originally developed for adsorption of gases onto solids, is predicated on three assumptions: (1)adsorption energy is constant and independent of surface coverage; (2)adsorption occurs on localized sites with no interaction between adsorbate molecules; and (3)maximum adsorption occurs when the surface is covered by a monolayer of adsorbate.[16]

The Freundlich equation can be expressed as follows:

$$\frac{x}{m} = k\,C^{\frac{1}{n}}$$

where:

- x = mass of adsorbate, mg
- m = mass of dry adsorbent, g
- k = constant, indicative of adsorption capacity
- C = equilibrium solution concentration, mg/l
- 1/n = constant, indicative of adsorption intensity.

Data for the Freundlich equation are usually fitted to the logarithmic form of the equation:[1]

$$\log\left(\frac{x}{m}\right) = \log k + \left(\frac{1}{n}\right)\log C$$

This expression is a straight line with a slope of 1/n and an intercept equal to the log k when C = 1 (log 1 = 0).

The Freundlich equation generally shows good agreement with both the Langmuir equation and experimental data, over moderate ranges of concentration, C.

In its linearized form, the Langmuir equation can be expressed as follows:

$$\frac{C}{\frac{x}{m}} = \frac{1}{ab} + \frac{1}{a} C$$

where: x = mass of adsorbate, mg

m = mass of dry adsorbent, g

C = equilibrium solution concentration, mg/l

a = solid phase concentration corresponding to complete coverage of available adsorption sites (mass solute adsorbed/mass carbon; for complete monolayer)

b = constant related to the enthalpy of adsorption

The coefficients of the Langmuir equation can be calculated by performing a linear regression of the data or determined graphically, by plotting C/(x/m) versus C on arithmetic graph paper (slope = 1/a, intercept = 1/ab).

The adsorption data from both models are useful in estimating the relative effectiveness of adsorbents for a given application. However, care must be exercised in assessing performance when the wastestream contains a large number of competing adsorbates. Most users will be forced to rely on laboratory scale adsorption isotherm results and prior industrial experience to assess performance and appropriate system design for a specific wastestream.

8.1.1 Process Description

Although activated carbon adsorption has been shown to effectively treat some metal and cyanide containing waste streams, it is generally employed in the treatment of organic containing wastes. Consequently, the focus on applied technology and research has been on the treatment of organics. However, research has been conducted on unique systems used for the treatment of metal and cyanide containing wastes, as discussed later in this section.

A schematic of an activated carbon adsorption system for the treatment of hexavalent chromium is shown in Figure 8.1.1. After exhaustion of the adsorption capacity, the activated carbon is regenerated with sulfuric acid. The regeneration acid is pH adjusted to precipitate chromium, and the sludge is removed via filtration. Carbon is returned to the process. There is typically an accompanying loss in adsorption capacity as a result of a small but significant depletion in effective surface area. This can result from a build up of hard to remove adsorbate, attrition, and other mechanisms.

8.1.1.1 Pretreatment/Post-Treatment Requirements--

Pretreatment of the feed to carbon adsorption columns is often required to improve performance and/or prevent operational problems. As discussed in Reference 9, there are four primary pretreatments which may be required:

- equalization of flow and concentrations of primary waste constituents;
- filtration;
- adjustment of pH; and
- adjustment of temperature.

Generally, the flow to the adsorber columns and the concentration of the primary waste constituent are not constant. Since variations in either can have a detrimental impact on system performance, it is necessary to make provisions to equalize flow and minimize concentration surges. Flow equalization is accomplished by employing a surge tank prior to the column. Concentration equalization is also accomplished somewhat by employing surge tanks, however, this may have to be supplemented by mechanical agitation. Mixing prevents concentration surges which can lead to premature column leakage and breakthrough or conversely, low concentration swings resulting in premature regeneration of an underloaded adsorber column.

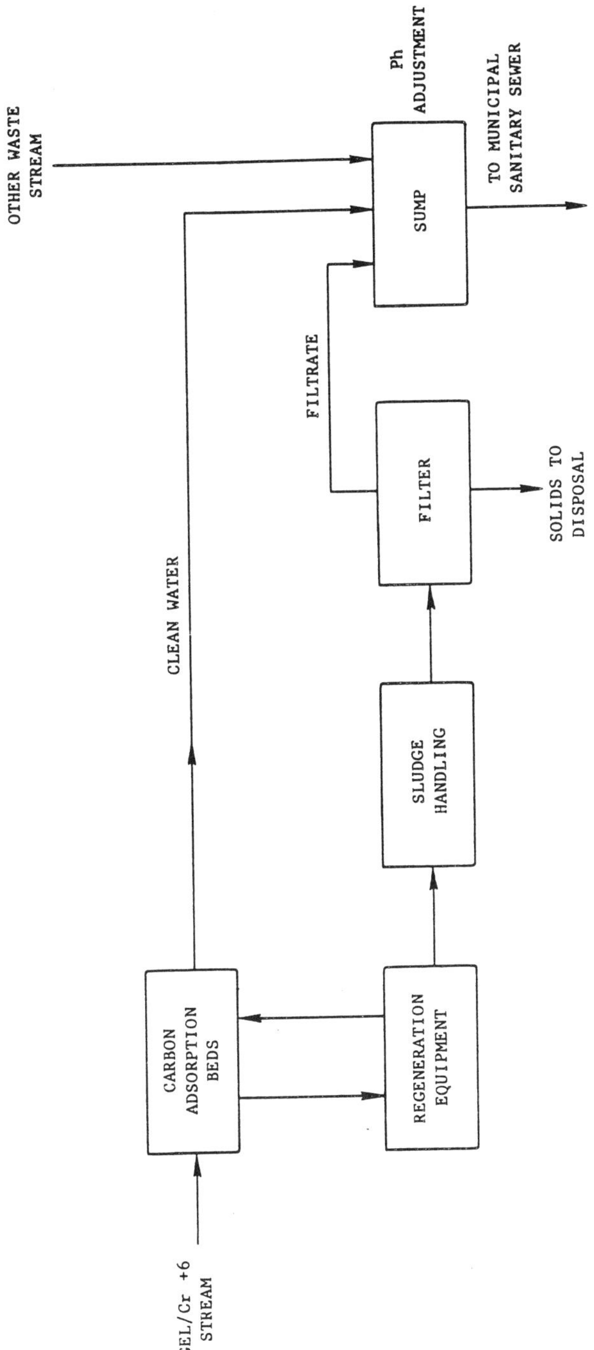

Figure 8.1.1. Regenerative carbon adsorption system.

It is a general requirement that the feed to column be low in suspended solids. It is difficult to set an upper limit on the absolute level of acceptable suspended solids because the physical nature of the solids is as important as their concentration. For example, finely divided, silty solids tend to pass through the bed, but coarse material of varying particle size can rapidly form a mat on top of the bed, thereby constricting flow. In general, if the column feed is turbid or the suspended solids level is greater than 10 mg/L, pretreatment for solids removal will be required.

In addition to suspended solids, many waste contaminants can interfere with carbon adsorption. For example, if calcium or magnesium are present in concentrations greater than 500 mg/L, they may precipitate and plug or foul the column.[17] Oil and grease in excess of as little as 10 mg/L has been reported to interfere with column operation.[18] The presence of many other compounds can influence adsorption of the contaminant of concern through competition for available adsorption sites.

Removal of suspended solids and other waste contaminants noted above may be achieved by pretreatment with multi-media pressure filters. Such filters complement fixed bed adsorption processes and can be readily integrated into a total design. Other filtration options include membrane filtration, when a highly clarified feed is desired, and ultrafiltration, if high molecular weight contaminants are encountered (over 1,000).

Activated carbon adsorption systems for metals are sensitive to changes in pH, particularly when the contaminants to be removed are either weakly acidic or weakly basic. Control can be easily achieved by installing pH measurement and acid/base reagent addition systems in the surge tank to maintain the desired pH feed to the adsorption columns. Finally, provisions for feed heating may be required since adsorption of metals has been shown to vary with temperature, as discussed above.

Under proper design and operating conditions, the treated water will generally be suitable for discharge to surface waters. Other aqueous streams such as backwash, carbon wash and transport waters are recycled or sent to a settling basin. Acidic regenerants, which are typically used for metals, may be treated through neutralization and precipitation.

8.1.1.2 Operating Parameters--

Optimal process design of both the adsorption and regeneration, or desorption, systems is dependent on the waste's physical, chemical, and flow characteristics. Isotherms, determined in a laboratory, measure the affinity of activated carbon for the "target" adsorbates in the process liquid. This provides data for determining the type and amount of carbon which will be required to treat the full scale process stream. Carbon requirements will be based on a limiting constituent for which attainment of acceptable effluent concentration is the most difficult.

Table 8.1.1 gives properties of some commercially available granulated activated carbons.[19] Properties of a typical powdered activated carbon are shown in Table 8.1.2.[20] Adsorption properties of the two types of carbon are generally comparable, the principal difference being the particle size. The fine size of the PAC makes it unsuitable for use in the contacting and regeneration equipment used for adsorber applications but makes it ideal for flow through processes (e.g., biological treatment) equipped with filtration systems for carbon removal.

A typical continuous adsorption system consists of multiple columns filled with activated carbon and arranged in either parallel or series. Total carbon depth of the system must accommodate the "adsorption wavefront"; i.e., the carbon depth must be sufficient to purify a solution to required specifications after equilibrium has been established. Bed depths of 8-40 feet are common.[8] Minimum recommended height-to-diameter ratio of a column is 2:1. Ratios greater than 2:1 will improve removal efficiency but result in increased pressure drop for the same flow rate. Optimum flow rate must be determined in the laboratory for the specific design and carbon used. For most applications, 0.5 to 5 gpm per square foot of carbon is common.[8]

Optimal adsorber configuration will be based on influent characteristics, flow rate, type of carbon, effluent criteria, and economics. Figure 8.1.2 illustrates several arrangements typically used for adsorber systems.[21] There are two basic modes of operation for columns; namely, fixed beds and moving or pulsed beds. In the fixed bed mode, the entire bed is removed from service when the carbon is reactivated. In the moving or pulsed bed, only the exhausted (inlet) portion of the bed is removed as new adsorbent is simultaneously added to maintain bed volume.

TABLE 8.1.1. PROPERTIES OF SEVERAL COMMERCIALLY AVAILABLE CARBONS

	ICI America Hydrodarco (lignite)	Calgon Filtrasorb 300 (bituminous)	Westvaco Nuchar WV-L (bituminous)	Witco 517 (12x30) (bituminous)
PHYSICAL PROPERTIES				
Surface area, m^2/g (BET)	600 - 650	950 - 1050	1000	1050
Apparent density, g/cm^3	0.43	0.48	0.48	0.48
Density, backwashed and drained, lb/g^3	22	26	26	30
Real density, g/cm^3	2.0	2.1	2.1	2.1
Particle density, g/cm^3	1.4 - 1.5	1.3 - 1.4	1.4	0.92
Effective size, mm	0.8 - 0.9	0.8 - 0.9	0.85 - 1.05	0.89
Uniformity coefficient	1.7	1.9 or less	1.8 or less	1.44
Pore volume, cm^3/g	0.95	0.85	0.85	0.60
Mean particle diameter, mm	1.6	1.5 - 1.7	1.5 - 1.7	1.2
SPECIFICATIONS				
Sieve size (U.S. std. series)[a]				
Larger than No. 8 (max. %)	8	8	8	--
Larger than No. 12 (max. %)	--	--	--	5
Smaller than No. 30 (max. %)	5	5	5	5
Smaller than No. 40 (max. %)	--	--	--	--
Iodine No.	650	900	950	1000
Abrasion No. minimum	b	70	70	85
Ash (%)	b	8	7.5	0.5
Moisture as packed (max. %)	b	2	2	1

[a]Other sizes of carbon are available on request from the manufacturers.

[b]No available data from the manufacturer.

-- Not applicable to this size carbon.

TYPICAL PROPERTIES OF 8 X 30-MESH CARBONS

	Lignite carbon	Bituminous coal carbon
Total surface area, m^2/g	600 - 650	950 - 1,050
Iodine number, min	500	950
Bulk density, lb/ft^3 backwashed and drained	22	26
Particle density wetted in water, g/cm^3	1.3 - 1.4	1.3 - 1.4
Pore volume, cm^3/g	1.0	0.85
Effective size, mm	0.75 - 0.90	0.8 - 0.9
Uniformity coefficient	1.9 or less	1.9 or less
Mean particle dia., mm	1.5	1.6
Pittsburgh abrasion number	50 - 60	70 - 80
Moisture as packed, max.	9%	2%
Molasses RE (Relative efficiency)	100 - 120	40 - 60
Ash	12 - 18%	5 - 8%
Mean-pore radius	33 A	14 A

Source: Reference 19.

TABLE 8.1.2. TYPICAL PROPERTIES OF POWDERED ACTIVATED CARBON (PETROLEUM BASE)

Surface Area m^2/g(BET)		2,300 - 2,600
Iodine No.		2,700 - 3,300
Methylene Blue Adsorption (mg/g)		400 - 600
Phenol No.		10 - 12
Total Organic Carbon Index (TOCI)		400 - 800
Pore Distribution (Radius Angstrom)		15 - 60
Average Pore Size (Radius Angstrom)		20 - 30
Cumulative Pore Volume (cm^3/g)		0.1 - 0.4
Bulk Density (g/cm^3)		0.27 - 0.32
Particle Size Passes:	100 mesh (wt%)	97 - 100
	200 mesh (wt%)	93 - 98
	325 mesh (wt%)	85 - 95
Ash (wt%)		1.5
Water Solubles (wt%)		1.0
pH of Carbon		8-9

Source: Reference 20.

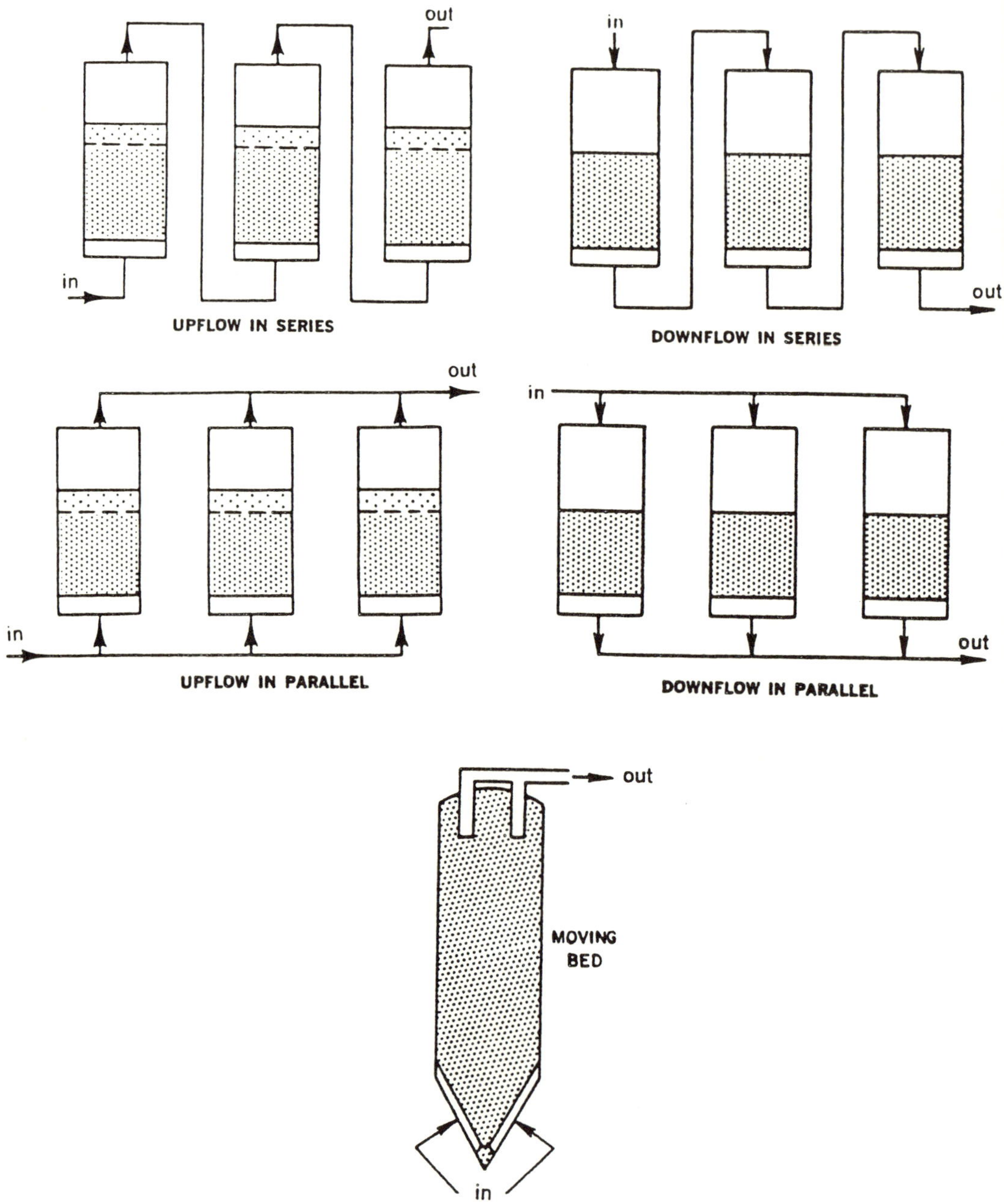

Figure 8.1.2. Carbon Bed Configurations.
Source: Reference 21.

Arrangement of columns in series permits the first column to become saturated with impurities while a solution of required purity is obtained through the second, or polishing, column. Upon reaching saturation, the first column is emptied and refilled with fresh or regenerated carbon. Fluid flow is redirected to the second column so that the replenished column is now in the downstream position, resulting in a variation of countercurrent flow between the waste stream and the adsorbent.[7]

Adsorption beds can be operated in either upflow or downflow mode. A downflow mode must be used where the adsorber is relied upon to perform the dual role of adsorption and filtration. Although lower capital costs can be realized by eliminating the need for pretreatment filters, operating costs escalate since more efficient and frequent backwashing of the adsorbers is required. Application rates of 2-10 gpm/ft^2 are employed, and backwash rates of 12-20 gpm/ft^2 are required to achieve bed expansions of 20-50 percent.[8] The use of a supplemental air scour can be used to increase efficiency of the backwashing.

While pre-filtration is normally required to prevent blinding of upflow-expanded beds with solids, smaller particle sizes of adsorber can be employed to increase adsorption rate and decrease adsorber size. Application rates can also be increased in the upflow-expanded mode, even to the extent that the adsorbent may be in an expanded condition.[9]

The design, flow, and configuration arrangements discussed above offer the following advantages and limitations:[21]

Adsorbers in Parallel	- For high volume applications - Can handle higher than average suspended solids (65-70 ppm) if downflow - Relatively low capital costs - Effluents from several columns blended, therefore, less suitable where effluent limitations are low
Adsorbers in Series	- Large volume systems - Easy to monitor breakthrough at tap between units - Effluent concentrations relatively low - Can handle higher than average suspended solids (65-70 ppm) if downflow - Capital costs higher than for parallel systems

Moving Bed	- Countercurrent carbon use (most efficient use of carbon) - Suspended solids must be low (<10 ppm) - Best for smaller volume systems - Capital and operating costs relatively high - Can use such beds in parallel or series
Upflow-expanded	- Can handle high suspended solids (they are allowed to pass through) - High flows in bed (>15 gpm/ft^2)

The above systems are not generally used with the much finer powdered activated carbons. The PAC systems now used involve mixing the PAC with the waste stream to form a slurry which usually can be separated later by methods such as filtration or sedimentation. A novel technique where powdered activated carbon is used to make activated carbon beads, based on a suspension-polymerization technique, has proven effective for treating some metal containing waste streams.[22]

8.1.2 Experimental Data and Demonstrated Performance

Information gathered from activated carbon manufacturers and industry indicates that few activated carbon systems are being used specifically for the treatment of metal and/or cyanide bearing waste streams. Thus, data for full scale applications are incomplete and essential operating parameters or pollutant removal characteristics have either not been generated or are considered to be proprietary information. However, the literature includes a number of efforts where the feasibility of activated carbon for metal and/or cyanide removal has been demonstrated. Specifically, some degree of success has been reported for the adsorption of arsenic, cadmium, chromium, mercury, and cyanide, as described below.

Arsenic--A number of different adsorbents were tested for their abilities to remove arsenic from a variety of aqueous solutions.[23] However, the results show that activated carbon was not the best adsorbent tested.

Three types of activated solids were chosen for the study including, activated alumina, activated bauxite, and activated carbon. Experiments were carried out with freshwater, seawater diluted ten times, undiluted seawater, and a 0.67 M sodium chloride solution. In experiments for As^{+3} adsorption, all reaction flasks were flushed with nitrogen gas to prevent the oxygenation of As^{+3} to As^{+5}.

The results demonstrate that As^{+5} is far more adsorbable than As^{+3}, and that As^{+5} was removed from solution much faster by activated alumina than by any other adsorbent (see Figure 8.1.3). In general, the rate of adsorption and extent of arsenic removal decreased with increasing salinity for all adsorbents tested. The effect of pH on As^{+5} adsorption by the three adsorbents was determined by varying pH from 2 to 12. Activated carbon adsorbs better in the acidic pH range (i.e., between 3 and 5) than at higher or lower pH values. Alumina and bauxite both displayed adsorption maxima for pH values between 4 and 7. Even at the pH of maximum adsorption by activated carbon, alumina and bauxite demonstrated superior performance. Figure 8.1.4 depicts the effect of pH on the performance of activated carbon.

Adsorption equilibrium were described adequately by both the Langmuir amd Freundlich isotherm models. The effect of solution composition on adsorption equilibria for activated carbon is shown using the Langmuir model in Figure 8.1.5. With regard to the ionic strength of the solutions and their effect on adsorption, it was determined that the rates of adsorption were slowest in seawater, yet the extent of adsorption was reduced by no more than 5 percent relative to freshwater.[23] However, it must be noted that the isotherm plots are based on only three data points.

Cadmium--In one study, adsorption of cadmium was batch tested using four brands of activated carbon (GAC and PAC), as shown in Table 8.1.3.[15] Stock solutions were prepared to represent cyanide and fluoroborate plating baths as follows:

1. cyanide bath: 10^{-2} M CdO + 10^{-1} M NaCN (molar ratio of Cd: CN = 1:10); and
2. fluoroborate bath: 10^{-1} M Cd $(BF_4)_2$ + 7.0×10^{-2} M $NH_4 BF_4$ + 5.0×10^{-2} M H_3BO_3 (molar ratios of $Cd:BF_4:NH_4:H_3BO_3$ = 1:2.7:0.7:0.5).

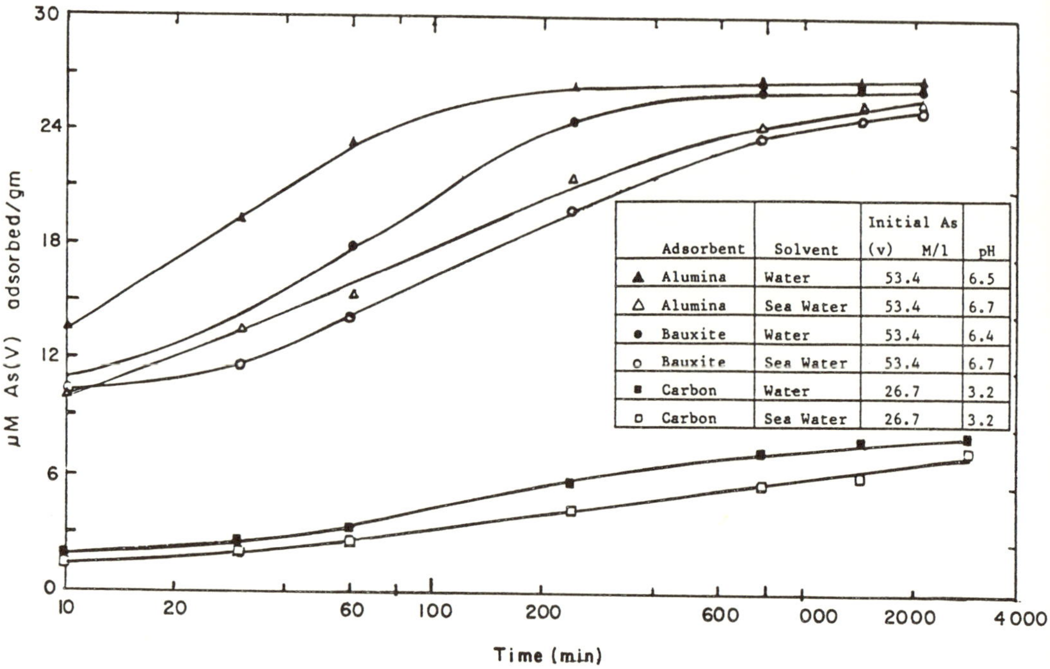

Figure 8.1.3. Adsorption of As(V) by different absorbents at optimum pH values.
Source: Reference 23.

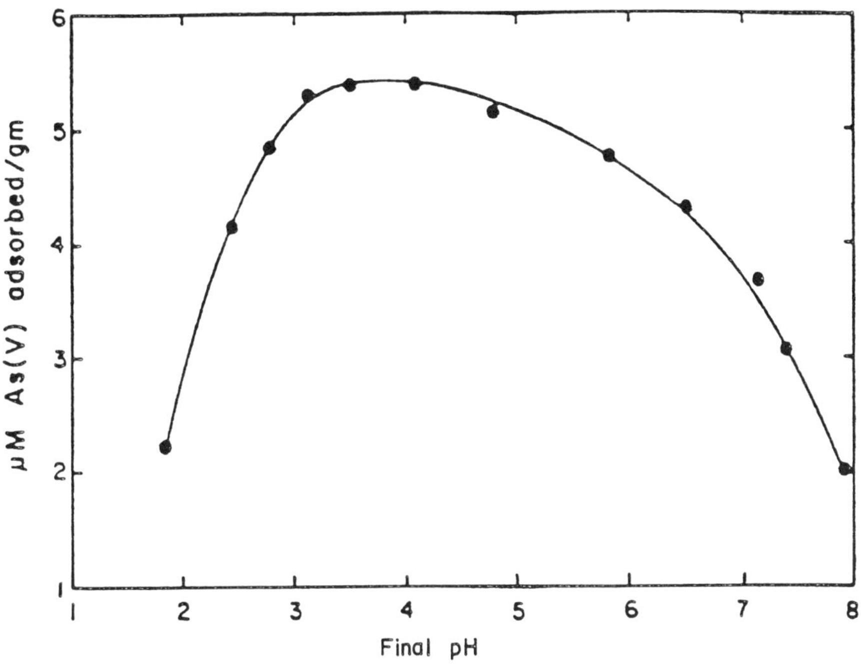

Figure 8.1.4. pH effect on adsorption of As(V) by activated carbon. Source: Reference 23.

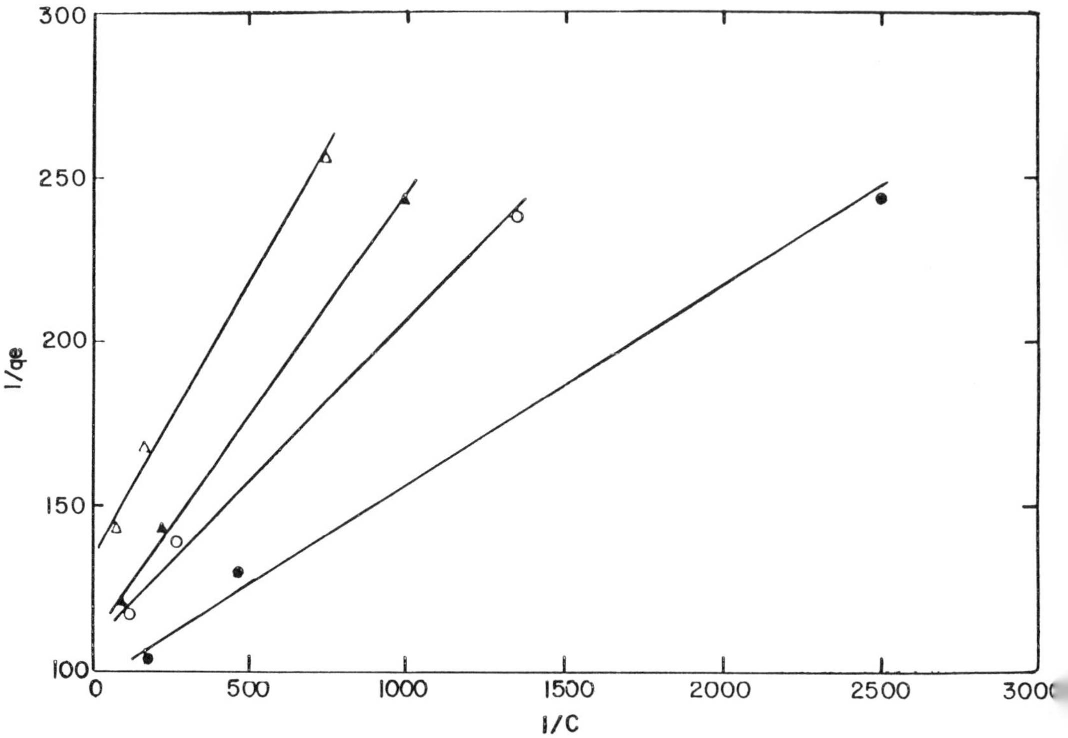

Figure 8.1.5. Langmuir isotherms of adsorption of arsenic(V) on activated carbon [q.=mM As(V) adsorbed/g of solids, C=mM of As(V)]. Source: Reference 23.

TABLE 8.1.3. TYPICAL SURFACE PROPERTIES OF ACTIVATED CARBONS USED IN THIS STUDY[a]

Carbon type	Specific surface area (m^2/g)	Particle size (mm)	pH_{zpc}
Nuchar S-N	1400 - 1800	~0.04 - 0.10	5.84[b]
Nuchar S-A	1400 - 1800	~0.03 - 0.10	3.80[b]
Darco HD 3000	550 - 650	0.84 - 4.75	c
Filtrasorb 400	941[d]	0.90 - 1.00	7.10[d]

[a]Information provided by manufacturer unless indicated otherwise (Reference 14).

[b]Determined in this study.

[c]Value not determined, but estimated to be 5-7 based on comparison with values of other Darco brand granular carbons and observation of pH drift in various solutions.

[d]From Reference 15.

The ionic strength of the waste water was shown to have a minimal effect on adsorption, with the rate increasing only slightly with decreasing ionic strength. However, as expected, the kinetics of cadmium adsorption onto PAC is faster than that onto GAC since pore diffusion is probably the rate limiting step for GAC. Conversely, external surface area contributes significantly to total surface area for PAC, therefore pore diffusion is less critical. Powdered activated carbons, in particular Nuchar S-N and Nuchar S-A, had larger cadmium adsorption capacities than the granular forms tested (Darco HD 3000 and Filtrasorb 400). Figure 8.1.6 shows that Nuchar S-A achieves 90-95 percent Cd^{+2} removal in the neutral pH range which is approximately three times that of granular carbons.

Since differences in specific surface area between the diferent carbon types are not large and the hydrated radius of a Cd^{+2} ion is estimated to be much smaller (4 angstroms) than the lower pore size values (10 to 1000 angstroms), the performance difference can be attributed to surface chemistry. Powdered carbon has a low pH at zero charge (pH_{ZPC}) and excellent adsorption capacity for cationic metal ions. Granular carbon, having a high pH_{ZPC}, is rather poor for metal ion adsorption. The pH_{ZPC} value reflects the nature of surface functional groups.

The distribution of cadmium species in solution is a function of pH. The hydrogen ion concentration of the wastewater solution plays a critical role in the extent of Cd^{+2} adsorption. For both the fluoroborate and cyanide wastewaters, the adsorption density was found to approach its maximum level in the neutral pH range. This is a positive feature of carbon treatment when compared to alternatives such as precipitation, which requires a pH adjustment to 10 or 11 for effective removal. Nuchar carbon was found to be particularly effective in this regard, with little or no pH adjustment required after addition to solution.[14]

Following these experiments, a suspension-polymerization technique was then used to aggregate one of the PACs, Nuchar S-A, to sizes suitable for column packing.[22] The beaded carbon was compared to a number of other activated carbons as listed in Table 8.1.4. All metal solutions were prepared from reagent grade chemicals. The cadmium was a synthetic cadmium fluoroborate, $Cd(BF_4)_2$, plating wastewater. Strong acids such as H_2SO_4, HCl, and $HClO_4$ were used to regenerate the Cd(II)-laden activated carbon beads.

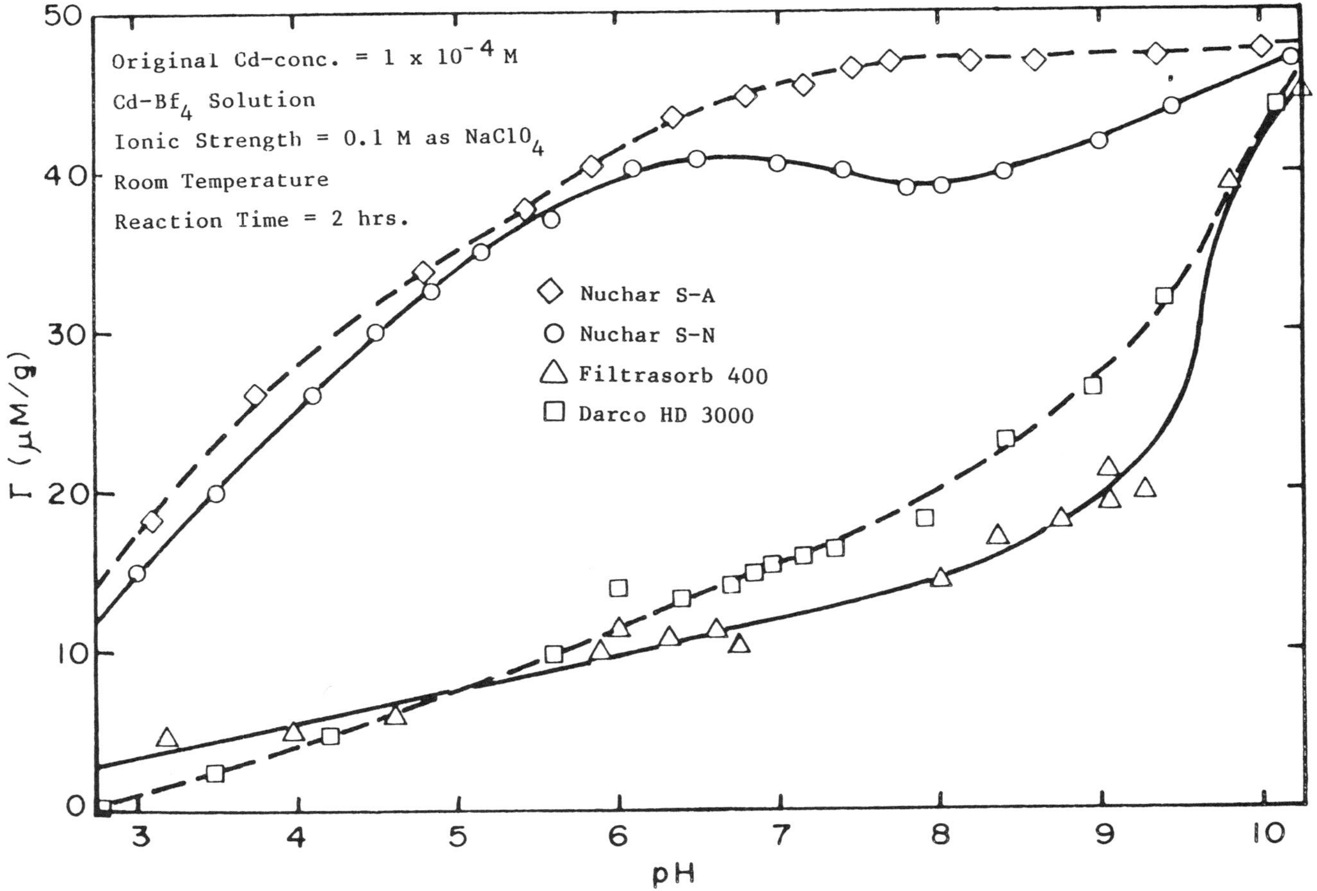

Figure 8.1.6. Comparison of Cd(II) adsorption capacity by granular and powdered activated carbon, as affected by pH. Source: Reference 14.

TABLE 8.1.4. COMPARISON OF ADSORPTION CAPACITY BY VARIOUS TYPES OF ACTIVATED CARBON[a]

Carbon	% Cd(II) Removed	Carbon	% Cd(II) Removed
Filtrasorb 100	20.5	Darco 12 x 20	26.0
Filtrasorb 200	20.5	Darco 20 x 40	25.5
Filtrasorb 300	20.5	Nuchar 722	30.0
Filtrasorb 400	17.0	Nuchar WV-G	23.0
Darco 12 x 40	26.0	Nuchar WV-L	22.0
Pittsburgh HGR	5.5	Nuchar WV-W	10.0
Darco Granules (HD 3000)	25.0	Nuchar S-A[b]	83.0
		Nuchar S-N[b]	67.0

[a]Batch adsorption conditions: $Cd(BF_4)_2$ 10^{-4} M; Carbon 1 g/L; pH = 7.00; I = 0.01 M $NaClO_4$; reaction time - overnight.

[b]The only two powdered activated carbons.

Reference 22.

Preliminary batch studies were performed for a total of fifteen different types of activated carbons. Table 8.1.4 shows the results of the preliminary runs. The PACs, Nuchar S-A and Nuchar S-N, exhibited a greater Cd(II) removal capacity than the granular carbons. Column experiments performed with the PAC beads demonstrated that the superior electrophoretic properties of powdered activated carbon could be combined with the manageability of GAC enabling use of the same contact equipment.[22]

Chromium--One study found that the removal of chromium from solution by activated carbon occurred through two major interfacial reactions: adsorption and reduction.[6] This study investigated chemical factors, such as pH and Cr^{+6} concentration, that affect the magnitude of Cr^{+6} adsorption. This study used a commercial activated carbon, Calgon Filtrosorb 400, in a continuous mixed batch system.

The adsorption density of Cr^{+6} increases with increasing pH to a maximum value and then declines rather rapidly with further increase in pH (see Figure 8.1.7). When the pH becomes greater than 10, no appreciable adsorption is observed. The extent of adsorption also increases with Cr^{+6} concentration.

Figure 8.1.8 demonstrates that Cr^{+6} is also removed by reduction to Cr^{+3} in the presence of activated carbon. In the absence of activated carbon, the Cr^{+6} added remained in the hexavalent state. However, based on the absence of Cr^{+3} in the supernatent, the researchers concluded that reduction only occurs at pH less than 6. This conclusion is valid since Cr^{+3} is adsorbed to a lesser extent by activated carbon than Cr^{+6}.

The results shown in Figure 8.1.9 demonstrate that cyanide is removed by activated carbon with a maximum value occuring around pH 8.[6]

Polaroid Corporation reported the successful application of activated carbon for the removal of hexavalent chromium from an aqueous waste stream generated by a slide film production facility. Several alternatives were considered including ion exchange, electrochemical treatment, sodium metabisulfite reduction, ferrous sulfate reduction, and carbon adsorption. A feasibility study and economic analysis resulted in selection of the activated carbon system.[24]

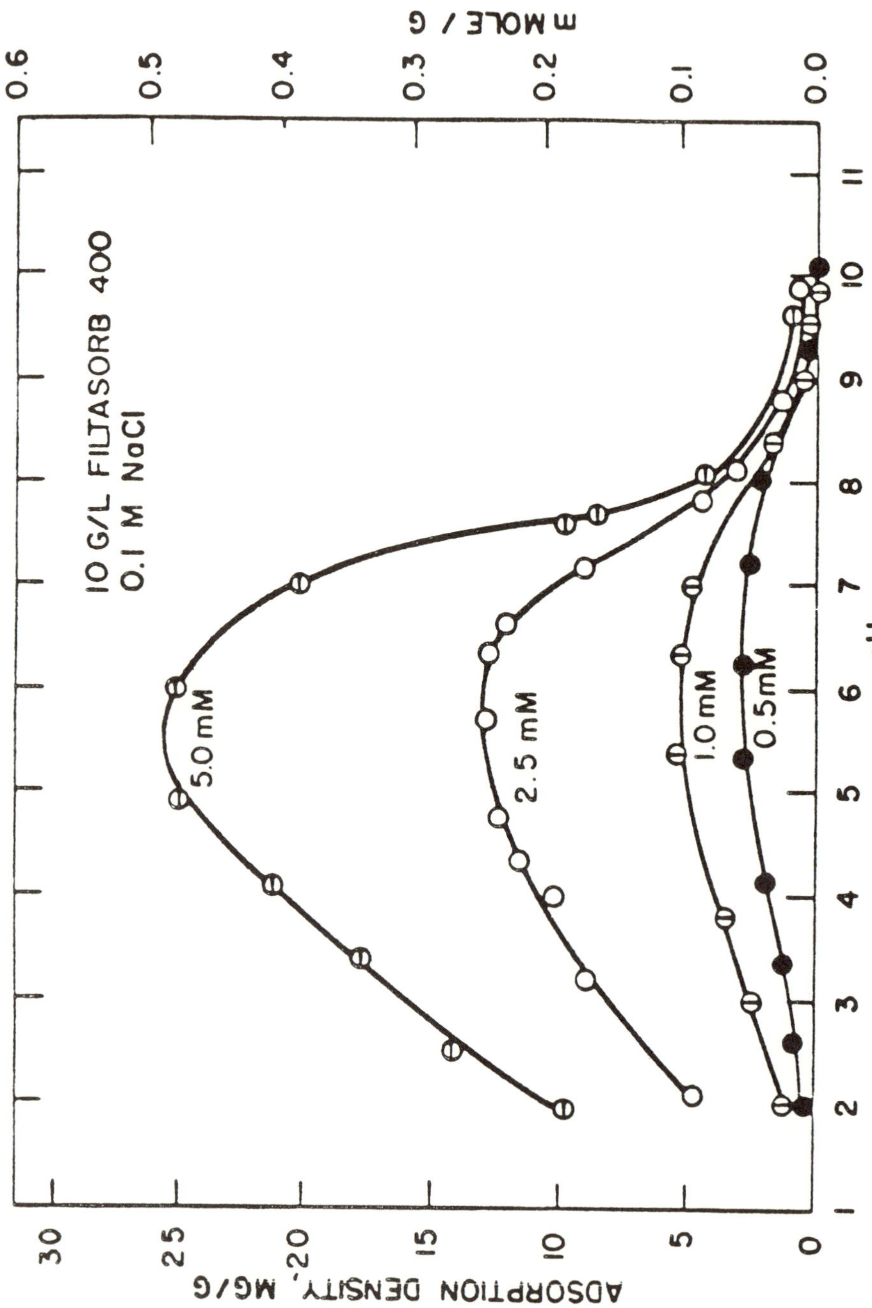

Figure 8.1.7. The effect of pH and total Cr(VI) on the adsorption of Cr(VI). Source: Reference 6.

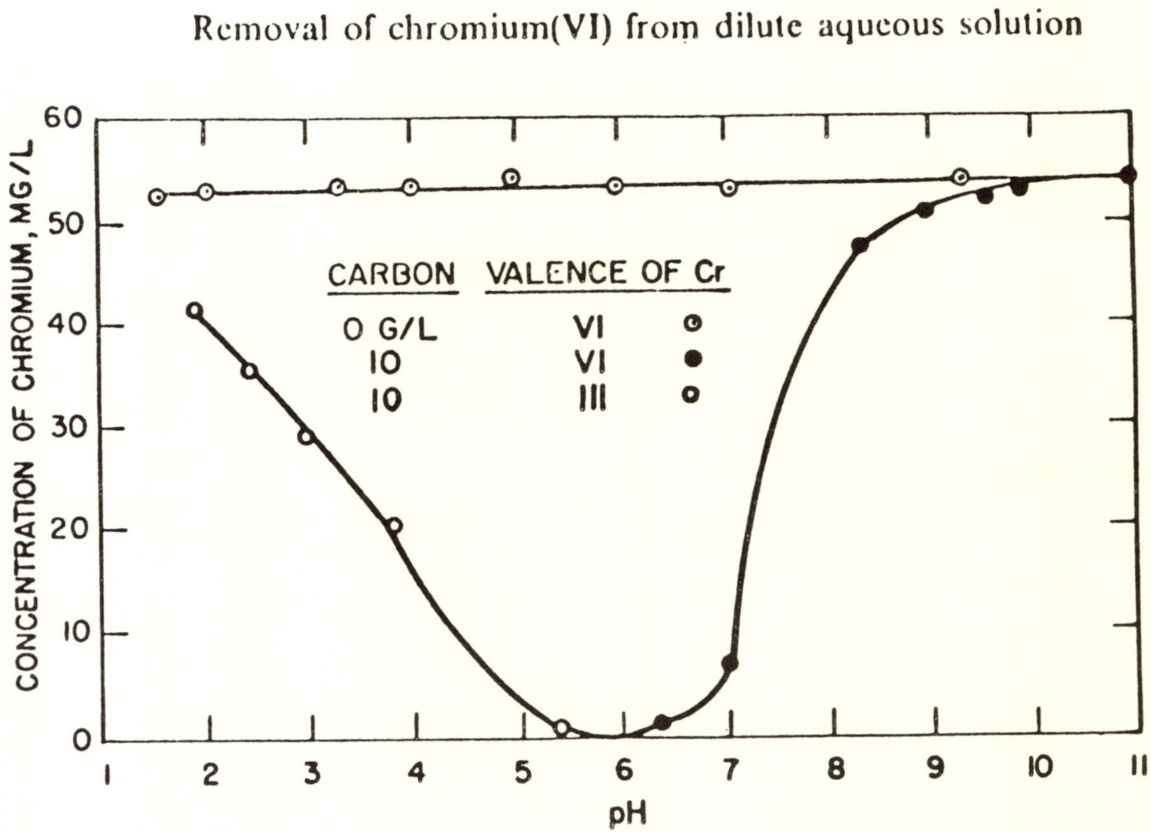

Figure 8.1.8. The effect of pH on the state of chromium in the presence and absence of activated carbon.
Source: Reference 6.

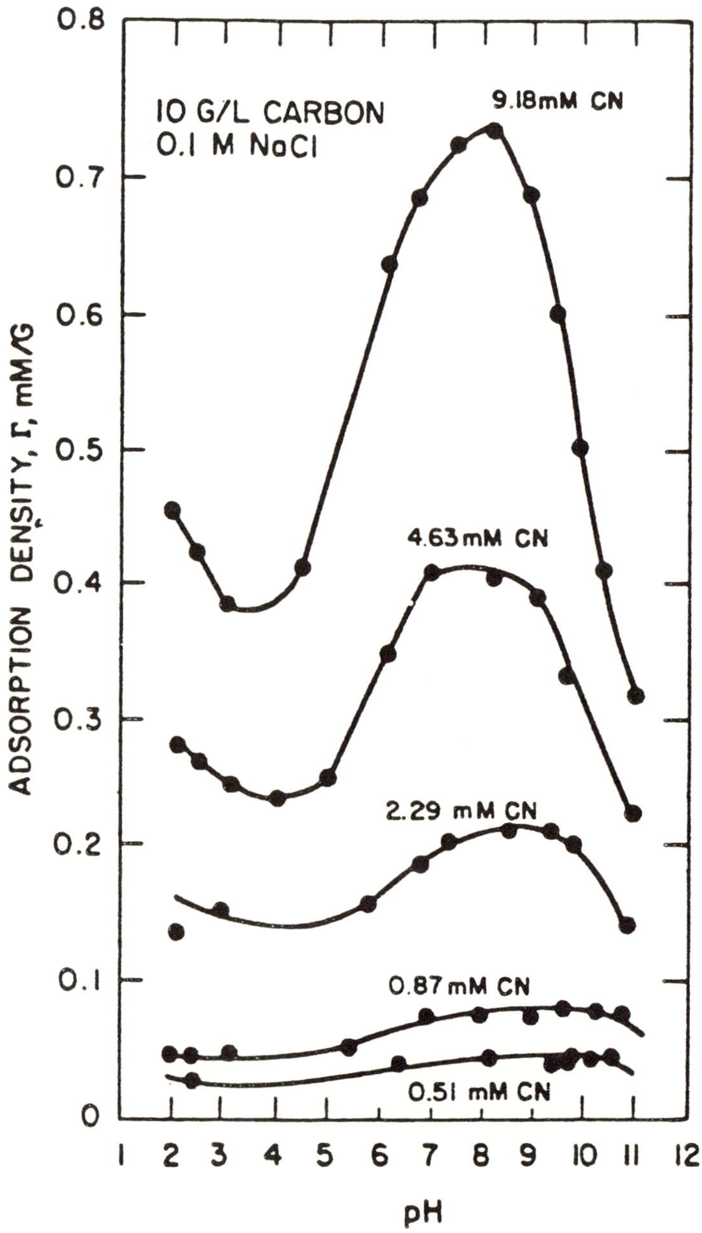

Figure 8.1.9. The effect of pH and total CN on the adsorption of CN. Source: Reference 6.

A schematic of the system is shown in Figure 8.1.10. The system utilizes carbon for adsorption of the chromium, which is believed to occur by reduction with subsequent adsorption of Cr^{+3} (Note: this mechanism is different than that postulated in the previous example since different carbon types are used).[24] After exhaustion, the carbon is regenerated by treatment with sulfuric acid which is then pH adjusted and filtered to remove precipitated chromium.[24]

During the pilot study, the feed pH and pretreatment (filtration) were found to have a major effect on successful operation of the carbon system. Adjustment of pH was necessary to extend the life and capacity of the carbon. If pH remained above 5.0, bed breakthrough occured 5 to 6 times more quickly than with adjustment.[6] Pre-filtration of the feed was required to prevent hydraulic fouling of the bed since the film production effluent contained a gelatinous component that easily plugged the carbon column.

In addition to technical success, Polaroid determined that a carbon system would also be more economical than the other technologies considered. Using the carbon on a once-through disposal mode (see Figure 8.1.10), as opposed to a regenerative mode, resulted in the lowest capital and operating costs in this particular application. A Carbon Service Agreement was selected as the optimal arrangement. Under this agreement, the carbon manufacturer leases an adsorber system, supplies carbon, disposes spent carbon, and provides maintenance support.

Mercury--Experiments were conducted to evaluate the enhancement of mercury adsoption that could be realized through preliminary chelation of mercury ions and chemical treatment of activated carbon.[25] The experiments included laboratory evaluations of process variables (i.e., pH, chelate type and dose, and carbon dose), batch capacity and isotherm tests, and continuous flow column studies. Synthetic solutions of mercuric chloride, which simulated wastewaters from chlor-alkali industries, were used throughout the studies.

Literature reviews performed by the investigators revealed the following:

- chelation of the mercury improves carbon removal capacity,
- reduction of Hg^{+2} to the elemental state may proceed subsequent to adsorption,

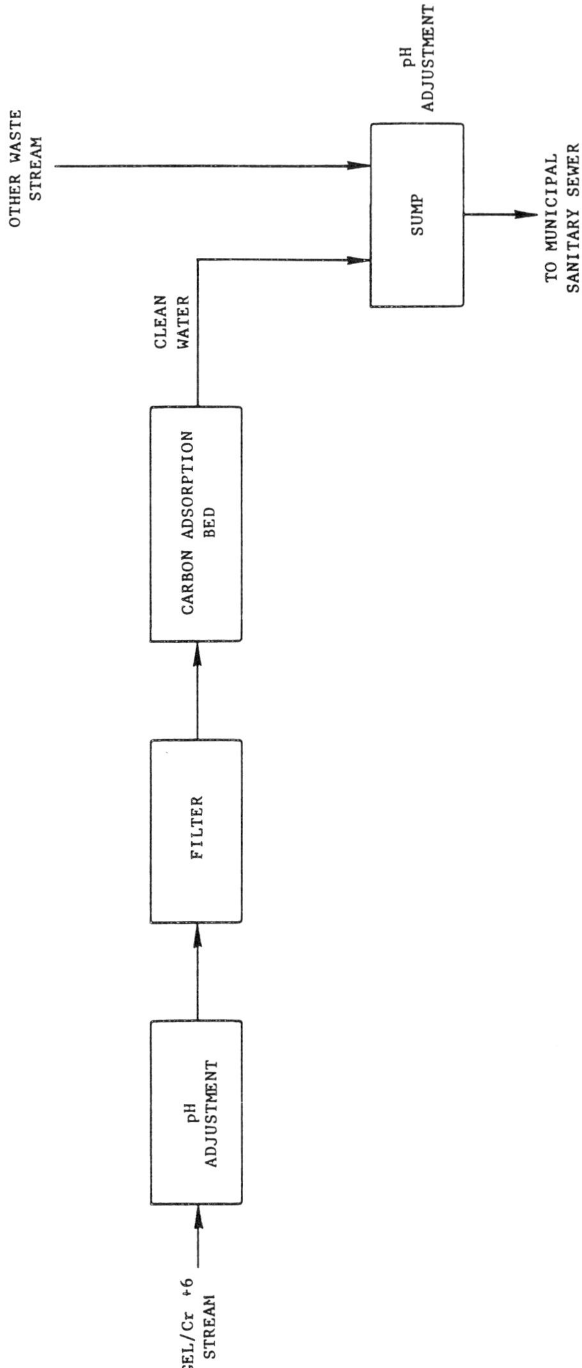

Figure 8.1.10. Single usage carbon adsorption system as installed. Source: Reference 24.

- o sulfurizing agents such as CS_2 can be used to improve mercury removal, and
- o regeneration of activated carbon laden with adsorbed mercury may be facilitated at high pH since its adsorption is enhanced by low pH.

Calgon Filtrasorb 300 effectively removed chelated Ammonium pyrrolidine dithiocarbonate (APDC) mercury from dilute mercuric chloride solutions at both pH 4 and pH 10, as shown in Figures 8.1.11 and 8.1.12, respectively. However, at the lower pH, carbon capacities were increased seven-fold in isotherm tests and nearly 14-fold in column tests. The pronounced pH effect may be explained in terms of changes in the carbon surface charge. At the mercuric chloride concentration in these experiments, sufficient chloride ion was available to complex the mercuric ion to its neutral or negatively charged chloride complexes. Any soluble metal oxides present at high pH would be converted to mercuric chloride species as the pH decreased.[25] A shift in pH would change the nature of the carbon surface. This carbon is an H-type carbon and possess basic surface oxide groups. Since the H-type carbons readily adsorb hydronium ions, the surface oxides could be neutralized at low pH, allowing pore diffusion of the mercury-chloride complexes. At higer pH, the basic surface groups repel the neutral to negative forms of the complexes.

Carbon disulfide (CS_2) greatly increased the removal of mercury by activated carbon at pH 10 when the carbon was soaked with CS_2 and dried prior to adsorption. At mercury concentrations of 1 ppm, CS_2 treatment resulted in a 50-fold increase in carbon capacity.[25]

Column tests with granular activated carbon showed improved operation at low pH. Figure 8.1.13 and 8.1.14 show the breakthrough curves for pH 10and pH 4, respectively. Relatively poor results were obtained at the high pH; breakthrough occurred within one day. At pH 4, a substantial improvement in performance was observed, with no breakthrough evident after 5 days of operation.[25]

In summary, the data indicate that removal of Hg^{+2} by activated carbon treatment is feasible. Carbon removals from alkaline wastes may be enhanced by chelation of the mercury with APDC, treatment of the carbon with CS_2, or by lowering the pH of the wastewater.[25]

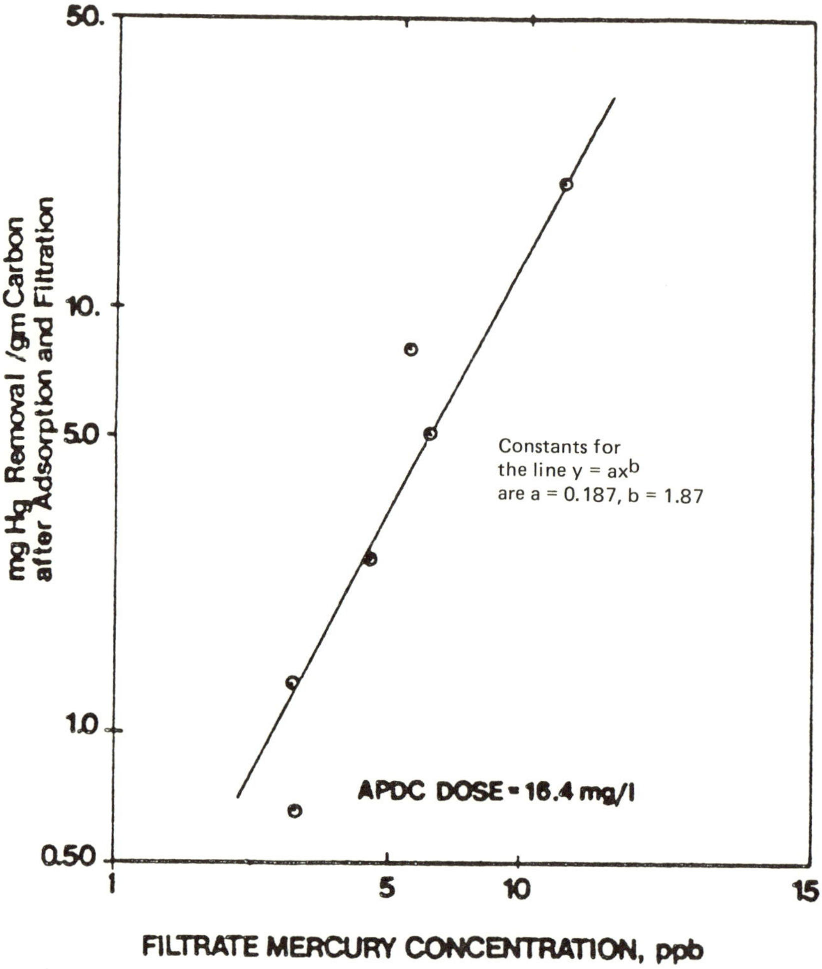

Figure 8.1.11. Freundlich isotherm demonstrating removal of mercury(II) by APDC and powdered activated carbon at pH 4 and 20°C. Source: Reference 25.

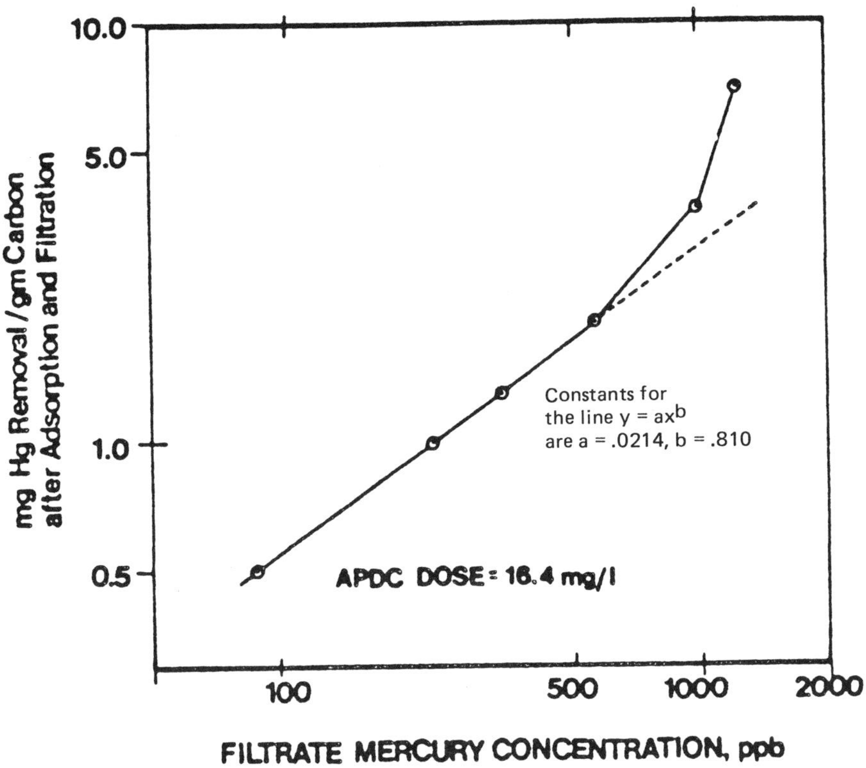

Figure 8.1.12. Freundlich isotherm demonstrating removal of mercury(II) by APOC and activated carbon at pH 10 and 20°C.

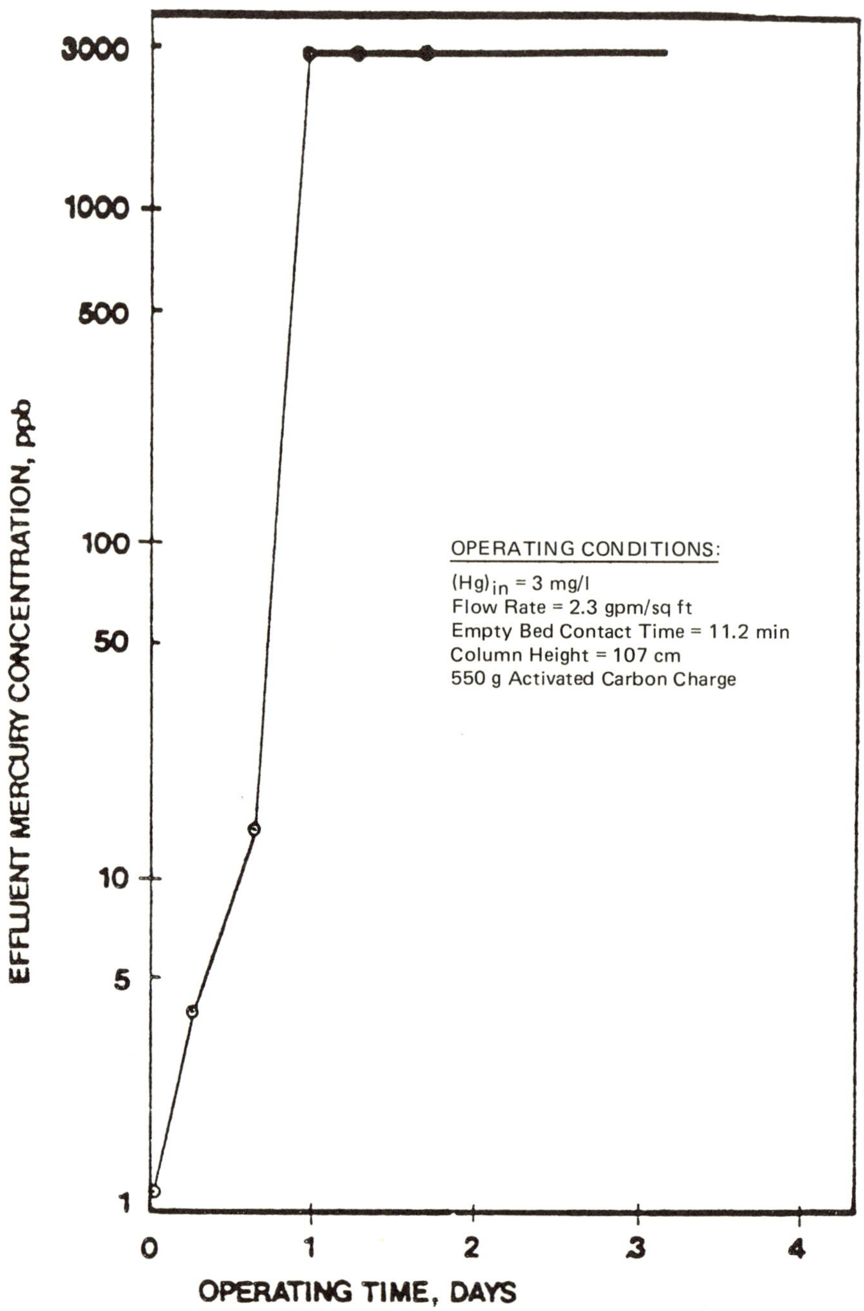

Figure 8.1.13. Column run at pH 10 and 25°C - Carbon-only system. Source: Reference 25.

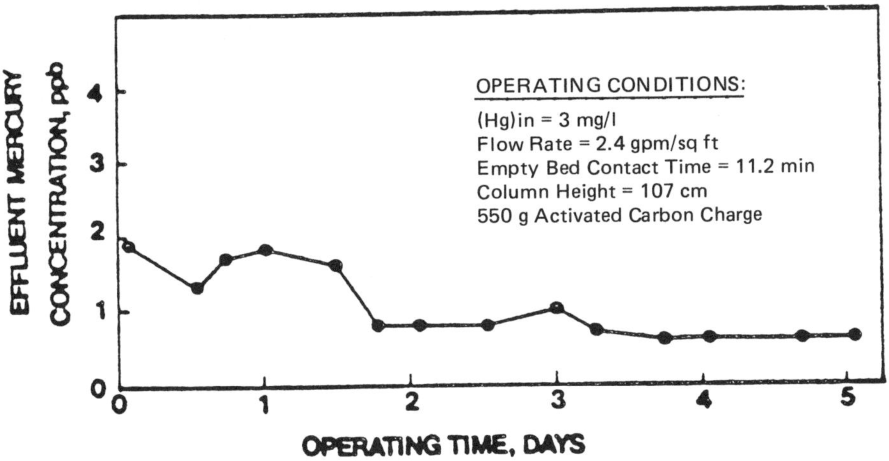

Figure 8.1.14. Column run at pH 4 and 25°C - Carbon-only system.
Source: Reference 25.

In another series of laboratory experiments, columns packed with Nuchar 722 activated carbon were used to determine the mechanism responsible for mercury removal.[26] A 17 percent caustic solution originating from mercury electrolysis cells was fed to one of the columns, while an aqueous preparation containing methyl mercuric chloride was percolated through the second. About 80 percent of the influent mercury was removed in the first column and no mercury was detected in the discharge from the second. The investigators concluded from the results obtained that organic mercury is readily adsorbable, both on an absolute basis and relative to other forms of mercury. They also postulated that filtration was the dominant mechanism in the observed removal of finely divided metallic mercury from the caustic stream.

Successful application of a full-scale activated carbon treatment system has been reported.[27,28] This particular system was devised for handling small volume, pesticide manufacturing discharges containing organic mercury compounds. In the process, suspended solids are removed by coagulation and flocculation with iron salts and polyelectrolytes prior to carbon adsorption in a series of packed beds. Mercury loadings of 0.05 kg per kg carbon are readily attained and the spent carbon sorbent is thermally regenerated.

Cyanide--Kuhn[29] patented a process where activated carbon is utilized as a catalyst for cyanide oxidation. The process involves mixing air and cyanide-bearing waste, at an alkaline pH, and pulsating the solution through a bed of activated carbon. Calgon Corporation extended this cyanide detoxification method by adding cupric ions to the wastewater along with oxygen prior to passing the cyanide-bearing waste through GAC columns.[30] The cupric ions accelerate and increase the efficiency of the catalytic oxidation of cyanide by granular activated carbon. The presence of cupric ions results in the formation of copper cyanides, which are adsorbed more readily than copper or cyanide alone.

Calgon demonstrated that the capacity of the granular carbon was limited to 2-3 mg of cyanide adsorbed per gram of adsorbent when no copper was used. However, the addition of copper increased the adsorption capacity to 25 mg/g. In the presence of dissolved oxygen, adsorption sites were continuously regenerated through the oxidation of the cyanide.[30]

Based on Calgon's studies, a study was undertaken to investigate the feasibility of a low-cost activated carbon treatment process for petroleum refinery wastewater.[7] The conceptualized process evaluated in this study involves the addition of powdered activated carbon (PAC) and cupric chloride directly into an activated sludge unit which is commonly used for secondary treatment at petroleum refineries. Some of the potential benefits of adding PAC to an activated sludge system include:[31]

- improvement in BOD and COD removals;
- improved solids settling, decreased effluent solids and increased sludge solids;
- adsorption of dyes and toxic components that are either not treated biologically or are poisonous to the biological system;
- prevention of sludge bulking over broader ranges of feed to microorganism;
- effective increases in plant capacity at little or no additional capital investment; and
- more uniform plant operation and effluent quality, especially during periods of widely varying organic or hydraulic loads.

From the results of batch tests, the following five parameters were considered to be the major variable affecting the cyanide treatment using PAC in activated sludge units: pH, mode of copper addition, carbon type, and carbon and copper dosages.[7]

Increased cyanide removal rate was observed at lower pH, as shown in Figure 8.1.15. However, although low pH favors increased cyanide removal, effluent copper levels were unacceptable. Further study demonstrated that at pH values near neutral, 95 percent cyanide removal was achieved while maintaining effluent copper levels of 0.05 mg/l or below.[7]

Copper salts can be introduced by two different techniques, either directly into the aeration basin or by being adsorbed onto the carbon prior to addition. The results did not demonstrate a significant difference between these methods.

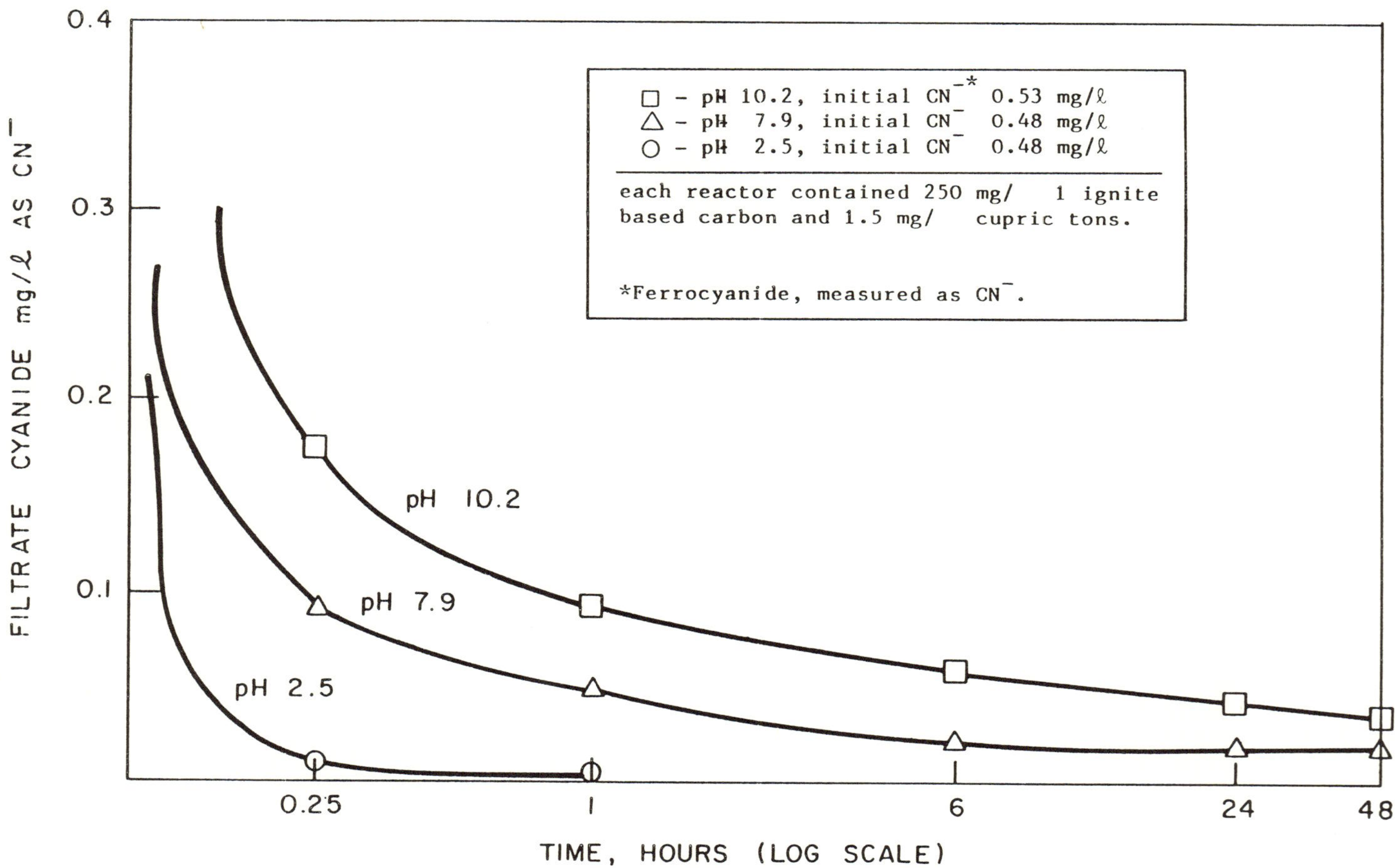

Figure 8.1.15. pH effect on rate of cyanide removal.
Source: Reference 7.

Two type of PAC, lignin-based (Aqua-Nuchar) and lignite-based (Hydarco C) were evaluated throughout a range of carbon and copper concentration. Typical results are shown in Figure 8.1.16. When utilizing potassium ferrocyanide, the lignite-based carbon was superior at all carbon dosages tested, however the overall improvement diminished as copper dosage was increased. Similar tests performed with potassium ferricyanide as the cyanide source showed the lignin-based PAC to be more effective. In both tests, the equilibrium soluble cyanide level was reduced as the carbon concentration was increased.[7]

Copper dosage was found to have the greatest influence on cyanide removal. The data presented in Table 8.1.5 demonstrate that as the carbon dosage increases, there is greater copper removal in addition to greater cyanide removal. Hence, concerns over excessive copper effluent levels can be addressed by increases in carbon dosage.[7]

The above results demonstrate that the addition of PAC/$CuCl_2$ directly into petroleum refinery activated sludge aeration basins can enhance cyanide removal without any detrimental effect on the microorganisms, provided that the copper concentration in the influent is maintained at less than 1 mg/l. The addition of PAC also improves the removals of BOD, COD and TOC.

8.1.3 Cost of Carbon Adsorption

The cost of carbon adsorption treatment can be described in terms of direct and indirect capital investment, operation and maintenance costs. For the small-scale system, direct capital investment costs include the purchase of a waste storage tank, a pre-filter, carbon columns, a waste feed pump, piping and installation. For the large-scale system, additional direct capital investment costs include storage tanks for spent and regenerated carbon and automatic controls.[32] A model has been developed by ICF, Inc. for calculating carbon adsorption costs.[33] Table 8.1.6 presents equations used to calculate direct capital costs as a function of carbon consumption rate and storage volume.

Indirect capital costs include engineering and construction costs, contractor's fee, startup expenses, spare parts inventory, interest during construction, contingency and working capital. These costs are expressed as percentages as summarized in Table 8.1.7. Direct and indirect capital costs are assumed to be incurred in year zero.

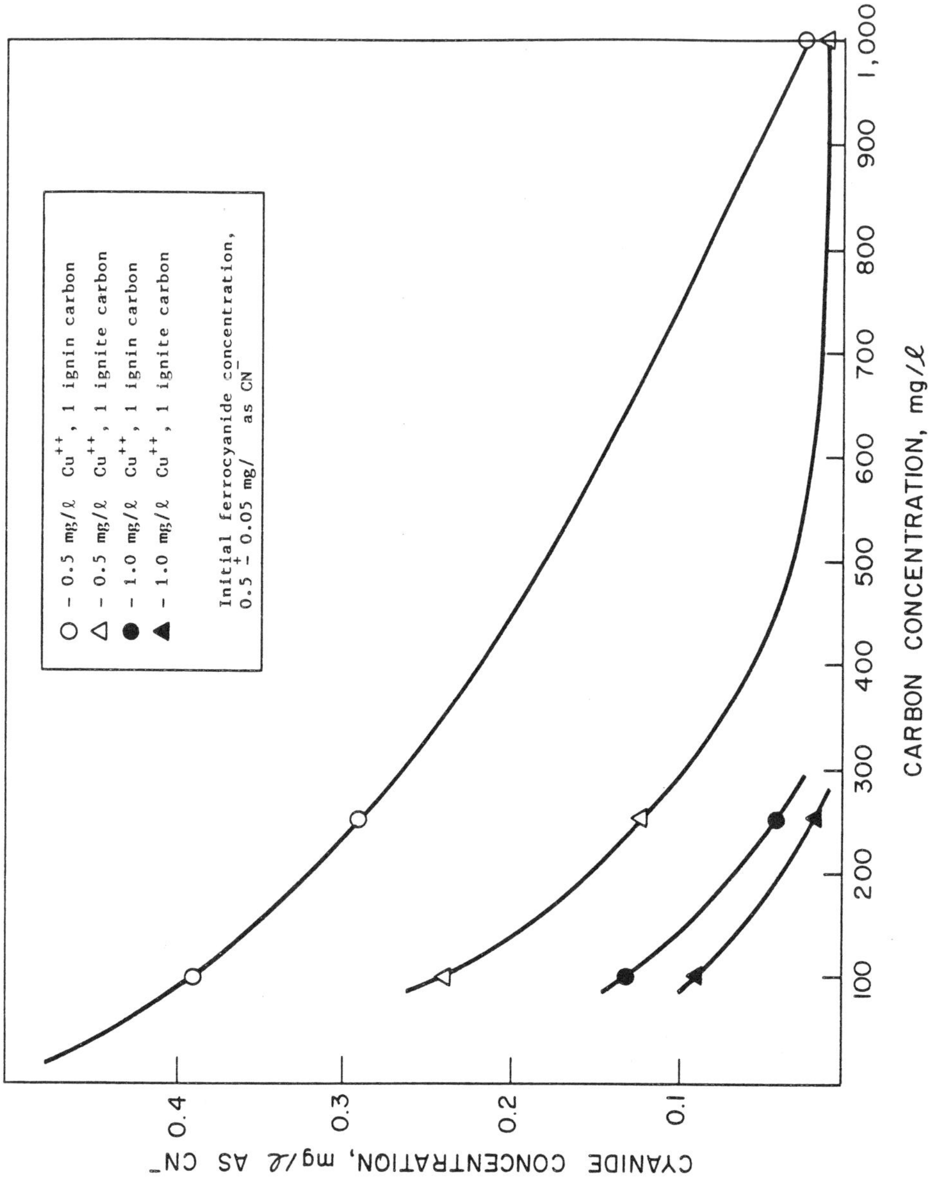

Figure 8.1.16. Equilibrium cyanide levels as a function of carbon concentration.
Source: Reference 7.

TABLE 8.1.5. FILTRATE COPPER LEVELS

Initial copper concentration	Average Filtrate Copper Levels		
	100 mg/L carbon	250 mg/L carbon	1,000 mg/L carbon
0.5	0.06	0.05	<0.02
2.0	0.08	0.05	0.02
1.5	0.19	0.09	--

Reference 7.

TABLE 8.1.6. DIRECT COSTS FOR CARBON ADSORPTION[a]

Carbon consumption rate (lbs/day)	Direct capital costs ($)	Direct operation and maintenance cost[b] ($/yr)
less than 400	$1,256(c)^{.603} + 140(s)^{.54}$	$29(c)^{.6} + 350(c)(cp) + 619(c)^{.168}(h) + 5(c)(p)$
greater than 400	$14,231(c)^{.522} + 140(s)^{.54}$	$58(c)^{.657} + 35(c)(cp) + 105(c)^{.455}(h) + 25,012^{.383}(c)(p) + 1.49\ 10^{6}(c)(f)$

where: c = carbon consumption rate in pounds per day

s = storage volume in gallons

cp = carbon price in dollars per pound ($0.8/lb)

h = hourly wage rate in dollars per hour ($14.56/hr)

p = power price in dollars per kilowatt-hour ($0.05/KWh)

f = fuel price (natural gas) in dollars per Btu ($6x10^{-6}/Btu)

[a]Cost estimates were developed for three model treatment systems (three small scale and three large scale systems). The cost estimates for these systems were then used to develop a cost equation in the form of a power curve.

[b]The power requirement is derived from the equipment specifications.

Source: Reference 33.

TABLE 8.1.7. INDIRECT COSTS FOR CARBON ADSORPTION

Item	Percent of direct capital costs	Percent of the sum of direct and indirect capital costs	Percent of total annual cost[a]
Indirect Capital Costs			
Engineering and Supervision	12	0	0
Construction and Field Expenses	10	0	0
Contractors Fee	7	0	0
Startup Expenses	5	0	0
Spare Parts Inventory	2	0	0
Interest During Construction	10	0	0
Contingency	0	15	0
Working Capital	0	18	0
Indirect Operation and Maintenance Costs			
Insurance, Taxes, General Administration	0	5	0
System Overhead	0	5	10

[a]The total annual cost is defined as the sum of the total capital cost multiplied by the capital recovery factor and the total operation and maintenance costs.

Reference 33.

Operation and maintenance costs also consist of direct and indirect costs. Direct operation and maintenance costs include operating labor, electricity, and carbon consumption. Table 8.1.8 also contains the equations used in the model to calculate direct operation and maintenance costs. All costs are presented for four flow rates ranging from 100 to 2,500 gal/hr.

8.1.4 Overall Status of Process

Activated carbon is a widely used technology for treating waste streams containing organic compounds. In contrast, the application of activated carbon technology to the treatment of metal and cyanide containing wastestreams is limited. However, the ability of activated carbon to treat these wastestreams has been demonstrated at bench, pilot, and full-scale levels. Full-scale systems have been used commercially to treat chromium and mercury wastestreams, but these applications are few in number. Performance data is difficult to acquire due to confidentiality agreements between activated carbon manufacturers and their customers.

Environmental impacts can occur when the exhausted activated carbon must be regenerated or disposed. The regeneration of activated carbons used for the treatment of metals or cyanides is accomplished using a strong acid or base. Regeneration is usually not performed unless there is an economic incentive to recover the adsorbed metals.[34] Consequently, the spent carbon is typically disposed of in a secure landfill.

TABLE 8.1.8. CARBON ADSORPTION COSTS[a]

	Quantity processed (gal/hr)			
	100	400	1,000	2,500
Capital Expenditures				
Capital Cost Including Installation[b] ($1,000)	59	561	904	1,462
Annual Operation and Maintenance ($1,000)[c]				
Energy	2	11	27	68
Labor	23	35	53	80
Carbon	7[e]	27	67	168
Other	1	5	10	18
Capital Recovery	10	99	160	259
Total Annual Cost	42[e]	177	317	593
Cost per 1,000 gal[d]	210[e]	221	159	119

[a]Costs are based on the RCRA RISK-COST ANALYSIS MODEL.[33]

[b]Capital costs for the 100 gal/hr system include waste storage tank, prefilter, carbon columns, waste feed pump, piping and installation; the other flow levels (400, 1,000, 2,500) include these units plus storage tanks for spent and regenerated carbon, a multiple hearth furnace and automatic controls.

[c]These costs are based on the following data:

carbon price = $0.8/lb
hourly wage rage = $14.56/hr
power price = $0.05/kwh
fuel price (natural gas) = $6 x 10^{-6}/Btu
capital recovery factor = 0.177

[d]Unit costs are based on 2000 hours of operation per year.

[e]Modified to reflect a direct relationship between carbon requirement and quantity processed.

*Note: 1984 dollars.

REFERENCES

1. Weber, W. J., Physicochemical Processes for Water Quality Control. John Wiley & Sons. 1972.

2. Lyman, W. J., Applicability of Carbon Adsorption to the Treatment of Hazardous Industrial Wastes. In: Cheremisinoff, et al., Carbon Adsorption Handbook, Ann Arbor Science. 1980.

3. U.S. EPA, An Investigation of Techniques for Removal of Cyanide from Electroplating Wastes. U.S. Environmental Protection Agency Water Pollution Control Research Series No. 12010 EIE 11/71. 1971.

4. Smithson, G.R., Jr. An Investigation of Techniques for the Removal of Chromium from Electroplating Wastes. U.S. Environmental Protection Agency, Water Pollution Control Research Series. No. 12010 EIE 03/71. 1971.

5. ICF Inc. Survey of Selected Firms in the Commercial Hazardous Waste Management Industry: 1984 update. Final report to U.S. EPA, Section II. OSW Washington, D.C. 1985.

6. Huang, C.P. and M.H. Wu. The Removal of Chromium(VI) from Dilute Aqueous Solution by Activated Carbon. Water Research 11:673-679. Pergamon Press. 1977.

7. Huff, J.E., E.G. Fochtman and J.M. Bigger. Cyanide Removal from Refinery Wastewater Using Powdered Activated Carbon. In: Cheremisinoff and Ellerbusch, Carbon Adsorption Handbook, Ann Arbor Science. 1980.

8. Wilk, Lisa et al. Alliance Technologies Corporation. Technical Resource Document for Treatment of Corrosive Wastes. Prepared for U.S. EPA HWERL under Contract No. 68-02-3997. October, 1986.

9. Slejko, F.L., Applied Adsorption Technology, Chemical Industry Series Volume 19. Marcel Dekker, Inc. NY, NY. December 1985.

10. Huang, C.P., Chemical Interactions Between Inorganics and Activated Carbon. In: Cheremisinoff and Ellerbusch, Carbon Adsorption Handbook, Ann Arbor Science. 1980.

11. Jevtitch, M.M. and D. Bhattacharyya. Separation of Heavy Metal Chelates by Activated Carbon: Effect of Surface and Species Charge. Chem. Eng. Commun. 23:191-213. 1983.

12. Steenberg, B. Adsorption and Exchange of Ions on Activated Charcoal. Almquist and Wilksell, Uppsala. 1944.

13. Benjamin, M.M. and J.O. Leckie. Conceptual Model for Metal-Ligand-Surface Interactions During Adsorption. Environmental Science & Technology 15:1050-1057. 1981.

14. Huang, C.P. and E.H. Smith. Removal of Cd(II) from Plating Waste Water by an Activated Carbon Process. In: Chemistry of Water Reuse, Ed: Copper, W. Ann Arbor Science. 1981.

15. Huang, C.P. and F.B. Ostovic. The Removal of Cadmium(II) from Dilute Aqueous Solution by Activated Carbon Adsorption. J. Environ. Eng. Div., ASCE 104(EE5). 1978.

16. Langmuir, I. The Adsorption of Gases on Plane Surfaces of Glass, Mica, and Platinum. Jour. Am. Chem. Soc., 40:1361-1403. 1918.

17. U.S. EPA Background Document for Solvents to Support 40 CFR Part 268, Land Disposal Restrictions, Volume II. January 1986.

18. Rizzo, J.L. Calgon Corporation, Letter to Paul Frillici, Alliance Technologies Corporation. May 5, 1986.

19. U.S. EPA. Activated Carbon Treatment of Industrial Wastewater-Selected Papers. EPA-600/2-79-177. Robert S. Kerr Environmental Research Laboratory. August 1974.

20. IT Enviroscience, Incorporated. Survey of Industrial Applications of Aqueous-Phase Activated-Carbon Adsorption for Control of Pollutant Compounds from Manufacture of Organic Compounds, Prepared for U.S. EPA IERL; PB-83-200-188. April 1983.

21. Lyman, W.J. Carbon Adsorption, In: Unit Operations for Treatment of Hazardous Industrial Wastes. Pollution Technology Review No. 47, Noyes Data Corporation, Park Ridge, New Jersey. 1978.

22. Huang, C.P. and P.K. Wirth. The Development of an Activated Carbon Adsorption Process for the Treatment of Cadmium(II)-Plating Waste. Heavy Metals in the Environment, Amsterdam, The Netherlands, 15-18. September 1981.

23. Gupta, S.K. and K.Y. Chen. Arsenic Removal by Adsorption. Journal WPCF. March, 1978.

24. Praino Jr., R.F. and R. O'Gorman. Technology Evaluation, Installation and Performance of a Chromium Removal System for Aqueous Discharges. Hazardous Waste 1:469-487. Publ: Mary Ann Liebert, Inc. 1984.

25. Humenick, Jr., M.J. and J.L. Schnoor. Improving Mercury(II) Removal by Activated Carbon. J. Environ. Eng. Div., ASCE 100E6:1249-1262. 1974.

26. Smith, S.B.; Hyndshaw, A.Y.; Laughlin, H.F. and S.C. Maynard (1971) Mercury Pollution Control by Activated Carbon: A Review of Field Experience. Rept. M1002.01, Westvaco Corp., Covington, VA.

27. Rosenzweig, M.D. (1975) Mercury Cleanup Routes - I. Chem. Eng., Vol. 82, No. 2, pp. 60-61.

28. TNO (undated) Purification of Mercury-Containing Waste Waters. Apeldoorn, The Netherlands.

29. Kuhn, R.G. Process for Detoxification of Cyanide Containing Aqueous Solutions. U.S. Patent 3,586,623. June 22, 1971.

30. Bernardin, F.E. Cyanide Detoxification Using Adsorption and Catalytic Oxidation on Granular Activated Carbon. J. Water Poll. Control Fed. 34:221-231. 1973.

31. Adams, A.D. Improving Activated Sludge Treatment with Powdered Activated Carbon. Presented at the 28th Annual Purdue Industrial Waste Conference, Purdue University, Lafayette, Indiana. May 1-3, 1973.

32. U.S. EPA. Development Document for Effluent Limitation Guidelines and Standards for Petroleum Refining Point Source Category. EPA-440/1-82-014. 1982.

33. ICF, Inc. RCRA Risk-Cost Analysis Model, Phase III U.S. EPA, OSW. March 1984.

34. Roy A., Calgon Corporation. Telephone conversation with D. Sullivan, Alliance Technologies Corporation. March 5, 1987.

8.2 ION EXCHANGE

Ion exchange has been used commercially to recover metal-containing wastes from the metal finishing, electroplating, and fertilizer manufacturing industries.[1] These wastes contain dissolved metal salts which dissociate to form metal ions. In conventional ion exchange, metal ions from dilute solutions (e.g., plating rinses) are exchanged for ions which are held by electrostatic forces to charged functional groups on the surface of the exchange resin. An alternative design, the acid purification unit, adsorbs acids from concentrated solutions (e.g., etchants) and allows metal contaminants to pass through the system. In both cases, the adsorbed constituent is subsequently removed by contacting the resin with a regenerant, resulting in a potentially recoverable by-product stream which is highly concentrated in the adsorbed constituent.

The major applications of ion exchange are water purification (deionization) and selective removal of toxic heavy metal and metal-cyanide complexes from dilute wastewater streams. Rinse water is reused and metal contaminants are concentrated in the regenerant stream, allowing more economical treatment and enhancing their recovery potential. As an end of pipe application, ion exchange resins have been applied for selective removal of toxic compounds, while allowing nontoxic dissolved ionic solids to remain in solution. The acid purification unit (APU) has been successfully applied commercially for recovery of steel pickling, aluminum anodizing, etchants, and rack stripping operations.[1]

8.2.1 Ion Exchange Process Description

General System Description--

The ion exchange system may be operated in a batch or flow-through (column) mode, the latter being generally preferred due to greater exchange efficiencies. With the batch mode of operation, the ion exchange resin and the waste solution are mixed in a batch tank. Upon completion of the exchange reaction (i.e., equilibrium is reached), the resin is separated from the treated solution by filtering or settling, regenerated, and reused. Unless the resin has a very high affinity for the contaminant ion, the batch mode of operation is chemically inefficient and thus has limited applications.

Flow-through operation involves the use of a bed or packed column of the exchange material (resin). These systems are typically operated in cycles consisting of the following steps:

1. Service (exhaustion) - Waste solution is passed through the ion exchange column or bed until the exchange sites are exhausted.

2. Backwash - The bed is washed (generally with water) in the reverse direction of the service cycle in order to expand and resettle the resin bed.

3. Regeneration - The exchanger is regenerated by passing a concentrated solution of the ion originally associated with it through the resin bed or column; usually a strong mineral acid or base.

4. Rinse - Excess regenerant is removed from the exchanger; usually by passing water through it.

A flow-through (column) system can be designed with cocurrent or countercurrrent flow of the waste and regenerant (steps 1 and 3 above). In cocurrent systems, the feed and the regenerant both pass through the resin in a downflow mode. Figure 8.2.1 illustrates the cocurrent flow process.[2] Each ion exchange unit consists of a cylindrical vessel having distributors or collectors at the top and bottom. Resin is loaded into approximately half of the vessel to accommodate resin expansion during the backwash cycle. Cocurrent systems are only cost-effective for weak acid or base exchangers which do not require highly concentrated regenerant solutions. However, regeneration of strong exchangers (high exchange capacity) requires strong acid and base solutions which can be more costly.

Often it is too costly to fully regenerate a bed. In order to avoid carry over of contaminants into the next service run, two or more sets of fixed columns arranged in parallel series can be used. Similarly, to avoid excessive downtime during the regeneration cycle, dual sets of fixed columns can be used. While one set of columns is being regenerated, the second set of columns will be switched on line permitting continuous operation of the system. Improved regenerant efficiency can also be accomplished by reusing

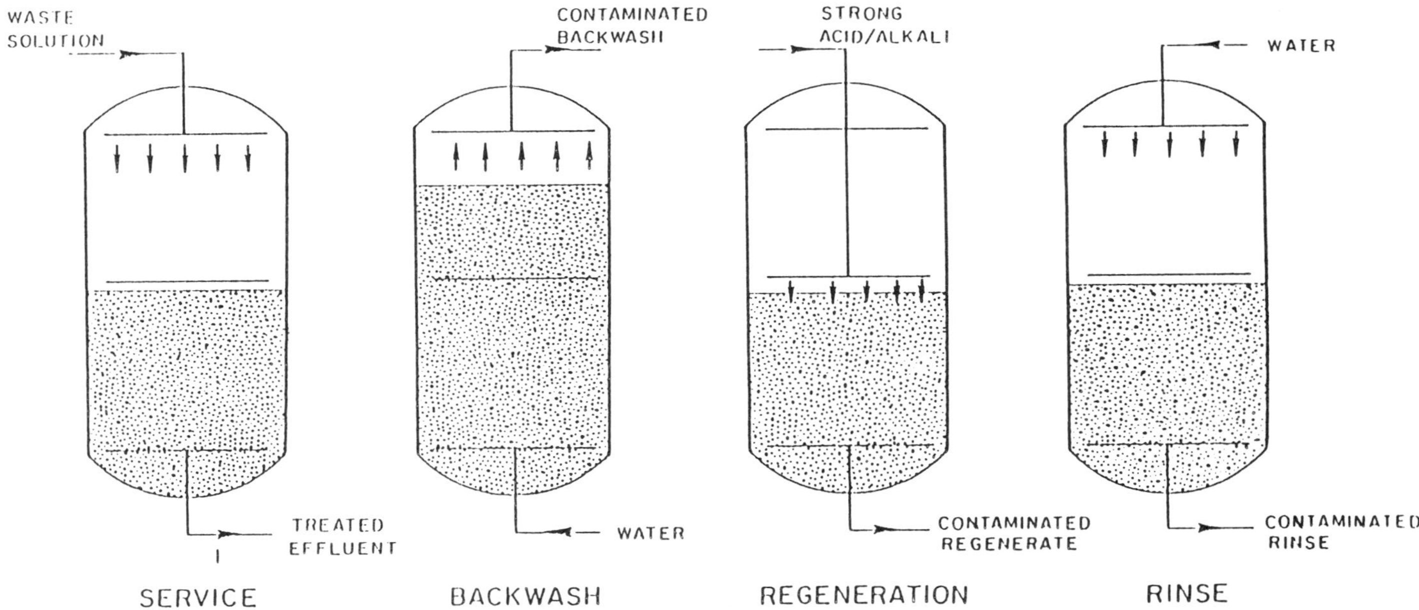

Figure 8.2.1. Cocurrent ion exchange cycle.

Source: Reference 2.

the last portion of the regenerant solution.[3] For example, if 5 lb/ft^3 (80 g/L) of regenerant were used for the system shown in the figure, the first 50 percent of spent regenerant would only contain 29 percent of the original acid concentration, whereas the remaining regenerant would contain 78 percent of the original acid.[3] If the last portion of the regenerant is reused in the next cycle before the resin bed is contacted with fresh HCl, the exchange capacity would increase from 60 to 67 percent at equal chemical doses.[3]

In addition to cocurrent designs, countercurrent systems are available which result in a more efficient use of regenerant chemicals. They also achieve a higher concentration of contaminant in the regenerant stream thereby enhancing potential for further recovery. A widely used countercurrent design for chemical recovery from plating rinses is the reverse or reciprocating flow ion exchanger (RFIE), as depicted in Figure 8.2.2. Another variation of this design, which uses countercurrent flow through a fluid bed, is called the acid purification unit (APU), developed by Eco-Tech, Ltd. Instead of adsorbing metallic species, the resin adsorbs acid which is then regenerated by flushing the bed with water (see Figure 8.2.3).

Resin Selection--

The most significant design parameter in an ion exchange system is the selection of an appropriate resin. Resin selection is based on the type of ion exchanger, flow volume and the resin's strength, exchange capacity and selectivity. Resins can be classified by functional (reactive) groups and the type of exchangeable ions present. Exchanger categories include strong and weak cation, strong and weak anion, and chelating ion exchangers. Some of the more common reactive groups are:[4]

Reactive Group	Exchangeable Ions
Strong acid (sulfonic)	Cations in general
Weak acid (carboxylic)	Cations in general
Weak acid (phenolic)	Cesium and polyvalent cations
Strong base (quaternary amine)	All anions, esp. used for anions of weak acids (cyanide, carbonate, silicate, etc.)

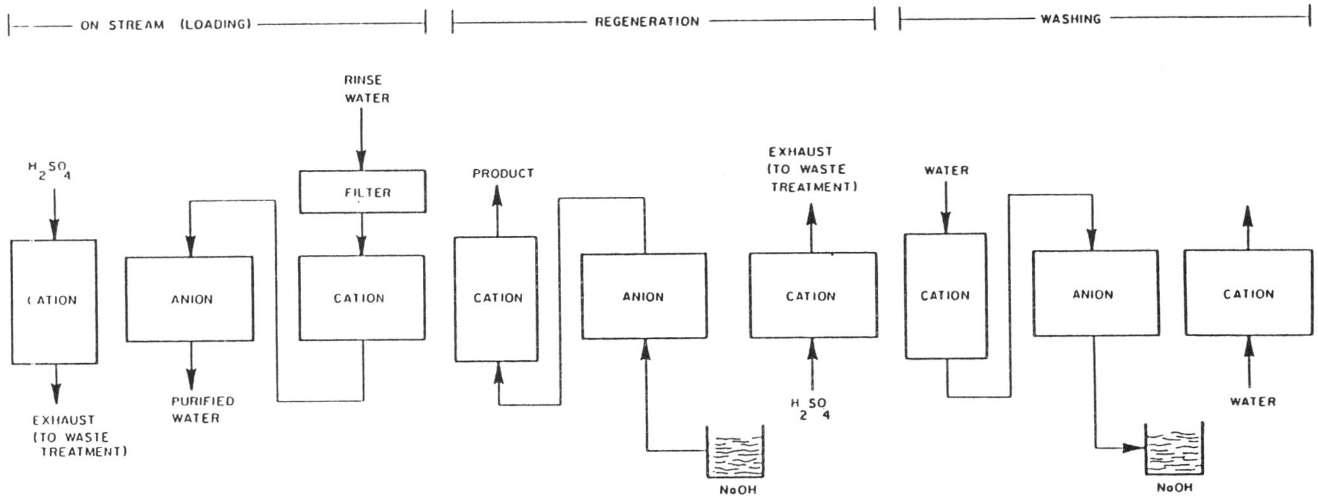

Figure 8.2.2. Schematic of a fixed bed reverse flow ion exchange (RFIE) system for the recovery of chromic acid from a dilute solution.

Source: Reference 3.

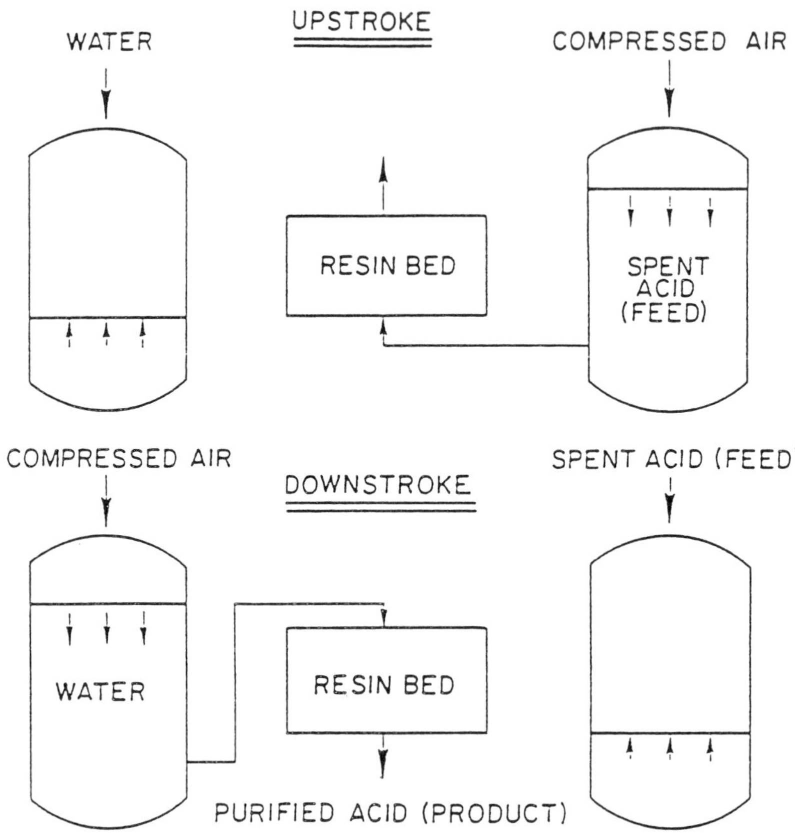

Figure 8.2.3. Basic operation of the acid purification unit (APU) using a continuous bed RFIE system.

Source: Reference 1.

Weak base (tertiary and secondary amine)	Anions of strong acids (sulfate, chloride, etc.)
Chelating (varied, may be imino-diacetate or oxime groups)	Cations, especially transition and heavy elements

Cation exchangers have positively charged, mobile ions for exchange. Strong acid cation resins are those containing functional groups derived from a strong acid. Their behavior is similar to that of a strong acid in that they can convert a metal salt to the corresponding acid. Both the hydrogen form (used for deionization) and sodium forms (used for water softening) are highly ionized. Due to the highly dissociated nature of these resin types, their exchange capacity is independent of solution pH.

Weakly acidic cation exchangers are resins derived from a weak acid. These resins behave like weakly dissociated organic acids. The degree of dissociation is strongly influenced by solution pH and they tend to demonstrate limited capacity below a pH of 6[2]. Due to the pH limits, weak acid resins are unsuitable for deionizing acidic wastes.

Strong and weak anion exchange resins behave in a fashion analogous to cation exchangers. Strongly basic anion exchangers are highly ionized and can be used over a wide pH range. Weakly basic anion resins are strongly influenced by solution pH and exhibit limited exchange capacity above a pH of 7[3].

Chelating resins behave similarly to weak acid cation resins, but are highly selective for heavy metal cations. This type of resin forms an essentially non-ionized complex with divalent metal ions. Consequently, once an exchanger group is converted to the heavy metal form, it is relatively unreactive with other similarly charged ions in solution, regardless of concentration. Chelating resins will effectively remove heavy metal cations from solutions of pH 4 and above.

The exchange capacity of a resin is generally expressed as equivalents per liter (eg/L), where an equivalent is equal to the molecular weight of the ion, in grams, divided by its electrical charge or valence.[3] For example, a resin with an exchange capacity of 1 eq/L could remove 37.5 g of divalent zinc (Zn^{+2}, molecular weight = 65 g) from solution.

As noted above, solution pH can have a significant effect on exchange capacity for weakly acidic, anionic, and chelating resins. For example, the effect of pH on the exchange capacity of Rohm and Haas Amberlite IRC-718, a chelating resin specifically designed for selective heavy metals removal, is quite dramatic.[5] Because of the resin's affinity for hydrogen ions, the capacity for most other ions falls off sharply below pH 4. Figure 8.2.4 compares the capacity of Amberlite IRC-718 when used to remove nickel from a waste containing calcium chloride at pH 2 and pH 4. The data show that good removal is realized for 200 bed volumes (1,500 gal/ft^3 of resin) when treating the stream at pH 4 whereas breakthrough occurred at pH 2 in less than 50 bed volumes.[5]

Figure 8.2.5 illustrates the effect of pH on the capacity of Amberlite DP-1 (a weak acid cationic resin) and Amberlite IRC-718 when these resins are used for removal of cadmium at pH 2.1 and pH 8.0.[5] In this example, Cd^{+2} was present at a concentration of 50 ppm with 1,000 ppm calcium chloride. At a flow rate of 8 bed volumes/hour (1 gpm/ft^3) and a pH of 2.1, both resins showed sharp breakthrough curves with end points less than 100 bed volumes. Conversely, at pH 8.0, Amberlite IRC-718 showed less than 0.1 ppm leakage for 350 bed volumes while leakage from Amberlite DP-1 remained under 0.1 ppm for 520 bed volumes.[5]

The metal removing performance of an ion exchange resin is also influenced by its ionic form. For example, Amberlite IRC-718 is available in hydrogen and sodium forms. Figure 8.2.6 demonstrates the difference in exchange capacity for this resin for the removal of copper from a stream containing 50 ppm Ca^{+2} and 1,000 ppm $CaCl_2$ at pH 4.[5] Breakthrough occurs much sooner for the sodium form of Amberlite IRC-718 despite the fact that the two resins demonstrated comparable removal efficiencies.

Ion exchange reactions are stoichiometric and reversible. A generalized form of an ion exchange reaction can be described as follows:

$$R\text{-}A^+ + B^+ \rightleftharpoons R\text{-}B^+ + A^+$$

where R is the resin, A^+ is the ion originally associated with the resin, and B^+ is the ion originally in solution.

The degree to which the exchange reaction proceeds is dependent on the preference, or selectivity, of the resin for the exchanged ion. The

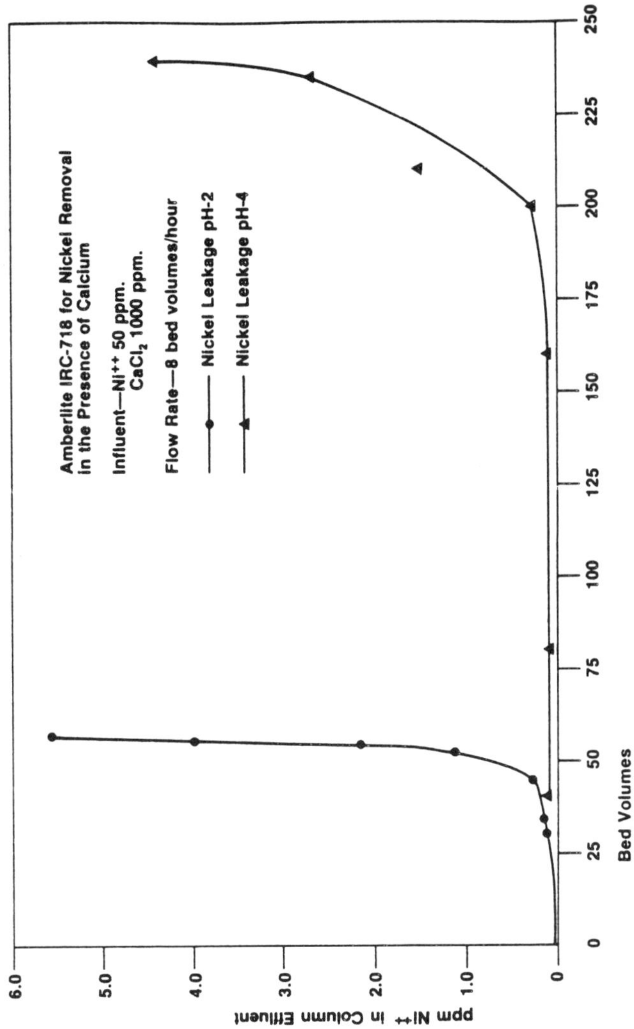

Figure 8.2.4. Breakthrough curve demonstrating variable pH performance of Amberlite IRC-718.

Source: Reference 5.

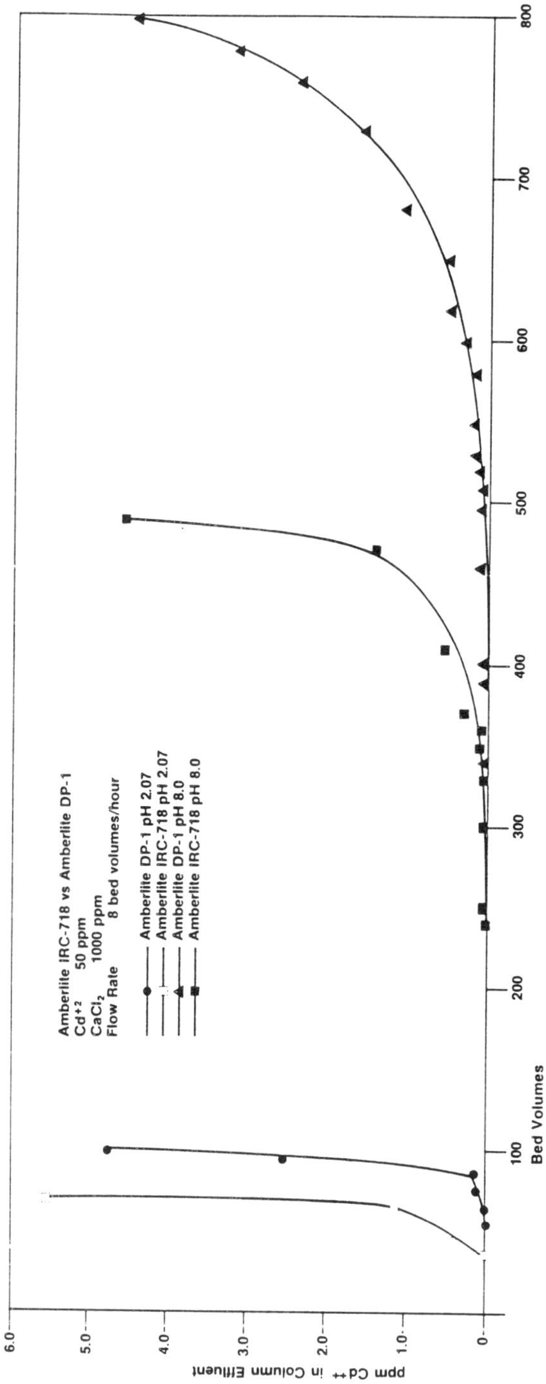

Figure 8.2.5. Breakthrough curves comparing pH variable performance of chelating (Amberlite IRC-718) and weak acid (Amberlite DP-1) resins.

Source: Reference 5.

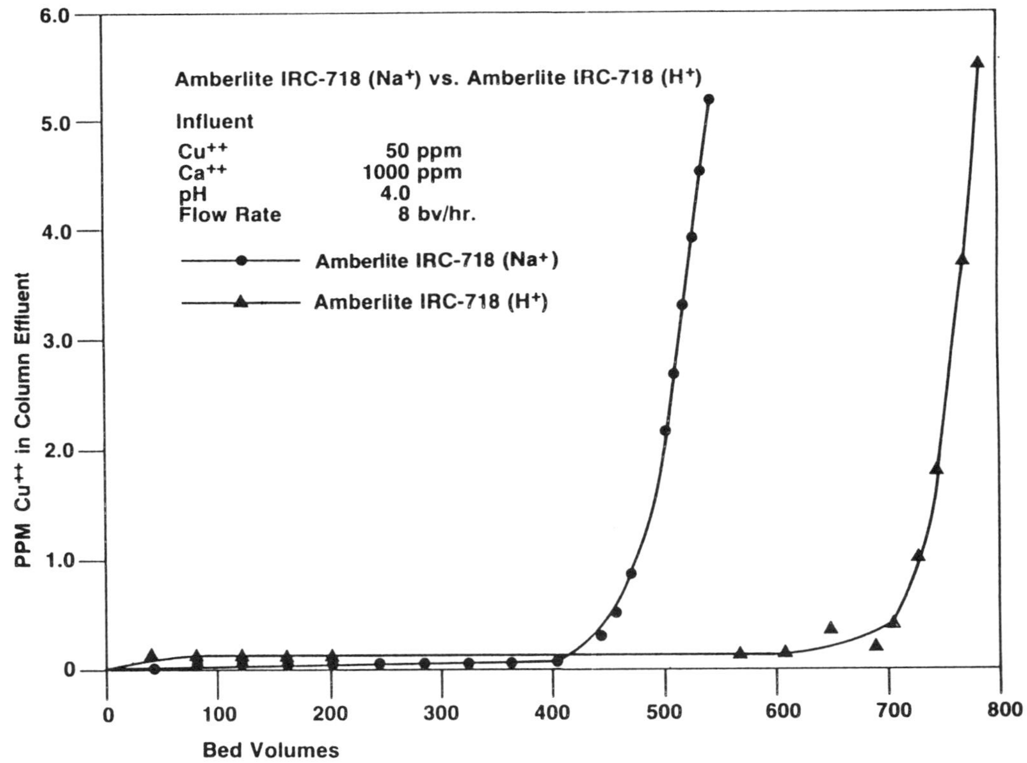

Figure 8.2.6. Comparison of chelating resin performance (Hydrogen form vs. Sodium form) using Amberlite IRC-718.

Source: Reference 5.

selectivity coefficient, K, expresses the relative distribution of ions when a charged resin is contacted with solutions of different, but similarly charged ions. For example, in the generalized ion exchange reaction presented above, the selectivity coefficient (K) is defined as follows:

$$K = \frac{[B^+] \text{ in resin}}{[A^+] \text{ in resin}} \times \frac{[A^+] \text{ in solution}}{[B^+] \text{ in solution}}$$

The selectivity coefficient of a resin will vary with changes in solution characteristics and the strength of the resin. Table 8.2.1 summarizes the selectivities of strong acid and strong base resins for various ionic species.[3,6] Resin selectivity is dependent upon ionic charge and size. The force with which an ion is attracted is proportional to its ionic charge and, therefore, the counter ion of higher valence is more strongly attracted into the exchanger.[3] The preference of exchange resins for counter ions of highest charge increases with dilution of the external electrolyte and is strongest with exchangers of high internal molarity.

With regard to ionic size, ions of smaller radius are preferentially adsorbed. When the resin is in a polar solvent, such as water, the fixed ions within the exchanger and mobile ions in both the resin and the solution tend to hydrate, causing the resin to swell. Hydration of the ions exerts a swelling pressure within the resin which is resisted by the cross-linked polymer matrix holding the resin particle together. As a result, the resin prefers the ion of smallest hydrated radius, since smaller ions can most readily enter the matrix of the resin and react with its functional groups. In general, multivalent hydrated ions are smaller in size than an equivalent charge unit of ions of lower valence and are therefore preferentially adsorbed. Within a given series of ions, the hydrated radius is generally inversely proportional to the unhydrated ionic radius.[7]

Another factor affecting resin selectivity is the interaction of ions within the exchanger and in bulk solution. The exchange resin prefers counter-ions which associate most strongly with the fixed ionic groups. If the groups are similar in structure to precipitating or complexing agents for a particular ion, the resin will prefer that ion. As a result of this phenomenon, many resins containing chelating functional groups show pronounced selectivities for transition group metal ions.[7]

TABLE 8.2.1. SELECTIVITIES OF ION EXCHANGE RESINS IN ORDER OF DECREASING PREFERENCES[a]

Strong acid cation exchanger	Strong base anion exchanger	Weak acid cation exchanger	Weak base anion exchanger	Weak acid chelate exchanger
Barium (+2)	Iodide (-1)	Hydrogen (+1)	Hydroxide (-1)	Copper (+2)
Lead (+2)	Nitrate (-1)	Copper (+2)	Sulfate (-2)	Iron (+2)
Mercury (+2)	Bisulfite (-1)	Cobalt (+2)	Chromate (-2)	Nickel (+2)
Copper (+1)	Chloride (-1)	Nickel (+2)	Phosphate (-2)	Lead (+2)
Calcium (+2)	Cyanide (-1)	Calcium (+2)	Chloride (-1)	Manganese (+2)
Nickel (+2)	Bicarbonate (-1)	Magnesium (+2)		Calcium (+2)
Cadmium (+2)	Hydroxide (-1)	Sodium (+2)		Magnesium (+2)
Copper (+2)	Fluoride (-1)			Sodium (+1)
Cobalt (+2)	Sulfate (-2)			
Zinc (+2)				
Cesium (+1)				
Iron (+2)				
Magnesium (+2)				
Potassium (+1)				
Manganese (+2)				
Ammonia (+1)				
Sodium (+1)				
Hydrogen (+1)				
Lithium (+1)				

[a]Valence number is given in parentheses.

Source: References 3 and 6.

The major disadvantage of a high degree of selectivity in an exchange reaction is the reluctance of the resin to release the ion during regeneration.[7] Figure 8.2.7 illustrates the elution curves for zinc from a chelating resin and a weak acid cation resin with a 10 percent HCl regenerant and a flow rate of 8 bed volumes/hour (1 gpm/ft^3).[5] The weak acid cation resin, Amberlite DP-1, gives a sharper elution curve, demonstrating the relative ease with which it is regenerated. Conversely, the highly specific chelating resin, Amberlite IRC-718, requires nearly twice as much regenerant. Although chelating resins clearly offer superior selectivity for metals removal, a weakly acidic cation exchange resin in the sodium form can sometimes exhibit equal or superior capacity and regeneration efficiency when treating heavy metal waste streams.[5]

Operating Parameters--

Operating parameters vary considerably depending on the particular application. The following factors will influence the selection of a resin type, pretreatment requirements, flow rates, cycle times, and the sizing of a system for a particular application:

- Types and concentrations of constituents present in the feed;
- rate of metal salt accumulation in the bath;
- flow rate; and
- number of hours of operation.

The types and concentrations of constituents present in the spent solution will determine the type of resin selected. Weak cation exchangers can be used for spent solutions containing low concentrations of metal ions. For solutions containing high concentrations, a strong anion exchanger may be preferred. The constituent concentrations and waste volume will also determine the resin volume needed to treat the stream. Commercially available systems are able to process wastes at throughput rates ranging from 38 to 6,700 liters/hour.[8] Cycle times for RFIE systems generally range from 5 to 15 minutes[8,9], whereas for cocurrent systems, they can be as much as 1 to 2 hours because of the time needed to regenerate the column.[3] As a result, dual sets of columns are typically used in cocurrent systems to avoid excessive downtime.

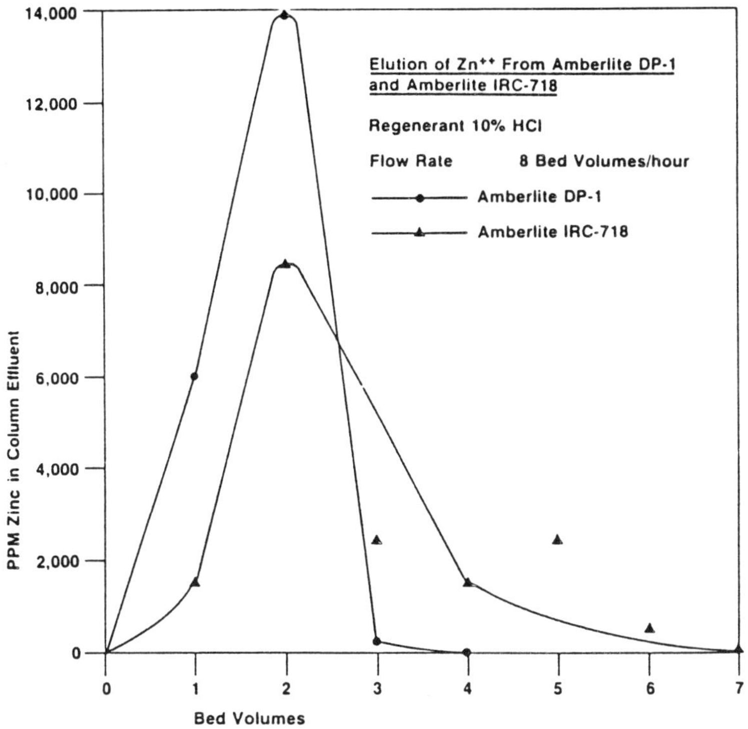

Figure 8.2.7. Regeneration performance of Amberlite DP-1 (weak acid resin) versus Amberlite IRC-718 (chelating resin).

Source: Reference 5.

Pretreatment Requirements--

Pretreatment of the waste stream (usually via filtration) is often necessary to remove many constituents which would otherwise adversely affect the resin. Certain organics (e.g., aromatics) become irreversibly sorbed by exchange resins, and oxidants, such as chromic or nitric acid, can damage the resin. Sodium metabisulfite, which converts hexavalent chrome to its trivalent state, can be added to the solution to prevent damage to the resin.[10] Eco-Tech claims that resin degradation is less of a problem with the RFIE process due to the short duration of contact (e.g., 1.5 min) between the acid and the resin.[12,13]

High concentrations of suspended solids, which can foul the resin bed, are typically pretreated through some form of filtration; e.g., activated carbon, deep bed, diatomaceous earth precoat, and resin filters. The filters eventually become clogged with particulates, and are replaced when overall cycle time increases to unacceptable levels due to excessive head loss.[7] For large volume systems which require frequent changing of filter cartridges, it may be more cost-effective to use a multimedia sand filter with a backwashing regeneration system. Although initial capital costs are higher, significant savings in filter replacement costs can be realized.[8]

The use of weak acid or base exchangers for treating wastes will require additional pretreatment. The exchange capacity of weak acid exchangers is generally limited below pH 6.0, and weak base exchangers are not effective above pH 7.[4] Therefore, a pH adjustment system must be incorporated prior to feeding the waste stream to the exchanger.

Ion exchange using cocurrent flow is not economically suitable for removal of high concentrations of exchangeable ions; i.e., above 2,500 mg/L, expressed as calcium carbonate equivalents. Above this level, the resin material is rapidly exhausted during the exchange process and regeneration becomes prohibitively expensive.[3,13] However, the reverse is true for acid purification units since they are capable of recycling the regenerate. In addition, higher acid concentrations in the waste feed solution will improve APU removal efficiencies.

Post-Treatment Requirements--

Overall savings in treatment and disposal costs can be realized through the use of ion exchange since, being a separation process, the total volume of wastes generated is reduced. Waste streams from ion exchange include spent regenerant solution, wash, and filtered solids. Cocurrent ion exchange generates an additional waste stream as a result of the need to backwash and expand the resin bed prior to regeneration (see Figure 8.2.1). Spent solutions from cocurrent operations are generally combined and managed through neutralization, precipitation, and disposal of the resulting metal sludge. Recovery of the regenerant or metals may not be economically justified since regeneration is conducted relatively infrequently in applications where this process is typically used; e.g., polishing treated effluents. Conversely, RFIE units generate more highly concentrated regenerant solutions which are more ammenable to metal recovery (e.g., electrolytic recovery) and regenerant reuse.

The only waste product generated from an APU is a moderately acidic metallic salt sludge. This also may be amenable to metal recovery techniques. The recovered acid stream is generally reused as make-up in the processing bath which was being treated.

For all units, filtrate from prefiltering systems can generally be land disposed without further treatment.[14] Otherwise, these can be managed through dewatering and solidification prior to landfilling. The quantity of sludge generated will depend on the types and concentrations of suspended solids present in the waste solution.

8.2.2 Process Performance

The performance of an ion exchange system will be predominantly influenced by the characteristics of the waste stream being treated including: types and concentrations of constituents present, acidity of the spent stream relative to that of the fresh stream, and required effluent quality. Factors which must be considered when evaluating system performance include: the quantity of residuals generated, cycle time, product concentration, process modifications required, attainable flow rate, system size, and overall processing costs.

A comparison of ion exchange systems is shown in Table 8.2.2.[14,15] Cocurrent flow units have the lowest capital costs but also the highest operating costs per unit of contaminant removal. RFIE units are generally more cost-effective than cocurrent fixed-bed systems for wastes with appreciable contaminant content (e.g., plating bath rinses). They use smaller resin volumes, minimizing capital costs and space requirements, have lower operating costs as a result of regenerant reuse, are capable of handling higher volume flows, and generate more consistent effluent quality. Examples of industrial applications for the various types of ion exchange units, including APUs, are discussed below.

Effluent Polishing: Plating Facility--

Polishing of effluents from conventional treatment systems using ion exchange has been applied successfully at a number of commercial installations. As an example, the Mogul Corporation designed a 2-stage, fixed bed polishing system for a client that could not meet effluent standards for Zn, Ni, Cu, and Cr.[4] The plating facility originally used sodium bisulfite chromium reduction and hydroxide precipitation to batch treat four segregated heavy metal plating waste effluents. Ion exchange was selected as the ideal choice to polish the combined waste discharge at this facility for several reasons. With a centralized treatment facility in place, no additional chemical destruction systems were needed to treat regenerant solutions. Similarly, no investment was needed for sophisticated pH control systems, flocculant feed systems, clarifiers and other process equipment. Finally, ion exchange units are compact and easy to automate and, therefore, were not difficult to incorporate into the existing system.

Rohm and Haas Company recommended using Amberlite XE-318 cation exchange resin for this application since, as a result of its strong chelating functionality, it is selective for removing transition metals in the presence of alkali or alkaline earth cations. Laboratory tests indicated that optimal removal was obtained by using the resin in both its hydrogen and sodium forms in a two-stage system. Since the selectivity of the resin is less for calcium ions than for sodium ions, lime was substituted for sodium hydroxide in the first-stage treatment. Resin column breakthrough tests were then performed to determine the quantity of resins needed to handle the 500 GPD volume of plating wastes.

TABLE 8.2.2. COMPARISON OF ION EXCHANGE OPERATING MODES

Criteria	Cocurrent fixed bed	Countercurrent fixed bed	Countercurrent continuous
Capacity for high feed flow and concentration	Least	Middle	Highest
Effluent quality	Fluctuates with bed exhaustion	High, minor fluctuations	High
Regenerant and rinse requirements	Highest	Somewhat less than concurrent	Least, yields concentrated regeneration waste
Equipment complexity	Simplest; can use manual operation	More complex; automatic controls for regeneration	Most complex; completely automated
Equipment for continuous operation	Multiple beds, single regeneration equipment	Multiple beds, single regenertion equipment	Provides continuous service
Relative costs (per unit volume:			
Investment	Least	Middle	Highest
Operating	Highest chemicals and labor; highest resin inventory	Less chemicals, water, and labor than concurrent	Least chemicals and labor; lowest resin inventory

Source: References 14 and 15.

Figure 8.2.8 shows a schematic of the upgraded treatment system. The primary system will continue to batch treat the segregated wastes which will then pass through the ion exchange system. Zinc pit wastes were judged to require pretreatment in a second H^+-form resin "roughing" column before entering the 2-stage columns (H^+-form followed by Na^+-form resin). This "roughing stage" column will require regeneration after each use. However the other wastewaters (Ni, Cr) only require treatment in the 2-stage column system. This system has the capacity to handle a full week's volume of wastewater before each column must be regenerated.

Resin regeneration procedures are briefly described as follows:

1. Backwash each column with city water for a minimum of 5 minutes at 60 gpm to reclassify the resin bed. Backwash water will dump into the chrome floor sump.

2. Regenerate the resins fully using 300 gallons of 10 percent sulfuric acid on the H^+-form columns, and a two-step regeneration of 100 gallons of 10 percent sulfuric acid followed by 50 gallons of a 5 percent solution for the Na^+-form column.

3. A final 5 minute rinse of 30 gpm city water for each column.

The effluent from the ion exchange system flows by gravity into a 15,000 gallon underground retention tank. A 24-hr composite sample of the effluent is collected daily by the sampling pump and an aliquot checked for compliance with the effluent limits. Table 8.2.3 compares the final effluent levels achieved by the upgraded treatment system with prior discharges and permissible legal limits.

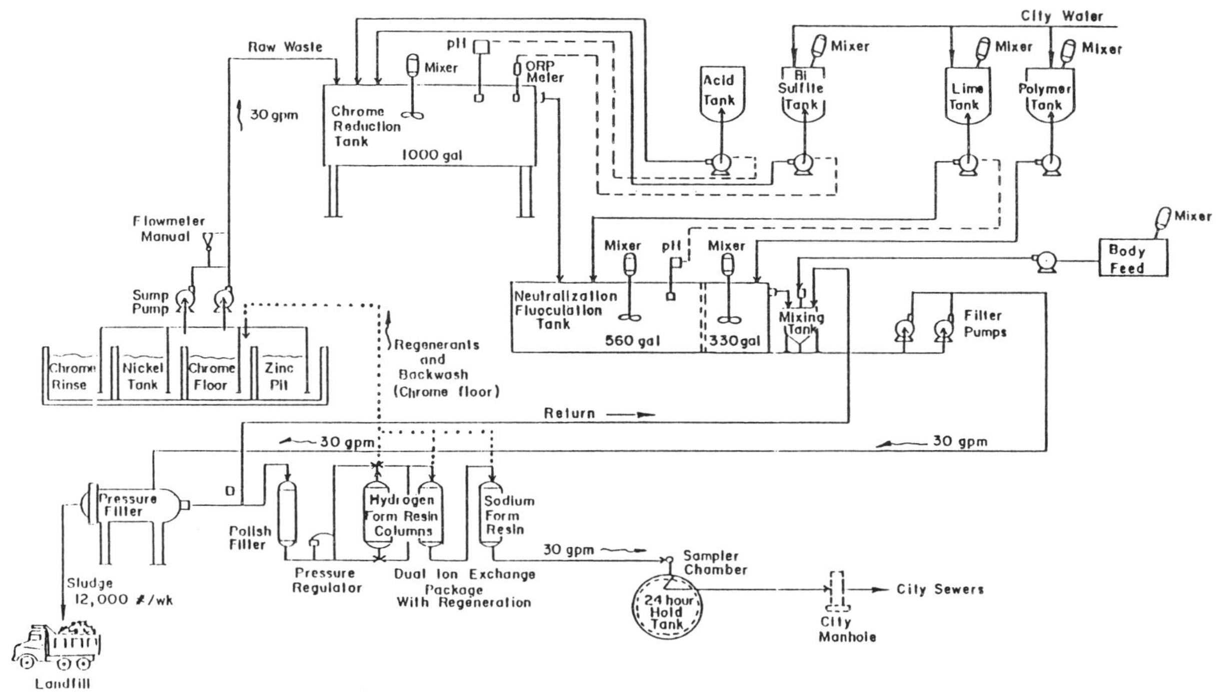

Figure 8.2.8. Schematic of the enhanced treatment system.

Reference: 4

TABLE 8.2.3. EFFLUENT QUALITY COMPARISON FOR UPGRADED ION EXCHANGE POLISHING

Parameter	State Limits (Avg.-Max.)	Effluent Range	
		Original System	With Ion Exchange Upgrade
pH	6.5 - 8.5	8.0 - 9.0	6.9 - 11.6
Color, pt-co	12	40 - 200	0
TSS, mg/L	15 - 20	15 - 150	<1.0
COD, mg/L	20 - 30	200 - 500	210 - 928
Cadmium, mg/L	0.01 - 0.01	<0.01	<0.005
Chromium (T), mg/L	0.05 - 0.10	0.02 - 1.8	<0.02
Copper, mg/L	0.05 - 0.10	0.1 - 20.0	<0.05
Iron, mg/L	0.50 - 1.0	0.2 - 0.3	<0.05
Lead, mg/L	0.05 - 0.10	<0.2	<0.05
Nickel, mg/L	0.10 - 0.30	2.0 - 4.5	<0.05
Zinc, mg/L	0.10 - 0.30	1.0 - 125.0	<0.02

Adapted from Reference No. 4.

The sorption filter design was slightly more expensive than the multi-media system, but would produce a better quality effluent. Since this system requires a proprietary media, it has a potentially high built-in uncontrollable operating cost tied to a single supplier. In addition, the solid product contains far more filter media than metal hydroxide.

The precoat filtration option would utilize three 150 gpm automatic diatomaceous earth precoat filters in place of the multi-media filters as described above. The sludge from this system would be mainly diatomaceous earth containing metal hydroxides.

The ultrafiltration design consisted of pretreatment followed by four 100 gpm trains in parallel, each consisting of a 1,500 gpm pump and 40 membrane modules. Ultrafiltration essentially replaces the multi-media filters with ultrafiltration units, while everything else remained virtually unchanged. However, this system required considerable pumping with its corresponding power costs and maintenance, and the membranes are susceptable to organic contamination. Ultrafiltration units are often used upstream of ion exchange systems to reduce the process load. In this case, however, the low inlet metal concentrations eliminated the need for both systems.

The sorption filtration design was a proprietary polishing system, yielding low metal effluent concentrations. It consisted of pH control, sodium sulfide addition, and filtration through a "sorption filter" precoated with diatomaceous earth. The advantage of this filter was that the media could be hydraulically pumped from the filters and reinstalled several times for each fresh charge, thereby reducing media consumption.

The proposed ion exchange system consisted of twin carbon towers for removal of trace organics, followed by twin sets of dual bed ion exchange columns manifolded for two-pass flow. The two-pass arrangement insured against breakthrough of poor quality water and capacity to handled variations in process load. Since the ion exchange regenerate solutions could average between 30 and 60 gpm, a second concentrate treatment system was required.

The ion exchange system was the least expensive to operate, but its capital cost was nearly twice that of the multi-media or sorption filter systems. However, since the exchange system recycles 80 to 90 percent of the processed water, it realizes significant savings in water/sewer charges. In addition, the recycled water is already warm and does not have to be heated from 40°F to 70°F, as does once-through water. Another advantage of the ion exchange system is that it concentrates the metals, thereby decreasing the size of the treatment system, and increasing the efficiency of the reducing agents. Concentrates, however, often require an additional stage of treatment to reach the low effluent levels mandated by environmental regulations.

Effluent Polishing: Chlor-Alkali Plant--

Akzo Chemicals Company of the Netherlands developed a process for the removal of mercury using Rohm and Haas Duolite GT-73, a weak acid cation exchange resin with a high degree of specificity for mercury.[17] The Duolite GT-73 resins utilize thiol (-SH) functional groups, which tend to form very strong bonds with ionic mercury.[17] The process, as installed at a chlor-alkali plant, involved the following steps:

- Oxidation--Since the resin reacts only with ionic mercury, metallic mercury must be converted to the ionic form. To accomplish this, an oxidation step is required, with solution pH maintained at 3 to prevent iron precipitation. To prevent clogging, metal hydroxides and unreacted mercury are filtered with sand or cloth filters.
- Dechlorination--The resin's thiol groups are readily oxidized, therefore, removal of chlorine is essential to retain resin activity. The Akzo process employs a two-stage dechlorination step. First, the stream is reacted with $NaHSO_3$, Na_2SO_3 or SO_2, and then it is passed through an activated carbon column.
- Ion Exchange--Two resin beds are used in series operating in a counter-current mode. One bed acts as a roughing stage and the second unit as a scavenger.

Figure 8.2.9 presents a schematic of the Akzo process for mercury removal in the treatment of chlor-alkali wastewater. It has been demonstrated to produce a mercury concentration well below 5 ppb at a flow rate of 1.25 gpm/ft^3. Figure 8.2.10 presents a typical breakthrough curve for chlor-alkali brine at pH 2 and a feed concentration of 20 to 50 mg/L mercury.

The performance of the Duolite GT-73 resin can be further demonstrated by its Freundlich isotherm, as shown in Figure 8.2.11. Figure 8.2.12 illustrates the capacity of the resin as a function of mercury concentration in the feed. Finally, Figure 8.2.13 presents a typical elution curve for Duolite GT-73, showing that it is readily regenerated using concentrated HCl.

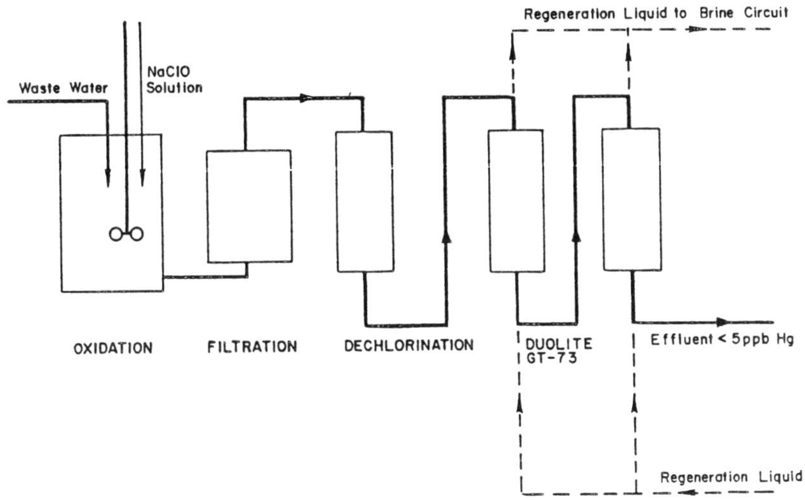

Figure 8.2.9. Akzo process for mercury removal.

Source: Reference No. 17.

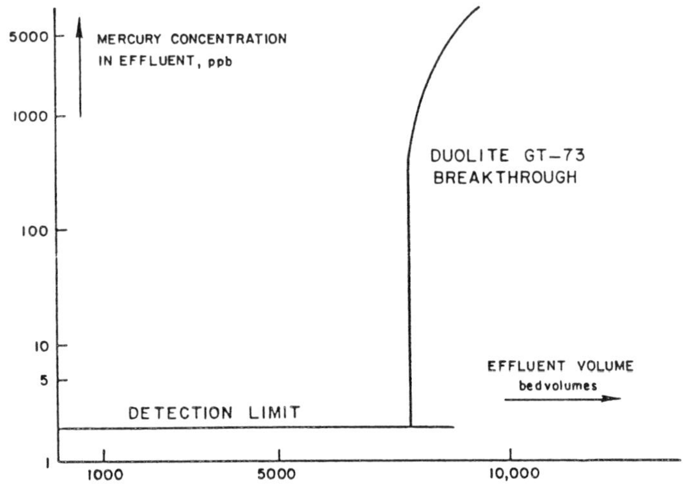

Figure 8.2.10. Typical breakthrough curve chlor-alkali plant brine, pH 2, mercury concentration in feed 20-50 mg/l.

Source: Reference No. 17.

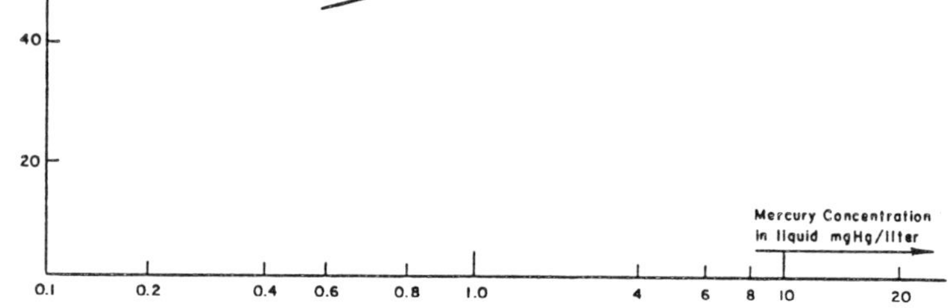

Figure 8.2.11. Freundlich Isotherm of Duolite GT-73.

Source: Reference No. 17.

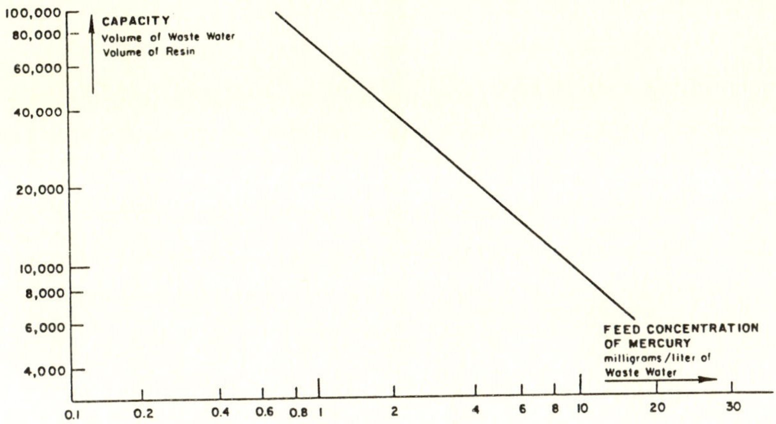

Figure 8.2.12. Capacity of Duolite GT-73 relative to the concentration of mercury in the feed.

Source: Reference No. 17.

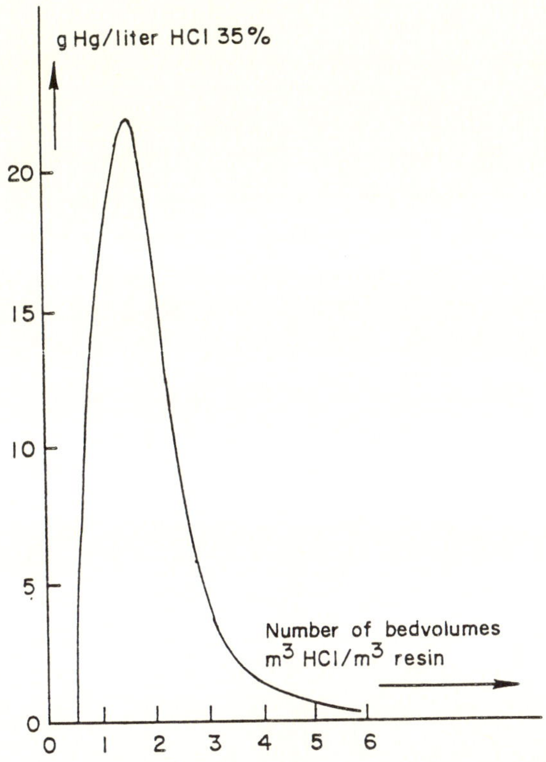

Figure 8.2.13. Regeneration of Duolite GT-73 with concentrated HCl. Regeneration rate 1 m^3 HCl/m^3 resin hour mercury concentration of resin prior to regeneration 35 g Hg/liter resin.

Source: Reference No. 17.

Effluent Polishing: Printed Circuit Board Manufacturer--

Honeywell Corporation upgraded its printed circuit board facility wastewater treatment plant to accomodate production increases and reduce water consumption.[16] The existing treatment system consisted of neutralization followed by automatic precoat pressure filtration with ion exchange as a rinse water recycling step. Several alternatives were considered to polish the effluent sufficiently to meet discharge levels while simultaneously reducing rinse water consumption from a projected 300 to 400 gpm to less than 75 gpm. Rinses which could be recirculated after treatment included those from combined rinses, alkaline etchants, ammonium persulfate deoxidizer, sulfuric acid and copper plate, and tin-lead and solder strip. Concentrates to be treated included the residuals created by rinse treatment, plating bath dumps, and spills.

Technologies considered for process enhancement included multimedia filtration, sorption filtration, automatic precoat filtration, ultrafiltration, and ion exchange. All process options would require 6,000 ft^2 of floor space, except the multi-media system which needed 5,000 ft^2. All of the systems met the effluent discharge limits of less than 1 ppm Cu and Ni and 0.5 ppm Pb. These alternatives are described below and their economics are summarized in Table 8.2.4.

The multi-media design would consist of two 400 gpm lined carbon steel pressure vessels charged with various filtration media. The rinse water was to be pH controlled and reducing agents added. The filter would require backwashing every 4 to 8 hours which would be pumped to a sludge conditioning tank. The final product would be a metal hydroxide sludge that contained little or no filter aide. The dewatering filter presses used for the concentrates would need to be increased in size since they also had to handle the main filter backwash.

The multi-media filtration system is the simplest, most compact and least capital intensive of the evaluated systems. Second only to the ion exchange system, it would have been the least expensive to operate. This type of filter can withstand higher feed metal concentrations than the other systems, and should produce an effluent between 0.4 and 1.0 ppm copper. The large amounts of backwash would require a large sludge conditioning tank and dewatering filter press. However, no filter aide is necessary.

TABLE 8.2.4. ECONOMIC COMPARISON - WASTEWATER POLISHING ALTERNATIVES*

	Ion exchange	Multi-media	Sorption filter	Precoat filter	Ultra-filter
Water consumption million gal/yr	10-20	108	108	108	108
Concentrate treatment	$ 150	$ 150	$ 150	$ 150	$ 150
Combined rinse treatment	1,102	200	402	650	630
Existing ion exchange	250	350	350	350	350
Control	100	75	75	75	75
Misc	250	250	100	100	250
Capital:	$1,852	$1,025	$1,077	$1,325	$1,455
Depreciation (10 years)	$ 185	$ 103	$ 108	$ 133	$ 146
Water ($1.87/k gal)	37	201	201	201	201
Heat water to 70°F[a]	0	378	378	378	378
Labor[b]	250	250	250	250	250
Electric ($0.10/kwh)	43	22	22	22	87
Media/resin[c]	220	68	261	162	98
Chemicals[d]	81	96	50	96	98
Sludge disposal[e]	10	10	60	120	10
Annual operating cost:	$ 826	$1,128	$1,330	$1,362	$1,268
$1,000 gallon	$ 7.65	$ 10.44	$ 12.32	$ 12.61	$ 11.74

[a]250 Btu/gal, $15/M Btu.
[b]Supervisor, foreman, and three operators.
[c]Resin life of 3 years, regenerating every day.
[d]NaBH4 at $1,500/drum.
[e]$150/ton.
*All values in $1,000.
Source: Reference No. 16.

Acid Purification Unit: Pickling Liquor Recovery--

Acid purification systems using RFIE have been commercially demonstrated to be effective in the recovery of acids from aluminum anodizing solutions, acid pickling liquors, and rack-stripping solutions.[1] Acid purification systems are the most effective form of ion exchange for recovering acids which have high concentrations of metal ion contaminants.

An APU was installed at the Continuous Colour Coat, Ltd. plant in Rexdale, Ontario, to recover sulfuric acid from a steel pickling process.[18] As an alternative to neutralization and disposal of spent baths, the plant employed an APU to remove iron build-up so that the solution could be recycled.

At a flow rate of 19 gal/min and a temperature of 119 F, iron concentration in the reclaimed acid was reduced by 80 percent and acidity losses were minimal.[8] Occasional replenishment of the bath was necessary, but draining of the tank (an expensive process) was no longer required. Also, improvements in product quality were noted due to more uniform bath consistency. Savings were realized in reduced neutralization and disposal requirements and net reductions in labor requirements for bath maintenance. An economic evaluation of the system (see Table 8.2.5) showed an estimated payback period for the unit of less than 2 years.[11]

Acid Purification Unit: Aluminum Anodizing Solution Recovery--

Another common application in which the APU works effectively is the recovery of acids from aluminum anodizing solutions. An APU system was installed at the Modine Manufacturing Company in Racine, Wisconsin to recover nitric acid from an aluminum etching process.[19] The APU was connected directly on-line, which allowed for continuous process operation. It generated a more concentrated solution of recovered nitric acid than it was fed; thus a slightly lesser volume of acid was returned to the tank.

Table 8.2.6 presents a summary of the results and operating parameters of the APU at the Modine plant. Improvements in product quality and savings in neutralization, disposal, and fresh acid makeup were noted by Modine personnel.[19] An economic evaluation of this system (Table 8.2.7) shows nitric acid recovery with an APU to be very cost-effective.

TABLE 8.2.5. ECONOMIC EVALUATION OF THE APU INSTALLED AT CONTINUOUS COLOUR COAT, LTD.

Item	Cost
CAPITAL COSTS[a] (includes costs for equipment & installation)	$100,000
OPERATING COSTS	
Resin Replacement (every 4 years at $58/liter)	$3,770/year
Utilities (0.5 KW x 16 hrs/day x 250 days/yr x $0.055/KWH)	$ 110/year
Taxes and Insurance (1% of TIC)	$1,000/year
TOTAL OPERATING COSTS	$4,880/year
COST SAVINGS	
Reduction in Acid Purchase @ $92.40/ton	$25,875/year
Reduction in Neutralization Costs (Lime) @ $80/ton	$18,000/year
Reduction in Sludge Disposal Costs	$20,000/year
TOTAL COST SAVINGS	$63,875/year
NET COST SAVINGS (Gross Savings - Operating Costs)	$58,995/year
PAYBACK PERIOD	1.7 years

[a]Capital Equipment included APU Model No. AP30-24, multimedia sand filter, water supply tank, and piping.

Source: Reference 9 (Ecotech cost quote, August 1986).

TABLE 8.2.6. TYPICAL OPERATING PARAMETERS AND RESULTS DURING TESTING OF THE APU FOR RECOVERY OF NITRIC ACID AT MODINE MANUFACTURING COMPANY IN RACINE, WISCONSIN

Parameter	Result
Feed to APU from etch tank	6.2 N
Product returned to etch tank	6.5 N
By-product going to waste treatment	0.6 N
Level of aluminum contamination:	
Coming into APU from etch tank	793 mg/L
Returning to the etch tank	231 mg/L
Average cycle time	12.7 min
Volume of water removed from etch tank/APU cycle	0.89 gal
Mass balance	
Equivalents of nitric acid into APU from etch tank	251
Equivalents returned to etch tank and waste	257

Source: References 19 and 20.

TABLE 8.2.7. ECONOMIC EVALUATION OF THE APU INSTALLED AT MODINE MANUFACTURING COMPANY IN RACINE, WISCONSIN FOR THE RECOVERY OF NITRIC ACID

Item	Cost
CAPITAL COSTS (includes costs for equipment and installation)	$37,234
COST SAVINGS	
Reduction in nitric acid purchase	$20,064/year
Reduction in neutralization costs	$ 6,276/year
Reduction in disposal costs	$ 7,236/year
Reduction in labor	$ 2,400/year
TOTAL COST SAVINGS	$35,976/year
OPERATING COSTS	
Resin replacement (every 4 years at $58/liter)	$ 1,305/year
Utilities (0.5 KW x 16 hrs/day x 300 days/yr x $0.055/KWH)	$ 132/year
Taxes and insurance (1% of TIC)	$ 372/year
TOTAL OPERATING COSTS	$ 1,809/year
NET COST SAVINGS (Gross savings - Operating costs)	$34,167/year
PAYBACK PERIOD	1.1 years

Source: References 8 and 19 using August 1986 cost data.

Another full-scale APU demonstration of aluminum anodizing solution recovery was performed at Springfield Machine and Stamping, Inc. of Warren, Michigan.[20] Typical operating parameters and results during the 6 month testing period are summarized in Table 8.2.8. The system proved to be cost-effective for recovery of the sulfuric acid solution due to high aluminum removal efficiency, retention of acid strength, and reductions in raw material purchase, disposal and labor costs.

TABLE 8.2.8. TYPICAL OPERATING PARAMETERS AND RESULTS FOR THE APU INSTALLED AT SPRINGFIELD MACHINE & STAMPING, INC. IN WARREN, MICHIGAN FOR SULFURIC ACID RECOVERY

Parameter	Feed	Product	Byproduct
Flow rate (liters/hr)	298	296	175
Sulfuric acid concentration (g/L)	183.8	175.0	13.0
Aluminum concentration (g/L)	12.2	4.2	12.0

Source: Reference 21.

Acid Purification Unit: Electroplating Pickle Liquor Recovery--

A pilot-scale unit for recovering hydrochloric acid from an electroplating pickling liquor was tested at Electroplating Engineering, Inc. in St. Paul, Minnesota.[10] The results were not as successful as with the cases presented previously since reduced metal removal efficiencies and high acidity losses were experienced.

Primary contaminants in the spent solution included iron (1,650 mg/l), zinc (4,283 mg/l), and nickel, copper, and chromium in the ppm range. A different system configuration was required for this application because zinc was present in the form of chloride complexes. As described in Section 8.2.1, the resin used in an APU shows a preferential affinity for acid anions as opposed to metal cations, which causes metals to pass through the resin while the acid is retained. However, instead of passing through the resin, zinc chloride complexes are also retained by the anion resin.

In order to remove both the zinc chloride complex and the iron contaminants, it was necessary to operate the system in two stages. Initially, the spent solution was passed through one resin to remove zinc. This is termed the inverse mode since the acid ions are not retarded. Then the solution is passed over a second resin in the normal mode of operation, retaining the acid while allowing iron ions to pass through the resin. As with typical APU processes, the acid is recovered during the regeneration cycle. Figures 8.2.14 and 8.2.15 illustrate these two modes of operation.

Three different HCl pickling liquors were used to test the performance of the APU. Analysis of these spent solutions yielded the following:

Parameter	Range of Concentrations
Acidity ($CaCO_3$ equivalents)	77,000 - 284,000 mg/L
Zinc content	640 - 52,000 mg/L
Iron content	1,100 - 7,000 mg/L

Several test runs were performed using the two-stage APU system, as summarized in Table 8.2.9. The results show that good zinc removal efficiencies (99.3 percent) were achieved during the inverse mode of operation with minimal losses in acidity (3.5 percent). However, during the normal mode of operation, an average of only 60 percent iron removal was achieved and acidity losses were high (averaging 38 percent). The results showed that increased iron removal could only be achieved at the expense of greater reductions in acidity. It was determined that the ratio of iron to acidity in the feed had to approach 1:15 in order to achieve effective performance. The iron to acidity ratio for the feed used during these tests was 1:67, which contributed to the poor performance results.[10]

It is possible that the intermediate byproduct solution generated after the inverse mode may be of sufficient quality to be returned to the pickling bath.[10] Iron content of the byproduct solution was comparable to the iron concentrations measured in the bath during its intermediate solution stage. Additional testing would be required in order to determine whether bath quality would be acceptable under these conditions. Based on the results of these pilot-scale tests, it was determined that the APU could only be cost-effective if a large volume of spent solution is processed.

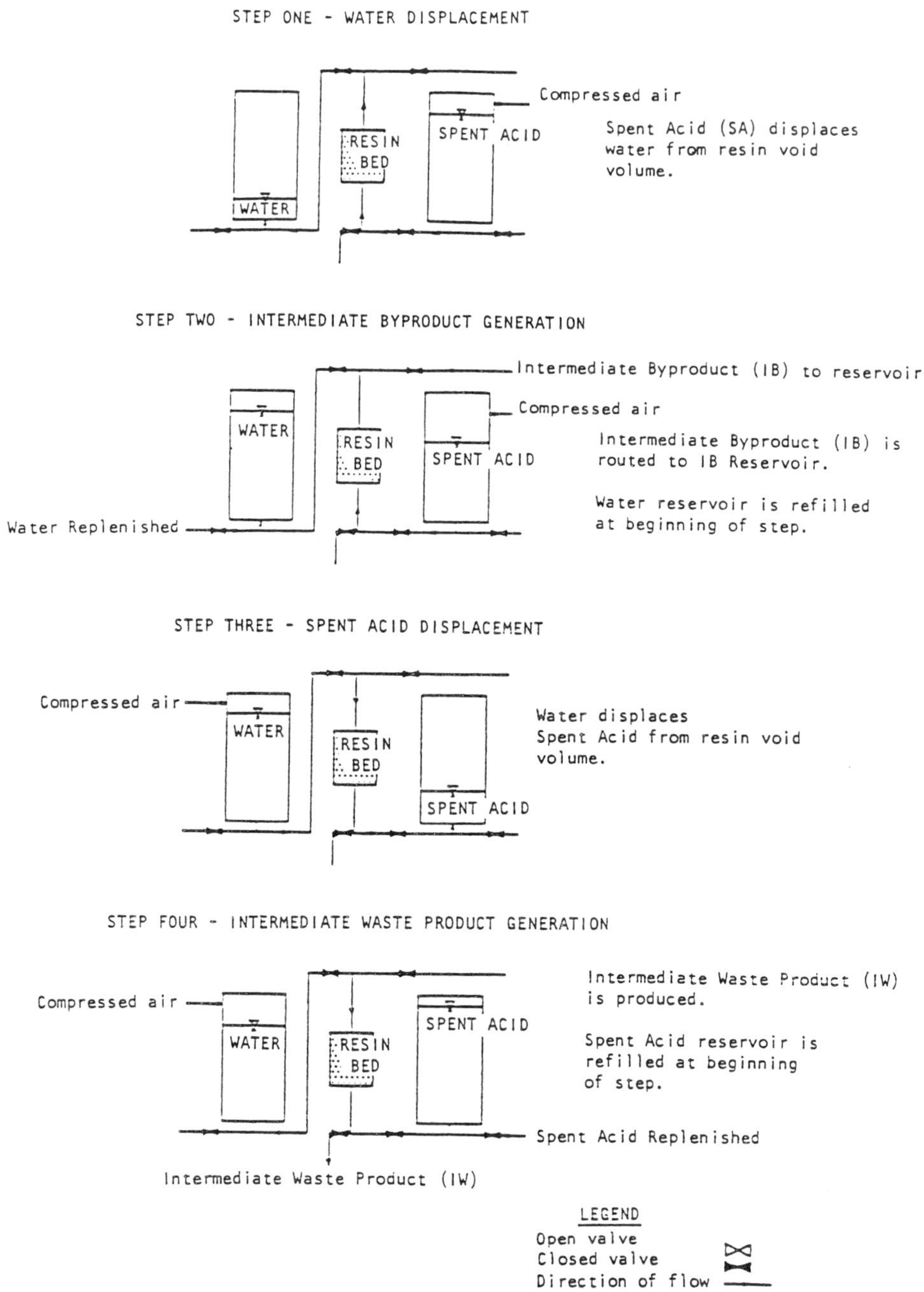

Figure 8.2.14. Schematic of inverse mode of operation.

Source: Reference 10.

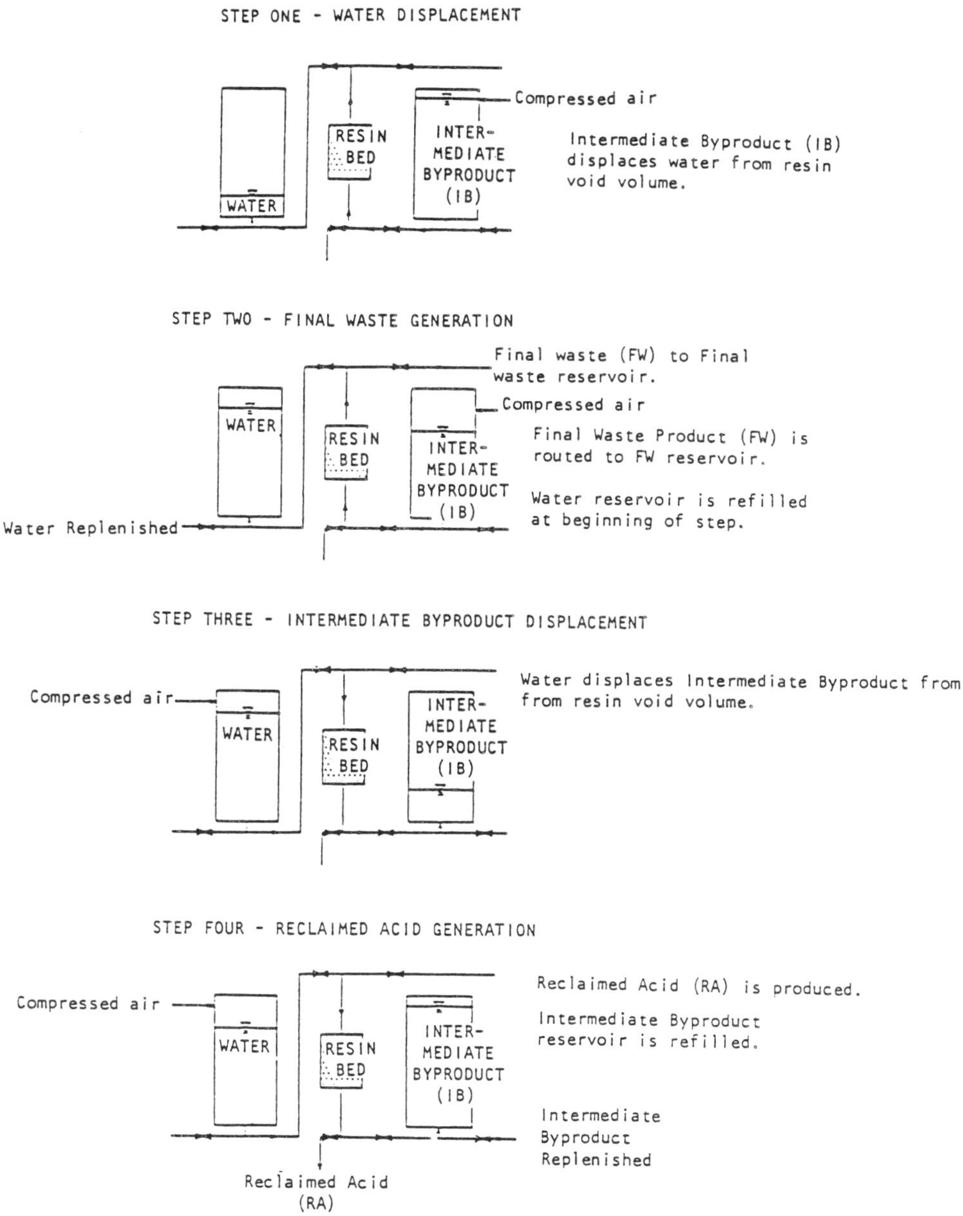

Figure 8.2.15. Schematic of normal mode of operation.

Source: Reference 10.

TABLE 8.2.9. SUMMARY OF APU RESULTS ON HCL PICKLING LIQUOR RECOVERY PERFORMED AT ELECTROPLATING ENGINEERING, INC.[a]

Parameter	Preliminary runs (No.2,3,4)[b]	Preliminary runs (No.5A,5B,6)[c]	Final runs (No. 7A-7P
VOLUME TREATED/GENERATED (LITERS):			
Spent acid	98	174	189
Intermediate by-product	86	174	189
Reclaimed acid	70	162	4.6
Intermediate waste product	64	96	106
Final waste product	67	121	3.3
INVERSE MODE LOADINGS TO RESIN:			
Zinc (grams/cycle)	34.2	4.4	2.6
Volume (bed volumes/cycle)	4.05	4.05	4.05
Feed rate (liters/hour)	11.4	11.4	11.4
NORMAL MODE LOADINGS TO RESIN:			
Acidity (grams $CaCO_3$/cycle)	43.4	35.2	14.6
Volume (bed volumes/cycle)	0.40	0.40	0.34
Feed rate (liters/hour)	6.0	6.0	5.5
STREAM CONCENTRATIONS OF ACIDITY: (expressed as g/L $CaCO_3$ equivalents)			
Spent acid	242	156	77
Intermediate by-product	217	151	74
Reclaimed acid	116	94	46
Intermediate waste product	39	15	5.9
Final waste product	90	57	30
STREAM CONCENTRATIONS OF ZINC (mg/L):			
Spent acid	40,333	1,100	640
Intermediate by-product	34,667	8	1.4
Reclaimed acid	19,233	18	5.47
Intermediate waste product	13,000	2,000	1,200
Final waste product	7,366	0.61	0.25

(continued)

TABLE 8.2.9 (continued)

Parameter	Preliminary runs (No.2,3,4)[b]	Preliminary runs (No.5A,5B,6)[c]	Final runs (No. 7A-7P)[d]
STREAM CONCENTRATIONS OF IRON (mg/L):			
Spent acid	4,700	2,600	1,100
Intermediate by-product	4,400	2,433	1,100
Reclaimed acid	1,450	920	439
Intermediate waste product	1,277	357	120
Final waste product	3,367	1,733	728
STREAM CONCENTRATIONS OF CHROMIUM (mg/L):			
Spent acid	43	2.7	NA
Intermediate by-product	42	2.6	NA
Reclaimed Acid	12	1.6	NA
Intermediate waste product	4.6	0.51	NA
Final waste product	28	3.4	NA

[a]The results of Run 1 were discarded due to improper installation.

[b]The zinc loadings for Runs 2, 3, and 4 were above the recommended 18 grams/cycle maximum loading recommended for the system.

[c]The objective for Runs 5A, 5B, and 6 were to process a sufficient quantity of acid for reuse and to optimize loadings.

[d]The objective for Runs 7A through 7P were to optimize loadings for the normal mode of operation.

Source: Reference No. 10.

In summary, available performance data suggest that the technical and economic feasibility of acid purification systems will mainly depend on the types and concentrations of metal ions present. These systems work well in recovering solutions with highly positively charged contaminant ions (e.g., aluminum, iron) because these ions pass rapidly through the strong base anion exchanger resin. Solutions containing low concentrations of contaminant ions are not efficiently recovered using the APU. Recommended minimum concentrations for efficient results are presented in Table 8.2.10. Although lower concentrations may be treated, removal efficiencies will be low unless larger systems are employed, compromising cost-effectiveness due to increased capital costs. A summary of demonstrated applications is provided in Table 8.2.11.

8.2.3 Process Costs

An economic evaluation of countercurrent (RFIE) systems is presented in this section since this is the only ion exchange system which is directly applicable to the treatmnet of California List metal/cyanide wastes. Cocurrent flow methods will only be technically and economically feasible for the treatment of California List wastes which have been diluted by mixing with spent rinse water.

Factors that affect the costs of RFIE units include: quantity and quality of constituents recovered, production rates, volume of spent solution to be treated, concentration of metal salts present in the spent solution, rate of build-up of metal ions in the bath, concentration of the bath, and number of hours of process operation.

Capital costs, which include equipment, installation, and peripheral costs, increase with system size. These costs are offset by savings which are realized through reduced volumes of wastes requiring post-treatment (e.g., neutralization) and disposal, and reduced purchase requirements for bath reagents. Operating costs will include replacement of filter cartridges, resin replacement (approximately every 5 years), and utilities.

TABLE 8.2.10. RECOMMENDED MINIMUM CONCENTRATIONS (g/L) FOR EFFICIENT METALS REMOVAL USING THE ECO-TECH APU

Solution	Iron	Zinc	Aluminum	Copper	Total metals
Hydrochloric acid	30-50	130-150	-	-	-
Sulfuric acid	30-50	-	5	20	-
Nitric/hydrofluoric acid	-	-	-	-	30
Nitric acid rack stripping	-	-	-	-	75-100

Note: The APU can be used for solutions with lower concentrations of these metals, but the metal removal efficiencies will be lower unless a larger unit is used. Metal removal efficiencies average 55% for typical systems.

Source: Reference No. 14.

TABLE 8.2.11. DEMONSTRATED APPLICATIONS OF ECO-TECH ACID PURIFICATION UNIT USING RFIE

Application/ bath components	Typical bath concentration (g/L)	Typical product concentration (g/L)	Typical by-product concentration (g/L)
Sulfuric acid	190	182	13
Aluminum	10	5.5	6
Sulfuric acid	127	116	10
Iron	36	10.5	21
Nitric acid	514	581	10
Nickel and copper	99	47.5	70.8
Sulfuric acid	128	113	18
Hydrogen peroxide	41	35	7
Copper	13.3	5.9	9.2
Hydrochloric acid	146	146	10
Iron	34	25	15
Nitric acid	150	139	4.5
Hydrofluoric acid	36	28.8	7.2
Iron	29	8.7	20.3
Nickel	7.02	2.1	4.9
Chromium	7.33	2.2	5.1
Sulfuric acid	61.3	54.9	5.88
Sodium	7.8	0.8	5.56

Source: Reference No. 11 (Based on July 1986 Ecotech cost data).

Capital costs for acid purification systems typically range from $15,000 to $180,000 depending on the throughput, as shown in Table 8.2.12. These costs include installation, equipment and peripherals, and a prefilter system. Capital costs presented in this table are for the recovery of sulfuric acid from aluminum anodizing solutions and may be slightly higher for other applications[22].

Typical operating costs are presented in Table 8.2.13. Finally, Table 8.2.14 presents an economic evaluation of several hypothetical systems.

8.2.4 Process Status

Cocurrent ion exchange systems are generally not employed for direct treatment of concentrated metal wastes. Cocurrent systems using weak exchangers have inefficient exchange capacities for these wastes and are generally only used as polishing systems following other treatment operations. Cocurrent systems using strong exchangers are technically feasible for the treatment of metal-containing rinses and other wastes, but they are not typically cost-effective because of the high costs for column regeneration.

Ion exchange systems, using the reverse or reciprocating flow mode (countercurrent), have been shown to be effective in the treatment of metal wastes. The process has been demonstrated commercially for chemical recovery from acid copper, acid zinc, nickel, cobalt, tin, and chromium plating baths, as well as for purification of spent acid solutions (i.e., the APU).

Chemical recovery systems using fixed bed RFIE have been used to recover chromic acid and metal salts. It has also been used to deionize mixed-metal rinse solutions for recovering process water and concentrating the metals for subsequent treatment.[3] Commercial units are available from several vendors.

Acid purification systems using continuous RFIE have been used to remove aluminum salts from sulfuric acid anodizing solutions, to remove metals from nitric and rack-stripping solutions, and to remove metals from sulfuric and hydrochloric acid pickling solutions.[1] The APU is primarily used for recovering aluminum anodizing solutions.[3] Acid purification systems are more cost-effective for removing high concentrations of contaminants than

TABLE 8.2.12. TYPICAL CAPITAL COSTS FOR ECO-TECH APU

Item	Small unit	Medium unit	Medium unit	Large unit
APU Model No.	AP-6	AP-24	AP-54	AP-72
Flow rate	38 L/hr	500 L/hr	800 L/hr	6700 L/hr
Capital cost	$14,000	$37,000	$116,000	$184,000

Notes: Capital Costs include equipment, installation, peripherals, and cartridge-type prefilter system.

Costs presented in this table are for application to recovery of a sulfuric acid anodizing solution. Costs for other applications may be slightly higher.

Twelve different size units are available from Eco-Tech, Ltd. The model numbers, which indicate bed diameters, for these units are: AP-6, AP-12, AP-18, AP-24, AP-30, AP-36, AP-42, AP-48, AP-54, AP-60, AP-66, and AP-72.

Source: Reference No. 22 (Ecotech quote July 1986).

TABLE 8.2.13. TYPICAL OPERATING COSTS FOR ACID PURIFICATION USING CONTINUOUS COUNTERCURRENT ION EXCHANGE (RFIE)

Item	Cost
Filter cartridges for prefilter system	\$10.00/month
Utilities:	
(0.5 KW x 16 hrs/day x 20 days/month x 0.055 \$/KWH)	\$8.80/month
Resin replacement (specific cost depends on system size)	\$58/liter every 4 years

Source: References 8 and 10 (Based on August 1986 cost data).

TABLE 8.2.14. ECONOMIC EVALUATION OF ACID PURIFICATION PROCESS

Description	30,000 gpy throughput	100,000 gpy throughput	500,000 gpv throughput
Case 1 - Purification of Sulfuric Acid Anodizing Solution: Previous approach used caustic acid neutralization: New approach uses APU with caustic neutralization.			
Approx. APU Cost	$ 6,000	$11,000	$ 25,000
Previous treatment cost	$ 9,690	$32,300	$161,500
Previous acid cost	$ 2,349	$ 7,830	$ 39,150
Annual savings	$ 8,427	$28,891	$140,455
Payback (months)	9	5	2
Case 2 - Purification of Sulfuric Acid Anodizing Solution: Previous approach used lime neutralization: New approach uses APU with lime neutralization.			
Approx. APU cost	$ 6,000	$11,000	$ 25,000
Previous treatment cost	$ 2,250	$ 8,500	$ 38,500
Previous acid cost	$ 2,349	$ 7,830	$ 39,150
Annual savings	$ 3,216	$10,731	$ 53,655
Payback (months)	22	12	6
Case 3 - Purification of Sulfuric Acid Anodizing Solution: Previous approach used waste haulage: New approach uses APU with caustic neutralization.			
Approx. APU cost	$ 6,000	$11,000	$ 25,000
Previous treatment cost	$ 3,000	$10,000	$ 50,000
Previous acid cost	$ 2,349	$ 7,830	$ 39,150
Present treatment cost	$ 2,907	$ 9,690	$ 48,450
Annual savings	$ 1,737	$ 5,791	$ 28,955
Payback (months)	41	23	10
Case 4 - Purification of Sulfuric Acid Anodizing Solution: Previous approach used waste haulage: New approach uses APU with lime neutralization.			
Approx. APU cost	$6,000	$11,000	$ 25,000
Previous treatment cost	$3,000	$10,000	$ 50,000
Previous acid cost	$2,349	$ 7,830	$ 39,150
Present treatment cost	$ 675	$ 2,250	$ 11,250
Annual savings	$3,969	$13,245	$ 66,155
Payback (months)	18	10	5
Case 5 - Nitric Acid Recovery: Previous approach used caustic neutralization: New approach uses APU with caustic neutralization.			
Approx. APU cost	$9,400	$11,300	$ 18,400
Previous treatment cost	$7,575	$30,300	$ 50,500
Previous acid cost	$8,775	$35,100	$ 58,500
Total previous cost	$16,350	$65,400	$109,000
Annual savings	$ 9,810	$39,240	$ 65,400
Payback (months)	11	3	3

Source: References 11, 12, 20, and 21. (Based on July 1986 Ecotech cost data).

other ion exchange systems. Although the use of ion exchange for acid purification is currently under investigation by several ion exchange vendors (e.g., Alpha Process Systems; Illinois Water Treatment Company; Ionics, Inc.; etc.), Eco-Tech, Ltd. is the only vendor with commercial units currently in operation.[10,22,24]

REFERENCES

1. Wilk, Lisa et al. Alliance Technologies Inc. Technical Resource Document: Treatment Technologies for Corrosive-Containing Wastes. Prepared for U.S. EPA HWERL, Cincinnati, OH. under Contract No. 68-02-3997. October, 1986.

2. GCA Technology. Industrial Waste Management Alternatives And Their Associated Technologies/Processes. Prepared for the Illinois Environmental Protection Agency, Division of Land Pollution Control, Springfield, Illinois. GCA Contract No. 2-053-011 and 2-053-012. GCA-TR-80-80-G. February 1981.

3. U.S. EPA, Industrial Environmental Research Laboratory, Cincinnati, Ohio. Summary Report: Control and Treatment Technology for the Metal Finishing Industry - Ion Exchange. EPA-625-8-81-007. June 1981.

4. Yeats, A.R. Ion Exchange Selectively Removes Heavy Metals From Mixed Plating Wastes. In: Ind. Waste Cont. Proc. 32, 467-76. 1978.

5. Waitz, Jr., W.H., Rohm & Haas Company. Ion Exchange in Heavy Metals Removal and Recovery. Amber-hi-lites No. 162. 1979.

6. U.S. EPA, Industrial Environmental Research Laboratory, Cincinnati, Ohio. Sources and Treatment of Wastewater in the Nonferrous Metals Industry. EPA-600/2-80-074. April 1980.

7. Weber, W.J. Physicochemical Processes for Water Quality Control. John Wiley & Sons. 1972.

8. Fontana, C., Eco-Tech, Ltd. Telephone Conversation with L. Wilk, GCA Technology Division, Inc. Re: Acid Purification Unit. August 21, 1986.

9. Dejak, M. Acid Recovery Proves Viable in Steel Pickling. Finishing, 10(1): 24-27. January 1986.

10. Pace Laboratories, Inc. Final Report: Reclamation and Reuse of Spent Hydrochloric Acid, Hazardous Waste Reduction Grant. Prepared for the Minnesota Waste Management Board on behalf of Electro-Plating Engineering Company, Inc. February 14, 1986.

11. Eco-Tech, Ltd. Product Literature: Acid Purification Unit (APU). Bulletin No. ET-4-84-5M. Received July 1986.

12. Eco-Tech, Ltd. Product Literature: Ion Exchange Systems. Bulletin No. ET-11-83-3M. Received July 1986.

13. GCA Technology. Corrective Measures for Releases to Ground Water from Solid Waste Management Units. Prepared for U.S. EPA-OSW Land Disposal Branch, under EPA Contract No. 68-01-6871, Work Assignment No. 51. GCA-TR-85-69-G. August 1985.

14. Fontana, C., Eco-Tech, Ltd. Telephone Conversation with L. Wilk, GCA Technology Division, Inc. Re: Acid Purification Unit. August 26, 1986.

15. U.S. EPA, Office of Research and Development, Washington, D.C. Treatability Manual, Volume III: Technologies for Control/Removal of Pollutants. EPA-600-8-80-042c. July 1980.

16. Van Dyke, Jr., B.H., Conoby, J.F., and C. Alderuccio. Innovative Hazardous Waste Stream Reduction Alternatives. In: Proc. of the Third Annual Hazardous Materials Management Conference. June 1985.

17. Rohm & Haas Company. Duolite GT-73 Ion Exchange Resin Product Bulletin. August 1986.

18. Chemical Processing Staff. Spotlight: Pickling Acid Recovery Unit Saves $40,000/year, Purifies Spent Sulfuric Acid. Chemical Processing, 49(3): 36-38. March 1986.

19. Robertson, W.M., James, C.E., and J.Y.C. Huang. Recovery and Reuse of Waste Nitric Acid From An Aluminum Etch Process. In: Proceedings of the 35th Industrial Waste Conference at Purdue University. May 13-15, 1980.

20. Brown, C.J., Davy, D., and P.J. Simmons. Recovery of Nitric Acid from Solutions Used for Treating Metal Surfaces. Plating and Surface Finishing. February 1980.

21. Brown, C.J., Davy, D., and P.J. Simmons. Purification of Sulfuric Acid Anodizing Solutions. Plating and Surface Finishing. January 1979.

22. Fontana, C., Eco-Tech, Ltd. Telephone Conversation with L. Wilk, GCA Technology Division, Inc. Re: Acid Purification Unit. July 7, 1986.

23. Parcy, E., Ionics, Inc. Telephone conversation with J. Spielman, GCA Technology, Inc. August 14, 1986.

24. Jain, S.M., Ionics, Inc. Telephone conversation with J. Spielman, GCA Technology Division, Inc. August 12, 1986.

8.3 DEVOE-HOLBEIN TECHNOLOGY

8.3.1 Process Description

DeVoe-Holbein Technology uses coordinating compounds covalently bonded to the surface of an inert carrier material to capture metal ions. In waste treatment applications, the reactants are used in equipment similar to that employed for ion exchange resins.

The technology was originally developed by DeVoe-Holbein as an adaptation of biological mechanisms in which living cells selectively extract a variety of metal nutrients (e.g., Na, K, Mg, Ca, Cu, Zn, Co, Fe, Se, and Mn) from their environment. Cells can acquire target metals by means of specialized molecular sites on their surfaces that recognize and bind only that species. Examples of such selective reactants are the nonprotein iron-binding molecules, collectively known as siderophores.[2]

Siderophores generally fall into two classes of molecules, hydroxomates and phenolate-catecholates. DeVoe-Holbein covalently linked microbial siderophores which belong to each of the two classes, Enterobactin and Desferrioxamine (see Figures 8.3.1 and 8.3.2), to porous glass bead supports.[1] In subsequent biological experiments, these particulate compositions were used successfully as a fixed-bed, iron-retrieval system.

DeVoe-Holbein has since synthesized a series of metal-capturing compositions with catechol, or substituted catechols, as the active component. Such compositions have similar properties to those of Enterobactin. Catechol was covalently bound to solid surfaces with bifunctional linking agents of defined lengths (Figure 8.3.3). Highly porous glass is the solid substrate which has been found to be most practical for the composition synthesis.[1]

According to DeVoe-Holbein, the resulting compositions proved to be highly efficient, typically achieving 99 percent or higher removal rates, and are selective for individual or groups of metals. The rapid adsorption kinetics minimizes required contact time and the compounds are mechanically and chemically stable. In addition, the compositions are regenerable, requiring only small volumes of regenerant.[3]

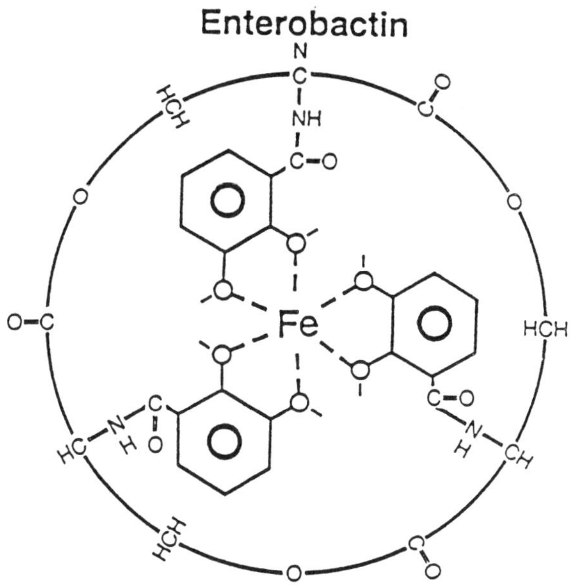

Figure 8.3.1. Microbial siderophore Enterobactin. Reference 1.

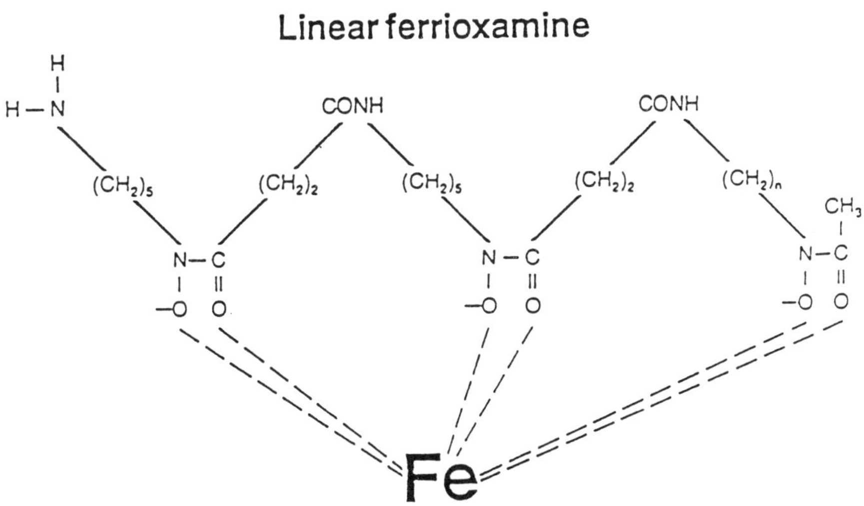

Figure 8.3.2. Microbial siderophore Ferrioxamine. Reference 1.

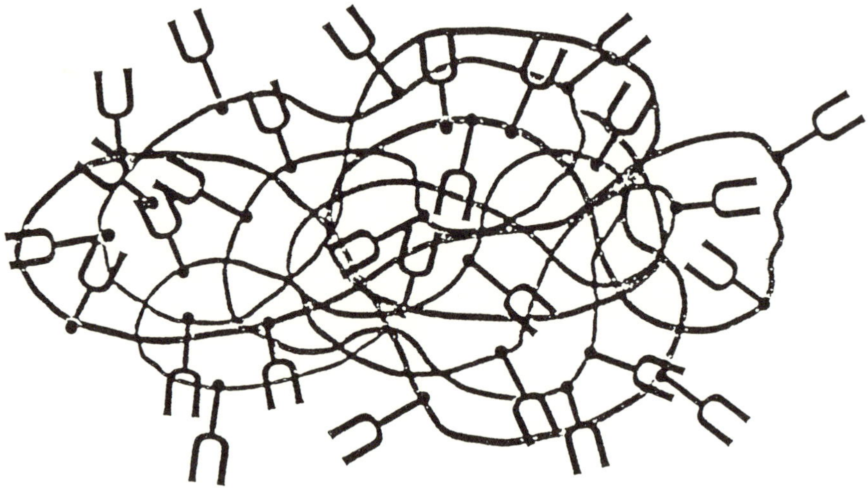

Figure 8.3.3. Graphic display of metal-capturing composition. Fork-like symbols represent siderophores immobilized through bridging agents to a solid surface represented by the continuous line. Reference 1.

The synthesized compounds were employed for waste treatment applications in a manner similar to classical ion exchange, as shown in Figure 8.3.4.[4] The media is contained in a fixed bed, and the metal-laden solution is passed through the bed during the service cycle. Following saturation of the media with metals, the bed is backwashed and the bound metal is displaced by an appropriate regenerant; e.g., 2 N HCl.[4] Co-current and counter-current fixed bed systems have been developed. The basic modular system which is now commercially available, can be expanded or realigned to correspond to end users' varying throughput requirements and spatial limitations.[4]

DeVoe-Holbein adsorption units are only able to treat contaminants in solution. Similar to ion exchange, high concentrations of suspended solids which can foul the adsorbent bed are typically pretreated through some form of filtration. Waste streams from the adsorption process include: contaminated regenerant and filtered solids from the pretreatment system. Filtrate from the pre-filtering system can generally be land disposed without further treatment. The regenerant may require treatment (e.g., neutralization, precipitation, dewatering) and disposal if not amenable to recycling.

One of the reported advantages of the DeVoe-Holbein system is that it is capable of yielding a more highly concentrated regenerant than ion exchange. Several options for downstream utilization of the concentrated metal regenerant are therefore possible. When it is compatible with the parent solution bath and metal concentrations are sufficiently high, the regenerant stream may be reused directly. If higher metal concentrations are required, an intermediate recovery step can be employed. For example, metal may be recovered from the regenerant electrolytically, recycling the regenerant to the adsorption process and selling the metal as scrap.[5]

8.3.2 Process Performance

The performance of a DeVoe-Holbein system will be influenced by the characteristics and quantity of the waste stream being treated. Parameters which need to be considered when evaluating the applicability of the system for a particular waste stream include: types and concentrations of constituents present in the waste stream, required effluent quality, and

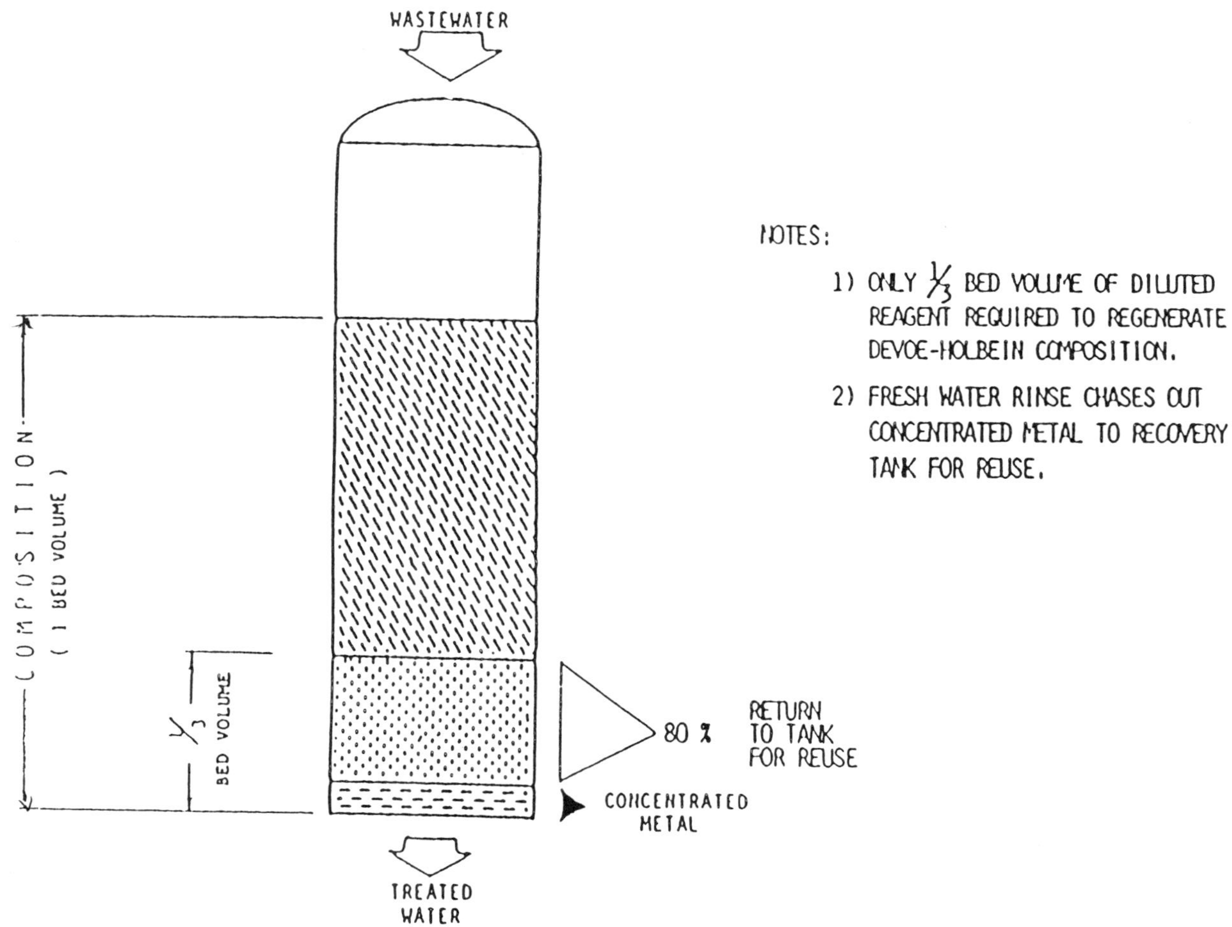

Figure 8.3.4. Wastewater process unit column showing regeneration. Reference 4.

options for managing the regenerant stream. DeVoe-Holbein says its compositions, trademarked Vitrokele, meet the following criteria:

- The ability to capture all or virtually all of a specific, target metal, even in the presence of very low concentrations of that metal or in the presence of competing metals.
- The ability to withstand harsh physical and chemical treatment without losing structural or functional integrity.
- The ability to allow easy displacement of the metal, permitting metal concentration and regenerant solution volume reduction, reuse of the composition and, possibly, reuse of the captured metal.
- The ability to capture substantial quantities of metal per unit of composition while maintaining high capture efficiency.
- The use of non-toxic agents; i.e., the process will not add trace toxic components to the solution from which the metal is being captured.
- The capability for being produced at a low cost, enhanced further by regenerability.

DeVoe-Holbein compositions all display very high metal capture efficiencies.[3] Table 8.3.1 illustrates the high capture efficiencies obtainable in the laboratory with test metal solutions of importance to metal finishing and hydro-metallurgical operations.[3] High capture efficiencies are demonstrated up to the capacity of the particular compound, with a sharp breakthrough curve occurring after saturation. A typical breakthrough curve is shown in Figure 8.3.5, where the DeVoe-Holbein DH-520 has been used for Cu removal from a relatively concentrated metal solution.

Selectivity and specificity of a particular composition are, in part, functions of the pretreatments used with the particular composition, and the conditions under which the metal solution or wastewater are treated. In many instances, selectivity and specificity can be altered (broadened or narrowed) to meet specific requirements of metal extraction.[4,6] Figure 8.3.6 demonstrates the selectivity of DeVoe-Holbein composition DH-506 (F-1) for iron relative to sodium, cadmium, and cobalt.

TABLE 8.3.1. THE EFFICIENT CAPTURE OF SOME TOXIC HEAVY METALS OF IMPORTANCE TO THE HYDROMETALLURGICAL AND METAL FINISHING INDUSTRIES BY DEVOE-HOLBEIN COMPOSITION

Toxic metals	DeVoe-Holbein composition	Influent concentration	Effluent concentration[a]	Capture efficiency (%)[b]
Cadmium	DH-516	674 ppm	<1.0 ppb	≥99.99
Chromium	DH-524	694 ppm	<0.01 ppm	≥99.99
Copper	DH-520	38 ppm	<1.0 ppt	≥99.99
Lead	DH-501	42 ppm	<42.0 ppb	≥99.99
Mercury	DH-573	12 ppm	<1.0 ppb	≥99.99
Nickel	DH-507	0.10 ppm	<1.5 ppt	≥99.99
Zinc	DH-508	6.5 ppm	<0.8 ppb	≥99.99

[a]Effluent concentration at or below normal detection limits using either radioactive tracer or atomic absorption spectrophotometric determinations.

[b]Capture efficiency determined as percent reduction in influent concentration; values are greater or equal to those shown due to detection limits of effluent metal concentration.

Reference 3.

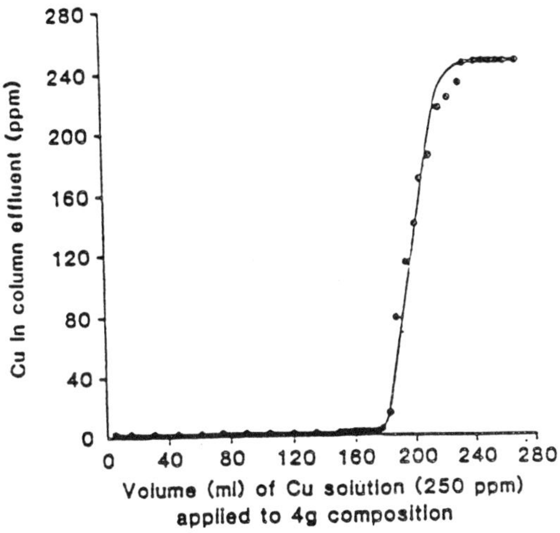

Figure 8.3.5. Copper removal breakthrough curve. Reference 1.

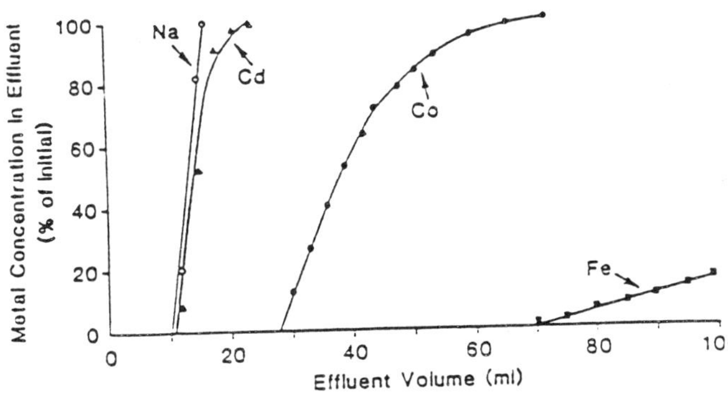

Figure 8.3.6. Selectivity of DH-506 (F-1) for iron. Reference 3, 6.

The specificity of DeVoe-Holbein compounds can be demonstrated by the capture of a test metal from a complex solution containing a number of other metals. Figure 8.3.7 demonstrates the ability of DeVoe-Holbein composition 506 (F-1) to selectively remove iron from sea water. The figure shows differences between the highly selective DeVoe-Holbein composition and competitive adsorbants, in this case, a selective ion exchange resin and a strong cationic exchange resin.[3]

Resin regenerability is of prime importance in determining overall economic viability of metals adsorption processes. Figure 8.3.8 illustrates the regenerability of DeVoe-Holbein compositions with a comparable cation exchange resin. This test was performed with identical bed sizes and similar materials were captured. Regeneration of the cation exchange resin required nearly five times the regenerant solution volume to achieve comparable regeneration.

Case studies of DeVoe-Holbein adsorption applied to metal/cyanide wastes, as adapted from DeVoe-Holbein, are presented below.

Case Study #1--

DeVoe-Holbein technology was evaluated for removal of chromium from three representative chromium wastewater streams wastewaters: boiler blowdown water, chrome plating waste precipitator effluent, and cooling tower wastewater. Waste stream metal concentrations are summarized in Table 8.3.2. DeVoe-Holbein composition DH-524 was used for this analysis.[3]

High recovery efficiency was demonstrated by using composition DH-524, as shown in Table 8.3.3. Both chromium removal and regeneration efficiency were essentially complete. In addition, the regenerant volume reqired was only 3 bed volumes and showed no loss in efficiency over 15 cycles.

Case Study #2--

The DeVoe-Holbein treatment process was employed to treat a 5 gpm counter-current rinse effluent from a zinc chloride electroplating line. It demonstrated high zinc removal efficiencies in the presence of other cations.

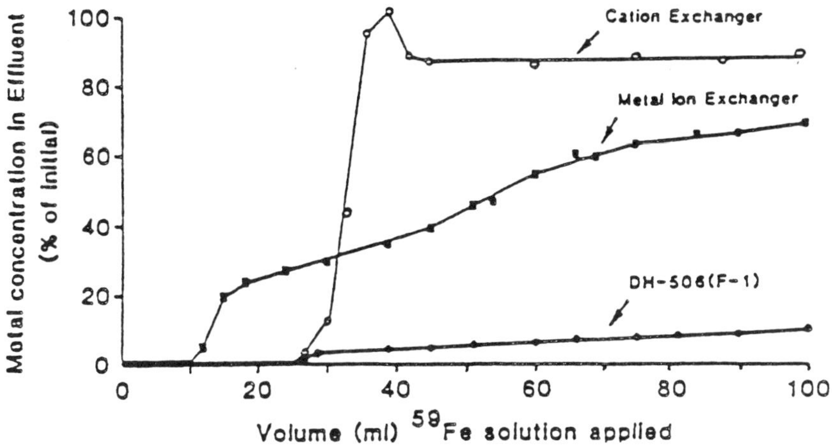

Figure 8.3.7. Demonstrated ability to DeVoe-Holbein DH-506 (F-1) to remove iron from seawater compared to cation and metal ion exchange resins. Reference 3.

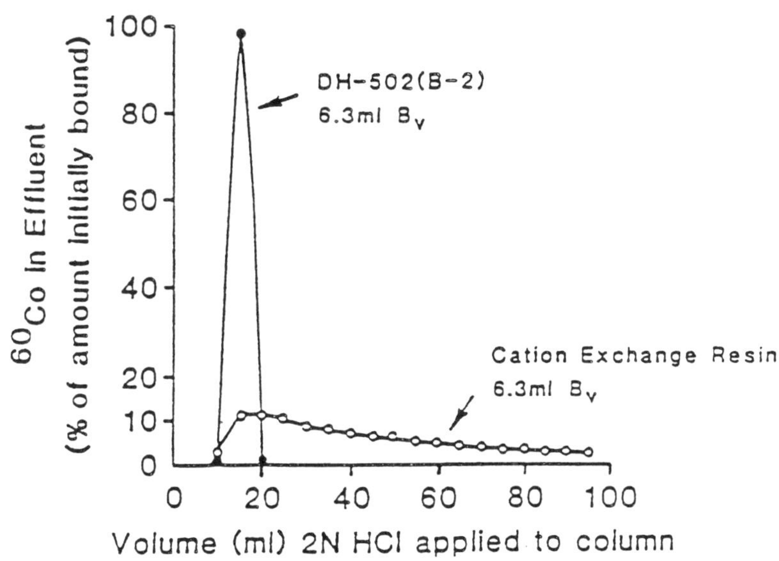

Figure 8.3.8. Regenerability performance of DeVoe-Holbein DH-502 (B-2) and cation exchange resin. Reference 3.

TABLE 8.3.2. PARTIAL ANALYSES OF DIFFERENT CHROMIUM WASTEWATERS

Element[a]	Boiler blowdown water (ppm)	Chrome plating[b] waste effluent (ppm)	Cooling tower bleed (ppm)
Chromium (VI)	694.3	3.8	7.6
Sodium	432.0	14.2	57.4
Calcium	155.3	88.4	176.4
Magnesium	64	6.8	42.1
Silicon	5.2	10.6	35.2
Strontium	0.8	0.3	0.9
Zinc	9.9	0.1	0.8
Boron	0.4	--	0.3
Iron	--	--	0.5

[a]Elemental analyses carried out by Inductively Coupled Plasma Emission Spectrometry.

[b]Wastewater following conventional chemical reduction and precipitation of chrome plating rinse water.

Reference 3.

TABLE 8.3.3. RECOVERY OF Cr FROM VARIOUS WASTEWATERS WITH DH-524

Type of wastewater	Influent CR VI (ppm)	Treated[a] effluent Cr VI (ppm)	Cr removal efficiency (%)	Capacity mg Cr VI per kilogram composition	Regeneration efficiency (%)[b]
Boiler blowdown	694.3	ND[c]	>99.99	≅ 20,000	100
Plating waste precipitator effluent	3.8	ND	>99.99	≅ 20,000	100
Cooling tower	7.6	ND	>99.99	≅ 20,000	100

[a]Flow rate of 20 bed volumes/hour in a fixed bed of DH-524.

[b]100 percent of bound Cr displaced in approximately 3 bed volumes of regenerant. Fully regenerable over 15 cycles of use so far tested.

[c]ND = not detectable by atomic absorption spectrophotometry.

Reference 3.

The wastewater process was designed to operate on a 16 hour feed cycle, with the influent zinc concentration ranging from 50 to 300 ppm. At the end of the operating period, the process unit is regenerated and reconditioned. Figure 8.3.9 shows the results of the operation of the wastewater process unit over a 2-week period. Depending on plating activity, the inlet zinc concentration to the process unit varied significantly, from 10 ppm to as high as 280 ppm. Yet zinc concentration in the treated effluent consistently remained below 1 ppm.[3]

The adsorption unit was regenerated daily following the 16 hour processing period. Less than one-third of a bed volume of regenerant was applied, at a flow rate of 0.5 bed volumes/hour, followed by a similar volume of rinsewater. The regenerant, with typical metal concentrations of 50,000 ppm (as high as 100,000 ppm) is directed to a storage tank for further recovery.[3]

Case Study #3--

In another example, a large job shop operating four different processes, and using at least eight different metals, recently installed a DeVoe-Holbein treatment system. Prior to this, spent baths, acids and soaps were sent to a landfill and aqueous effluents were pH adjusted in a neutralization pit and discharged directly to the sewer.

The new wastewater process successfully treats both the individual and combined wastewater effluents which contain nickel, zinc, brass, chrome, precious metals, and possibly cyanide. Although combined effluents are treated efficiently, treatment of individual rinse lines offers several advantages over treatment of the combined effluent. Smaller treatment systems are required for individual rinsewater effluents, and these can be operated in a closed loop cycle, reusing treated rinsewater and recovered metals.[4] To demonstrate the efficiency of the DeVoe-Holbein wastewater treatment system, a treatment study of the nickel rinse effluent from an automatic rack plating operation was undertaken.

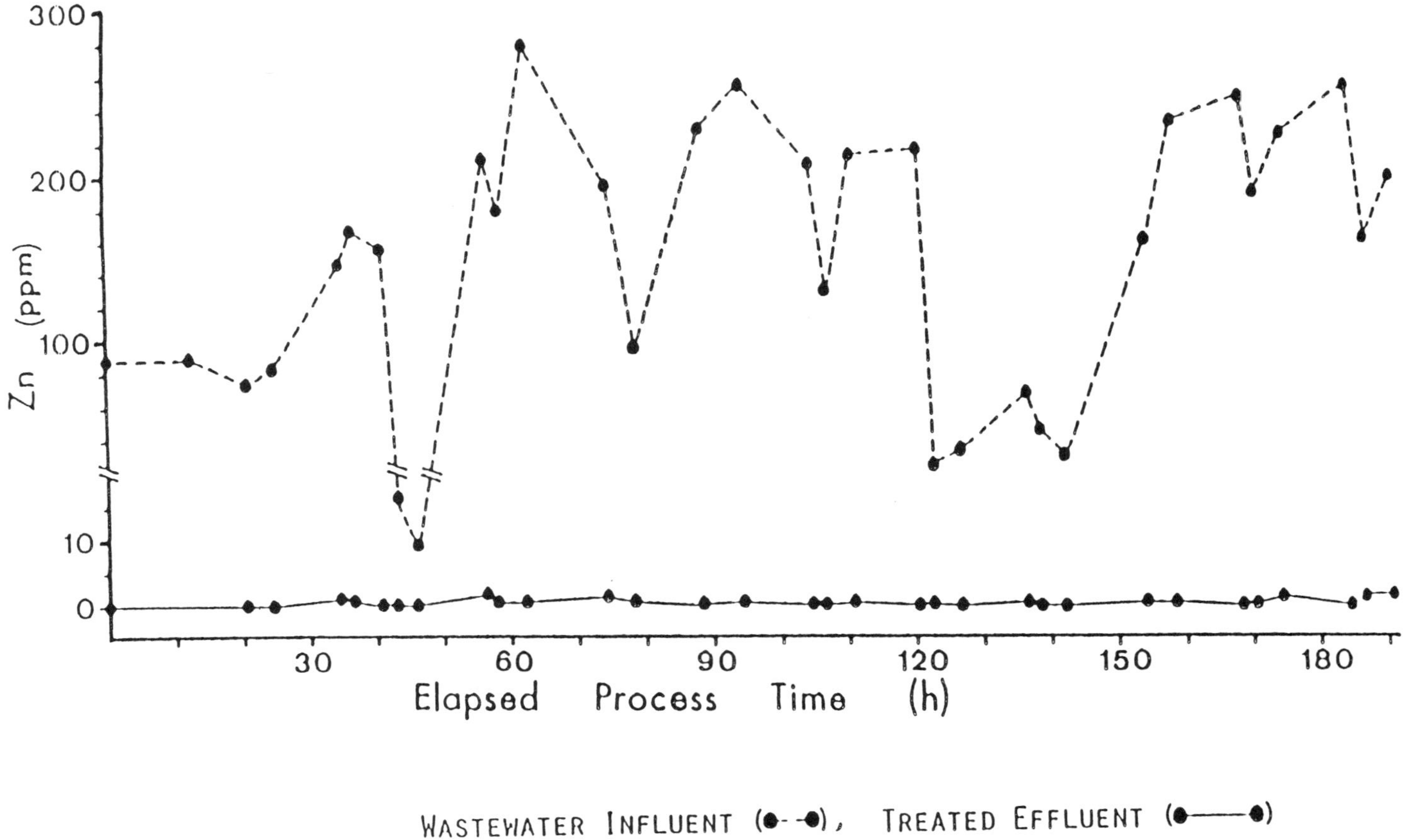

Figure 8.3.9. Removal of Zn from electroplating wastewater. Reference 4.

Figure 8.3.10 shows the results of operation of the adsorption treatment system over two complete cycles of loading and regeneration. The nickel concentration in the rinsewater was reduced from 130 to 180 ppm in the feed to an average of less than 1 ppm nickel in the treated effluent. When the nickel concentration in the effluent reached 5 ppm, the unit was regenerated. Less than two bed volumes of regenerant was required, resulting in a stream that was highly concentrated in nickel (7,000 to 8,000 ppm).[4]

Case Study #4--

For cyanide complexes, DeVoe-Holbein media have demonstrated high efficiency but, in some cases, result in relatively low capacity. This has been successfully overcome by pretreating cyanides using a destruct process. For example, following cyanide oxidation with chlorine, both cadmium and zinc were efficiently captured (99 percent) with capacities in excess of 12 grams of zinc or cadmium per liter of DeVoe-Holbein media.[5]

Silver-contaminated effluents containing thiosulfate or cyanide complexes do not require destruction of the cyanide prior to extraction of silver. Using a selective and regenerable DH media, silver can be efficiently (99.99 percent) removed from solutions to concentrations less than 10 ppb. Although the capacity of this medium for silver is somewhat dependent on the nature of the effluent, capacities as high as 20 grams of silver per liter of media have been realized.[7]

8.3.3. Process Costs

DeVoe-Holbein offers several different VITROKELE compositions, each of varying selectivity and metal capture capacity (depending upon the environment in which the metal must be captured and recovered).[8] The costs associated with DeVoe-Holbein treatment systems is subject to the particular VITROKELE composition sought and the volume ordered, and total system costs are assessed on an individual basis.[9] As a general guide per liter of D-H composition, the prices range from \$10 to \$50.[10] System costs, however, were unobtainable in conversations with DeVoe-Holbein representatives.

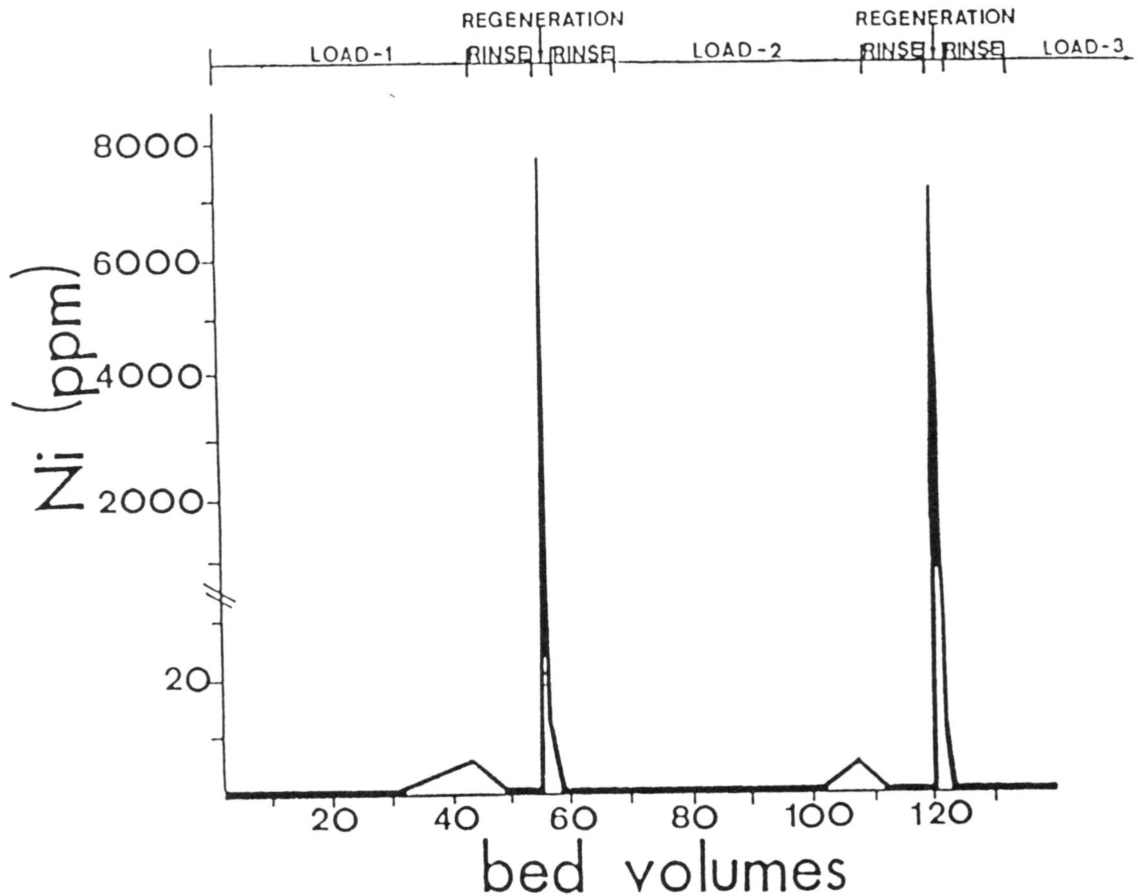

Figure 8.3.12. Removal of Ni from electroplating rinsewater, and regeneration of wastewater treatment unit.

Source: Reference 4.

8.3.4. Process Status

DeVoe-Holbein technology is protected under U.S. Patent No. 4,530,963 and a number of pending patent applications throughout the world. DeVoe-Holbein International N.V. holds the worldwide rights to the technology and is commercializing various aspects through subsidiaries and joint ventures.[11] The process appears most applicable to the selective removal of valuable metals (e.g. silver) from waste streams. Although the process appears promising, further information concerning selectivity and capacity is needed.

REFERENCES

1. DeVoe, I.W., and B.E. Holbein. "A New Generation of Solid-State Metal Complexing Materials: Models and Insights Derived from Biological Systems." In: Newer Methods for the Removal of Trace Metals from Aqueous Solution. Amual. Chem. Congress, Royal Society of Chemistry, University of Warwick, England. April 1986.

2. J.B. Neilands. Ann. Rev. Biochem. 50:715. 1981.

3. Holbein, B.E., DeVoe, I.W., Neirick, L.G., Nathan, M.F., and R.N. Arzonetti. "DeVoe-Holbein Technology: New Technology for Closed-Loop Source Reduction of Toxic Heavy Metal Wastes in the Nuclear and Metal Finishing Industries. Presented at: Massachusetts Hazardous Waste Source Reduction Conference Proceedings. October 17, 1984.

4. Brener, D., Greer, C.W., and S. G. Nelson. "Novel, Synthetic Compounds for the Efficient Removal of Nickel and Zinc From Plating Effluent." Presented at: 7th American Electroplaters Society, U.S. Environmental Protection Agency Conference on Pollution Control for the Metal Finishing Industry. 1986.

5. Greer, C. W. and A. C. Huber. Synthetic Compositions: Effective Treatment of Heavy Metal Containing Wastewaters from the Metal Finishing Industry. DeVoe-Holbein Canada, Inc., Montreal, Quebec, Canada.

6. DeVoe, I.W., and B.E. Holbein. "New Technology for the High Affinity Capture of Radioactive Metals from Water. Presented: 4th Annual Conference Canadian Nuclear Society. June 1983.

7. Neirick, L.G., and B.E. Holbein. "Removal of Heavy Metals from Waste Streams with Novel High-Affinity Selective and Regenerable Media. Presented at: Annual Meeting American Electroplaters Society, Detroit, Michigan. 1985.

8. DeVoe-Holbein Technical Brochure.

9. O. D'Sylva. Telephone conversation with David Sullivan, Alliance Technologies Corporation. May 4, 1987.

10. T. Resch, DeVoe-Holbein. Personal communication with David Sullivan, Alliance Technologies. April 1, 1987.

11. U.S. Patent No. 4,530,963. Insoluble Chelating Compositions. I.W. DeVoe and E. Holbein. July 23, 1985.

9. Electrolytic Processes

9.1 PROCESS DESCRIPTION

The electrolytic cell is the basic device used in electroplating operations. The cell consists of an anode and a cathode immersed in an electrolyte. When current is applied, dissolved metals in the electrolyte are reduced and deposited on the cathode. This process is attractive for pollution control because of its ability to remove specific contaminants from the waste stream without the addition of chemicals which produce large quantities of sludge. In addition, it is often possible to reuse the metal which is removed from solution, thereby making the technology a recovery process as opposed to an end-of-pipe treatment process.[1,2]

A commonly used configuration for electrolytic recovery is to connect the electrolytic unit to the dragout tank that follows metal plating or etching baths and precedes the running rinse (see Figure 9.1.1). The solution in the dragout tank, which contains diluted plating chemicals, is circulated through the electrolytic reactor and back into the dragout tank.[3] In this way, the concentration of metals in the dragout tank is maintained at a low level. Instead of being carried into the running rinse and eventually into the wastewater treatment system, the metals are recovered by the electrolytic reactor.

Electrolytic treatment is not effective in removing all contaminants. It is most effective in removing the noble metals such as gold and silver. These metals have high electrode potentials (see Table 9.1.1) and are easily reduced and deposited on the cathode. Other metals, such as aluminum and magnesium, cannot be removed by this type of process because their electrode potentials favor oxidation rather than reduction. Compounds such as cadmium, tin, lead and copper can be removed, but a greater amount of current is required, particularly when the metal concentration is low; e.g. less than 1,000 ppm.

In addition to the type of metal, the type of solution also has an effect on the practicality of ectrolytic recovery. Extremely corrosive solutions (e.g., certain etchants) may pose problems for electrolytic recovery because

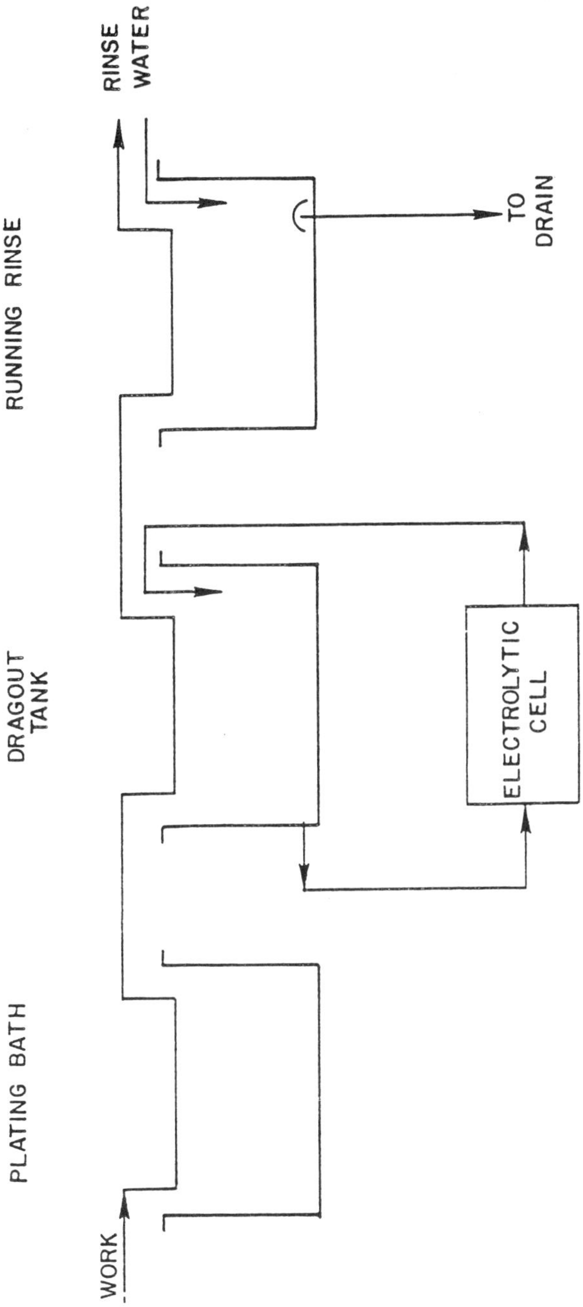

Figure 9.1.1. Typical electrolytic recovery system configuration.

TABLE 9.1.1. ELECTRODE POTENTIALS AT 25°C

Metal	Cathode reaction		Electrode potential (volts)
Gold	Au^{3+} + 3e-	Au(s)	1.5
Platinum	Pt^{2+} + 2e-	Pt(s)	1.42
Silver	Ag^{+} + e-	Ag(s)	0.8
Copper	Cu^{2+} + 2e-	Cu(s)	0.345
Lead	Pb^{2+} + 2e-	Pb(s)	-0.126
Tin	Sn^{2+} + 2e-	Sn(s)	-0.136
Nickel	Ni^{2+} + 2e-	Ni(s)	-0.25
Cadmium	Cd^{2+} + 2e-	Cd(s)	-0.40
Steel or Iron	Fe^{2+} + 2e-	Fe(s)	-0.44
Zinc	Zn^{2+} + 2e-	Zn(s)	-0.76
Aluminum	Al^{3+} + 3e-	Al(s)	-1.68
Magnesium	Mg^{2+} + 2e-	Mg(s)	-2.37

Source: Reference No. 4.

the metal that is plated on the cathode is etched off as quickly as it is plated. In addition, solutions with chelated metals, such as electroless copper plating solutions, may be more difficult for electrolytic recovery than solutions containing free metal ions such as acid copper electroplating solutions.[1]

For dilute metal-containing solutions, electrolytic recovery can be extremely difficult, particularly when using standard flat plate electrodes. One of the primary limitations of this type of electrode is that high mass-transfer rates are difficult to achieve. When plating metals from a solution, the layer of solution next to the cathode becomes depleted in metal ions. Since there are fewer ions present in dilute solutions, diffusion into and across the depleted layer is much slower and the layer becomes thicker and more depleted[5]. Mass transfer rates can be enhanced both by agitation and by increasing the effective surface area of the electrodes, particularly the cathode. Both of these actions will increase the rate of movement of metal ions to the cathode, which is equivalent to an increase in the current passed between the electrodes.

Since most rinsewaters requiring treatment contain metals at concentrations of less than 1,000 ppm, a number of electrolytic reactors have been designed with electrodes that either enhance mixing or have large surface areas.[6] Some of the electrode designs are:

- Concentric cylinder;
- Parallel, porous plates;
- Rotating cylinder;
- Packed bed;
- Fluidized-bed; and
- Carbon fiber.

The electrodes used in these reactors may be more effective in removing metals from solution, but their design may also make it difficult to remove the metal once it has been plated onto the cathode. For example, the use of a reactor with parallel stainless steel cathodes generally allows for the production of a compact, adherent layer of metal. This can be mechanically removed and sold as scrap. Conversely, the use of a reactor with a high electrode area results in the deposition of metal within pores of a cathode which may be comprised of carbon fibre, carbon granules, metal mesh, or metal sponge. In this case, mechanical removal of the metal is generally not feasible. Therefore, recovery of the metal must be accomplished by leaching the deposited metal out of the cell as a concentrate, by either corrosion or anodic dissolution. Alternatively, the cathode material may be disposed, or in the case of precious metals, sent to a refiner.

Both of these methods have drawbacks. Disposal does not allow reuse of the cathode, and leaching may not be practical in all situations. For example, precious metals such as silver, gold, and platinum are difficult to dissolve by corrosion, and aggressive solutions may damage cell components. Anodic dissolution involves reversing the polarity of the electrolytic cell which may also damage electrode materials.[6] Therefore, increasing mass transfer by using high surface area electrodes may come at the expense of reuse of the deposited metal and/or cathode.

Electrolytic cells have also been used to treat plating solutions and rinsewaters containing cyanide. In this case, cyanide is oxidized at the anode forming cyanate as an intermediate product and carbon dioxide, nitrogen, and ammonia as the end products.[7] In some cases, it is even possible to concurrently deposit metals at the cathode and oxidize cyanide at the anode. Optimal destruction usually requires temperatures between 150 and 200°F. Sodium chloride can be added to provide a source of chlorine which also acts as an oxidant to enhance cyanide destruction.[7,8] However, with conventional reactors, it is difficult to treat solutions containing low concentrations of cyanide; e.g., less than 1,000 mg/L. Therefore, another treatment process, such as alkaline chlorination, is often used after electrolytic oxidation to destroy residual cyanide.

A pilot scale electrolytic device called a Trickle Tower Electrochemical Reactor, has been developed which is reportedly capable of acheiving low cyanide levels.[9] The column is comprised of alternating layers of an electrolytically conductive packing (e.g., carbon Raschig rings) separated by thin sheets of insulating mesh. Voltage applied to plate electrodes at either end of the column results in each of the conductive layers of the column becoming a bipolar electrode; i.e. one face positive and the other negative, leaving a neutral center zone. The tower is regenerated by filling with anodic solution and reversing the polarity. The concentrated metal solution which developes can be recycled to the plating bath.

Copper is catalytic for the oxidation of cyanide due to the intervention of the $Cu^{+}(CN^{-})_2$ complex in the solution phase. Cyanide is oxidized as it passes the anodes and, when excess cyanide has been eliminated, copper is deposited on the cathodic surfaces of the bipolar layers. Cyanide levels have been reduced from over 200 ppm to near zero in 60 minutes during bath treatment of copper/cyanide solutions. In comparison, electrolytic oxidation of free cyanides (i.e., without metals present) takes roughly twice as long.[9]

9.2 PRETREATMENT AND POST-TREATMENT REQUIREMENTS

Electrolytic processes are generally used at the source of waste generation. The aqueous effluent is then either reused directly (e.g., bath make-up) or further treated to remove other contaminants or to be neutralized. Therefore electrolytic recovery is itself somewhat of a pretreatment process. However, in many cases, it is necessary to filter the wastewater prior to feeding it through the electrolytic reactor.[8,10] This is particularly true with reactors that utilize porous or packed bed electrodes since particulates can potentially clog the reactor.

Adjustment of pH is also sometimes necessary as a pretreatment measure since the waste pH affects metal speciation.[10] At a low pH, free metal ions predominate. These exhibit a higher mass-transfer rate to the cathode than do metals at higher pH. However, when treating wastewater containing chelated metals, the pH of the solution will not have a significant effect unless pH goes below 3.[1]

As discussed previously, post-treatment may be required to recover metals from the cathodic regenerent solution if this cannot be reused directly as bath make-up. Recovered wastewater will also require eventual disposal or treatment due to build-up of organics and other impurities present in the bath. Finally, stripped metals and metal laden cathodes can be shipped offsite to smelters or reclamation facilities.

9.3 PERFORMANCE OF ELECTROLYTIC RECOVERY SYSTEMS

The performance of several electrolytic reactors on specific metal/cyanide wastes is summarized in Table 9.3.1. However, these data reflect performance of a particular electrolytic reactor on a specific waste stream and, thus, should not be taken as a general indicator of performance.

Performance can be assessed in terms of rate of metal removal from solution, or current efficiency. Rate of metal removal can be determined either by doing a metal mass balance on inlet and outlet streams or by weighing the amount of metal which has deposited on the cathode. Current efficiency compares the actual amount of metal (or other contaminant) removed to the amount that could be theoretically removed for a given current. In practice, a high current efficiency is not necessarily equivalent to a high rate of removal since removal rate increases with current.

The paragraphs below discuss, in detail, the case studies of electrolytic treatment that were summarized in Table 9.3.1.

9.3.1 Concentric Cylinder Reactor[1]

The reactor employed in this study consists of a central post-type anode surrounded by a cylindrical cathode with a diameter of 8 inches and a height of 6 inches. The waste solution is rapidly recirculated through the annular space between the electrodes to provide a constant supply of metal ions. The cathode material is stainless steel and the anode material depends on the type of solution being treated. For acid copper solutions, the material is titanium; for corrosive fluoborate solutions used in solder (tin/lead) plating niobium is preferred.

At this facility, the reactors were being used to treat rinsewater from copper and solder electroplating. Four reactors were connected in parallel to the copper plating dragout tank, and rinse solution was circulated through each of the reactors at a rate of about 16 gal/min. This setup is shown schematically in Figure 9.3.1. The four reactors removed approximately 20 grams of copper per hour at an average current of 5 amps. The

TABLE 9.3.1. SUMMARY OF PERFORMANCE DATA

Reactor type	Flow	Current	Voltage	Waste type (contaminant)	Removal efficiency or rate	Current efficiency
Concentric cylinder	16 gpm	5 amps	7.5 volts	Acid Copper 300-400 mg/L Cu	5.7 g/hr	80 - 90 %
Carbon Fiber	46 gpm	--	--	Cadmium Cyanide- 300 mg/L Cd	89-98% 0.8-2.4 g/min	----
	46 gpm	--	--	800 mg/L cyanide	30-94% 1.47-5.34 g/min	----
Parallel Porous Plate	2.0 gpm	--	7.5 volts	Electroless Copper- 60-120 mg/L Cu	80-85%	----
Fluidized Bed	--	300 amps	--	Cadmium Cyanide- 200 mg/L Cyanide	40-50 kg Cadmium per year	20-30%
Packed Bed	10 gpm	175 amps	7.7 volts	Chromate - 17 mg/L of heaxavalent chromium	99.7% conversion to trivalent chromium	----
	10 gpm	560 amps	40 volts	Cyanide at 80 mg/L	65%	----

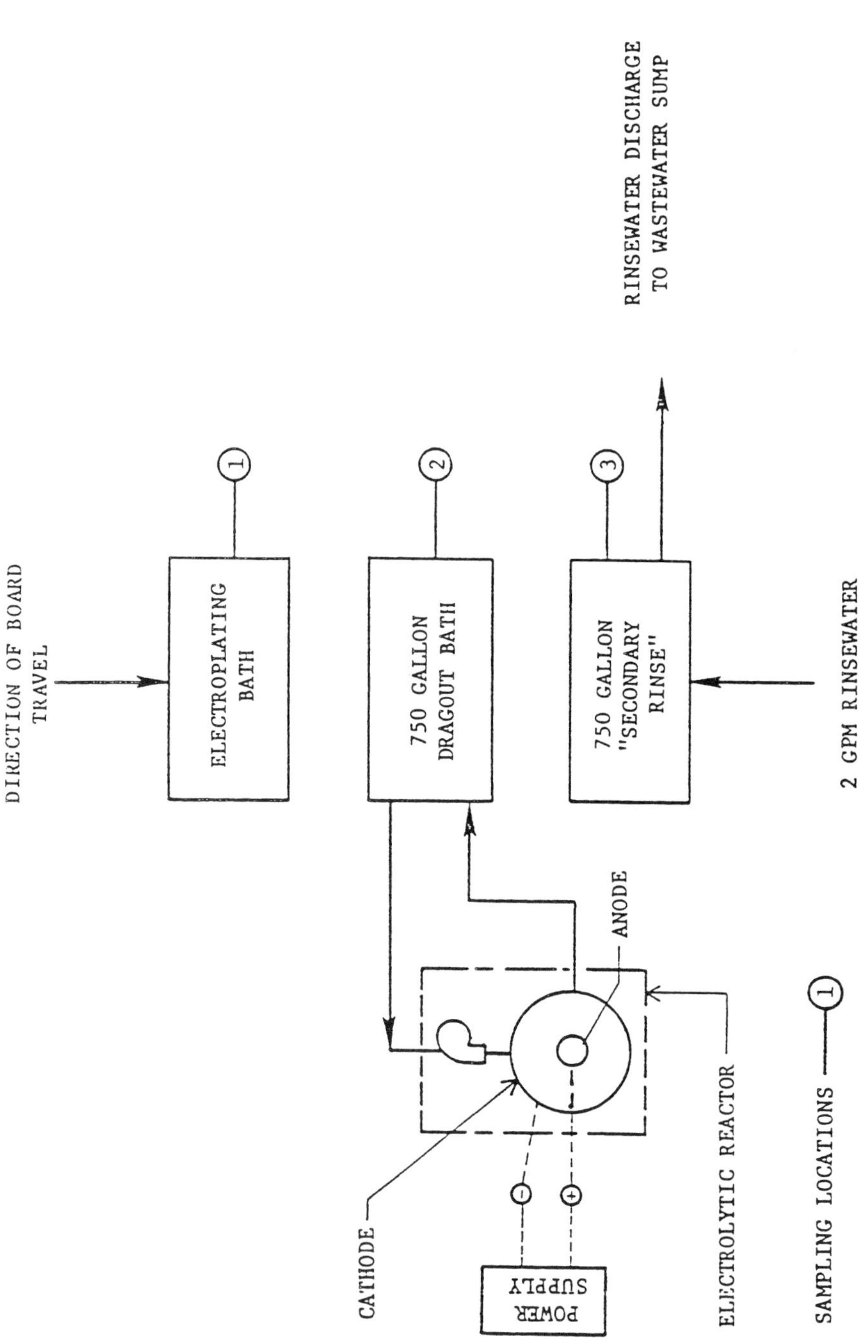

Figure 9.3.1. Schematic of electrolytic recovery system using concentric cylinder reactors. Source: Reference 12.

concentration of copper in the rinse discharge was reduced from 3,000 to 4,000 mg/L to between 50 to 100 mg/L to yield a current efficiency of 90 percent. Copper was recovered from the stainless steel cathodes once each week. Removal was easily accomplished by hand, producing a metal foil that could be sold as scrap metal.

Three reactors were connected to the dragout tank that followed the solder electroplating bath. However, initial tests were not successful in recovering tin or lead. Personnel at the facility indicated that these metals are sometimes recovered, but occasionally problems result due to the corrosivity of the solder plating solution. This corrosivity may cause etching of tin and lead from the cathode, thereby negating any electrolytic removal. To overcome this effect, it would be necessary to increase current to the reactor. The corrosivity may also cause occasional breakdown of the pump that is used to circulate the solution through the reactor. For these reasons, this type of solution is not always amenable to electrolytic recovery.[12,13]

9.3.2 Carbon Fiber Cell[8]

The data presented in Table 9.3.1 for the carbon fiber cell were generated using a unit developed by HSA Reactors Ltd. of Rexdale, Ontario and currently marketed by Metal Removal Systems, Inc. of Melville, NY. The electrolytic cell module contains a carbon fiber cathode and an anode of titanium coated with a rare earth oxide. The cathode's carbon fibers have a diameter of only 5 to 15 microns and 1 gram of fiber has a surface area of 2.6×10^6 cm^2. This is reportedly 1,000 times greater than the surface area of other porous electrode materials.[6]

Electrolytic recovery was used to recover cadmium and destroy cyanide in a cadmium-cyanide barrel-plating line. A reactor was operated in a closed-loop circuit connected to a dragout/recovery tank. Plating solution drag-out was carried into the recovery tank in volumes of 1.5 to 8 liters for each barrel plated. The contaminated rinse solution was then pumped through a 700 liter electrolytic reactor at a rate of 175 L/min.

As shown in Figure 9.3.2, the rinsing of a barrel raises the cadmium concentration in the rinse from 80 to 300 mg/L. Within approximately two minutes, the electrolytic reactor restores this concentration to its original level. The rate of removal then levels off, so that by the time another barrel is ready for rinsing, the concentration is approximately 60 mg/L. This curve clearly shows the dependency of rate of removal on metal concentration. In this case, removal drops off sharply when the cadmium concentration falls below 100 mg/L. Overall, between 89 and 98 percent of the cadmium washed out in the dragout tank was removed by the electrolytic reactor, corresponding to a recovery of up to 24 g/min of cadmium. As a result, the average cadmium concentration in the facility's effluent decreased from 4.0 to 1.0 mg/L.

An electrolytic reactor was also used to destroy cyanide in the drag-out tank. To enhance cyanide destruction, 80 g/L sodium chloride was added to the system. The sodium chloride acted both as a source of chlorine for cyanide destruction and as an electrolyte. Initially, cyanide destruction was achieved in the same electrolytic reactor as was used for metal removal. To increase destruction rate, a holding tank and a second, smaller electrolytic reactor were added to the system. This unit provided extended retention time for increased cyanide destruction, resulting in a 93 percent destruction efficiency, even when cyanide loadings were high (5.74 g/min). This destruction rate maintained the concentration of cyanide in the recovery tank at just over 1,000 mg/L. Before the modifications, the cyanide concentration averaged 4,000 to 8,000 mg/L.

Operation of the electrolytic reactor requires a 19 hour cycle of which 16 hours is for cadmium removal and cyanide destruction and the other 3 hours are for removal of the cadmium from the carbon fiber cathode. The latter is accomplished by pumping a high-strength cyanide solution through the reactor, allowing cadmium oxide to form and dissolve in solution. This product is then reused as cadmium-cyanide plating solution make-up.

9.3.3 Parallel Porous Plate Electrodes[1]

The RETEC heavy metal recovery system used in this application consists of a parallel plate electrode configuration, in which the cathodes are a metal sponge-like material (as shown in Figure 9.3.3) formed by depositing copper on a polyester foam. The result is a porous, flow-through cathode with a much

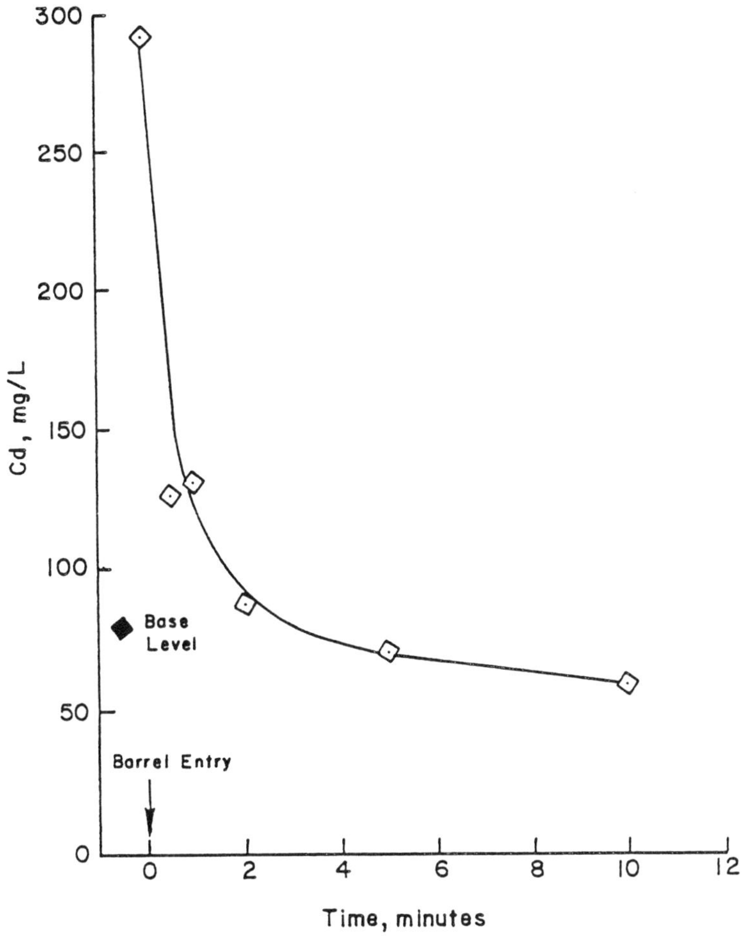

Figure 9.3.2. Reduction Curve for Cadmium.
Source: Reference No. 6.

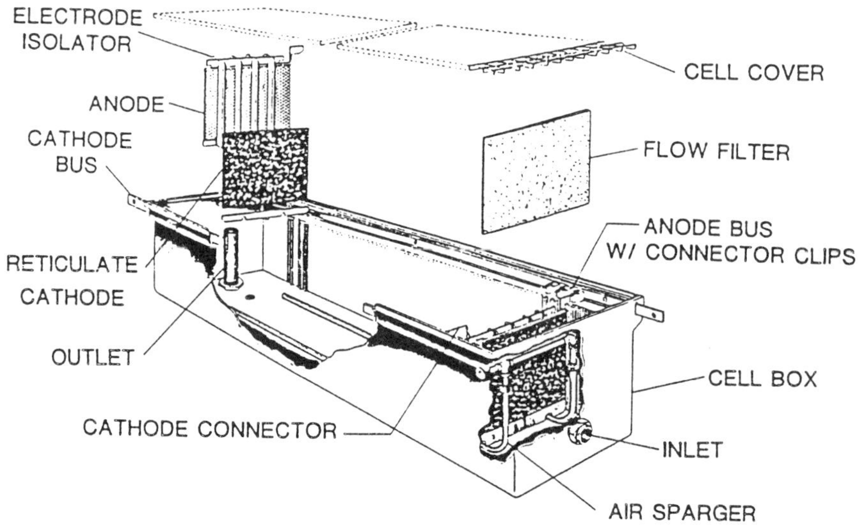

Figure 9.3.3. Retec cell.
Source: Reference No. 14.

higher surface area than a standard flat plate, stainless steel cathode. The applicability of this system to different types of metal containing solutions is summarized in Table 9.3.2. The majority of commercial applications have been in acid or electroless copper plating rinses.

A recent study evaluated the performance of this unit on an electroless copper rinse solution from a printed circuit board manufacturing facility.[1] The RETEC unit contained 25 copper-plated polyester "sponge" cathodes alternated by 26 titanium-coated anodes, each measuring 45 by 38 cm, spaced 0.64 cm apart. Influent copper concentration ranged from 8 to 1,100 mg/L, and flow rate ranged from 0.5 to 2.5 gpm. Testing of the unit yielded the following conclusions:[1]

- Copper removal efficiency averaged 80 to 85 percent.
- pH had little effect on copper removal over the pH range of 3 to 11.
- Removal efficiency was independent of influent copper concentration above 50 mg/L but removal efficiency decreased significantly below this concentration.
- Removal efficiency was best at low flow rates - removal efficiency increased by 15 percent when the flow rate was decreased from 25 to 0.5 gpm.
- Recirculation did not affect removal efficiency.

At a second facility, an identically designed but larger unit (50 cathodes as opposed to 25) was used to treat 10 gpm of combined copper plating and etching rinses. Inlet copper concentration was 100 to 200 ppm and the discharge was from the RETEC unit was below 1 ppm copper.[15]

9.3.4 Fluidized Bed

The Chemlec Cell, a fluidized bed developed by the Electricity Council Research Centre in England (see Figure 9.3.4), consists of a set of apertured expanded metal-mesh electrodes immersed in a bed of small glass beads. The bed is fluidized to about twice its packed depth by pumping rinsewater upwards through a distributor and the bed. The glass beads impinge on the electrodes and provide a simple means of agitation and mixing. The electrodes are

TABLE 9.3.2. RETEC APPLICATIONS

Metal	Electrolyte	Comments/conditions
Cadmium	Alkaline	Above pH 10.5 cadmium is soluble. Theoretically it can be removed at efficiencies and to levels equal to cadmium cyanide.
Cadmium	Acid	Cadmium can be reduced with a RETEC unit but not to Federal compliance. Cadmium can be reduced to approximately 50 ppm in dilute acid media (pH 2-5) (lab data). Strong acid would require bulk cathodes to achieve the high cathode potentials needed to protect the substrate.
Cadmium	Cyanide	Cadmium can be removed easily from a cyanide solution to compliance levels. 300 ppm at 3 gpm and 500 amp should operate at approximately 30% efficiency (field and lab results).
Chromium	Any	Neither trivalent $(Cr^{+3})n$ or Hexavalent (Cr^{+6}) is removable at 2 grams/liter (gpl.) In theory chrome could be removed from a more concentrated bath but this has not been achieved with RETEC (lab and field).
Copper	Dilute Acid	Copper can be removed easily from acid copper plating, bases (labe and field).
Copper	Electroless	Copper can be removed from electroless copper rinses less easy than acid baths but still at acceptable levels. High levels of chelate in the rinse will reduce anode life (lab and field).
Copper	Strong Acids Bright Dip	Soluble copper can be removed from almost any acid if the current density is raised sufficiently to counter the corrositivity of the acid (cathodically protect). This generally means elevated 1000 amp current for acidity .05 N. Bulk cathodes and/or elevated current for

(continued)

TABLE 9.3.2 (continued)

Metal	Electrolyte	Comments/conditions
Copper (continued)		acidity < 5N. For acidity > 5N a RETEC-50 cell would have to be modified to carry sufficient current. Higher current densities could be achieved in a RETEC Jr. Due to the lower surface area, efficient metal removal cannot be achieved with bulk cathodes with metal concentrations < 500 ppm. Long recycle times would be necessary, if the acid could not be reused with this metal loading. If the acid is not going to be reused neutralization is usually a more efficient alternative.
Copper	Ammoniacal Etches	Copper can be removed from ammoniacal etches at a lower efficiency than in acid media. Acidification will help efficiency but it will not be as efficient as straight acid. In order to remove copper from ammonicacal baths, it is necessary to raise the current density (and the cathode potential) to counter the action of the etch out. Complete copper removal from very concentrated ammoniacal baths cannot be achieved (lab).
Copper	Cyanide	Copper can be removed easily from copper cyanide rinses. Efficiencies are less than in acid baths (field and lab).
Lead	Acid Baths	Lead can be removed efficiently from acid baths. lead fluoborate will disassociate to fluoride and attack DSA anodes. Lead sulfamate baths are preferred. Lead and lead complexes are not very soluble. Lead above pH 4 will form insoluble lead oxide

(continued)

TABLE 9.3.2 (continued)

Metal	Electrolyte	Comments/conditions
Lead (continued)		and in the presence of sulfate (from copper sulfate plating for instance) will form insoluble lead sulfate. (lab and field).
Mercury	Aqueous	Mercury can be removed electrolytically but will not adhere, it will roll off the cathode and deposit in the bottom of the cell (beaker testing).
Nickel	Watts Rinse NiCl Rinse Ni Acetate	Removal at extremely low efficiency at low current densities from conc. solution. Inefficient removal from dilute solution with bulk cathodes at approximately 1 gpl (lab).
Palladium	Pd Cl-NH_4	Efficient removal from Pd bath (lab).
Ruthenium	Alc-Acid	Good removal from HCl alcohol solution (lab).
Selenium	Dilute Tab	Removal to approximately 20 ppm. Better results theoretically achievable.
Silver	Cyanide	Removable electrochemically at high efficiency (lab).
Silver	Thiosulfate	Removable electrochemically at high efficiency. May react with copper on cathode to cause premature cathode deterioration (lab and field).
Tin	Acid	Removable to approximately 40 ppm in RETEC cell (lab and field).
Tin	Alkaline	Should be removable to approximately 1 ppm.

(continued)

TABLE 9.3.2 (continued)

Metal	Electrolyte	Comments/conditions
Zinc	Acid	Not removable to compliance level. Zinc should be reduced to 100-200 ppm (lab).
Zinc	Cyanide	Zinc removal electrochemically to compliance levels (lab and field).
Zinc	Alkaline	Low levels should be achievable.

Source: Reference No. 14.

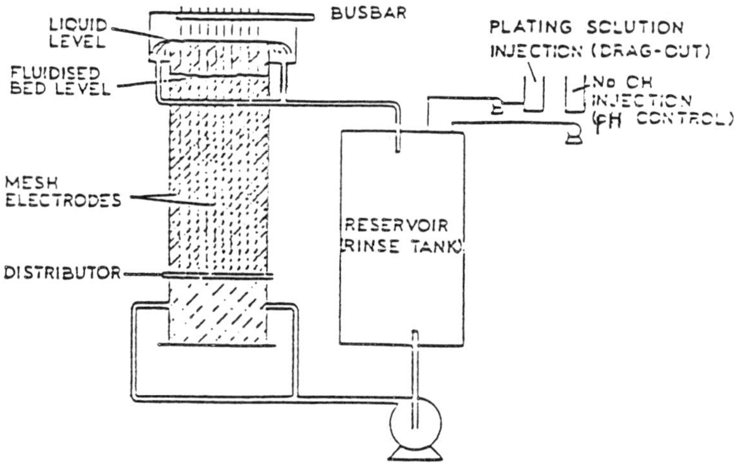

Figure 9.3.4. The Chemelec cell.
Reference No. 6.

commonly titanium, but when treating alkaline cyanide solutions, both anode and cathode may be made of mild steel. Also, since the metal is deposited on the cathode mesh rather than on the bed medium (e.g., as for the carbon fiber reactor), it is easy to remove the cathode from the cell and place it in the plating solution where it can be used as an auxiliary anode.

At one facility, the Chemelec cell was used to recover cadmium from an alkaline cyanide plating drag-out tank. The concentration of cadmium in the plating bath was 20 g/L and the Chemelec cell maintained the dragout tank concentration between 100 and 400 mg/L. The current efficiency ranged between 20 and 30 percent at an applied current of 30 amps. In the first 2 years of use, the cell recovered 33.7 and 59.7 kg cadmium, respectively, despite lengthy stretches of downtime due to low workloads. The maximum capacity of the 1 m^2 cell is estimated to be 200 to 300 kg/year. Power consumption for electrolysis for these 2 years was 13 and 11 kw-hr/kg cadmium recovered, respectively.

Under these operating conditions, it was necessary to remove the anodes from the Chemelec cell every 2 weeks to collect the cadmium deposit. The cathodes are lifted out of the cell and hung by copper hooks from the anode rail in the plating bath. After about one 8-hour shift, the cadmium from the coated cathodes has been dissolved. In this way, there is complete recycling of cadmium.[16]

9.3.5 Packed Bed

The packing in a packed-bed electrolytic reactor can be any of a number of different types of material. In the case discussed here, the packing material is inert carbon particles. These are loosely packed in a cell with the major electrodes in a parallel configuration, spaced 6 inches apart. The major electrodes act as baffles to create an overall bed length of 100 inches. The loose packing of the carbon particles imparts a semi-conductive nature to the bed. Potentials are maintained between particles and/or agglomerates of particles creating many anodic and cathodic sites within the major cell.[11]

In this application, the electrolytic reactors are used to reduce hexavalent chromium to the trivalent state and to oxidize cyanide. Chromium reduction is accomplished by four electrolytic cells used in parallel, each having a nominal capacity of 10 gpm. The system was designed to treat 120 mg/L of hexavalent chromium, although, the actual concentration fed to the reactors averaged only 17 mg/L. As a result, the length of travel through the cells was greater than required and the cell discharge averaged 0.06 mg/L chromium, well below the design target of 0.2 mg/L. This was equivalent to a 99.7 percent reduction efficiency. At higher feed concentrations (100 to 150 mg/L), the cell discharge was closer to the design value of 0.2 mg/L. Other conclusions of the testing are as follows:

- Control of waste pH entering the cell is essential; at hexavalent chromium concentrations of 50 mg/L, a pH of 1.8 to 2.1 is adequate; at 150 mg/L, a pH of 1.5 to 1.6 is preferred.
- Power consumption varies with chromium concentration; for a 20 gpm flow at 17 mg/L, 1.4 kw (175 amp, 7.7 volts) of power is consumed; at the same flow rate, but 156 mg/L, 2.0 kw of power was consumed.
- The deposition of chromium onto the carbon particles will result in a very high initial removal rate, however, after this initial period, a steady state will exist for electrolytic removal of hexavalent chromium.

It should be noted that the low and narrow pH range (1.5 - 1.6) observed during testing is extremely difficult to achieve on a consistent basis. Pilot plant testing with an emphasis on process control procedures is recommended prior to any full scale implementation.

The electrolytic treatment system for cyanide consisted of three cells used in parallel, each with a nominal capacity of 10 gpm. This system was designed to treat cyanide-bearing rinses at a hydraulic loading of 30 gpm and cyanide concentrations of up to 30 mg/L. However, actual cyanide concentration was found to exceed 30 mg/L 98 percent of the time, with an average concentration of 80 mg/L. As a result, it was only possible to achieve an overall destruction efficiency of 65 percent (to 28 mg/l), which was not sufficient to meet the design goal of 2 mg/L of cyanide in the cell discharge.

Further evaluations indicated that there was straight-line relationship between distance of travel along the packed bed and removal of cyanide; namely, 0.5 mg/l of cyanide is removed for each inch of travel. Therefore, to achieve the design goal of 2 mg/L of cyanide, it would be necessary to increase the bed length from 100 to 156 inches.

Other results of the study include:

- Cyanide was completely oxidized to CO_2 and N_2.
- Electrical power use averaged 22.4 kw (560 amps at 40 volts) which was calculated to be equivalent to 45 kw/kg of cyanide removal.
- It was not demonstrated that complexed metals were destroyed by the electrolytic process.

9.4 SYSTEM COSTS

The major cost associated with electrolytic treatment is usually the cost of the reactor itself. This can range from $3,500 for a reactor with a 1-ft^2 stainless steel cathode, up to $89,000 for a reactor with a high surface area, carbon fibre cathode.[8,17] Additional capitol costs will include items such as a rectifier and electrical connections, pumps and plumbing, and installation labor. These items may represent 15 to 25 percent of the cost of the reactor.[8,11] Operating costs for electrolytic recovery include electricity, maintenance or replacement of electrodes, labor, and chemicals for oxidation of cyanide; e.g., NaCl. Chemicals may also be used in some cases to strip metals from cathodes.

The best way to illustrate the cost of electrolytic treatment is to compare its costs with those of other treatment methods; e.g., precipitation. Electrolytic treatment can be very cost effective since it usually permits recovery of metal from the waste solution and also precludes the generation of metal-bearing sludges that require subsequent management as hazardous waste.

Cost comparisons of electrolytic recovery versus other treatment methods are presented in Tables 9.4.1 and 9.4.2. Table 9.4.1 shows the costs of using a carbon fibre electrolytic reactor versus the costs for alkaline chlorination and precipitation for a waste stream containing cadmium and cyanide. The annual operating costs for electrolytic recovery are $25,000 less than for the chemical treatment alternative. This would permit a payback time for the higher cost reactor of less than a year.[8]

The chemical costs for electrolytic treatment shown in Table 9.4.1 are for sodium chloride (NaCl), which is used to aid cyanide destruction, and sodium cyanide (NaCN) and oxygen, which are used to remove cadmium from the

TABLE 9.4.1. COSTS FOR CARBON FIBER ELECTROLYTIC TREATMENT VERSUS CHEMICAL TREATMENT

	Electrolytic treatment (carbon fiber)	Alkaline chlorination/ precipitation
Annual operating costs ($)		
Electricity	900	NA
Chemicals	11,150	33,800
Sludge disposal	0	5,700
Labor	1,150	4,600
Maintenance	5,950	0
Total:	19,150	44,100
Capital costs ($)		
Reactor	87,500	NA
Electrical and plumbing	21,900	NA
Installation	1,200	NA
Total:	110,600	90,000

NA = Not available

Source: Reference No. 8.

TABLE 9.4.2. COSTS FOR ELECTROLYTIC TREATMENT USING ARETEC CELL VERSUS CHEMICAL TREATMENT

	Electrolytic treatment (Retec cell)	Sulfide precipitation
Annual operating costs ($)		
Electricity	875	NA
Labor	1,250	NA
Maintenance	7,750	NA
Recovered copper	(625)	0
Total Operating Cost ($)	9,250	52,000
Total Capital Costs:	44,000	140,500

NA = Not available.

Source: Reference No. 15.

carbon fiber cathode. The maintenance costs are for replacement of the anodes, which is required every 2 years, and for filter cartridges which are used to remove particulate matter from the waste influent.[8]

Table 9.4.2 compares the costs of using electrolytic treatment versus sulfide precipitation to treat a copper-bearing waste stream generated from the manufacture of printed circuit boards. In this case, two RETEC electrolytic reactors are required to treat the 10 gpm waste stream which contains 200 mg/L copper. The major operating cost of electrolytic recovery is the maintenance cost to replace the cathodes. However, the overall operating costs are only one-sixth of that incurred using sulfide precipitation. The latter costs were not itemized in the reference but probably result from the purchase of sodium sulfide and the cost for disposing the metal-bearing sludge that results from precipitation of cadmium.

The capital costs for electrolytic removal are one-third those for sulfide precipitation. The capital costs for sulfide precipitation include the purchase of tanks, pumps, a clarifier, and a filter, plus a building addition to hold this equipment. The equipment required for electrolytic treatment is much more compact and does not require the construction of additional building space in which to place tanks and clarifiers. If these assumptions are accurate, the capital costs for electrolytic treatment are significantly lower than for sulfide precipitation.

Figure 9.4.1 presents an operating cost comparison for treatment of electroplating wastewater containing hexavalent chromium with two different treatment technologies. One of the technologies is a packed carbon bed electrolytic reactor and the other is chemical reduction with sodium bisulfite. Both of these technologies are followed by addition of NaOH to precipitate trivalent chromium. Figure 9.4.1 shows that chemical reduction is less expensive than electrolytic reduction at low Cr^{+6} concentrations, but becomes more expensive as the concentration increases. Electrolytic reduction becomes more efficient at high metal concentrations; at a concentration of 17 mg/L Cr^{+6}, approximately 25 kwh of power are required to reduce 1 kg whereas, at a concentration of 156 kg/l, only 2 kwh are required. Conversely, the amount of sodium sulfite required for reducing Cr^{+6} increases almost linearly as the concentration of chromium in the wastewater increases from 17 to 156 mg/l.[6]

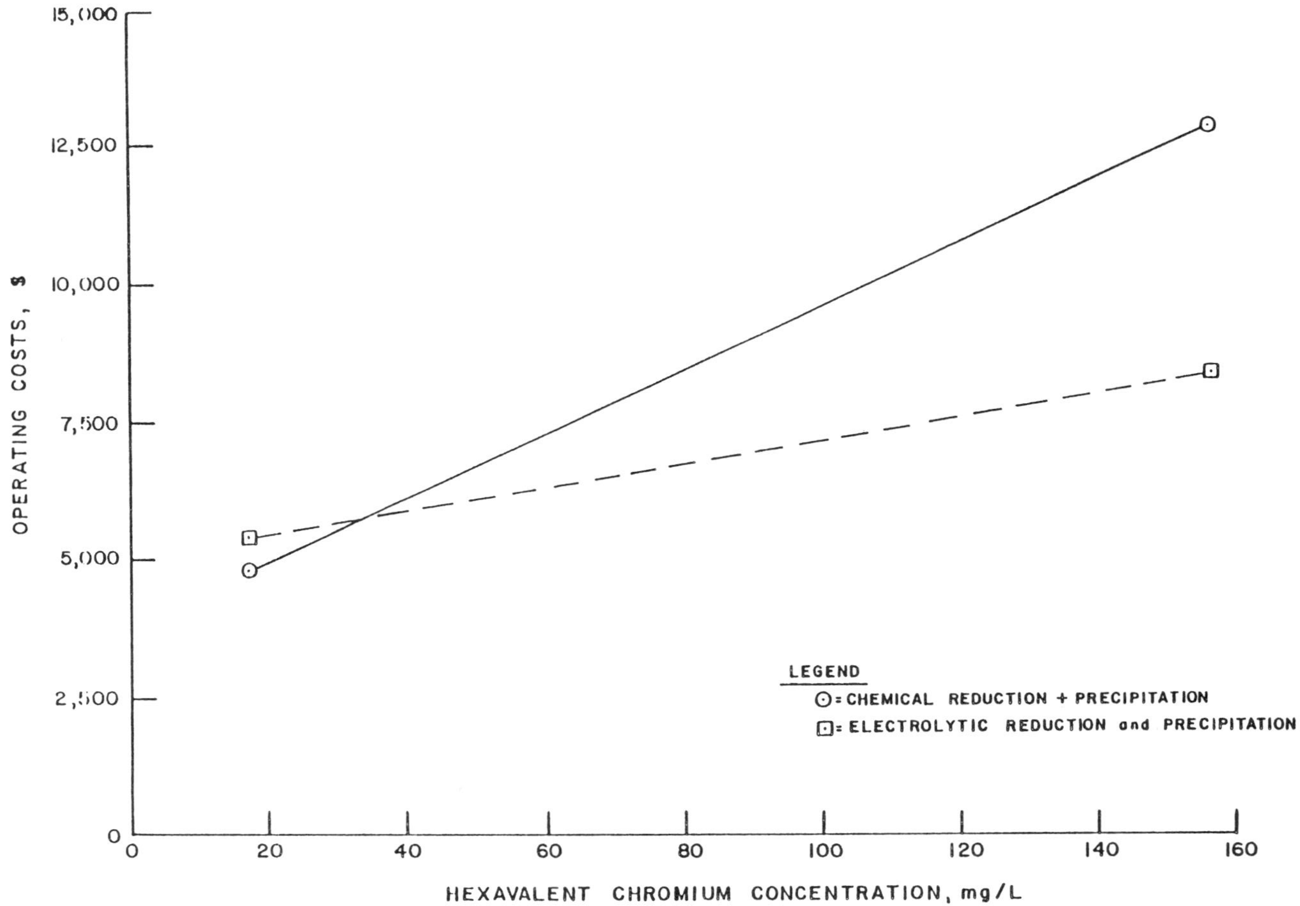

Figure 9.4.1. Operating Cost Comparison for Treatment of Hexavalent Chromium Rinse.
Source: Reference No. 11.

The case studies discussed above illustrate the fact that electrolytic recovery can be more economical than other treatment methods in certain situations. However, as indicated in the final case study above, the unit cost of reducing hexavalent chromium electrolytically is greatly affected by the inlet concentration. If the concentration is low, it may be more economical to use chemical reduction.

Another major factor to consider is the cost of disposing the metal-bearing sludge generated by most chemical treatment methods. The land disposal ban is likely to cause further increases in secure landfilling costs. Thus, situations for which recovery is not yet economically feasible may be more cost-effectively managed through recovery in the future.

9.5 PROCESS STATUS

Electrolytic recovery is applicable for certain metal/cyanide waste streams. It is a particularly attractive process for metal-bearing waste streams because it allows for metal recovery, thereby precluding the generation of metal-bearing sludge.

A number of different types of electrolytic reactors are currently manufactured. Simple, parallel-plate reactors can be used to recover noble metals such as gold and silver. More complex units with porous or granular electrodes may be required to remove metals such as copper, tin, and lead, particularly when these metals are present in low concentrations; e.g., less than 100 mg/L.

REFERENCES

1. Bishop, P.L., and R.A. Breton. Treatment of Electroless Copper Plating Wastes. In: Proceedings of the 38th Annual Purdue Industrial Waste Conference.

2. USEPA. Environmental Pollution Control Alternatives: Reducing Water Pollution Control Costs in the Electroplating Industry. EPA-625/5-85-016. 1985.

3. Centec Corporation. Navy Electroplating Pollution Control Technology. Prepared for Naval Facilities Engineering Laboratory. February 1984.

4. Snoeyink, V.L. and D. Jenkins. Water Chemistry. John Wiley & Sons, Inc. 1980.

5. Swank, C.A. Electrolytic Recovery from Rinse Waters. Printed Circuit Fabrication. 5(5). 1982.

6. Walsh, F.C., and D.R. Gabe. Electrochemical Cell Designs for Metal Removal and Recovery. In: Proceedings of the Symposium on Electroplating Engineering and Waste Recycle New Developments and Trends. Cleveland, OH. August-September 1982.

7. Patterson, J.W. Wastewater Treatment Technology. Ann Arbor Science Publishers, Ann Arbor, MI. 1975.

8. Vachon, D.T. et al. Evaluation of Electrochemical Recovery of Cadmium at a Metal Finishing Plant. Plating and Surface Finishing. pp. 68-73. April 1986.

9. Allen, R. Electrolytic Recovery from Dilute Solutions. Finishing Industries. September, 1982. pp 27.

10. Kim, B.M., and J.L. Weininger. Electrolytic Removal of Heavy Metals from Wastewaters. Environmental Progress, 1(2), pp. 121-125. May 1982.

11. Warner, B.E. Electrolytic Treatment of Job Shop Metal Finishing Wastewater. Prepared for U.S. EPA Industrial Environmental Research Laboratory. EPA-600/2-75-028. September 1975.

12. Alliance Technologies Corporation. Case Studies of Existing Treatment Applied to Hazardous Waste Banned from Landfill. Phase II - Case Study Report for Facility F. Prepared for U.S. EPA Hazardous Waste Engineering Research Laboratory. Contract No. 68-03-3243. June 1986.

13. Rosenbaum, Wayne. ETICAM, Inc. Warwick, RI. Telecon with Mark Arienti, Alliance Technologies Corporation. February, 1986.

14. Omnipore, Inc., Sugar Land, TX. Technical Literature for the Retec Heavy Metal Recovery System. March 1987.

15. Kula, D. The Cost Benefit of Metal Recovery in the Plating Shop Instead of Waste Treatment While Attaining Compliance. Presented at the Interconnections Packaging Circuitry (IPC) Fall Meeting, Denver, CO. September 1983.

16. Tyson, A.G. An Electrochemical Cell for Cadmium Recovery and Recycling. Plating and Surface Finishing. pp. 44-47. December 1984.

17. Pierce, R. International Circuit Technology (ICT), Lynchburg, VA. Telecons with Mark Arienti, Alliance Technologies Corporation. February 18-19, 1986.

10. Chemical Treatment/Removal Processes for Metals

The treatment processes discussed in this section are based on physical/ chemical methods of separation and removal of metallic contaminants in the waste feed stream. Processes discussed are:

10.1 Precipitation

10.2 Coagulation and Flocculation

10.3 Chemical Reduction

10.4 Flotation

All of these processes are used to some extent for the treatment of wastes, but differ in their applicability to various types of waste and their need for pretreatment and post-treatment procedures. The physical/chemical treatment processes (and the other treatment processes discussed in the following sections) are considered within the framework of four major areas: (1) Process Description including pretreatment and post-treatment requirements; (2) Demonstrated Performance in Field and Laboratory; (3) Cost of Treatment; and (4) Overall Status of the Technology.

10.1 PRECIPITATION

All precipitation processes operate under the same fundamental chemical principles and utilize similar types of equipment and process configurations. Additionally, pretreatment requirements and residual post-treatment options are comparable, regardless of the specific precipitation method under investigation. Therefore, similar aspects of precipitation systems will be addressed prior to discussion of specific reagent/waste combinations. Section 10.1.1 serves as introduction to the basic theory of precipitation chemistry and proceeds to identify considerations in pretreatment requirements, process equipment, process configurations, post-treatment and disposal of residuals. The remaining subsections (Sections 10.1.2 through 10.1.4) address specific precipitation reagents. These highlight the unique aspects of each, including compatable waste types, treatment costs, sludge generation and special considerations in equipment design and reagent handling practices. The reagents are:

- Hydroxides;
- Sulfides;
- Carbonates.

Each reagent subsection covers the following topics:

- General process description including typical operating characteristics;
- Performance data which identifies operating parameters, processing equipment, and system configurations;
- Capital and operating costs;
- Status of the technology.

10.1.1 General Considerations

10.1.1.1 Precipitation Theory--

The principal mechanism of precipitation involves the alteration of the ionic equilibrium of a metallic compound to produce an insoluble precipitate. Typically, an alkaline reagent is used to lower the solubility of the metallic constituent and thus, bring about precipitation. In certain cases, chemical reduction (Section 10.3) may be needed to change the characteristics of the metal ions (i.e., valence state) in order to achieve precipitation. In general, precipitation reactions form a salt and an insoluble metal complex, as illustrated in the following reaction between nickel sulfate and sodium hydroxide (caustic soda):

$NiSO_4$	+	2 NaOH	=	Na_2SO_4	+	$Ni(OH)_2$ (s)
nickel sulfate		sodium hydroxide		sodium sulfate		nickel hydroxide

Chemical precipitation normally depends on several variables[1]:

- Maintenance of an alkaline pH throughout the precipitation reaction and subsequent settling.
- Addition of a sufficient excess of treatment ions to drive the precipitation reaction to completion.
- Addition of an adequate supply of sacrificial ions (such as iron or aluminum) to ensure precipitation and removal of specific target ions.
- Effective removal of precipitated solids.

Control of pH is essential for precipitation of many metals, as illustrated by the solubility curves for selected metal hydroxides and sulfides shown in Figure 10.1.1. Hydroxide precipitation is effective in removing arsenic, cadmium, chromium (+3), copper, iron, manganese, nickel, lead, and zinc. Sulfide treatment is superior to hydroxide (and carbonate treatment) for removal of several metals. As shown by theoretical solubilities of hydroxides and sulfides of selected metals (Table 10.1.1), sulfide precipitation is highly effective in removal of cadmium, cobalt, copper, iron, mercury, manganese, nickel, silver, tin, and zinc. Estimated achievable maximum 30-day average concentrations of several heavy metals under different chemical precipitation and solids removal technologies are shown in Table 10.1.2. The estimated achievable concentrations are based on the performance data reported in literature.[1]

Another factor that effects precipitation reagent performance is the presence in solution of chelator/complexing agents. A list of common agents, together with their structures, is given in Table 10.1.3. These chelator/complexing agents prevent the complete precipitation of heavy metal hydroxides by competing with the hydroxyl ion for possession of the heavy metal, e.g.,

$$Zn(NH_3)_4^{++} + 2OH^- = Zn(OH)_2(s) + 4NH_{3\,(aq)}$$

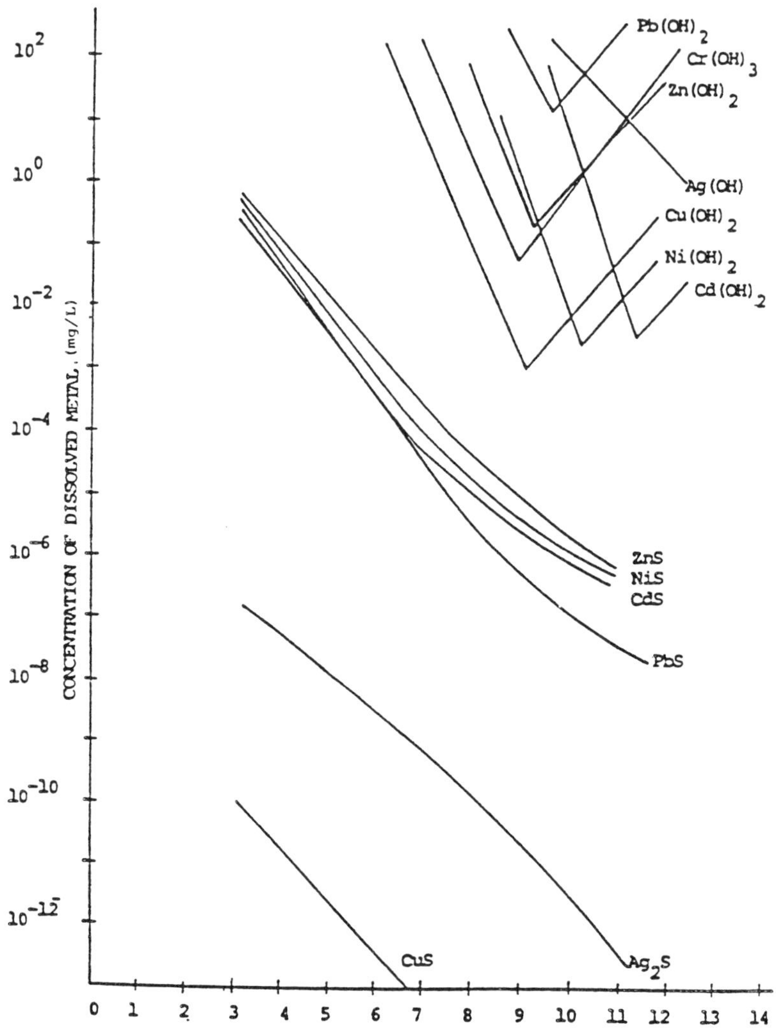

Figure 10.1.1. Solubility of metal hydroxides and sulfides as a function of pH.

Source: Reference 1.

TABLE 10.1.1. THEORETICAL SOLUBILITIES OF HYDROXIDES AND SULFIDES OF SELECTED METALS IN PURE WATER

Metal	Solubility of metal ion, mg/L		
	As hydroxide	As carbonate	As sulfide
Cadmium (Cd^{++})	2.3×10^{-5}	1.0×10^{-4}	6.7×10^{-10}
Chromium (Cr^{+++})	8.4×10^{-4}	-	No precipitate
Cobalt (Co^{++})	2.2×10^{-1}	-	1.0×10^{-8}
Copper (Cu^{++})	2.2×10^{-2}	-	5.8×10^{-18}
Iron (Fe^{++})	8.9×10^{-1}	-	3.4×10^{-5}
Lead (Pb^{++})	2.1	7.0×10^{-3}	3.8×10^{-9}
Manganese (Mn^{++})	1.2	-	2.1×10^{-3}
Mercury (Hg^{++})	3.9×10^{-4}	3.9×10^{-2}	9.0×10^{-20}
Nickel (Ni^{++})	6.9×10^{-3}	1.9×10^{-1}	6.9×10^{-8}
Silver (Ag^{+})	13.3	2.1×10^{-1}	7.4×10^{-12}
Tin (Sn^{++})	1.1×10^{-4}	-	3.8×10^{-8}
Zinc (Zn^{++})	1.1	7.0×10^{-4}	2.3×10^{-7}

Source: Reference 1.

TABLE 10.1.2. ESTIMATED ACHIEVABLE MAXIMUM 30-DAY AVERAGES FOR THE APPLIED TECHNOLOGIES

	Final concentrations (mg/L)					
	Lime ppt followed by sedimentation	Lime ppt followed by filtration	Sulfide ppt followed by filtration	Ferrite coprecipitation followed by filtration	Soda ash addition followed by sedimentation	Soda ash addition followed by filtration
Antimony, Sb	0.8 - 1.5	0.4 - 0.8				
Arsenic, As	0.5 - 1.0	0.5 - 1.0	0.05 - 0.1			
Beryllium, Be	0.1 - 0.5	0.01 - 0.1				
Cadmium, Cd	0.1 - 0.5	0.05 - 0.1	0.01 - 0.1	0.05		
Copper, Cu	0.05 - 1.0	0.4 - 0.7	0.05 - 0.5	0.05		
Chromium, Cr(+3)	0.0 - 0.5	0.05 - 0.5		0.01		
Lead, Pb	0.3 - 1.6	0.05 - 0.6	0.05 - 0.4	0.20	0.4 - 0.8	0.1 - 0.6
Mercury, Hg(+2)			0.01 - 0.05	0.01		
Nickel, Ni	0.2 - 1.5	0.1 - 0.5	0.05 - 0.5			
Silver, Ag	0.4 - 0.8	0.2 - 0.4	0.05 - 0.2			
Selenium, Se	0.2 - 1.0	0.1 - 0.5				
Thallium, Tl	0.2 - 1.0	0.1 - 0.5				
Zinc, Zn	0.5 - 1.5	0.4 - 1.2	0.02 - 1.2	0.02 - 0.5		

ppt = precipitation

Source: Reference 1.

TABLE 10.1.3. STRUCTURES OF CHELATING AGENTS SEPARATED

NTA

(Nitrilo triacetic acid)

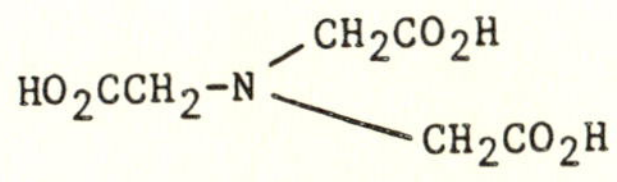

EDTA

(Ethylene d

(HO_2CCH_2

EBTA

(Ethylene bis(oxyethylenenitrilo)tetraacetic acid)

$(HO_2CCH_2)_2NCH_2CH_2OCH_2CH_2OCH_2CH_2N(CH_2CO_2H)_2$

CDTA

1,2 diamino cyclohexane tetraacetic acid

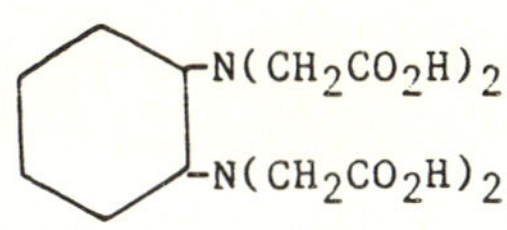

Source: Adapted from Reference 3.

The equation indicates that solutions which contain dissolved ammonia tend to drive the reaction to the left, thereby preventing removal of zinc as the hydroxide. Calculations show that for a solution containing 100 ppm of dissolved NH_3, at a pH of 8.0, nearly 3.0 ppm of Zn^{++} will remain unprecipitated.[2] All complexing agents solubilize certain heavy metals in a fashion similar to that given in the above example.[2]

10.1.1.2 Pretreatment Requirements--

Pretreatment of metals containing wastes prior to precipitation typically consists of gross solids removal (e.g., filtration), flow equalization, neutralization, or treatment of individual waste streams prior to combination with other process wastes. These treatments of segregated wastes result in economic benefits from reduced reagent costs and smaller equipment sizing. Other common pretreatment processes include cyanide destruction, chromium reduction, and oil removal.

Cyanide destruction--Cyanide wastes cannot be mixed with metal-containing wastes due to the formation normally stable organo-metallic complexes or the possible evolution of toxic hydrogen cyanide gas. Instead, cyanide is typically oxidized to carbon dioxide and nitrogen gas through a chemical oxidation process. In two-stage chlorination, pH is typically maintained around 11.0 in the first reaction vessel and 8.0 to 8.5 in the second vessel through addition of NaOH, as required (see Section 14.1).[3]

Chromium reduction--Chromic acid wastes may contain hexavalent chromium which must be reduced to the trivalent form prior to precipitation. Reduction typically occurs at pH 2.0 to 3.0 through addition of acid (e.g., sulfuric) and a reducing agent (e.g., sulfur dioxide, ferrous sulfate, sodium metabisulfite, or sodium bisulfite). However, alkaline reduction (pH 7 to 10) using ferrous iron has also been demonstrated. It has proven to be cost-effective for highly buffered alkaline waste and the treatment of mixed metal wastes containing less than 10 mg/L of hexavalent chromium (see Section 10.3).[3]

Neutralization--Neutralization consists of adjusting an acidic or alkaline waste stream with the appropriate reagent to a final pH of 6 to 9, which meets surface water discharge requirements established under the Clean Water Act. However, it is sometimes only necessary to adjust the pH to approximately 5 to 6 (i.e., partial neutralization) to achieve certain treatment objectives. In other applications it may be necessary to neutralize an acid to pH 9 or higher to precipitate metallic ions or to completely clarify a waste for acceptable discharge. These techniques are called under- and over-neutralization, respectively.[4]

Table 10.1.4 identifies several of the more prevalent neutralization reagents and their characteristics. The selection of the appropriate reagent for wastewater neutralization processes is site-specific and dependent on the following considerations: wastewater characteristics, reagent costs and availability, speed of reaction, buffering qualities, product solubility, costs associated with reagent handling, and residual quantities and characteristics. Typically, the first step in reagent selections is to characterize the wastewater. General parameters of interest include flow (rate, quantity), pH, pollutant loading, physical form of waste, and waste/reagent compatibility. These characteristics narrow the range of reagents and treatment configurations available for consideration.

Following the selection of candidate reagents, the quantity of reagent required to neutralize the waste to the desired end point must be determined. Reagent quantity is usually calculated by developing a titration curve for each candidate reagent using representative wastewater samples.[5,6] These data determine the quantity of reagent required to bring the sample volume of wastewater to the desired pH.

The next step in the experimental procedure is the preparation of reaction rate curves and development of kinetic rate equations for each candidate reagent. Reagent reactivity is an important factor in determining retention time and consequently the size of the treatment facility, the final effluent quality, and the ease or difficulty of process system control. These parameters, in turn, will affect both capital and operational costs associated with the wastewater treatment system. Reaction rate curves for various quantities of residual reagent (i.e., excess above stoichiometric

TABLE 10.1.4. ACID/ALKALINE NEUTRALIZATION AGENT CHARACTERIZATION

Reagant	Molecular formula	Chemical name	Molecular weight	Common form and commercial strength	Bulk density Kg/m^3	Solubility g/100 g water	Typical use	Advantages	Disadvantages	Shipment form	Equivalent basicity factor[a]	Approximate cost/ton ($)	Cost/ton basicity[a] ($)	Equivalent weight
High Calcium Limestone	$CaCO_3$	Calcium Carbonate	100.1	Powder 95% $CaCO_3$ granules	2000-2800	0.0014^{25} (@ 25°C)	Acid Neutralization	Relatively inexpensive	Contains impurities slow reacting	Bulk	0.489	6	12.27	50.5
High Calcium Hydrated Lime	$Ca(OH)_2$	Calcium Hydroxide	74.1	Powder 72-74% CaO	400-640	0.15^{30} (@ 30°C)	Acid Neutralization	Relatively inexpensive	Contains impurities slow reacting	Bulk	0.710	46	64.79	37.05
High Calcium Quicklime	CaO	Calcium Oxide	56.1	Pebble 93-98% CaO	770-1120	Converted to $Ca(OH)_2$	Acid Neutralization	Relatively inexpensive	Requires dry storage	Bulk	0.941	39	41.45	28.05
Carbon Dioxide	CO_2	Carbon Dioxide	54.0	Gas liquified under pressure	--	--	Alkali Neutralization	May be available from flue gas	--	Tank car	---	200	---	27
Dolomitic Hydrated Lime	$Ca(OH)^2$ MgO	Normal Dolomitic	114.4	Powder 46-48% CaO 33-34% MgO	416-666	0.15^{30} (@ 30°C)	Acid Neutralization	Relatively inexpensive	Less reactive than high calcium lime	Bagged	0.912	46	50.44	33.9
Dolomitic Quicklime	CaO-MgO	Calcium Magnesium	96.4	Pebble 55-58% CaO 38-41% MgO	801-1165	Converted to $Ca(OH)_2$ and $Mg(OH)_2$	Acid Neutralization	Relatively inexpensive	Less reactive than high calcium lime	Bulk	1.110	39	35.14	24.9
Soda Ash	Na_2CO_3	Sodium Carbonate	106.0	Powder 58% Na_2O	560-1041	27.6^{30} (@ 30°C)	Acid Neutralization	Highly reactive soluble	Higher costs than calcium reagents	Bulk	0.507	83	163.71	53.0
Caustic Soda	NaOH	Sodium Hydroxide	40.0	Liquid 73% NaOH	----	347^{100} (@ 100°C)	Acid Neutralization	Highly reactive, easy handling	High cost, requires heated storage	Tank car	0.687	205	298.4	40
Magnesia	MgO	Magnesium Oxide	40.31	Powder	1017.9	0.0086^{30} (@ 30°C)	Acid Neutralization	Highly reactive	High Cost	Bagged	0.929	365	392.89	20.16
Sulfuric Acid	H_2SO_4	Sulfuric Acid	98.1	Liquid 77% H_2SO_4 93% H_2SO_4	1704-1834	Complete	Alkali Neutralization	Highly reactive, inexpensive	Forms calcium sulfate sludge with calcium	Tank car	---	59 71	---	49.05
Muriatic Acid	HCL	Hydrochloric acid	36.5	Liquid 20°Be 22°Be	1157-1177	Complete	Alkali Neutralization	Highly reactive	More expensive than sulfuric acid	Tank car	---	66 74	---	36.5

[a]Reference 3.

requirements) are determined by plotting pH as a function of time. Other variables which should be monitored include temperature rise, agitator speed, density, viscosity, color, sludge volume, and settleability.

From the titration and reaction rate curves, kinetic rate equations can be developed. Methods for determining the kinetic rate equation (e.g., integral or statistical analysis) are discussed in the literature.[7] The reader is also referred to standard engineering texts for reactor and costing methodologies based on flow parameters and kinetic rate equations.[8]

In the final selection, the optimal reagents and reagent/waste feed ratio will be those which incur the least overall cost, including not only the cost of the reagent itself, but also the cost of purchasing and maintaining the reagent and neutralization systems, and the costs associated with residual handling. The combination of all such factors may make a slightly more expensive reagent less expensive overall.

Oil removal--Removal of oil through emulsion breaking, dissolved air flotation, skimming or coalescing may also be needed prior to precipitation. Traditionally, emulsified oils have been treated at low pH (e.g., pH of 2.0) with alum. However, this form of treatment is giving way to the use of more effective emulsion breaking coagulants such as cationic polymers and other specialty chemicals.[3]

Flow equalization--The most prevalent form of pretreatment is flow equalization. It is generally used in facilities which experience wide variations in the wastewater flow or pollutant concentrations.[7] Figure 10.1.2 illustrates a number of ways that flow equalization can be achieved. In all methods of flow equilization, care must be exercised during the wastewater analysis to completely characterize any peak flows or concentrations that might overload the system. In addition, flexibility in system design should be provided for any future expansion, change in location, or deviation in flow rates.

10.1.1.3 General Precipitation Equipment--

The heart of a wastewater precipitation process is the pH control system, which must bring the wastewater pH to the level required to precipitate the optimal quantity of contaminant metal salt. As previously discussed, pH

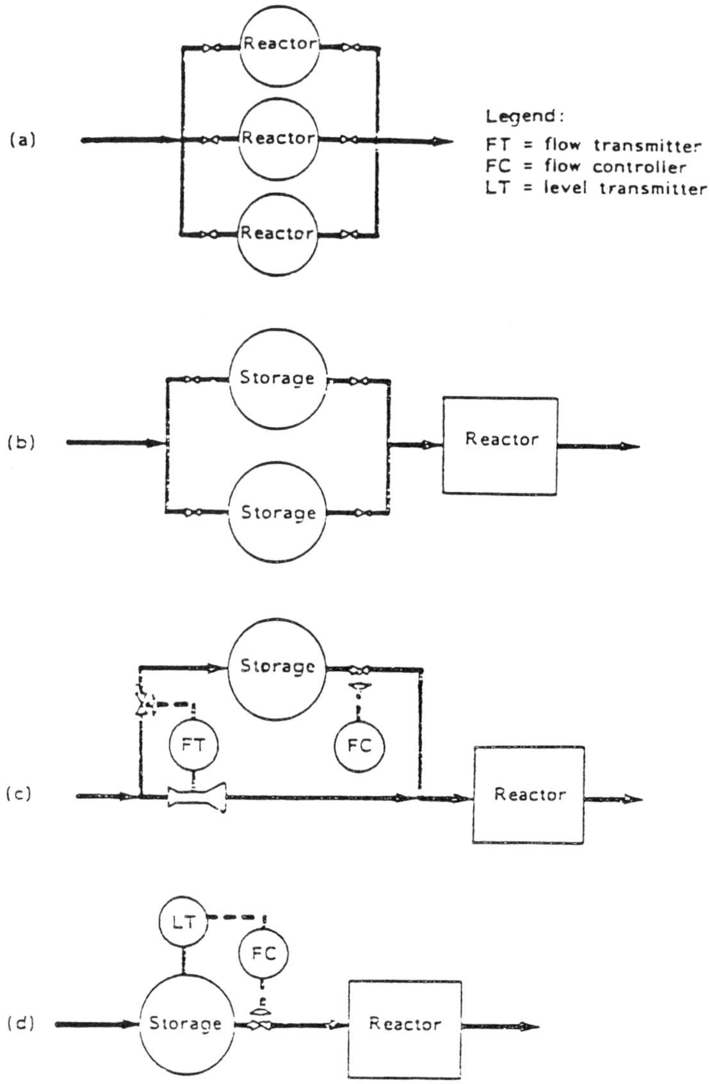

Figure 10.1.2. Alternative concepts for wastewater equalization: (a) batch reactor system, (b) batch equalization continuous processing, (c) side-stream equalization, and (d) flowthrough equalization.

Source: Reference 7.

control is achieved through the use of a neutralization system. Neutralization may occur either in the precipitation reactor or in a separate tank. A wide variety of treatment options and configurations are available. However, fully engineered component neutralization/precipitation systems generally consist of the following equipment:

- Neutralization/Precipitation System
  - Tank(s)
  - Mixer(s)
  - pH control instrumentation
- Chemical Feed System
  - Tank(s)
  - Mixer(s)
  - Level instrumentation
  - Metering equipment
- Miscellaneous
  - Flow monitoring
  - Effluent pH recorder
  - Electrical and mechanical fit-up
  - Incremental engineering requirements

In addition, there is a need for facilities and equipment to collect and segregate the wastewaters, transport the wastewaters to equalization sumps, pump the wastewaters to the treatment system, perform liquid/solid separation, and convey the treated wastewaters to the point of discharge.

Precipitation tanks are fabricated from a wide range of construction materials such as masonry, metal, plastic, or elastomers. Corrosion resistance can be enhanced with coatings or liners which prevent the premature decomposition of tank walls. For example, concrete reactors susceptible to corrosion can be installed with a two-layer coating of a 6.3 mm base surface (glass-reinforced epoxy polyamide) covered by a 1.0 mm coating of polyurethane elastomer to extend service lifetimes.[7]

Vessel geometrics can be either cubical or cylindrical in nature with agitation provided overhead in line with the vertical axis. While cubical tanks need no baffling, cylindrical vessels are typically constructed with suitable ribs to prevent swirling and maintain adequate contact between the reactants. A general rule of thumb in the design of precipitation reactors is that the depth of the liquid should be roughly equivalent to the tank diameter or width.[9]

Reactors can be arranged in either single- or multi-stage configurations and operate in either batch or continuous mode. Multi-stage continuous configurations are typically required to neutralize and precipitate concentrated wastes with variable feed rates. In these units most of the neutralization reagent is added in the first vessel with only final pH adjustments (polishing) and precipitation agent addition made in the remaining reaction vessels. This is particularly true when using reagents which require extensive retention time. Single-stage continuous or batch precipitation is suitable for most applications with highly buffered solutions or dilute wastewaters not subject to rapid changes in flow rate or pH.

An adequate retention time is required to provide time for the precipitation reaction to go to completion. This factor is especially critical where a dry feed (lime or ferrous sulfide) or slurry is used as the control agent. In these systems, the solids must dissolve before they react, increasing the required retention time and tank capacity. For example, liquid reagents used in continuous flow operations generally require 3 to 5 minutes of retention time in the first tank. Three minutes corresponds to the minimum time for adequate mixing. In comparison, solid-based reagent systems such as lime or ferrous sulfide typically require approximately 30 minutes of retention time.[4,5,10]

The pH control systems for batch precipitation processes can be quite simple with only on-off control provided via solenoid or air activated valves. Control system designs for continuous flow precipitation systems are more complicated because the wastewater feeds often fluctuate in both flow and concentration. Systems currently available include: proportional, cascade, feedforward, or feedback pH control. Each system has distinct advantages and disadvantages which are discussed in detail in the literature.[5,9,11,12] The pH control equipment usually consists of a pH probe, monitor, and recorder. In addition, there is typically a control panel with an indicator, starters and controls for metering pumps, all relays, high/low pH alarms, switches, and mixer motor starters.

Chemical feed apparatus consists of storage tanks, agitation, level instrumentation, and metering pumps. Storage tanks should be sized according to maximal feed rate, shipping time required, and quantity of shipment. The total storage capacity should be more than sufficient to guarantee a chemical supply while awaiting delivery. Storage containers must be suitable for the reagent being used. For example, hydroscopic reagents such as high calcium quicklime must be stored in moisture-proof tanks to prevent atmospheric degradation.

In addition to the chemical feed and neutralization/precipitation systems, both flow monitoring and effluent pH recording equipment are necessary to prevent discharge of insufficiently treated waste resulting from surges or upsets. Also, spare parts such as pH probes, pH controller circuit board, metering pump ball valves, o-rings, and strainers should be kept on hand to prevent any excessive downtime.

10.1.1.4 Clarification and Sludge Consolidation--

Clarification and sludge consolidation unit operations are typically applied as post-treatments to the majority of aqueous metals containing waste treatment systems. Figure 10.1.3 illustrates a general treatment/ post-treatment approach for aqueous metal/cyanide bearing waste streams.[13]

Usually, wastewaters undergo chemical treatment and enter a clarifier where the flow is decreased to a point that allows solids with a specific gravity greater than that of the liquid settle to the bottom. For liquid/solid mixtures with a slight density difference, an organic polymer (flocculant) can be added to allow the solids to agglomerate and improve the settling characteristics (see Section 10.2).[14] The supernatant in the overflow is drawn off and residual trace organics or solids are removed in a final polishing step such as carbon adsorption, ultrafiltration, or ion exchange. The solids in the underflow can then be discharged to a holding tank for subsequent dewatering.

In addition to differences in the quantity of sludge generated, each reagent imparts to the sludge variable settling characteristics, thereby affecting the sizing parameters of downstream equipment.[15] For example, lime neutralized sludge exhibits a granular nature that settles fairly rapidly and dewaters effectively (4 to 20 lb of dry solids/hr/ft^2 yielding a 3/16 to 3/8 in. cake). Conversely, sodium hydroxide sludge results in a fluffy gelatinous precipitate with low settling rates.[16] Figure 10.1.4 shows the

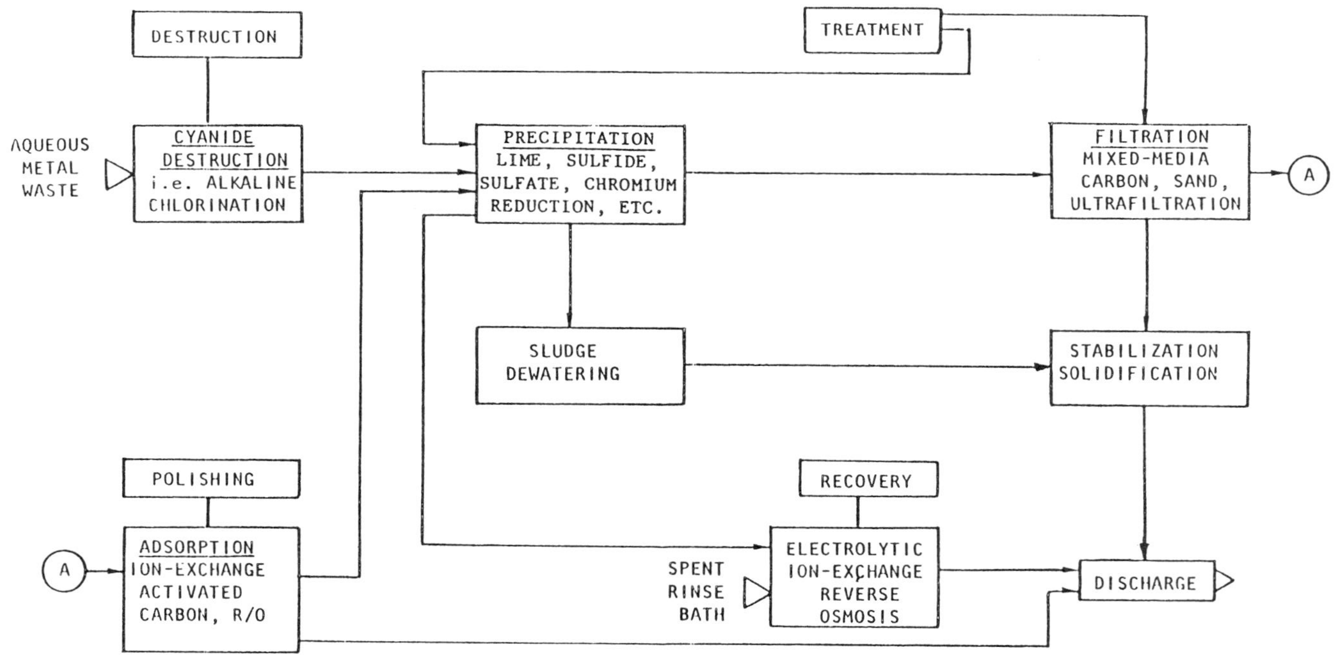

Figure 10.1.3. A general treatment approach for aqueous metal/cyanide bearing waste streams.
Source: Reference 13.

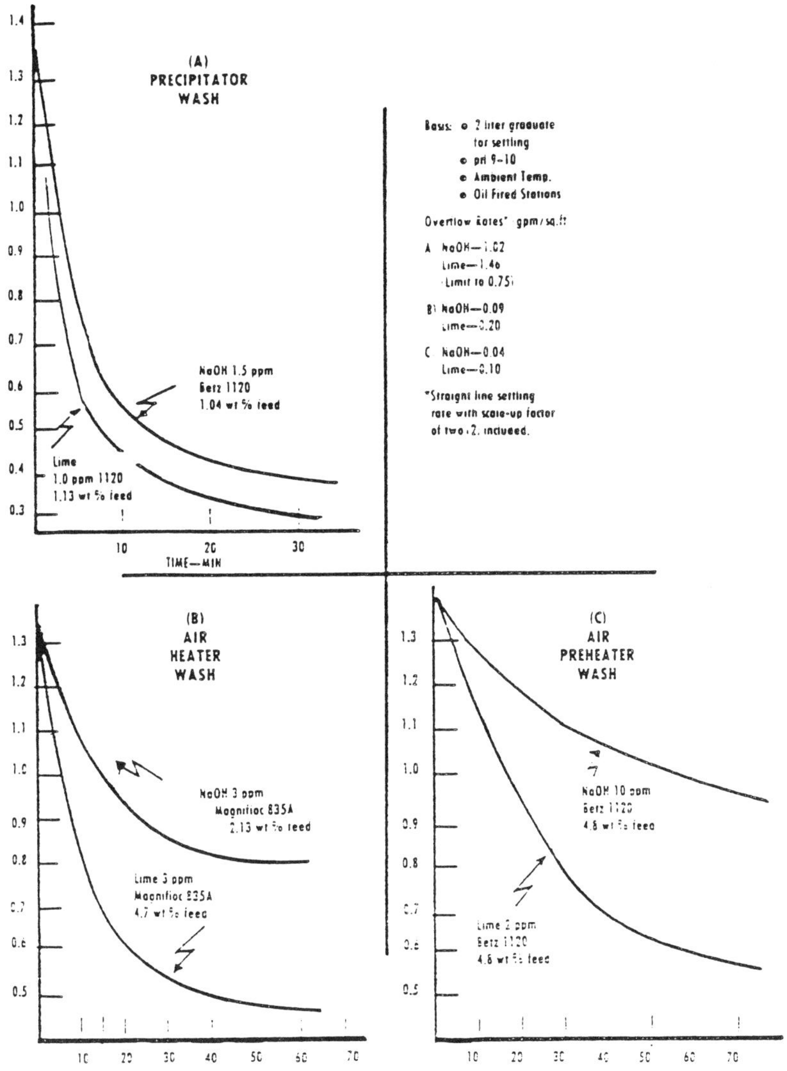

Figure 10.1.4. Settling rate curves.

Source: Reference 16.

results of three settling tests conducted on power plant effluents with both lime and sodium hydroxide. In all three cases sodium hydroxide settled more slowly, and in subsequent filtration tests, dewatered about half as effectively.[16] However, the use of lime or calcium carbonate generates greater sludge weight and volume. This is primarily due to insoluble acid salts and calcium sulfates formed when precipitating metal sulfate containing wastes such as acid plating baths. Therefore, as landfill and hauling costs become more significant, sodium hydroxide becomes more competitive with lime and limestone as a precipitation agent.

Few, if any, sludges settle at a rate sufficient to utilize only clarifiers or thickeners to accumulate sludge for disposal on land.[17] Therefore, the underflow from the clarifier is typically concentrated through the use of mechanical dewatering equipment such as centrifuges, rotary vacuum filters, belt filters, drying ovens, and recessed-plate filter presses. The obtainable degree of cake dryness can be determined by bench-scale tests by the equipment vendor to identify the suitability of a particular dewatering device (see Table 10.1.5). The low solids content of sodium hydroxide after sedimentation (3 to 10 percent) requires the use of a filter press.[18] Conversely, suspended solids removal from lime neutralized sludges can be accomplished through use of a wider range of equipment including rotary vacuum or continuous belt filters.

10.1.1.5 Land Disposal of Residuals--

Installation of a metals precipitation system inevitably results in the problem of sludge disposal. The cost of hauling the sludge to a licensed hazardous waste landfill will depend on the volume of sludge, the distance hauled, and the sludge composition. Sometimes it is possible to dispose of calcium-based reagent sludges through agricultural or acid pond liming. In one neutralization/precipitation application, over 200,000 lbs/acre of lime-treated waste pickle liquor sludge was applied onto Miami silt loam to improve overall crop yields.[19]

TABLE 10.1.5. SUMMARY OF SLUDGE DEWATERING DEVICE CHARACTERISTICS

Parameter	Gravity (low pressure)	Basket centrifuge	Solid bowl centrifuge	Vacuum filter	Belt filter press	Recessed filter press
Cake solids %	16 - 24	20 - 30	30 - 42	30 - 40	36 - 46	50 - 60
Operational variables	-Rate of sludge feed -Polymer concentration -Belt speed -Depth of sludge in cylinder	-Bowl speed -Time at full speed -Depth of skimming -Sludge feed rate	-Bowl/conveyor differential speed -Pool depth -Sludge feed rate	-Quantity of H_2O -Drum speed -Vacuum level -Conditioning chemicals -Filter media	-Belt speed -Belt tension -Washwater flow and pressure -Belt type -Polymer conditioner	-Feed pressure -Filtration time -Use of precoat -Cloth washing frequency -Filter cloth used
Advantages	-Low energy & capital cost -Low space requirements -Requires little operator skill	-Same machine for thickening & dewatering -Very flexible -Little operator attention	-Easy to install -Low space requirement -Either thickening or dewatering -High rate of feed -Can operate on highly variable feeds	-Continuous operation -Long media life -Low maintenance -Easy operation	-Only filter press produces drier cake -Low power -Low noise & vibration -Continuous operation	-High solids filter cake -High solids capture -Only mechanical device capable of meeting some landfill requirements
Disadvantages	-Limited capacity -Low solids concentration -Requires large quantity of conditioning chemicals	-Unit is not continuous -High ratio of capital cost to capacity -Requires complex controls -Requires noise control	-Requires prescreening -Very noisy with high vibration -High power consumption -Requires high maintenance skills	-High power requirement -Vacuum pumps are noisy -Requires at least 3% feed solids for operation	-Very sensitive to incoming feed -Short media life -Greater operational attention and polymer dosage	-High capital cost -Batch discharge -High polymer usage -Media replacement costs are high

Source: Adapted from Reference 18.

Another option is to treat the waste by immobilizing the waste constituents for as long as they remain hazardous. This method of treatment, based on fixation or encapsulation processes, is a possibility for some metals containing wastes. Certain of these residuals could be found hazardous; their heavy metal content may lead to positive tests for EP toxicity. In such cases, encapsulation may be needed to eliminate this characteristic.

The following discussions will summarize available information concerning immobilization techniques, namely solidification/fixation or encapsulation. Chemical fixation involves the chemical interaction of the waste with a binder; encapsulation is a process in which the waste is physically entrapped within a stable, solid matrix.

Solidification technologies are usually categorized on the basis of the principal binding media. These media include: cement-based compounds, lime-based pozzolanic materials, thermoplasts, and organic polymers (thermosets). The resulting stable matrix produces a material that contains the waste in a nonleachable form, is nondegradable, cost-effective, and does not render the land it is disposed in unusable for other purposes. A brief summary of the compatibility and cost data for selected waste solidification/stabilization systems is presented in Tables 10.1.6 and 10.1.7.

Cement-based systems--These systems utilize type I Portland cement, water, proprietary additives, possibly fly ash, and waste sludges to form a monolithic, rock-like mass. In an EPA publication,[20] several vendors of cement-based systems reported problems with organic wastes containing oils, solvents, and greases not miscible with an aqueous phase. For although the unreactive organic wastes become encased in the solids matrix, their presence can retard setting, cause swelling, and reduce final strength.[21] These systems are most commonly used to treat inorganic wastes such as incinerator generated wastes and heavy metal sludges from neutralization/precipitation processes.

TABLE 10.1.6. COMPATIBILITY OF SELECTED WASTE CATEGORIES WITH DIFFERENT WASTE SOLIDIFICATION/STABILIZATION TECHNIQUES

	Treatment Type						
Waste component	Cement based	Lime based	Thermoplastic solidification	Organic polymer (UF)*	Surface encapsulation	Self-cementing techniques	Classification and synthetic mineral formation
Organics:							
1. Organic solvents and oils	May impede setting, may escape as vapor	Many impede setting, may escape as vapor	Organics may vaporize on heating	May retard set of polymers	Must first be absorbed on solid matrix	Fire danger on heating	Wastes decompose at high temperatures
2. Solid organics (e.g., plastics, resins, tars)	Good--often increases durability	Good--often increases durability	Possible use as binding agent	May retard set of polymers	Compatible--many encapsulation materials are plastic	Fire danger on heating	Wastes decompose at high temperatures
Inorganics:							
1. Acid wastes	Cement will neutralize acids	Compatible	Can be neutralized before incorporation	Compatible	Can be neutralized before incorporation	May be neutralized to form sulfate salts	Can be neutralized and incorporated
2. Oxidizers	Compatible	Compatible	May cause matrix break down, fire	May cause matrix break down	May cause deterioration of encapsulating materials	Compatible if sulfates are present	High temperatures may cause undeable reactions
3. Sulfates	May retard setting and cause spalling unless special cement is used	Compatible	May dehydrate and rehydrate causing splitting	Compatible	Compatible	Compatible	Compatible in many cases
4. Halides	Easily leached from cement, may retard setting	May retard set, most are easily leached	May dehydrate	Compatible	Compatible	Compatible if sulfates are also present	Compatible in many cases
5. Heavy metals	Compatible	Compatible	Compatible	Acid pH solubilizes metal hydroxides	Compatible	Compatible if sulfates are present	Compatible in many cases
6. Radioactive materials	Compatible	Compatible	Compatible	Compatible	Compatible	Compatible if sulfates are present	Compatible

* Urea-Formaldehyde resin.

Source: Reference 22.

TABLE 10.1.7. PRESENT AND PROJECTED ECONOMIC CONSIDERATIONS FOR WASTE SOLIDIFICATION/ STABILIZATION SYSTEMS

Type of treatment system	Major materials required	Unit cost of material	Amount of material required to treat 100 lbs of raw waste	Cost of material required to treat 100 lbs of raw waste	Trends in price	Equipment costs	Energy use
Cement-based	Portland Cement	$0.03/lb	100 lb	$ 3.00	Stable	Low	Low
Pozzolanic	Lime Flyash	$0.03/lb	100 lb	$ 3.00	Stable	Low	Low
Thermoplastic (bitumen-based)	Bitumen Drums	$0.05/lb $27/drum	100 lb 0.8 drum	$18.60	Keyed to oil prices	Very high	High
Organic polymer (polyester system)	Polyester Catalyst Drums	$0.45/lb $1.11/lb $17/drum	43 lb of polyester-catalyst mix	$27.70	Keyed to oil prices	Very high	High
Surface encapsulation (polyethylene)	Polyethylene	Varies	Varies	$ 4.50*	Keyed to oil prices	Very high	High
Self-cementing	Gypsum (from waste)	**	10 lb	**	Stable	Moderate	Moderate
Glassification/mineral synthesis	Feldspar	$0.03/lb	Varies	--	Stable	High	Very high

* Based on the full cost of $91/ton.
** Negligible but energy cost for calcining are appreciable.

Source: Reference 22.

Lime-based (pozzolanic) techniques--Pozzolanic concrete is the reaction product of fine-grained aluminous siliceous (pozzolanic) material, calcium (lime), and water. The pozzolanic materials are wastes themselves and typically consist of fly ash, ground blast furnace slag, and cement kiln dust. The cementicious product is a bulky and heavy solid waste used primarily in inorganic waste treatment such as the solidification of heavy metal and flue gas desulfurization sludges.

Thermoplastic material--In a thermoplastic stabilization process, the waste is dried, heated (260 to 450°F), and dispersed through a heated plastic matrix. Principal binding media include asphalt, bitumen, polypropylene, polyethylene, or sulfur. The resultant matrix is relatively resistant to leaching and biodegradation, and the rates of loss to aqueous contacting fluids are significantly lower than those of cement or lime-based systems. However, this process is not suited to wastes that act as solvents for the thermoplastic material. Also, there is a risk of fire or secondary air pollution with wastes that thermally decompose at high temperature.[22]

Organic polymers (thermosets)--Thermosets are polymeric materials that cross-link to form an insoluble mass as a result of chemical reaction between reagents, with catalysts sometimes used to initiate reaction. Waste constituents could conceivably enter into the reaction, but most likely will be merely physically entrapped, within the cross-linked matrix. The cross-linked polymer or thermoset will not soften when heated after undergoing the initial set. Principal binding agents or reactants for stabilization include ureas, phenolics, epoxides, and polyesters. Although the thermosetting polymer process has been used most frequently in the radioactive waste management industry, there are formulations that may be applicable to certain precipitation sludges. It is important to note that the concept of thermoset stabilization, like thermoplastic stabilization, does not require that chemical reaction take place during the solidification process. The waste materials are physically trapped in an organic resin matrix that, like thermoplastics, may biodegrade and release much of the waste as a leachate.[23] It is also an organic material that will thermally decompose if exposed to a fire.

Encapsulation is often used to describe any stabilization process in which the waste particles are enclosed in a coating or jacket of inert material. A number of systems are currently available utilizing polybutadiene, inorganic polymers (potassium silicates), Portland concrete, polyethylene, and other resins as macroencapsulation agents for wastes that have or have not been subjected to prior stabilization processes. Several different encapsulation schemes have been described in Reference 24. The resulting products are generally strong encapsulated solids, quite resistant to chemical and mechanical stress, and to reaction with water.[25] Wastes successfully treated by these methods and their costs are summarized in Tables 10.1.8 and 10.1.9. The technologies could be considered for stabilizing precipitation sludges, but are dependent on the compatibility of the precipitation waste and the encapsulating material. EPA is now in the process of developing criteria which stabilized/solidified wastes must meet in order to make them acceptable for land disposal.[26]

10.1.2 Hydroxide Precipitation

10.1.2.1 Process Description--

Hydroxide precipitation for heavy metals removal from aqueous waste streams is both an effective and economical treatment technology. The treatment converts soluble metal ions into insoluble hydroxide compounds. The metals can then be separated from the liquid through sedimentation and/or filtration. The most commonly used precipitating agents are lime [CaO or $Ca(OH)_2$] and caustic soda ($NaOH$).

Hydroxide precipitation has been widely applied in treating industrial wastewaters. The following is a list of some of the industries which use this technology:

- nonferrous metal processing,
- ore mining and dressing,
- utility power generation,
- metals plating, and
- battery manufacturing.

TABLE 10.1.8. ENCAPSULATED WASTE EVALUATED AT THE U.S. ARMY WATERWAYS EXPERIMENT STATION

Code No.	Source of Waste	Major Contaminants
100	SO_x scrubber sludge, lime process, eastern coal	Ca, $SO_4^{=}/SO_3^{=}$
200	Electroplating sludge	Cu, Cr, Zn
300	Nickel - cadmium battery production sludge	Ni, Cd
400	SO_x scrubber sludge, limestone process eastern coal	Cu, $SO_4^{=}/SO_3^{=}$
500	SO_x scrubber sludge, double alkali process eastern coal	Na, Ca, $SO_4^{=}/SO_3^{=}$
600	SO_x scrubber sludge, limestone process, western coal	Ca, $SO_4^{=}/SO_3^{=}$
700	Pigment production sludge	Cr, Fe, CN
800	Chlorine production brine sludge	Na, Cl^-, Hg
900	Calcium fluoride sludge	Ca, F^-
1000	SO_x scrubber sludge, double alkali process, western coal	Cu, Na, $SO_4^{=}/SO_3^{=}$

Source: Reference 25.

TABLE 10.1.9. ESTIMATED COSTS OF ENCAPSULATION

Process Option	Estimated Cost
Resin Fusion:	
Unconfined waste	\$110/dry ton
55-Gallon drums	\$0.45/gal
Resin spray-on	Not determined
Plastic Welding	\$253/ton = \$63.40/drum (80,000 55-gal drums/year)

Source: Reference 25.

In the first step, the hydroxide precipitating agent is thoroughly mixed with the wastewater stream. The reactions which begin in the flash-mix tank and which result in formation of the insoluble metal hydroxides are given below where M^{++} is the metal cation removed.

for quicklime:

$$CaO + H_2O = Ca(OH)_2$$

$$M^{++} + Ca(OH)_2 = M(OH)_2 + Ca^{++}$$

for hydrated lime:

$$M^{++} + Ca(OH)_2 = M(OH)_2 + Ca^{++}$$

for caustic soda:

$$M^{++} + 2NaOH = M(OH)_2 + 2Na^{+}$$

Hydroxide precipitation is capable of removing certain metals found in acid wastewaters. Among the metal ions removed are arsenic, cadmium, copper, trivalent chromium, iron, manganese, nickel, lead, and zinc.[27] Table 10.1.10 presents reported residual concentrations to which hydroxide precipitation can remove these metals. This information is based upon application of hydroxide precipitation to various industry wastewaters. It is important to note in Table 10.1.10 that in some cases, e.g., lead, cadmium and zinc, the residual concentrations reported are lower than the theoretical solubilities of the pure element in water.[27] Several phenomena influence the effectiveness of precipitation, e.g., ionic strength, coprecipitation, and adsorption. These phenomena will ultimately determine the residual concentrations in specific applications, especially in solutions containing several metal ions.

As stated previously, the most commonly used precipitating agents are lime, hydrated lime, and sodium hydroxide. The following is a brief description of each reagent type.

<u>Lime slurry</u>--Lime slurry treatment of metal laden waste streams is one of the oldest and perhaps most prevalent of all industrial waste treatment processes.[10] It is used extensively as an alkaline reagent in the

TABLE 10.1.10. HYDROXIDE PRECIPITATION METAL REMOVAL EFFECTIVENESS

Metal	Inlet concentration (mg/L)	Residual concentration (mg/L)
Arsenic	0.2 - 0.5	0.03
Cadmium	ND	0.0007
Chromium, trivalent	1,300	0.06
Copper	204 - 385	0.2 - 2.3
Iron	10	0.1
Lead	0.5 - 25	0.03 - 0.1
Manganese	ND	0.5
Nickel	5	0.15
Zinc	16.1	0.02 - 0.23

Source: Reference 27.

precipitation of pickling wash waters, plating rinses, acid mine drainage, and process waters from chemical and explosive plants.[9,28,29] It is used in many applications as a low-cost alkali due to its pumpable form, and effectiveness in removing Ca salts from the process.[10] However, a major disadvantage of the process is the formation of a voluminous sludge product.

Since limes are formed by the thermal degradation of limestone (calcination), they are available in either high calcium (CaO) or dolomitic (CaO-MgO) form[29]. These pure, oxidized products are referred to as quicklime. Quicklime varies in physical form and size, but can generally be obtained in lump (63 to 255 mm), pebble (6.3 to 63 mm), ground (1.45 to 2.38 mm), or pulverized (0.84 to 1.49 mm) form.[29] Experimental evidence has shown an increase in dissolution as the size of a lime particle diminishes. For example, a 100 percent quicklime of 100 mesh (0.149 mm) will dissolve twice as fast as one of 48 mesh (0.35 mm).[29]

Although lime can be fed dry, for optimal efficiency it is slaked (hydrated) and slurried before use. Slaking is usually carried out at temperatures of 82 to 99°C with reaction times varying from 10 to 30 minutes. Following slaking, a wet plastic paste is formed (lime putty) and then slurried with water to a concentration of 10 to 35 percent.[29]

While most lime is sold as quicklime, small lime consumers often cannot economically justify the additional processing step that slaking entails. Therefore, high calcium and dolomitic lime are also available in hydrated form (either $Ca(OH)_2$ or $Ca(OH)_2$ MgO).[30] This product is made by the lime manufacturer in the form of a fluffy, dry, white powder. It is supplied either in bulk or in 23 kg (50 lb) bags. Hydrated lime is suitable for dry feeding or for slurrying and the resulting purity and uniformity are generally superior to slaked lime prepared onsite. High calcium hydrate is far more reactive than dolomitic hydrate, Dolomitic hydrate, which possesses greater basicity (approximately 1.2 times), is a much slower reactant, although heat and agitation can accelerate its inherently slow reactivity.[28]

Both quicklime and hydrated lime deteriorate in the presence of carbon dioxide and water (air-slaking). Therefore, lime is generally stored within moisture-proof containers and consumed within a few weeks after manufacture. The storage characteristics of dry hydrated lime are superior to quicklime, but carbonation may still occur causing physical swelling, marked loss of chemical activity, and clogging of discharge valves and pipes.

Dry chemical feed systems consist of either manual addition of 50 lb bags or, in large operations where lime is stored in bulk, an automatic mixing and feeding apparatus. Two types of automatic feed systems are available. Volumetric feed systems deliver a predetermined volume of lime while gravimetric systems discharge a predetermined weight. Gravimetric feeders require more maintenance, are roughly twice as expensive, but can guarantee a minimum accuracy of 1 percent of set rate versus 30 percent for volumetric feeders.[31]

In a typical lime slurry system with storage and slaking equipment, slurry tank with agitator is used followed by a slurry recirculation line.[16] The process flow lines bleed off a portion of the recirculation slurry to the reactors. The process line is as short as possible (to prevent caking) and the control valves are located close to the point of application.

Lime neutralization/precipitation operations are typically conducted under atmospheric conditions and room temperatures.[32-35] The precipitation unit is usually a reinforced tank with acid-proof lining and some sort of agitation to maintain intimate contact between the metals-containing wastes and the lime (slurry) solution. Vertical ribs can be built into the perimeter to keep the contents from swirling instead of mixing.

During operations, adequate venting may have to be provided due to the possible evolution of heat and noxious gases. Table 10.1.11 presents a summary of process parameters gathered from various lime slurry precipitation systems. However, while these provide an indication of typical system design, testing under actual or simulated conditions is the only sound basis for the determination of individual waste treatment parameters.

Caustic soda--Pure anhydrous sodium hydroxide (NaOH) is a white crystalline solid manufactured primarily through the electrolysis of brine. Caustic soda is a highly alkaline, sodium hydroxide solution. It is used in the precipitation of heavy metals and in neutralizing strong acids through the formation of sodium salts.

Although available in either solid or liquid form, NaOH is almost exclusively used in water solutions of 50 percent or less.[36] The solution is marketed in either lined 55-gallon drums or in bulk; i.e., tank car or truck. As a solution, caustic soda is easier to store, handle and pump,

TABLE 10.1.11. SUMMARY OF TYPICAL LIME-SLURRY OPERATING PARAMETERS

Parameter	Unit(s)	Operating range	Optimum range
Type of Stone	% MgO	5 - 40	5
Stone Size	mm	0.149 - 255	0.149
Slaking Temperature	°C	82 - 99	Same
Slurry Solids	%	5 - 40	a
Retention Time	Min	5 - 15	5
Sedimentation Time	Min	15 - 60[b]	15 - 30
Mineral Acidity	Mg/L	10,000 - 100,000	20,000

[a]is dependent on site specific factors

[b]High calcium lime will settle in 15 minutes with 1-2% acid wash streams and 30-60 minutes with 3-10 percent acid streams. Dolomite will typically take 15-60 minutes.

Source: References 10, 28, and 29.

relative to lime. In comparison to lime slurries, caustic soda will not clog valves, form insoluble reaction products, or cause density control problems. However, when sodium hydroxide is stored in locations where the ambient temperature is likely to fall below 12°C, heated tanks should be provided to prevent reagent freezing.[36]

After lime, sodium hydroxide is the most widely used alkaline reagent for precipitation systems. Its chief advantage over lime is that, as a liquid, it rapidly dissociates into available hydroxyl (OH-) ions. Holdup time is minimal, resulting in reduced feed system and tankage requirements. Caustic soda's main disadvantage is reagent cost.[14] As a monohydroxide, in precipitating divalent metals such as nickel, two parts hydroxide are required per part of metal precipitated. In contrast, dihydroxide bases such as hydrated lime, only require one part hydroxide per part of divalent metal precipitated.

This increase in reagent requirements combined with a higher cost/mole (approximately five times that of hydrated lime), makes caustic soda more expensive on a precipitation equivalent basis. Generally, in high volume applications where reagent expenditures constitute the bulk of operating expenses, lime is the reagent of choice. However, in low volume applications where low space requirements, ease of handling, and rapid reaction rates are the deciding factor in reagent selection, caustic soda is clearly superior. Also, in any system where sludge disposal costs will be high, caustic soda will compete more favorably with lime.

The higher solubility of NaOH in water (approximately 100 times that of lime at 25°C) reduces or eliminates the need for complex slaking, slurrying, or pumping equipment. In a typical system, caustic is added through an air-activated valve controlled by a pH sensor.[16] Reagent is demanded as long as the pH of the waste stream remains below the controller setting required for precipitation. Agitation is provided by a mechanical mixer to prevent excessive lag time between the addition of the reagent and the first observable change in the effluent pH. The precipitated solution is then pumped to a large settling tank for liquid/solid separation. Table 10.1.12 provides sodium hydroxide sludge generation factors for seven metallic species commonly encountered in metal-containing wastes.

The precipitation reaction is typically carried out under standard operating temperatures and pressures. The reaction is almost instantaneous since caustic soda reacts vigorously with water. At concentrations of 40 percent or greater, the heat generated by dilution can raise the temperature above the boiling point. Handling precautions are required when performing dilution or other reagent handling since even moderate concentrations of NaOH solution are highly corrosive to skin.[36]

Process configurations for caustic soda treatment are a function of waste type, volume, and raw waste pH level and variability. For example, the precipitation of concentrated acidic metals-laden waste streams with low dead times depends on pH as follows: one reactor system for feeds with pH ranging between 4 and 10, a reactor plus a smoothing tank for feeds with pH fluctuations of 2 and 12, and two reactors plus a smoothing tank for feeds with pH less than 2 or greater than 12.[7] Retention times vary with the rate of reaction and mixing, however, 15 to 20 minutes appears to be optimal for complete neutralization/precipitation in most systems.[37] The interval between the addition of sodium hydroxide and the first observable change in effluent pH (dead time) should be less than 5 percent of the reactor residence time in order to maintain good process control.[7] A summary of typical operating parameters is provided in Table 10.1.13.

A typical caustic system is designed to add most of the reagent in a preliminary precipitation stage, while a second stage acts as a smoothing and finishing tank. In this manner, the second reactor is able to compensate for pH control overshoots or concentrated batch dumps which may temporarily overwhelm the primary precipitation system.[11]

Overshoot is due primarily to the lack of sodium hydroxide solution buffering capacity. For example, Figure 10.1.5 illustrates the titration curve for the neutralization of a ferric chloride etching solution (pH 0.5) with a 5 Molar caustic soda solution. The steep slope of the titration curve beginning at pH 2.0, combined with a strong demand for alkali prior to that point, often make over- or under-correction unavoidable. For continuous precipitation applications of greater than 20 gpm, pH control in the portion of the titration curve which is nearly vertical (between pH 2.0 and 9.0) is achieved in a second reactor to prevent excess reagent usage or effluent discharge violations.

TABLE 10.1.12. SODIUM HYDROXIDE SLUDGE GENERATION FACTORS

Metal ion	lb dry solids generated / lb of metal precipitated
Cr	1.98
Ni	1.58
Cu	1.53
Cd	1.30
Fe	1.61
Zn	1.52
Al	2.89

Source: Reference 14.

TABLE 10.1.13. SODIUM HYDROXIDE NEUTRALIZATION: SUMMARY OF TYPICAL OPERATING PARAMETERS

Parameter	Unit(s)	Operating range	Ideal range
Sodium hydroxide concentration	% NaOH	12 - 50	40 - 50
Dead time	% Retention time	3 - 10	3 - 5
Retention time	Minute	5 - 30	15 - 20
Batch treatment throughput	gal/min	1 - 20	20
Continuous treatment throughput	gal/min	15	20
Suspended solids	Weight %	3 - 10	10
Storage temperature (40-50% NaOH)	°C	12 - 20	16 - 20

Source: References 4, 7, and 38.

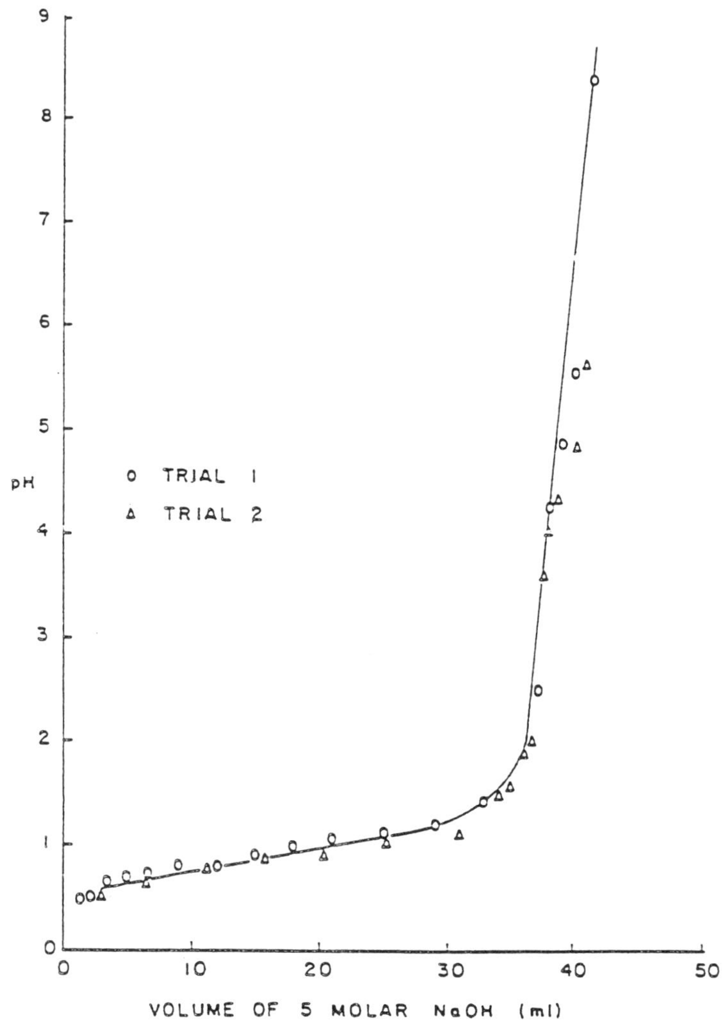

Figure 10.1.5. Neutralization of ferric chloride etching waste by sodium hydroxide.

Source: Reference 6.

10.1.2.2 Process Performance--

Chemical precipitation of metal hydroxides through the use of lime and sodium hydroxide is a classical waste treatment technology used by most industrial waste treatment systems. The performance of these technologies in removing metallic pollutants from industrial wastewaters is well documented in the literature. Tables 10.1.14 (lime precipitation) and 10.1.15 (NaOH precipitation) contain general performance indicators which incorporates effluent concentrations and removal efficiencies developed from plant-specific full-scale and pilot plant data bases.[1]

In recent years, research has centered around the evaluation of supplemental chemicals to the already well defined hydroxide precipitation sedimentation process. Organic and inorganic polyelectrolytes (see Section 10.2), acid,[39] and soda ash (see Section 10.1.4) have all been used in this capacity. The purpose of these supplementary chemicals is to improve the efficiency of liquid-solid separation break complexing/chelating agents, and take advantage of the lower solubility of carbonate complexes.

Process wastes containing complexing/chelating agents are often untreatable with established technologies. The difficulty arises due to the formation of a highly stable organo-metallic bonds formed between the metal ion and the complexing/chelating agent. Ammonia is an example of a complexing agent, with each molecule of ammonia bound to a metal species such as copper by a single bond. A chelating agent such as EDTA, on the other hand, forms more than one bond with each metal ion. Complexing and chelating agents are typically used to keep the metals in solution for plating. During rinsing, the complexed and chelated metals end up in the processing rinsewater. The major complexing agents found in metal waste streams are ammonia, cyanide, fluoborate, and pyrophosphate. The foremost chelating agents are EDTA, Quadrol, citrate, and tartrate.[40]

Established chemical methods for breaking chelator/complexes and removing metals to low concentrations are starch xanthate, sodium DTC, ferrous sulfate, waste acids, sulfide ions, sodium hydrosulfite, sodium borohydride, and high pH lime.[40] A typical process used by industry is the combination waste acid-high pH lime treatment method. For this type of waste treatment process

TABLE 10.1.14. PERFORMANCE SUMMARY FOR LIME PRECIPITATION OF HEAVY METALS

Metallic species	Effluent concentration (mg/L)	Removal efficiency (%)
Arsenic	ND - 110	20 - >99
Cadmium	ND - 80	20 - >99
Chromium	ND - 1,800	47 - >99
Copper	ND - 220	33 - >99
Cyanide	ND - 5,500	67 - >99
Lead	ND - 580	0 - >99
Mercury	0.1 - 43	69 - >96
Nickel	ND - 5,200	6 - >99
Selenium	ND - 87	40 - >99
Silver	ND - 90	99 - >99
Thallium	1.1 - 20	58 - >75
Zinc	13 - 26,000	25 - >99

Source: Reference 1.

TABLE 10.1.15. PERFORMANCE SUMMARY FOR SODIUM HYDROXIDE PRECIPITATION OF HEAVY METALS

Metallic species	Effluent concentration (mg/L)	Removal efficiency (%)
Cadmium	ND - 930	22 - >99
Chromium	18 - 3,000	53 ->99
Copper	1.0* - 5,900	36 - 98
Lead	-	>99
Nickel	ND - 210	>99
Silver	11 - 64	76
Zinc	44 - 560	80 - >99
Hexavalent chromium	ND - 25	73 - >99

*Approximate value.

Source: Reference 1.

the pH of the organo-metallic waste is first adjusted to approximately 2 with dilute mixed waste acid (sulfuric, nitric, hydrochloric, or chromic acid) and/or virgin hydrochloric acid. After the chelator/complex breaking step the pH of the waste solution is raised to approximately 11, resulting in the formation of insoluble metal hydroxides.[39] Table 10.1.16 presents alternate precipitation technologies for the removal of metals such as copper from complexed and chelated rinsewaters.

An alternate technology for the precipitation of metal hydroxides which has shown promise in recent years is magnesium oxide (MgO). Magnesium oxide is available in slurry form composed of 55 to 60 percent magnesium hydroxide, $Mg(OH)_2$. The slurry has a bulk density of 1.5 kg/L and due to its low solubility (0.0009 g/100 mL), must be mildly agitated during storage.[41]

The main advantage of magnesium hydroxide over comparable hydroxide precipitation technologies is that the precipitate formed is more particulate in nature (due to longer reaction times). The sludge formed has better handling and dewatering characteristics and sludge volumes are much less. Table 10.1.17 compares typical physical, chemical, and filtered sludge properties of magnesium hydroxide to those of caustic soda and hydrated lime. As can be seen, dewatering characteristics and filtration time for separating solids are considerably enhanced in the case of magnesium hydroxide.[41]

The main disadvantage of magnesium hydroxide is that it costs approximately three times as much as hydrated lime. In addition, operation of the magnesium hydroxide system is not as straightforward as comparable hydroxide systems. Reaction times are slower and it will be necessary to make modifications in waste treatment operating procedures and equipment.

Table 10.1.18 presents the results of a Bureau of Mines research effort into magnesium oxide precipitation of metals. The researchers found that when equal pH values are obtained, MgO leaves less dissolved metal and less suspended metal hydroxide of sedimentation is part of the process. MgO was able to remove any metal that is precipitated as an hydroxide. However, a threefold to fourfold stoichiometric excess was required to reach adequate pH values (8-9).

TABLE 10.1.16. CHEMICALS FOR Cu REMOVAL FROM COMPLEXED AND CHELATED RINSEWATERS

Precipitation chemical	Alkaline etchants	Ammonium persulfate etchants	Electroless Cu	Pyrophosphate Cu	Fluoborate Cu
Insoluble starch xanthate (ISX)	X	X	X	X	
Sodium dimethyl-dithiocarbamate (DTC)	X	X	X	X	
Ferrous sulfate	X	X	X	X	
Spent pickle liquor	X	X	X	X	
Ferrous sulfide	X	X	X		
Sodium hydrosulfite	X	X	X		
High-pH lime			X	X	X
Sodium borohydride	X		X		

Source: Reference 40.

TABLE 10.1.17. COMPARISON OF HYDROXIDE REAGENT PROPERTIES

Property	NaOH	$Ca(OH)_2$	$Mg(OH)_2$
Molecular weight	40.0	74.1	58.3
Hydroxide content (%)	42.5	45.9	58.3
Heat of solution (Kg-cal/mole)	9.94	2.79	0.0
Solubility (g/100 mL H_2O)	42.0[a]	0.185[a]	0.0009[b]
Reactive pH maximum	14.0	12.5	9.0
Weight equivalency	1.37	1.27	1.0
Freezing point	16.0[c]	0.0[d]	0.0[e]
Solids content of sludge (%)[f]	30.0	35.0	55.0
Sludge density (lb/ft^3)	80.0	85.0	100-110
Filtration time (hr)	7-8	7-9	1.5-2.0
Sludge volume (yd^3/10,000 lb)	15.0	12.5[g]	6.4

[a]Temperature, 0°C.

[b]Temperature, 18°C.

[c]50 percent solution.

[d]30 percent slurry.

[e]58 percent slurry.

[f]Sludge from a plate-and-frame filter press.

[g]Consists of metal hydroxide and gypsum.

Source: Reference 41.

TABLE 10.1.18. TEST RESULTS FROM TREATING METALS BEARING WASTEWATER WITH MgO AND LIME

	Chemical analysis, ppm								
	pH	Fe	Cu	Zn	Ni	Mn	Co	Cd	Pb
Beneficiation process water-CM									
Untreated water	5.4	5.7	0.63	0.55	ND[a]	9.9	ND	ND	ND
Treated with:									
0.1 g/L[b] MgO, filtered	8.6	0.2	0.1	0.1	ND	7.1	ND	ND	ND
0.2 g/L MgO, filtered	9.2	0.2	0.1	0.1	ND	3.4	ND	ND	ND
0.35 g/L MgO, filtered	9.4	0.2	0.1	0.1	ND	1.3	ND	ND	ND
Beneficiation process waste-BM									
Untreated water	6.4	0.2	0.1	12.7	0.2	17.5	ND	ND	ND
Treated with:									
0.16 g/L MgO, filtered	8.3	ND	ND	0.2	ND	8.3	ND	ND	ND
0.21 g/L MgO, filtered	8.7	ND	ND	0.2	ND	5.7	ND	ND	ND
0.31 g/L MgO, filtered	8.9	ND	ND	0.2	ND	1.9	ND	ND	ND
Mine drainage									
Untreated water	2.7	40	ND	39	ND	41	ND	ND	ND
Treated with 0.5 g/L MgO, filtered	8.9	0.2	ND	0.1	ND	15	ND	ND	ND
Prepared solution No. 1									
Untreated water	4.2	ND	8.7	ND	12.0	ND	11.0	ND	ND
Treated with:									
0.4 g/L MgO, filtered	8.9	ND	0.1	ND	0.2	ND	0.2	ND	ND
0.4 g/L MgO, settled	8.9	ND	0.5	ND	0.2	ND	0.2	ND	ND
0.1 g/L lime,[c] settled	9.4	ND	0.7	ND	1.2	ND	1.6	ND	ND
Prepared solution No. 2									
Untreated water	5.4	5.0	0.21	2.7	ND	4.4	ND	ND	ND
Treated with:									
0.125 g/L MgO, filtered	8.9	0.2	0.1	0.1	ND	0.2	ND	ND	ND
0.04 g/L lime,[c] filtered	8.9	0.2	0.1	0.2	ND	2.2	ND	ND	ND
Prepared solution No. 3									
Untreated water	4.0	ND	ND	4.2	ND	ND	ND	5.2	4.7
Treated with:									
0.2 g/L Mg, settled	9.0	ND	ND	0.1	ND	ND	ND	0.31	0.5
0.05 g/L lime,[c] settled	9.0	ND	ND	0.8	ND	ND	ND	1.4	1.6

[a]The unit g/L is the grams of MgO used per liter of water treated.

[b]ND = not determined, since initial concentrations were below the analysis limit of atomic absorption.

[c]Weight of lime is for CaO, not $Ca(OH)_2$.

Source: Reference 42.

The researchers concluded that when influent metals content is low, increased chemical costs will be balanced by savings from easier sludge dewatering, compactness, and stability. It is anticipated that as land disposal costs for metal hydroxide sludges continue to increase, the economics of this process will become more favorable. In addition, by mixing magnesium hydroxide with sodium hydroxide in a dual reagent system, sludge reductions of approximately 45 percent can be realized. Although alkali costs will increase, savings in sludge conditioning polymers and disposal costs will help to defray the added reagent expense.[43]

10.1.2.3 Process Costs--

The basic equipment train for an hydroxide precipitation system consists of a collection sump, piping system, precipitation reactor, feed system, flocculation/clarification unit, sludge storage tank(s), and plate and frame filter press. Figure 10.1.6 illustrates the treatment train design used for the remainder of this section.

The capital and annualized cost information contained in this section was adapted from costing methodology developed by Versar.[44] Versar calculated the direct costs, indirect costs, and working capital as a percentage of the purchased equipment and installation (PE&I) costs. The cost elements and assumptions made by Versar are summarized below:

Cost elements	Assumed value (% PE&I)
Direct costs (DC)	
• Instrumentation and controls	10
• Piping	21
• Electrical equipment and materials	13
• Buildings	26
• Yard improvements	7
• Service facilities	41
Total direct cost:	118

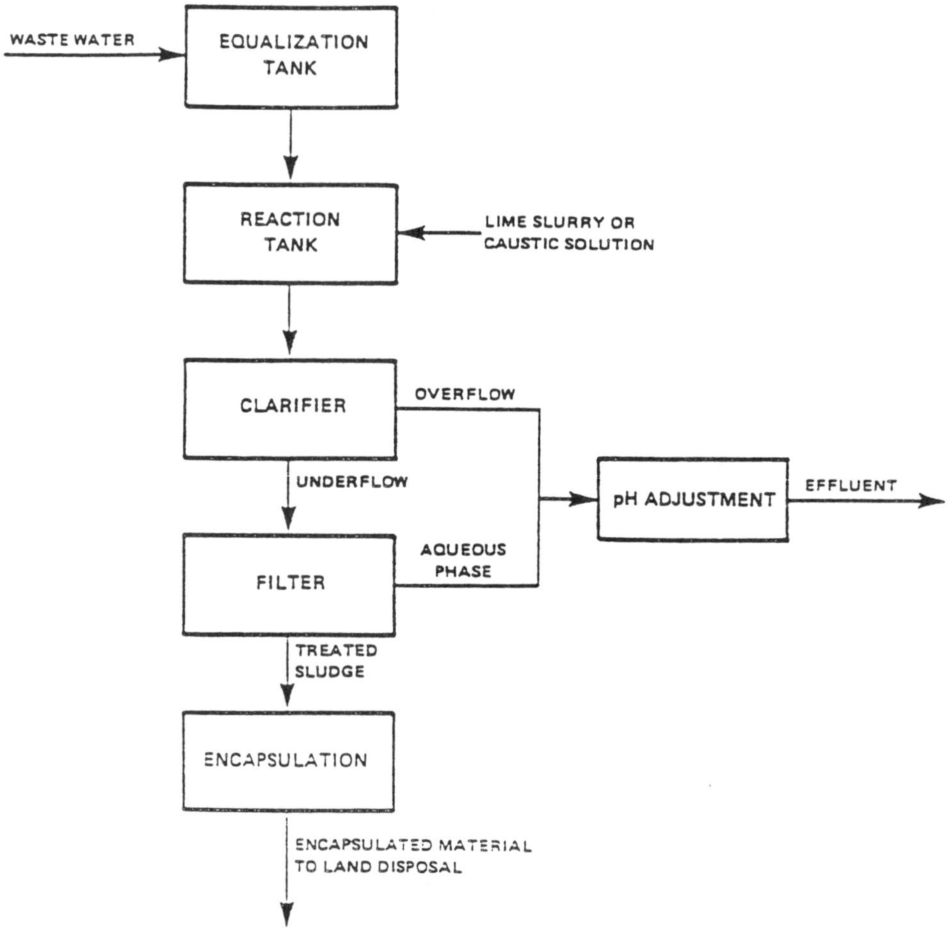

Figure 10.1.6. Lime|hydroxide precipitation.

Source: Reference 44.

Indirect costs (IC)	
• Engineering and supervision	29
• Construction expenses	32
• Contractor's fees	7
• Contingency	27
Total indirect cost:	95
Fixed capital investment (FCI)	PE&I + DC + IC
Working capital (WC)	47
Total capital investment (TCI)	FCI + WC = 360% PE&I

Annualized costs included variable costs, plant overhead costs, general and administrative expenses, and fixed costs. The variable costs included costs for labor, maintenance, materials, chemicals, and hazardous contracted sludge disposal. The fixed costs include taxes, insurance, and capital recovery costs.

The chemical requirements for each treatment were based on stoichiometric requirements. The cost for chemicals were obtained from the Chemical Marketing Reporter. In deriving annualized costs, a certain set of assumptions were made. These assumptions are listed below:[14,44-47]

- Plant overhead operating costs are 5.8 percent of the total capital investment costs.
- Taxes and insurance costs are 1 percent of total capital investment costs.
- Labor costs are based on 4 hrs/shift at $20/hr.
- Maintenance costs are at 6 percent of total capital investment costs.
- Power costs are at 2 percent of total capital investment.
- The nonhazardous contracted sludge disposal costs are based on $200/ton.
- The sludge transportation costs are based on $0.25/ton-mile and a transportation distance of 15 miles. It is further assumed that all hazardous solid wastes generated by the treatment processes would be encapsulated and disposed of as nonhazardous wastes.

In all cases, capital recovery was calculated at a 12 percent interest rate over a period of 10 years. The capital recovery factor (CRF) was estimated as follows:

$$CRF = \frac{i\ (1 + i)^n}{(1 + i)^n - 1} = 0.177$$

where:

i = interest rate
n = number of years.

Costs for items which were not in the size range of available information extrapolations were made using the following equation:[46]

$$\text{Cost A} = \text{cost B}\ \frac{\text{Capacity A}}{\text{Capacity B}}^{x}$$

The exponent, "x", was determined with available information and is presented where necessary in the footnotes to the cost summary tables.

The capital costs for the base equipment in hydroxide treatment train have been adapted from cost figures and tables contained in "Reducing Water Pollution Costs in the Electroplating Industry." The base system discussed in this section is designed to handle an aqueous waste stream containing 200 mg/L of heavy metal ions. The flow rates for this system were developed for three different sizes: 1,000, 10,000, and 100,000 gal/hr. Operation of this system is assumed to be 24 hrs/day, 300 days/year. Mixed reactor construction costs for the first stage flow/concentration equalization tank are presented in Figure 10.1.7. The equalization tank has been sized for 1-hour of retention time and was fabricated from reinforced concrete.[7]

The precipitation reaction tank consists of a continuous neutralization/ precipitation vessel equipped with pH control, reagent storage, and reagent feed systems. The reagent feed and storage system is sized for a 1-week supply and uses hydrated lime as a precipitation agent. Sulfuric acid capability is also included in case of pH overshoot. Figure 10.1.8 presents the mixed reactor construction and installation costs. The reactor has been sized for a minimum of 30 minutes of retention time to ensure a complete precipitation reaction.

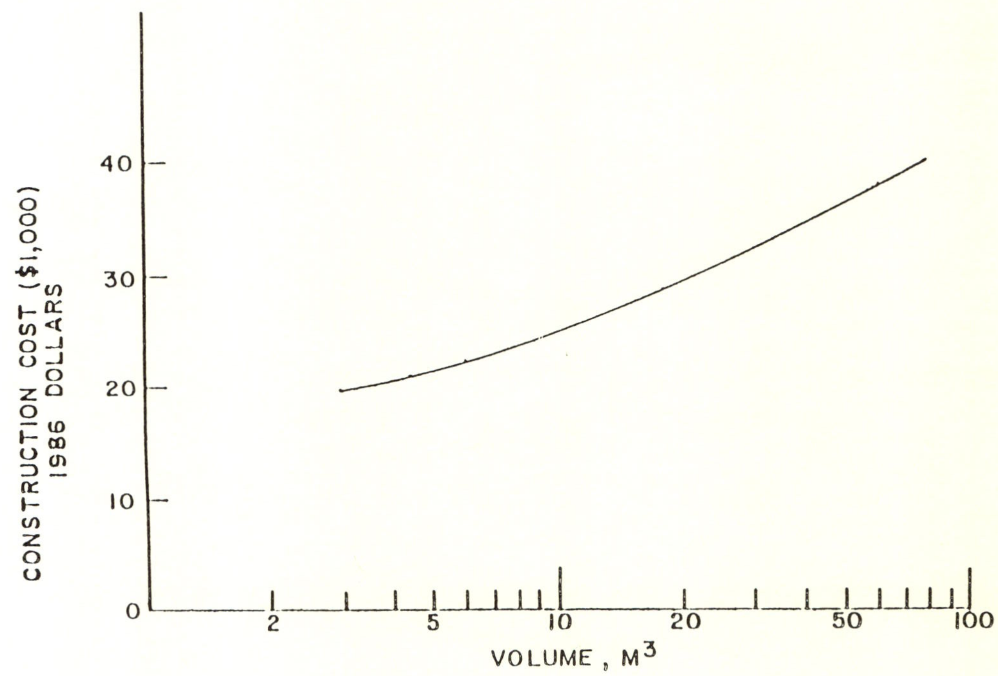

Figure 10.1.7. Construction costs for reinforced concrete equalization reactor.

Source: Reference 7.

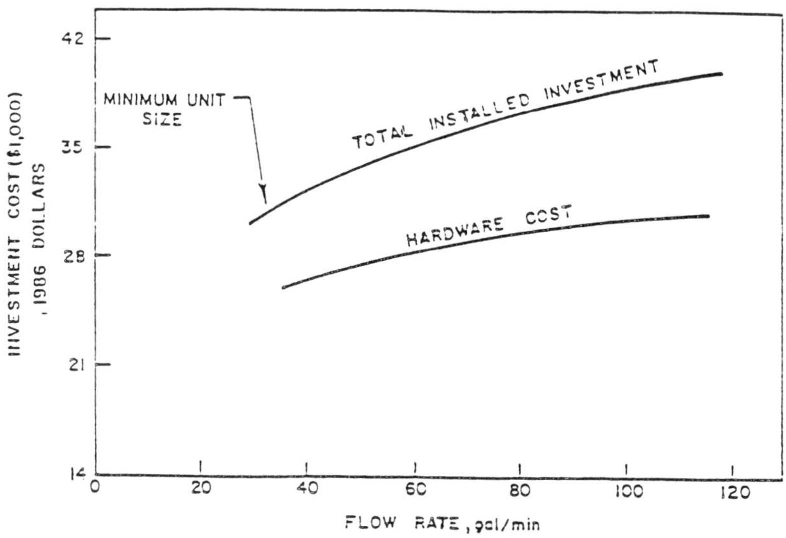

Figure 10.1.8. Investment cost for continuous single-stage precipitation.

Source: Reference 14.

Assuming complete reaction between the heavy metal ions in the waste stream and the hydroxide ions in the precipitation reactor, some sort of separation will be necessary to remove the metal hydroxides and other insoluble pollutants from the reactor effluent. Figure 10.1.9 shows the hardware and installed costs for a flocculation/clarification unit used to enhance the settling characteristics of the suspended solids. The unit is assumed to have a separate flocculation tank, a polymer feed system, a "lamella" or slant-tube separator, and a zone in which the sludge collects before being discharged. The costs like those of the equalization and precipitation reactors are a function of flow rate. The solids concentration of the underflow is assumed to be 2 percent, while the overflow is assumed to be solids-free.[14]

Typically, the underflow from the flocculation/clarification unit must be stored onsite in sludge holding tanks before the sludge is shipped to a disposal site or transferred to another dewatering stage. The investment cost for sludge tanks is presented in Figure 10.1.10. The tanks are of carbon steel construction and the cost is a function of tank volume. The sludge holding tanks for the hydroxide treatment base case have been sized for 10 hours of clarifier underflow.[14]

In many cases, further concentration of thickened sludge through the use of mechanical dewatering equipment is desirable in reducing sludge disposal costs. Figure 10.1.11 presents the unit costs for a recessed plate filter press as a function of the feed volume capacity (filter cake volume is also given). The feed solids concentration in this case is assumed to be 2 percent, the cake solids concentration is 20 percent, with an 8-hour press-cycle.[14] Items not included, but will contribute to the cost of installation include:

- High pressure feed pumps;
- Filtrate return lines (to clarifier); and
- Cake solids handling equipment.

Table 10.1.19 details the costs developed for the continuous hydrated lime precipitation system previously shown in Figure 10.1.10. The high cost of the 1,000 gph continuous system relative to the other two pretreatment

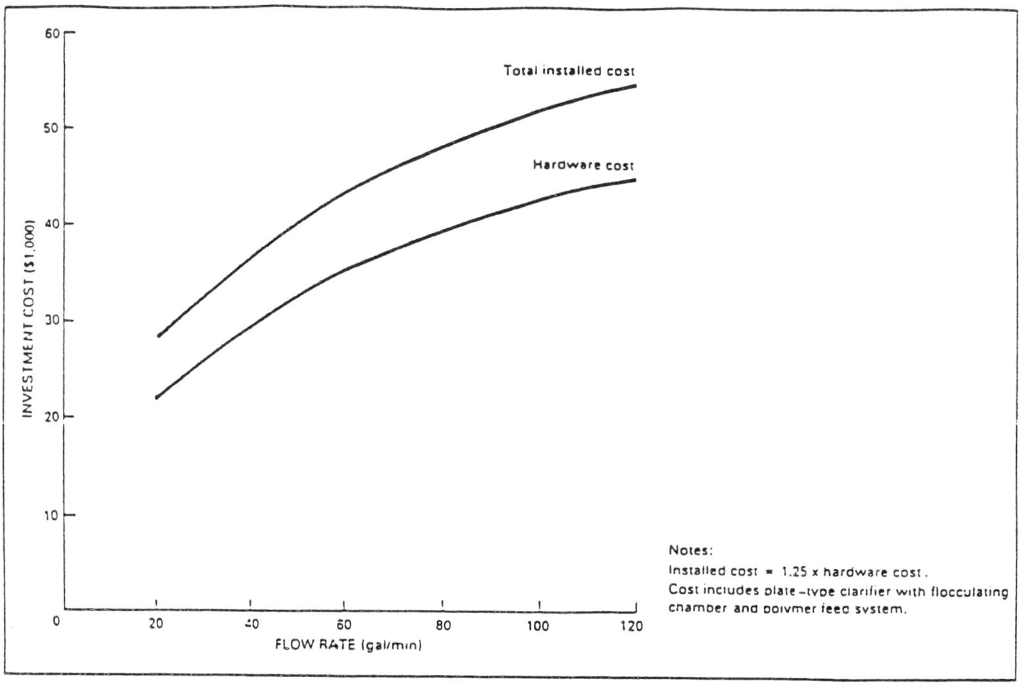

Figure 10.1.9. Investment cost for flocculation/clarification units.

Source: Reference 14.

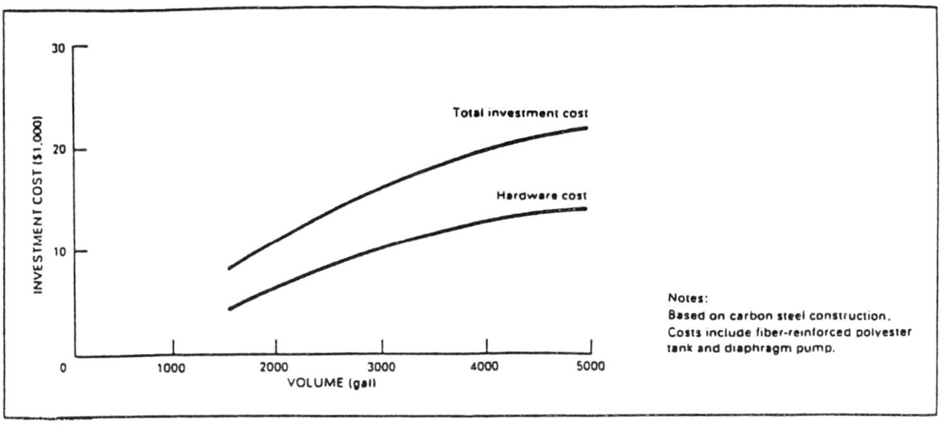

Figure 10.1.10. Investment cost for sludge storage/thickening units.

Source: Reference 14.

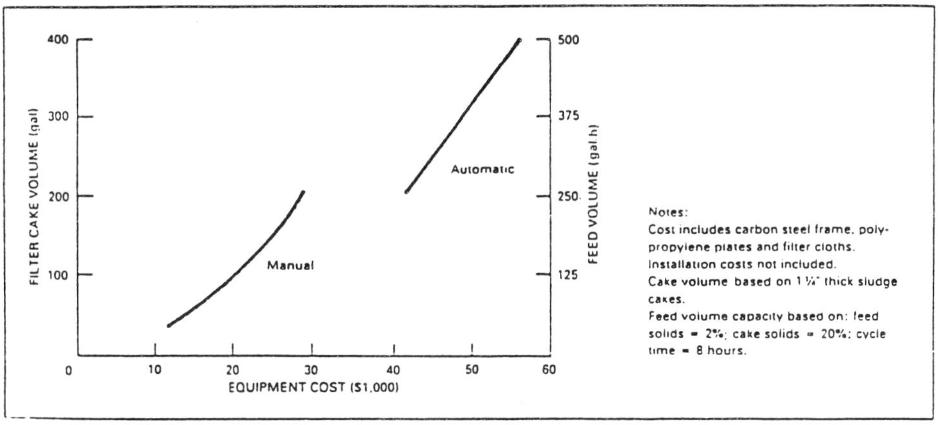

Figure 10.1.11. Hardware cost for recessed plate filter presses.

Source: Reference 14.

TABLE 10.1.19. CONTINUOUS HYDRATED LIME PRECIPITATION COSTS[a]

	Flow rate (gph)		
	1,000	10,000	100,000
Purchased equipment and installation (PE&I) ($)			
Equilization tank	17,000	29,000	50,000
Precipitation reactor	24,000	40,000	69,000
Flocculator/clarifier	18,000	50,000	140,000
Sludge molding tank(s)	3,000	6,000	48,000
Filter Press	10,000	25,000	100,000
	72,000	150,000	407,000
Total capital investment (360% PEI)	259,000	540,000	1,465,200
Operating costs ($)			
Operating labor ($20/hr)	72,000	72,000	72,000
Maintenance (6% TCI)	15,500	32,500	87,900
General plant overhead (5.8% TCI)	15,000	31,400	85,000
Utilities (2% TCI)	5,200	10,800	29,000
Taxes and insurance (1% TCI)	2,600	5,400	14,700
Chemical costs ($40/ton)	500	5,300	53,000
Sludge transportation ($0.25/ton-mile)	200	2,300	22,500
Sludge disposal ($200/ton)	12,000	120,000	1,200,000
Annualized capital (CFR = 0.177)	45,800	95,800	259,000
Total cost/year	168,800	375,500	1,823,100
Cost/1,000 gallon	23	5	2.5

Source: Reference 44.

[a]1987 Dollars.

processes illustrates why precipitation systems under 50,000 gpd (2,000 gph) are typically batch in nature to reduce equipment costs. In addition, the large costs attributed to sludge disposal in every system demonstrates the main drawback to hydroxide precipitation. As land disposal costs increase, treatment processes such as hydroxide precipitation, which generate large quantities of hazardous sludge will lose their cost advantage over the more expensive recovery technologies.

10.1.2.4 Process Status--

Hydroxide precipitation is a widely used and well developed technology for reducing metals effluent concentrations to acceptable levels. The process operates at ambient temperature and pressure and is well suited to automatic control. Its ability to treat a wide variety of industrial waste streams has been well demonstrated in bench, pilot, and full-scale systems. Environmental impacts can result from emissions during the precipitation process and the production of large volume of potentially hazardous sludge.[48] Exit gases can be scrubbed by using a control system, however, sludge reduction methods (seeding, dilution, vacuum filtration, etc.),[48,49] have only partially offset the problems associated with sludge generation. Therefore, new methods of sludge disposal and reduction and recycle/reuse options (such as agricultural liming) should be considered. The advantages and disadvantages of lime-precipitation and caustic soda precipitation are summarized in Tables 10.1.20 and 10.1.21, respectively.

10.1.3 Sulfide Precipitation

10.1.3.1 Process Description--

The basic principle of sulfide precipitation is similar to that of hydroxide treatment, in that the precipitation process converts soluble metal ions into insoluble (sulfide) compounds. Some advantages over hydroxide precipitation are that with sulfides, heavy metals can be removed to extremely low concentrations at a single pH. In addition, the use of sulfides allow precipitation of contaminants, even in the presence of chelating agents. Sulfide precipitation has been limited to relatively few applications, however, due to the toxicity and odor of hydrogen sulfide (H_2S) evolved from the precipitation process.[1]

TABLE 10.1.20. ADVANTAGES AND DISADVANTAGES OF LIME PRECIPITATION

Advantages

- Proven technology with documented neutralization efficiencies.
- No temperature adjustments normally necessary.
- Modular design for plant expansion.
- Can be used in different configurations.
- Able to coprecipitate a mixture of metal ions to achieve residual metal solubilities lower than that achieved by precipitating each metal at its optimum pH.
- Reagent is easy to handle, and has treatment effectiveness for wide range of dissolved materials.

Disadvantages

- The theoretical minimum solubilities for different metals occur at different pH values. For a mixture of metal ions, it must be determined whether a single pH can produce sufficiently low solubilities for the metal ions present in the wastewaters.
- Hydroxide precipitates tend to resolubilize if the solution pH is increased or decreased from the minimum solubility point; thus maximum removal efficiency will not be achieved unless the pH is controlled within a narrow range.
- The presence of complexing ions, such as phosphates, tartrates, ethylenediaminetetraacetic acid (EDTA), and ammonia may have adverse effects on metal removal efficiencies when hydroxide precipitation is used.
- Hydroxide precipitation usually makes recovery of the precipitated metals difficult because of the heterogeneous nature of most hydroxide sludges.

TABLE 10.1.21. ADVANTAGES AND DISADVANTAGES OF CAUSTIC SODA PRECIPITATION

Advantages

- Proven technology with documented neutralization efficiencies
- Strong alkali with rapid reaction rate
- Smaller tanks and retention times than comparitive reagents
- Inventory and storage handling procedures are less complicated due to liquid form
- Storage does not require continuous agitation to maintain homogeneity
- Does not require complex slaking or slurrying equipment
- Produces more soluble by products in low pH applications

Disadvantages

- Chemical costs are significantly higher ($205/ton vs. $46/ton for hydrated lime)
- Does not impart any buffering capacity to industrial waste streams
- Close attention must be given to the design of the pH control
- Caustic soda precipitation will result in a fluffy gelatinous floc increasing the size of the clarification chambers and sludge dewatering equipment.
- Cannot effectively precipitate sulfate waste streams due to solubility of sodium sulfate.

Source: Reference 3.

Sulfide precipitation is used to remove lead, copper, silver, cadmium, zinc, mercury, nickel, thallium, arsenic, antimony, and vanadium from wastewaters. Typically, the precipitation reaction is conducted under near neutral conditions (pH 7.0 to 9.0). Exceptions to this rule are arsenic and antimony which require a pH below 7 for optimum precipitation.[1] As with hydroxide treatment, cyanides are usually oxidized prior to precipitation.

The first step in the sulfide precipitation process is the preparation of a sodium sulfite solution. The solution is then added to the reaction tank (30 minutes retention time)[44] in excess to precipitate the pollutant metal as illustrated in the following reaction:

Na_2S	+	$NiSO_4$	=	NiS	+	Na_2SO_4	(1)
Sodium Sulfide		Nickel Sulfate		Nickel Sulfide		Sodium Sulfate	

The process is controlled by means of a feedback control loop employing ion-selective electrodes.[50] Physical separation of the metal sulfide takes place in thickeners or clarifiers, with reducing conditions maintained by excess sulfide ions. The final step is usually oxidation of the excess sulfide ions through aeration or hydrogen peroxide addition.[44] Currently, two methods of delivering sulfide ions to the process reactor are available. The first method utilizes soluble-sulfides such as a sodium sulfide (Na_2S) or sodium hydrosulfide (NaHS). A second method (Sulfex process) uses a sparingly soluble metal sulfide such as ferrous sulfide (FeS) as a source of sulfide ions. Each process will be discussed individually in the following subsections.

Soluble sulfides--Pure sodium sulfide (sodium sulfuret) is a white, crystalline solid (mp 1180°C, sp gr 1.856).[51] Commercial material is white to light yellow or pink. It crystallizes from aqueous solutions as the nonahydrate, $Na_2S.9H_2O$. In air, sodium sulfide slowly converts to sodium carbonate and sodium thiosulfate and is deliquescent. Reactions with strong oxidizing agents give elemental sulfur.[51]

Pure sodium hydrosulfide (sodium sulfhydrate, sodium hydrogen sulfide, sodium bisulfide) is a white, crystalline solid (mp 350°C, sp gr 1.79). It is highly soluble in water, alcohol, or ether. The commercial product occurs in different shades of yellow and is highly deliquescent. Exposure to air converts it to sodium thiosulfate and sodium carbonate. In the presence of organic matter, combustion can occur. Heating releases hydrogen sulfide, which is a toxic gas.

Sodium sulfide is marketed as 30 to 34 weight percent fused crystals and 60 to 62 weight percent flakes.[51] Each container has a corrosive label and a product label stating that the product causes severe burns to eyes or skin, and that contact with acid liberates poisonous hydrogen sulfide gas. The material is nonflammable, noncombustible, and nonexplosive. Sodium hydrosulfide is marketed as 70 to 72 weight percent flakes and 44 to 60 weight percent liquor in the high purity grades, and as 10 to 40 weight percent liquor from recovered caustic wash in the oil-refining desulfurization processes. Shipment labeling is the same as for sodium sulfide. The product is shipped either as flake in drums or as solutions in tank cars or tank trucks.

The lower freezing points of solutions of sodium hydrosulfide provide an advantage over those of sodium sulfide in shipping by tank truck and tank car. Recently, systems have been designed to enable customers to make their own sodium sulfide solutions by reaction of NaHS and NaOH.

The high solubilities of sodium sulfide and sodium hydrosulfide eliminate the need for slaking and slurrying apparatus. Reagent is added either from storage in the case of liquid reagents or from rapid-mix tanks when using solid reagents.[2,52] Reagent demand is determined through a specific-ion sulfide reference electrode pair, which is set to a preselected potential.[50] Normally, sulfide reagent demand depends on the total metal concentration contained in the effluent waste stream. For continuous processes where metals concentrations are constant, electrode set points can be set at the potential which corresponds to the maximum electrical potential-sulfide concentration gradient and where the wastewater solution has the least detectable odor.[2,50] For batch processes, a simple jar test prior to reagent addition can accurately determine optimal sulfide dosages.

Since sodium sulfide and sodium hydrosulfide have such high solubilities, dissolved sulfide concentrations are correspondingly high. This high concentration of dissolved sulfide causes a rapid precipitation of the metals dissolved in the water as metal sulfides. However, it often results in the generation of small particle fines and hydrated colloidal particles.[50] The rapid precipitation reaction tends more discrete particle precipitation than toward nucleation precipitation (the precipitation of a particle from solution onto an already existing particle). The resulting poor-settling or poor-filtering floc is difficult to separate from the wastewater discharges. This problem has been solved by the effective use, separately or combined, of coagulants and flocculants to aid in the formation of large, fast-settling particle flocs.[53]

One major disadvantage of the soluble sulfide precipitation method is the formation of hydrogen sulfide (H_2S) from dissolved sulfide ions. Figure 10.1.12 is a graph developed by Centec Corporation for determining the percentage of the dissolved sulfide in the form of H_2S as a function of the pH of the solution. According to Centec, the relationship shows that at a pH of 9, H_2S accounts for only 1 percent of the free sulfide in solution.[52] The rate of evolution of H_2S from a sulfide solution per unit of water/air interface will depend on the temperature of the solution (which determines the H_2S solubility), the dissolved sulfide concentration, and the pH. In practice, considering typical response lags of instruments and incremental reagent addition, control of the level of dissolved sulfide and pH would require fine tuning and rigorous maintenance to prevent an H_2S odor problem in the work area.[52] In currently operating treatment systems, the H_2S odor problem is eliminated by enclosing and vacuum evacuating the process vessels.

<u>Insoluble-sulfides</u>--The insoluble-sulfide (Sulfex) process precipitates dissolved metals by mixing the wastewater with an FeS slurry in a solid/liquid contact chamber. The FeS dissolves to maintain the sulfide ion concentration at a level of 2 mg/L.[2] Due to its instability, the ferrous sulfide has to be generated onsite from sodium sulfide and ferrous sulfate. The sulfide is released from ferrous sulfide only when other heavy metals with lower equilibrium constants for their sulfide form are present in solution (see Table 10.1.22).

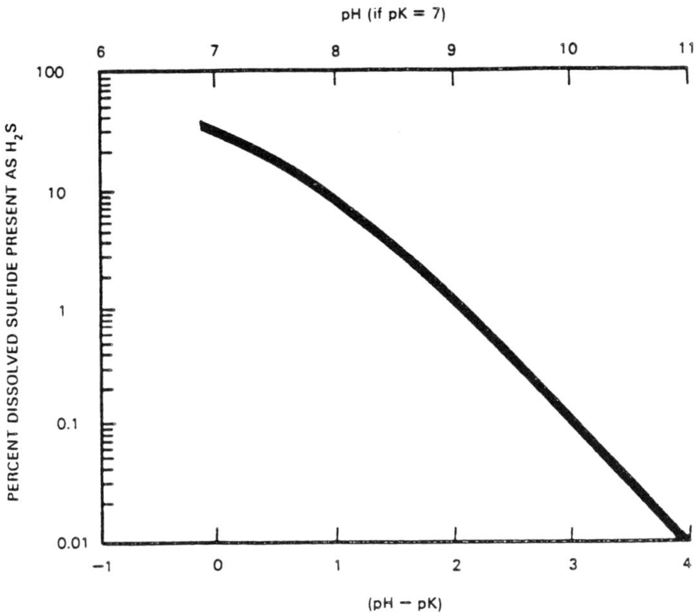

Note.—pK (logarithmic practical ionization constant) is used to measure the degree of dissociation of weak acids, in this case H_2S.

Specific electrical conductance of solution at 77° F (μohm/cm)	Value of pK		
	50° F	68° F	104° F
0[a]	7.24	7.10	6.82
100	7.22	7.08	6.80
1,000	7.18	7.04	6.76
50,000[b]	7.09	6.95	6.67

[a]Distilled H_2O.

[b]Seawater.

Figure 10.1.12. Percent of dissolved sulfide in the H_2S form.

Source: Reference 52.

TABLE 10.1.22. SOLUBILITIES OF SULFIDES

Metal sulfide	K_{sp} (64° to 77°F)[a]	Sulfide concentration (mol/L)
Manganous sulfide	1.4×10^{-15}	3.7×10^{-8}
Ferrous sulfide	3.7×10^{-19}	6.1×10^{-10}
Zinc sulfide	1.2×10^{-23}	3.5×10^{-12}
Nickel sulfide	1.4×10^{-24}	1.2×10^{-12}
Stannous sulfide	1.0×10^{-25}	3.2×10^{-13}
Cobalt sulfide	3.0×10^{-26}	1.7×10^{-13}
Lead sulfide	3.4×10^{-28}	1.8×10^{-14}
Cadmium sulfide	3.6×10^{-29}	6.0×10^{-15}
Silver sulfide	1.6×10^{-49}	3.4×10^{-17}
Bismuth sulfide	1.0×10^{-97}	4.8×10^{-20}
Copper sulfide	8.5×10^{-45}	9.2×10^{-23}
Mercuric sulfide	2.0×10^{-49}	4.5×10^{-25}

[a]Solubility product of a metal sulfide, K_{sp}, equals the product of the molar concentrations of the metal and sulfide.

Source: References 54 and 55.

When the pH is maintained between 8.5 and 9, the liberated iron will form a hydroxide and precipitate as well. The unreacted ferrous sulfide is filtered or settled out with the metal sulfide precipitate, while the effluent is practically sulfide free. Anionic polymers aid settling of metal sulfide precipitates. The sludge is easily dewatered by conventional techniques. In chelated systems, a 4-molar excess of ferrous sulfide is required to obtain maximum heavy-metal removal (see Table 10.1.23 for operating parameters).

The following reactions occur when FeS is introduced into a solution containing dissolved metals and metal hydroxide:

$$FeS = Fe^{+2} + S^{-2} \quad (2)$$

$$M^{+2} + S^{-2} = MS \quad (3)$$

$$M(OH)_2 = M^{+2} + 2(OH)^- \quad (4)$$

$$Fe^{+2} + 2(OH)^- = Fe(OH)_2 \quad (5)$$

The addition of ferrous ions to the wastewater and their precipitation as ferrous hydroxide [$Fe(OH)_2$] results in a considerably larger quantity of solid waste from this process than from a conventional hydroxide precipitation process.

When the Sulfex process was compared to hydroxide precipitation in a series of jar test studies and pilot plant demonstration tests, the following conclusions were reported.[2]

- When treating the same influent, the Sulfex process obtains lower residuals of copper, cadmium, nickel, and zinc than can be obtained with the hydroxide process.
- Satisfactory effluent quality is usually obtained with the Sulfex process within the 8.5 to 9.0 pH range which is within the 6.0 to 9.5 pH range permitted by EPA for discharge.
- The removal of a particular heavy metal is more effective when it is in a solution containing other heavy metals than when it is the only metal in solution.

TABLE 10.1.23. SUMMARY OF TYPICAL INSOLUBLE-SULFIDE PRECIPITATION OPERATING PARAMETERS

Parameter	Unit	Operating range	Optimum range
Reaction temperature	°C	Room	-
Reaction pH	S.U.	6.0 - 9.0	8.5 - 9.0
Reagent excess	%	0 - 400	100 - 300
Influent metal concentration	mg/L	1 - 500	20 - 50
Retention time	Minute	30 - 60	30
Sedimentation rate	gpm/ft^2	0 - 2	2[a]

[a]Tube settler.

Source: Reference 2.

- The Sulfex process can be applied in precipitators (and similar devices) at surface rates up to 2.0 gpm/ft^2 when tube settlers are used.

- The required dosage of ferrous sulfide reactant is dependent upon the type of waste being treated. It should normally vary from about 1.5 times theoretical requirement for wastes with no complexing agents to three or more times theoretical for wastes containing complexing agents.

- The concentration of settleable ferrous sulfide solids in the mixing zone, the pH of the process, and use of certain polyelectrolytes are important to obtaining satisfactory results in the Sulfex process.

- It may be more economically desirable to pretreat wastes containing high concentrations of dissolved heavy metals (i.e., a total heavy metal concentration greater than 50 mg/L) by hydroxide before polishing with Sulfex.

10.1.3.2 Process Performance--

While not as prevalent as hydroxide treatment, sulfide precipitation has seen increasing usage in recent years due to improvements in both reagent dosage and hydrogen sulfide emission controls. The following are illustrations of soluble, insoluble, and calcium sulfide precipitation technologies.

Soluble sulfide precipitation--At a 37 gallon/minute (gpm), industrial pretreatment facility a full-scale demonstration of the soluble sulfide precipitation process for the pretreatment of a metal finishing wastewater was performed. Soluble sulfide precipitation was selected because the lower solubility of metal sulfides was expected to result in better metal removal efficiency than conventional hydroxide precipitation.[56]

Three segregated wastes were treated separately. Cyanide-containing wastes were treated in a two-stage alkaline chlorination process for complete cyanide oxidation. Chromium-containing wastes were acidified to pH 2.5 and treated with sodium metabisulfite to reduce hexavalent chromium to the less soluble trivalent form. Following separate treatment, these wastes were combined with the acid/alkali and metals contaminating wastes for treatment by soluble sulfide precipitation.

This treatment system consists of pH adjustment with caustic soda, addition of ferrous sulfate and anionic polymer as coagulants, addition of sodium sulfide to precipitate metals, flocculation, parallel plate clarification, gravity sand filtration, and peroxide destruction of residual sulfide. Sludge processing consists of gravity thickening and dewatering in a plate and frame filter press. Table 10.1.24 compares limitations for a 6 month period.

Insoluble sulfide precipitation--In 1980, three plants using the Sulfex process to remove heavy metals from wastewater discharge were investigated to assess system performance. Two of the plants (Plants A and B) use the Sulfex process singularly, while the third (Plant C) uses the process as a polishing step after hydroxide precipitation and clarification.[52]

Plant A uses both electroless and electrolytic plating processes to plate plastic components. The heavy metals in the wastewater (copper, nickel, and chromium) are complexed/chelated with a variety of proprietary agents. Plant B manufactures carburetors for the automotive industry. Wastewater from the metal finishing portion of the process contains chromium, zinc, iron, phosphates, organic chelating agents, and assorted plating chemicals. Plant C treats wastewater from a barrel-dip, zinc-phosphating line.

Table 10.1.25 presents the chemical consumption and sludge generation rates for Plants A, B, and C. While Plants B and C were successful in lowering metallic contaminants to effluent discharge requirements, Plant A was unable to treat both hexavalent and total chromium. The poor performance in chromium removal was primarily due to an increase in the level of hexavalent chromium in the mixer/clarifier without a commensurate increase in the FeS feed to compensate for the increased demand. Consequently, the level of unreacted FeS in the sludge blanket was gradually depleted and eventually, insufficient FeS was present in the blanket to achieve the normal high level of removal. The FeS stored in the sludge blanket prior to reagent depletion was able to maintain the high removal efficiency.

TABLE 10.1.24. TOBYHANNA ARMY DEPOT WASTE AND TREATED EFFLUENT ANALYSIS (mg/L)

Parameter	Waste average	Effluent		Daily maximum		Daily average	
		Average	Maximum	Design	Permit	Design	Permit
Cadmium	1.34	0.09	0.25	1.2	0.69	0.5	0.26
Chromium	1.14	0.31	1.15	7	2.77	2.5	1.71
Copper	2.35	0.07	0.47	4.5	3.38	1.8	2.07
Lead	0.43	0.19	0.4	0.6	0.69	0.3	0.43
Nickel	1.61	0.08	0.35	4.1	3.98	1.8	2.38
Silver	-	0.01	0.02	-	0.43	-	0.24
Zinc	3.4	0.37	2.69[a]	4.2	2.61	1.8	1.48
Cyanide	1.08	0.04	0.12	0.8	1.2	0.23	0.65
Aluminum	6.67	4.3	18	1	-	0.5	-
Tin	0.003	0.01	1.0	2.5	-	1	-
Suspended solids	-	18.8	152[a]	-	60	-	31
Oil and grease	-	12.8	22	-	52	-	26

[a]Exceeds permit limit.

- Indicates data not available or no standard specified.

Source: Reference 56.

TABLE 10.1.25. WASTEWATER TREATMENT PROCESS CHARACTERISTICS FOR PLANTS A, B, AND C[a]

Characteristic	Value		
	Plant A	Plant B	Plant C
Wastewater			
Average flow rate (gal/min)	39	21	16
pH:			
Feed	2.0 - 4.0	4.5 - 6.0	2.5 - 3.0
Effluent	9.0 - 10.0	8.5 - 9.5	7.5 - 8.5
Average feed concentration (ppm):			
Nickel	31	NA	NA
Copper	28	NA	NA
Hexavalent chromium	76	27	0.07
Total chromium	88	39	8
Zinc	NA	48	24
Iron	NA	1.4	127
Phosphorus	NA	NA	289
Average effluent concentration (ppm):			
Nickel	0.54	-	-
Copper	0.03	-	-
Hexavalent chromium	0.10	0.005	0.02
Total chromium	0.20	0.13	0.10
Zinc	-	0.02	0.12
Iron	-	0.05	0.60
Phosphorous	-	-	0.3
Treatment chemicals			
Lime:[b]			
lb/h	8.8	2.0	8.1
Calcium chloride (for phosphate removal):[b]			
lb/h	NA	NA	17.0
Cationic polymer:[b]			
lb/h	0.1	0.17	0.02
Anionic polymer:[b]			
lb/h	NA	NA	0.01
Ferrous sulfide:			
lb/h	12/5[c]	4/5[d]	0.30[b]
Sludge generation factors			
Dry goods generation:			
lb/h	23.7	7.2	16.4
First stage	NA	NA	16
Second stage	NA	NA	0.4
lb/1,000 gal wastewater	10.1	5.7	17
Underflow volume (gal/h at 0.75% solids)	380	114	262
Filter cake volume (gal/h at 30% solids)	7.9	2.4	5.3

[a]All three plants use an ISP process to remove metals from wastewater, but Plant C uses ISP as a polishing system.

[b]Observed rates.

[c]Based on three times the stoichiometric requirement.

[d]Based on four times the stoichiometric requirement.

Source: Reference 52.

Calcium sulfide--Many of the problems associated with soluble and insoluble sulfide precipitation (i.e., excess reagent requirements and H_2S evolution) can be minimized with calcium sulfide as the sulfide source.[50,57] Solid CaS can be added to the wastewater as a slurry. The addition of CaS as a slurry produces easily settleable precipitates: calcium sulfide particles act as nuclei for production of metal sulfide particles, and the dissolved calcium ion functions as a coagulant. Since calcium, which is added as CaS, is mostly dissolved in the wastewater after reaction, the increase in the sludge volume is minimal. For the same reason, unlike FeS, the CaS requirement is near stoichiometric.

Calcium sulfide is stable only in dry solid form. In aqueous solution, it reacts with water to produce $Ca(HS)_2$ and $Ca(OH)_2$.

$$2CaS + 2H_2O = Ca(HS)_2 + Ca(OH)_2$$

or: $$S^= = HS^- + OH^-$$

The main reactions involved in the precipitation of metal sulfides after adding the CaS solutions are:

$$M^{++} + HS^- = MS + H^+$$

$$H^+ + OH^- = H_2O$$

$$M^{++} + S^= = MS$$

Research conducted by the General Electric Company, Schenectedy, New York, investigated the effectiveness of calcium sulfide as a precipitation agent. The investigation involved batch treatment of the wastewater by sulfide precipitation with the addition of lime until the pH reached 7 and next, 0.1 M CaS solution to a desired pH value, normally 9.0. The precipitates were flocculated with 2 mg/L Nalco 7763 polyelectrolyte. Vigorous mixing (600 rpm) of the solution for 2 minutes followed by moderate mixing (30 rpm) for 1 minute was sufficient for effective flocculation. The flocs were settled for 30 minutes before sampling supernatant liquid for analysis. The solution was further filtered with 0.2 um Acropor filter to remove any suspended solids.

Both actual and simulated metal finishing wastewaters were treated using hydroxide and calcium sulfide (a mixture of $Ca(OH)_2$ and $Ca(HS)_2$) solutions. In addition, both treatment methods were evaluated on wastewaters containing chelating agents to test their effects. The calcium sulfide preparation system produced the CaS solution from H_2S and $Ca(OH)_2$. The ratio of Ca and S was controlled by measuring the pH of the solution. Vessels for pH adjustment, sulfide precipitation, and flocculation were included in the system.

The results of these experiments, showed that sulfide precipitation is very effective for the removal of heavy metals such as Cd, Cu, Pb, Ag, and Zn. The method works in the presence of chelating agents and removes metals to extremely low concentrations. The calcium sulfide slurries, prepared by reacting lime with hydrogen sulfide or sodium hydrosulfide, are effective sulfide sources. The addition of calcium sulfide can be controlled simply, in most cases, by measuring the pH. The processes employing such techniques have been demonstrated in bench-scale experiments using wastewater samples. The two-stage process may be employed when the wastewater contains a large amount of iron and nontoxic suspended solids.

10.1.3.3 Process Costs--

Table 10.1.26 presents the cost data developed for a continuous soluble sulfide precipitation system. The purchased equipment and installation costs are based on the treatment process shown in Figure 10.1.6 and the assumptions made in Section 10.1.2. An additional aeration vessel consisting of a reinforced concrete reactor, 4-6 acid resistant spargers ($82/sparger), and 30 feet of 6 inch pipe ($2.40/ft) has been included in the treatment train to reduce the quantity of H_2S fumes evolved.

Operating labor requirements have been increased from 4 to 6 hours per shift due to the greater need for process control (to prevent excess sulfide dosing) associated with this process.[2,52,58] Maintenance, overhead, utilities, taxes and insurance have all remained constant; however, reagent chemical costs have increased dramatically. Since sodium sulfide flake costs $410/ton vs. $40/ton for hydrated lime, an equivalent influents metals concentration would result in a greater than 10 fold increase in reagent cost when comparing the two systems.

TABLE 10.1.26. CONTINUOUS SOLUBLE SULFIDE PRECIPITATION COSTS[a]

	Flow Rate (gph)		
	1,000	10,000	100,000
Purchased Equipment and Installation (PE&I) ($)			
Equalization Tank	17,000	29,000	50,000
Precipitation Reactor	24,000	40,000	69,000
Flocculator/Clarifier	18,000	50,000	140,000
Aeration Vessel	17,400	29,500	50,500
Sludge Holding Tank(s)	3,000	6,000	48,000
Filter Press	10,000	25,000	100,000
	89,400	179,500	457,500
Total Capital Investment (360% PE&I)	321,800	646,200	1,647,000
Operating Costs ($)			
Operating Labor ($20/hr)	108,000	108,000	108,000
Maintenance (6% TCI)	19,300	38,800	98,800
General Plant Overhead (5.8% TCI)	18,700	37,500	95,500
Utilities (2% TCI)	6,400	12,900	32,900
Taxes and Insurance (1% TCI)	3,200	6,500	16,500
Chemical Costs Na_2S ($410/ton)	4,050	40,500	404,800
$FeSO_4$ ($145/ton)	350	3,600	35,700
Sludge Transportation ($0.25/ton-mile)	300	2,600	25,600
Sludge Disposal ($200/ton)	13,800	138,100	1,380,000
Annualized Capital (CFR = 0.177)	57,000	114,400	291,500
Total Cost/Yr	231,100	502,900	2,489,300
Cost/1,000 gallon	32	7.0	3.5

[a]1987 Dollars.

Sludge disposal costs are roughly equivalent for sulfide and hydroxide precipitation, while annualized capital costs are slightly greater due to the requirement of an additional reaction vessel. Overall costs for sulfide precipitation, based on the assumptions presented in this section, are approximately 40 percent greater than those associated with hydrated lime. However, sulfide precipitation economics become more favorable when compared to the more expensive hydroxide reagents such as sodium hydroxide ($175/ton for 30 percent solution) or magnesium hydroxide ($0.78/lb). If lower dissolved heavy metal concentrations were desired, the most economically efficient use of sulfide precipitation would be as a final polishing step to, or in conjunction with (co-precipitation), hydroxide precipitation.

10.1.3.4 Status of Technology--

Sulfide precipitation has been demonstrated to be an effective alternative to hydroxide precipitation for removing various heavy metals from industrial wastewaters.[2,52,58-61] The major advantage of the sulfide precipitation process is that because of the extremely low solubility of metal sulfides, very high metal removal efficiencies can be achieved. The major limitations of the sulfide precipitation process are the evolution of toxic hydrogen sulfide fumes and the discharge of treated wastewaters containing residual levels of sulfide[52,57] (see Table 10.1.27 for summary of advantages and disadvantages of sulfide precipitation).

The use of ferrous sulfide (insoluble sulfide process) as a source of sulfide reduces or virtually eliminates the problem of hydrogen sulfide evolution.[2] The use of ferrous sulfide, however, requires reagent consumption considerably higher than stoichiometric and significantly higher sludge generation than either the hydroxide or soluble sulfide treatment processes.

The use of calcium sulfide as a source of sulfides reduces the problems (H_2S evolution and excess reagent requirements) associated with the previous two technologies.[57] However, as with ferrous sulfide, calcium sulfide precipitation results in high solids generation. These solids must be removed in a subsequent treatment step, such as sedimentation or filtration. Sulfide sludges are less subject to leaching than hydroxide sludges. Many landfills now require post-treatment of the residuals through such technologies as stabilization or encapsulation be performed prior to land disposal.

TABLE 10.1.27. ADVANTAGES AND DISADVANTAGES OF SULFIDE PRECIPITATION

Advantages

- The sulfide process has the ability to remove chromates and dichromates without preliminary reduction of the chromium to the trivalent state.
- The high reactivity of sulfides with heavy metal ions and the insolubility of metal sulfides over a broad pH range are attractive features compared with the hydroxide precipitation process.
- Sulfide precipitation, unlike hydroxide precipitation, is relatively insensitive to the presence of most chelating agents and eliminates the need to treat these wastes separately.

Disadvantages

- Sulfide reagent will produce hydrogen sulfide fumes when it comes in contact with acidic wastes. This can be prevented by maintaining the pH of the solution between 8 and 9.5 and may require ventilation of the treatment tanks.
- As with hydroxide precipitation, excess sulfide ion must be present to drive the precipitation reaction to completion. Since the sulfide ion itself is toxic, sulfide addition must be carefully controlled to maximize heavy metals precipitation with a minimum of excess sulfide to avoid the necessity of post-treatment. Where excess sulfide is present, aeration of the effluent stream would be necessary to oxidize residual sulfide to the less harmful sodium sulfate (Na_2SO_4).
- The cost of sulfide precipitants is high in comparison with hydroxide precipitant, and disposal of metallic sulfide sludges may pose problems.

Source: References 2, 52, and 57-61.

10.1.4 Carbonate Precipitation

10.1.4.1 Process Description--

Carbonate precipitation may be used to remove metals either by direct precipitation using a carbonate reagent such as calcium carbonate (limestone) or sodium carbonate or by converting hydroxides into carbonates using carbon dioxide. The solubility of most metal carbonates is intermediate between hydroxide and sulfide solubilities; in addition, carbonates form easily filtered precipitates.

Calcium Carbonate--Limestone is available in either high calcium ($CaCO_3$) or dolomitic ($CaCO_3$ $MgCO_3$) form.[30] Both types of limestone are available as either a powder or crushed stone. Crushed stone diameters are typically 0.074 mm (200 mesh) or less since both the reactivity and completeness of the reaction increase proportionately to the available surface area.[3] High calcium is most commonly used because of its greater reaction rate and its more widespread availability. Dolomitic limestone reactivity will increase if it is finely ground, and sludge production will be minimal due to the formation of soluble magnesium sulfate. However, its reactivity is generally too slow even with grinding, and hence not suitable for most applications.[30]

The inherent problem with calcium carbonate precipitation is that it is only effective for reducing metallic species such as trivalent chromium and iron in its operational pH range (5.0 to 7.0).[62] In addition, the inhibition of the stone particles in the presence of high quantities of sulfate and/or metallic ions make it less attractive than other reagents. Limestone is a solid-based reagent that liberates CaO for precipitation through surface dissolution. The inhibition of the particle surface through calcium sulfate precipitation increases retention times, reagent purchases, equipment sizing, and lowers waste throughput.[29,63,64] Improved reaction kinetics can be achieved by increasing the available solid surface area through greater limestone loading.[65] However, both reagent purchase and sludge disposal costs will increase proportionately with the excess limestone applied.

The primary advantage of limestone neutralization is that limestone is a low cost and widely available reagent. However, limestone is limited in its ability to neutralize over pH 6.0 or treat acid concentrations greater than 5,000 mg/L.[62,65] There have been attempts to use limestone in combination with lime in a dilute, dual alkali mode. The limestone is used as a pretreatment to raise the pH to approximately 3.0 or 6.0 with lime completing the process of precipitation.[15] The limestone/lime process is usually more complicated than a simple lime slurry process, resulting in higher projected costs and limited application. However, in high volume applications the savings in reagent (when used in pebble form) may offset any increase in capital expenditures.

Sodium Carbonate--Sodium carbonate (Na_2CO_3) is a highly reactive soluble alkali that is marketed most often as an anhydrous powder. Wet crystal bulk storage typically facilitates solution feeding. In dry form it is also easily fed from hoppers.[5] Positive provision for dissolution is desirable for dry feed applications. Suitable materials for handling the compound or its solutions include plastic, iron, rubber, and steel. Shipment is made in bags, barrels, or in bulk with transfer usually performed by pneumatic conveyor.

In the chemical trade the terms "ash," "soda ash," "soda," and "calcined soda" are used for the anhydrous salt, although soda ash is the most common name in English-speaking countries. Sodium carbonate is moderately soluble in cold water and soluble to approximately 30 percent of solution weight in hot water the solution is strongly alkaline.[30] (Melting point, 851°C; heat capacity at 25°C, 1043.01 J/(kg-K) [249.3 cal/(kg-K)]; heat of fusion, 315.9 kJ/kg (75.5 kcal/kg); density at 20°C, 2533 kg/m^3. Bulk densities of various commercial grades range from 576 to 1072 kg/m^3 (36-67 lb/ft^3).[30]

Ordinary light soda ash produced by calcining crude bicarbonate is satisfactory for many uses. Dense soda ash is most often manufactured by hydration of light ash to produce larger sodium carbonate monohydrate crystals followed by dehydration. Hydration may be accomplished by either feeding light ash and water to mixers or blenders or by adding light ash to a saturated solution of soda ash containing a slurry of monohydrate crystals. The monohydrate crystals are fed to a continuous dryer. The dehydrated product from the dryers needs only screening before packing and shipping. Most dense ash is shipped in bulk to large industries.

Sodium carbonate is an alternative to sodium alkali for acidic-metals wastestreams lacking buffering capacity such as deionized acid-bath rinsewaters. The use of sodium carbonate (a weak base) with strong acids, such as sulfuric, will impart a buffer to the wastewater stream, thereby facilitating pH control and precipitation within the neutral range. These buffering reagents will produce a smaller change in pH per unit addition than comparable unbuffered, strong bases such as high calcium lime or caustic soda.[37] This phenomena can be seen in Figure 10.1.13, which illustrates the neutralization of a 1 percent sulfuric acid solution with caustic soda and soda ash. A small incremental addition of caustic soda caused the pH to change from 2 to 11 standard units. Alternatively, approximately three times the quantity of soda ash resulted in a modest pH change from 6 to 9 units.[37]

Due to its carbonate-based reaction mechanism, the neutralization/ precipitation metals-containing acidic rinsewaters with soda ash (as with limestone) proceeds at a much slower pace than comparable hydroxide-based reagent systems such as lime or caustic soda. Accordingly, continuous flow reactors must be sized to provide a minimum of 45 minutes hydraulic retention in each stage.[37] In addition, soda ash is commercially available only in a dry form. Consequently, onsite batch mixing and solution preparation facilities, similar to those of hydrated lime, are mandatory when using this chemical as a neutralizing agent. The solubility of soda ash also limits its use since a chemical solution feed strength of only 20 percent by weight can be maintained at ambient temperatures without salt recrystallization. Continuous mixing of the prepared solution is recommended to maintain homogeneity.[5]

An advantage of soda ash is lower sludge generation since sodium-based end products are more soluble than calcium-based products. However, sodium-based sludges do not filter as readily or to as high solids content as calcium-based sludges. In addition, the clarified liquid effluent may not be as low in metals content or total dissolved solids as insoluble end product systems such as lime. All these factors must be carefully weighed before selecting sodium carbonate or any other alkaline reagent as a precipitating agent.

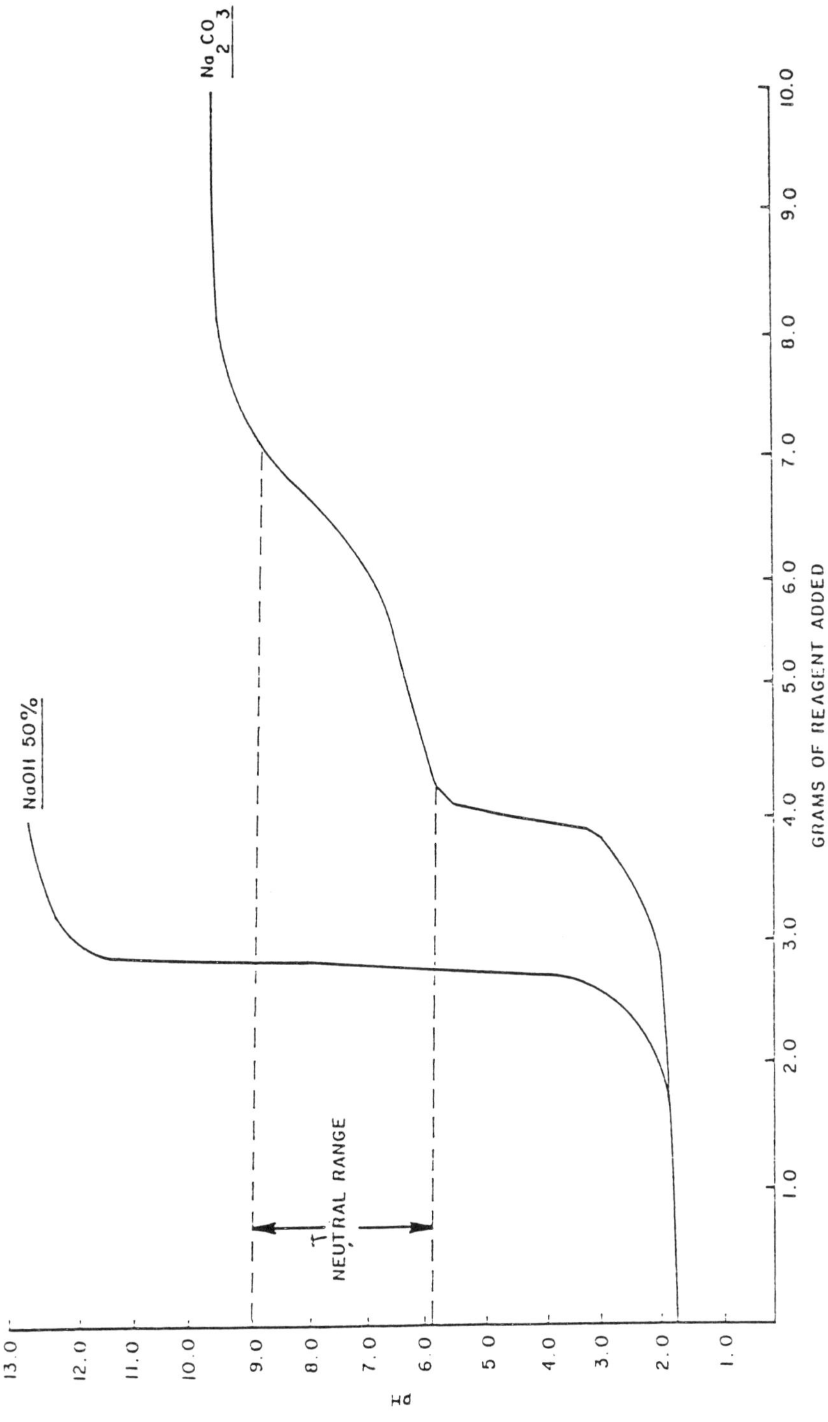

Figure 10.1.13. Titration curve for the neutralization of a 1% H_2SO_4 solution with sodium hydroxide and sodium carbonate.

Source: Reference 37.

Carbon Dioxide--Carbon dioxide is a relatively old but, as of yet, undeveloped technology for treating metals-containing wastestreams. Typically, carbonic acid is generated directly in the neutralization/precipitation chamber by injecting carbon dioxide into the wastewater solution. Upon hydration, the carbon dioxide will form carbonic acid and react with available hydroxides to form less soluble carbonates.

H_2CO_3	+	$Ni(OH)_2$	=	$NiCO_3$	+	$2H_2O$	(1)
Carbonic acid		Nickel hydroxide		Nickel carbonate			

H_2CO_3	+	$Ca(OH)_2$	=	$CaCO_3$	+	$2H_2O$	(2)
Carbonic acid		Hydrated lime		Calcium carbonate pH 9.4 (saturated)			

Compressed (liquid) carbon dioxide is stored and transported at ambient temperatures in cylinders containing up to 22.7 kilograms. Larger quantities are stored in refrigerated, insulated tanks maintained at -18°C and 20 atmospheres.[66] Transportation is by insulated tank truck and rail car.

The standard method of applying compressed carbon dioxide for precipitation is to vaporize carbon dioxide in a heat exchanger or across a flash valve. The pressurized gas is forced through porous diffuser tubes placed along the bottom of a batch treatment tank. Carbon dioxide gas is released from the diffusers as fine bubbles (15 microns) which are preferentially absorbed by the surrounding wastewater. This type of treatment requires a slow-moving effluent stream with a treatment tank of sufficient depth to ensure that the carbon dioxide is fully absorbed before reaching the surface.[67] Since hydration of carbon dioxide forms carbonic acid, it is recommended that the diffuser assembly be constructed of a corrosion-proof material.

The primary advantages of compressed carbon dioxide are minimal capital requirements, uncomplicated piping, and the inability to over-acidify the wastewater. Its primary disadvantages are a low dissolved oxygen content (4.5 percent) at the point of injection, and a high reagent cost on a neutralization equivalent basis (approximately \$200 to \$300/ton. However, for large volume users of 200 tons or more per year, the unit cost per ton of compressed carbon dioxide drops to \$90 to \$100/ton.[68,69]

10.1.4.2 Process Performance--

Carbonate precipitation technology is sometimes preferred over hydroxide precipitation because in some instances it provides superior precipitation properties; i.e., with cadmium it produces cadmium carbonate which is preferred cadmium hydroxide for recovery purposes. Also, nickel and lead precipitation with carbonate gives lower final levels than precipitation with hydroxide.[70]

Treatment of cadmium with sodium carbonate (soda ash) will give good levels of removal at a slightly lower pH than hydroxide, typically in the range of 9.5-10. Due to the value of cadmium, it is often desirable to send the precipitated sludge to a reprocessor for recovery of the cadmium, or to reuse it. Whether the cadmium is in the hydroxide or carbonate form may be important to the reprocessor plant operator.

Bench-scale tests conducted by Nassau Smelting and Refining Co.[71] studied lead precipitation by caustic, lime and caustic soda/soda ash. It was found that both lime and caustic soda/soda ash gave good results. The optimum pH was 9.0 to 9.5. Influent lead was 5 mg/L and final lead was 0.01 to 0.04 mg/L.

Figure 10.1.14 shows solubility levels of lead with different alkali agents. As can be seen, the soda ash/caustic soda systems produced slightly better results than the straight-line system. Separan AP 30 was used as a coagulant aid.[71]

Investigators at the Illinois Institute of Technology performed a series of solubility experiments.[72] In this series, precipitation experiments were performed over 24-hour periods at constant pH and C_T (total carbonate). Two levels of carbonate were evaluated from each metal: a low background

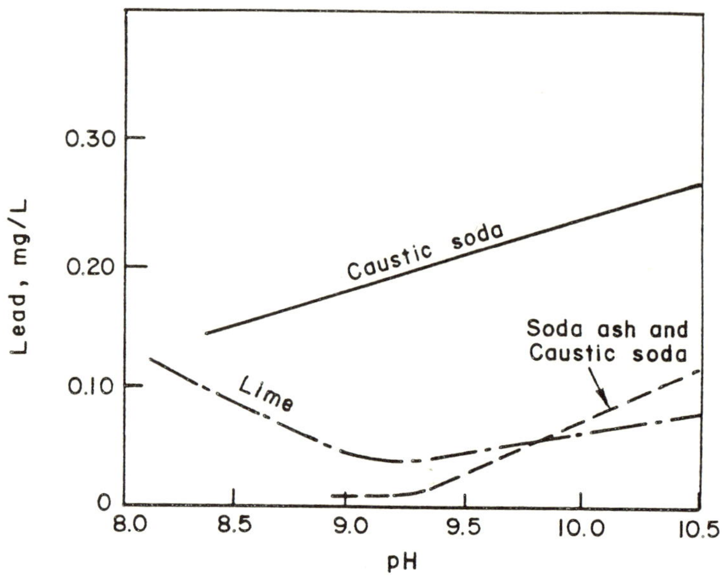

Figure 10.1.14. Lead solubility in three alkalies.

Source: Reference 71.

carbonate level of less than $10^{-3.8}$ M(2 mg/L inorganic carbon) and a carbonate level of approximately $10^{-3.2}$ M(7.6 mg/L inorganic carbon). Values of pH ranging from pH 6 to pH 13 were tested. In addition, hydroxide experiments were performed under the same conditions for cadmium, copper, lead, and zinc.

Cadmium Solubility--Minimum cadmium solubility of 0.08 mg/L was obtained at pH 10-10.5. In the pH range of 6.5 to 8.5, the carbonate system yielded a soluble cadmium concentration range of 81.0 mg/L to 0.66 mg/L. The hydroxide system, over the same pH range of 6.5 to 8.5, yielded a much greater soluble cadmium concentration range of 8 x 10^4 mg/L to 129 mg/L. Lower soluble cadmium concentrations are observed in the test system with carbonate present. At higher pH values of 9-10, the soluble metal concentrations are comparable for both systems. This suggests that both systems are controlled by hydroxide solubility at pH 9-10 rather than pH near 10. At pH above 10.0 there is a significant difference in soluble cadmium concentrations. Soluble cadmium concentrations for the carbonate system were much lower. This appears to be due to a slight increase in carbonate concentration at pH 9.5, which would decrease soluble cadmium concentrations.

Copper Solubility--The minimum soluble copper concentration attained was 0.005 mg/L at pH 8.9 to 9.3. Over the pH range 6.7 to 7.9, soluble copper concentrations were reduced from 3.5 mg/L to 0.016 mg/L.

Minimum solubility of 0.015 mg/L to 0.018 mg/L was obtained in the carbonate test system in the pH range of 8.6-10.4. From pH 7.5 to 9.5 the soluble copper concentration in the hydroxide test system ranged from 0.021 to 0.005 mg/L. With an increase in carbonate concentrations to C_T = $10^{-3.2}$ M, the soluble copper concentration in the carbonate system ranged from 0.061 to 0.016 mg/L.

Lead Solubility--It is apparent that the carbonate induces lower soluble lead concentrations than occur in the hydroxide system, at pH below 8. At pH 7.0 to 7.5, the carbonate system yielded a minimum soluble lead concentration of 0.025 mg/L while the hydroxide system produced a lead concentration of 0.131 mg/L. Above pH 8, carbonate functions as a ligand to increase lead solubility.

Zinc Solubility--Solubility patterns for both systems are similar, with data points for the carbonate system generally below those of the hydroxide system, but higher than the theoretical carbonate solubility curve. There is some evidence that the zinc carbonate precipitation system approaches equilibrium extremely slowly, perhaps requiring more than 10 days to near equilibrium solubility.[5] This has been postulated to result from the more rapid kinetics of zinc hydroxide precipitate formation, even in a system thermodynamically stable for zinc carbonate. The subsequent kinetics of zinc solubility then would be limited by the slow transformation of solid phase zinc hydroxide to zinc carbonate.

10.1.4.3 Process Costs--

Table 10.1.28 details the cost data developed for a continuous sodium carbonate precipitation system. The purchased equipment and installation costs are equivalent to those of the hydrated lime precipitation system except that a retention time of 1 hour instead of 30 minutes has been used to size the precipitation reactor (due to the slower reactivity of sodium carbonate). In addition, chemical reagent costs and useages are significantly higher for sodium carbonate when compared to hydrated lime. For example, approximately 2.9 lbs of sodium carbonate (at $120/ton) are required to precipitate 1 lb of heavy metal, while only 2.2 lbs of hydrated lime (at $40/ton) are required per lb of metal. However, due to the higher solubility of the sodium cation in the sodium carbonate complex sludge generation is only 7 percent higher (on a dry weight basis).

Overall costs for sodium carbonate, based on the cost data presented in this section, are only 1 to 18 percent greater than those presented for hydrated lime. The viability of this technology as an alternative to either hydroxide or sulfide precipitation is enhanced due to the lower pH requirements (usually 8-9) for carbonate precipitation. The lower pH requirement will result in lower alkali demand for neutralization and consequently less sludge generation. Therefore in any consideration of alternate precipitation technologies, influent pH should also be examined.

TABLE 10.1.28. CONTINUOUS SODIUM CARBONATE PRECIPITATION COSTS[a]

	Flow Rate (gph)		
	1,000	10,000	100,000
Purchased Equipment and Installation (PE&I) ($)			
Equalization Tank	17,000	29,000	50,000
Precipitation Reactor	24,000	60,000	150,000
Flocculator/Clarifier	18,000	50,000	140,000
Sludge Holding Tank(s)	3,000	6,000	48,000
Filter Press	10,000	25,000	100,000
	72,000	170,000	488,000
Total Capital Investment (360% PE&I)	259,000	612,000	1,756,800
Operating Costs ($)			
Operating Labor ($20/hr)	72,000	72,000	72,000
Maintenance (6% TCI)	15,500	36,700	105,400
General Plant Overhead (5.8% TCI)	15,000	35,500	101,900
Utilities (2% TCI)	5,200	12,200	35,100
Taxes and Insurance (1% TCI)	2,600	6,100	17,600
Chemical Costs ($120/ton)	2,100	20,600	206,400
Sludge Transportation ($0.25/ton-mile)	200	2,100	21,400
Sludge Disposal ($200/ton)	12,800	128,400	1,284,000
Annualized Capital (CFR = 0.177)	45,800	108,300	311,000
Total Cost/Yr	171,200	421,900	2,154,800
Cost/1,000 gallon	24	6	3

[a]1987 Dollars.

10.1.4.4 Status of Technology--

Carbonate precipitation has been demonstrated to be a viable alternative to either hydroxide or sulfide precipitation for removing various heavy metals from industrial wastewaters. The solubility of most metal carbonates is intermediate between hydroxide and sulfide solubilities. In addition, the reagent cost is also intermediate. The main advantages of carbonate technology are buffering capability, superior handling characteristics (i.e., little dust, good flow, and no arching in the feeder), and widespread availability. Main disadvantages are slow reaction time (typically a minimum of 45 minutes retention) and low solubility (20 percent by weight). Since carbonates are not particularly corrosive and soda ash generates less sludge than comparable calcium-based technologies, environmental impacts are few. See Table 10.1.29 for summary of advantages and disadvantages of carbonate precipitation.

TABLE 10.1.29. ADVANTAGES AND DISADVANTAGES OF CARBONATE PRECIPITATION

Advantages

- Carbonate reagents have a relative ease of handling and can be obtained in bulk by railcar or truck or in 100 lb bags
- Calcium carbonate forms easily filtered precipitates
- Sodium carbonate imparts buffering capacity and generates less sludge

Disadvantages

- Retention times are longer due to slower reacting carbonate-based chemistry
- Carbonates do not mix easily into solution and have the potential for evolving carbon dioxide which, without aeration, will slow reaction times further
- Calcium carbonate particles have the potential to become deactivated if calcium sulfate precipitates on particle surface
- Sodium carbonate sludges do not filter as readily or to as high solids content as calcium-based sludges
- Calcium carbonate is only able to achieve an operational pH range of 5-7

Source: References 62, 66, 70, and 72.

REFERENCES

1. U.S. EPA. Treatability Manual, Volume III, EPA-600/8-80-042. July 1980.

2. U.S. EPA-600/2-77-049. Treatment of Metal Finishing Wastes by Sulfide Precipitation. February 1977.

3. Wilk, L. et al., Alliance Technologies Corporation. Technical Resource Document: Treatment Technologies for Corrosive-Containing Wastes. Draft Final Report. October 1986.

4. Camp, Dresser, and McKee. Technical Assessment of Treatment Alternatives for Wastes-Containing Corrosives. Contract No. 68-01-6403. September 1984.

5. Cushnie, G.C. Removal of Metals from Wastewater: Neutralization and Precipitation. Pollution Technology Review, No. 107. Noyes Publications, Park Ridge, NJ. 1984.

6. Oberkrom, S.L., and T.R. Marrero. Detoxification Process for a Ferric Chloride Etching Waste. Hazardous Waste and Hazardous Materials, Vol. 2, No. 1. 1985.

7. MITRE Corporation. Manual of Practice for Wastewater Neutralization and Precipitation. EPA-600/2-81-148. August 1981.

8. Levenspiel, O. Chemical Reaction Engineering. 2nd. Edition, John Wiley & Sons, New York, NY. 1972.

9. Hoyle, D.L. Designing for pH Control. Chemical Engineering, November 8, 1976.

10. Lewis, C.J., and R.S. Boynton. Acid Neutralization with Lime for Environmental Control and Manufacturing Processes. National Lime Association, Bulleton No. 216. 1976.

11. Hoffman, F. How to Select a pH Control System for Neutralizing Waste Acids. Chemical Engineering. October 30, 1972.

12. Jungek, P.R., and E.T. Woytowicz. Practical pH Control. Industrial Water Engineering. February/March 1972.

13. Grosse, D.W. A Review of Alternative Treatment Processes for Metal-Bearing Hazardous Waste Streams. Journal of the Air Pollution Control Association. May 1986.

14. U.S. EPA. Reducing Water Pollution Costs in the Electroplating Industry. EPA-625/5-85-016. September 1985.

15. Hale, F.D., et al. Spent Acid and Plating Waste Surface Impoundment Closure. Management of Uncontrolled Hazardous Waste Site. Washington, D.C. October 31-November 2, 1983.

16. Mace, G.R., and D. Casaburi. Lime vs. Caustic for Neutralizing Power. Chemical Engineering Progress. August 1977.

17. Janson, C.E., et al. Treatment of Heavy Metals in Wastewaters. Environmental Progress. August 1982.

18. U.S. EPA. Remedial Action at Hazardous Waste Disposal Sites (Revised). EPA-625/6-85-006. October 1985.

19. Berger. Land Application of Neutralized Spent Pickle Liquor. 17th Industrial Waste Conference, Purdue University. 1962.

20. Guide to the Disposal of Chemically Stabilized and Solidified Waste, EPA SW-872. September 1980.

21. Environmental Laboratory, U.S. Army Engineer Waterways Experiment Station. Survey of Solidification/Stabilization Technology for Hazardous Industrial Wastes, EPA-600/2-79-056.

22. Alliance Technologies Corporation Technical Resource Document: Treatment Technologies for Dioxin-Containing Wastes. Contract No. 68-03-3243. August 1986.

23. Stabilizing Organic Wastes: How Predictable are the Results? Hazardous Waste Consultant. p. 18. May 1985.

24. Thompson, D.W., Malone, P.G., and L.W. Jones. Survey of Available Stabilization Technology in Toxic and Hazardous Waste Disposal. Vol. 1, R.B. Pojasek (Editor). Ann Arbor Science, Ann Arbor, MI. 1979.

25. Lubowitz, H.R. "Management of Hazardous Waste by Unique Encapsulation Processes." Proceedings of the 7th Annual Research Symposium. EPA-600/9-81-002b.

26. Wiles, C. Hazardous Waste Engineering Research Laboratory, U.S. EPA, private communication; and Critical Characteristics and Properties of Hazardous Waste Solidification/Stabilization, HWERL. U.S. EPA Contract No. 68-03-3186 (in publication).

27. U.S. EPA. Sources and Treatment of Wastewater in the Nonferrous Metals Industry. EPA-600/2-80-074. April 1980.

28. Besselievre, E.B. The Treatment of Industrial Wastes. McGraw-Hill Book Company, New York, NY. 1967.

29. Boynton, R.S. Chemistry and Technology of Lime and Limestone. Interscience Publishers, New York, NY. 1966.

30. Kirk-Othmer Encyclopedia of Chemical Technology. Vol. 14, 3rd. Edition, John Wiley & Sons, New York, NY. 1966.

31. U.S. EPA. Process Design Manual: Sludge Treatment and Disposal. EPA-625/1-79-011. September 1979.

32. Hugget et al. Automatic Continuous Acid Neutralization. 23rd. Industrial Waste Conference, Purdue University. 1968.

33. Houk, R.D. et al. Lime Treatment of Waste Pickle Liquor. Industrial and Engineering Chemistry. February 1947.

34. Mooney, G.A., et al. Two-Stage Lime Treatment in Practice. Environmental Progress, Vol. 1, No. 4. November 1982.

35. Hsu, D.Y., et al. Soda Ash Improves Lead Removal in Lime Precipitation Process. 34th. Industrial Waste Conference, Purdue University. 1978.

36. Kirk-Othmer Encyclopedia of Chemical Technology. Vol. 1, 3rd. Edition, John Wiley & Sons, New York, NY. 1981.

37. Mabbett, Cappacio & Associates. Industrial Wastewater Pretreatment Study: Preliminary Engineering Design Report. January 1982.

38. Leedberg, T., Honeywell Corporation. Telephone conversation with Steve Palmer, Alliance Technologies Corporation. August 13, 1986.

39. Metcalf & Eddy. Facility Test Report for Frontier Chemical Waste Process, Inc. November 1985.

40. Wing, R.E. Complexed and Chelated Copper-Containing Rinsewaters. Plating and Surface Finishing. July 1986.

41. Teringo, J. Magnesium Hydroxide for Neutralizing Acid Waste-Containing Metals. Plating and Surface Finishing. October 1986.

42. Schiller, J.E. et al. Mineral Processing Water Treatment Using Magnesium Oxide. Environmental Progress. May 1984.

43. State of Environmental Affairs Report. Plating and Surface Finishing. April 1987.

44. Versar, Inc. Technical Assessment of Treatment Alternatives for Wastes-Containing Metals and/or Cyanides. Contract No. 68-03-3149, U.S. EPA/OSW (Draft). October 1984.

45. Duffey, D.P. et al. A Survey of Metal Finishing Wastewater Treatment Costs. Plating and Surface Finishing. April 1987.

46. Peters, M.S., and K.D. Timmerhaus. Plant Design and Economics for Chemical Engineers. McGraw-Hill Book Company, New York, NY. 1980.

47. U.S. EPA. Economics of Wastewater Treatment Alternatives for the Electroplating Industry: Environmental Pollution Control Alternatives. EPA-625/5-79-016. 1979.

48. Levine, R.Y. et al. Sludge Characteristics for Lime Neutralized Pickling Liquor. 7th. Industrial Waste Conference, Purdue University. 1952.

49. Dickerson, B.W. et al. Neutralization of Acid Wastes: Industrial Engineering Chemistry 42:599-605. 1950.

50. Kim, B.M. Treatment of Metal-Containing Wastewater with Calcium Sulfide. AICHE Symposium Series Water. 1980.

51. Kirk-Othmer Encyclopedia of Chemical Technology, Vol. 22, 3rd. Edition, John Wiley & Sons, New York, NY. 1981.

52. U.S. EPA. Control and Treatment Technology for the Metal Finishing Industry. Sulfide Precipitation. EPA-625/8-80-003.

53. Sundstrom, D.W., and M.D. Khei. Wastewater Treatment. Prentice-Hall, Inc., Englewood Cliffs, NY. 1979.

54. Weast, R.C. Handbook of Chemistry and Physics, 50th. Edition, West Palm Beach, FL. 1969.

55. Meites, L. Handbook of Analytical Chemistry, New York, NY, McGraw-Hill Book Company, New York, NY. 1963.

56. CH_2MHill, Industrial Processes to Reduce Generation of Hazardous Waste at DOD Facilities. Phase 2 Report. July 1985.

57. Kim, B.M. et al. Calcium Sulfide Process for Treatment of Metal-Containing Wastes. Environmental Progress. August 1983.

58. Higgins, T.E., and S.G. TerMaath. Treatment of Plating Wastewaters by Ferrous Reduction, Sulfide Precipitation, Coagulation, and Upflow Filtration. Proceedings of the 36th. Industrial Waste Conference, Purdue University. 1981.

59. Resta, J.E. et al. Soluble-Sulfide Precipitation Treatment of Metal Finishing Wastewater. Proceedings of the 16th. Mid-Atlantic Industrial Waste Conference. 1984.

60. Talbot, R.S. Co-Precipitation of Heavy Metals with Soluble Sulfides Using Statistics for Process Control. Proceedings of the 16th. Mid-Atlantic Industrial Waste Conference. 1984.

61. Peters, R.E., et al. The Effect of Chelating Agents on the Removal of Heavy Metals by Sulfide Precipitation. Proceedings of the 16th. Mid-Atlantic Industrial Waste Conference. 1984.

62. Gloves, H.G. The Control of Acid Mine Drainage Pollution by Biochemical Oxidation and Limestone Neutralization Treatment. 22nd. Industrial Waste Conference, Purdue University. 1967.

63. Gehm, H.W. Neutralization of Acid Wastewaters with Upflow Expanded Limestone Bed. Sewage Works Journal 16:104-120. 1944.

64. Tully, T.J. Waste Acid Neutralization. Sewage and Industrial Wastes 30:1385. 1985.

65. Volpicelli, G., et al. Development of a Process for Neutralizing Acid Wastewaters by Powdered Limestone. Environmental Technology Letters, Vol. 3, pp. 97-102. 1982.

66. Kirk-Othmer Encyclopedia of Chemical Technology. Vol. 4, 3rd. Edition, John Wiley & Sons, New York, NY. pp. 725-741. 1981.

67. Griffith, M.J. et al. Carbon Dioxide Neutralization of an Alkaline Effluent. Industrial Waste. March 1980.

68. Ponzevik, D. Liquid Air Products. Telephone conversation with Stephen Palmer, Alliance Technologies Corporation. September 6, 1986.

69. Berbick, D., Cardox Corporation. Telephone conversation with Stephen Palmer, Alliance Technologies Corporation. September 25, 1986.

70. Lanoutte, K.H. Heavy Metals Removal. Chemical Engineering. October 17, 1977.

71. Day, R.V., Lee, E.T., and E.S. Hochuli. Bell System's Metals Recovery Plant. Industrial Waste. July-August 1974.

72. Patterson, J.W. Effect of Carbonate Ion on Precipitation Treatment of Cadmium, Copper, Lead, and Zinc. 36th. Industrial Waste Conference. Purdue University. 1981.

10.2 COAGULATION AND FLOCCULATION

Chemical coagulation and flocculation are two terms often used interchangeably to describe a process whereby a chemical addition is made to enhance sedimentation (removal of solid particles from suspension by gravitational settling) operation. Coagulation and flocculation are often used to remove the insoluble and colloidal heavy metal complexes formed by precipitation. In this text, chemical coagulation is defined as particle agglomeration brought about by the reduction of electrostatic particle surface charges. Flocculation is a time-dependent physical process of aggregation of fine particles into solids large enough to be separated.

The coagulation process involves the destabilization of the suspension by neutralizing or decreasing the repulsive forces on the particles, so that the particles will approach each other and agglomerate.[1,2,3]

The charge on organic, inorganic and biocolloids is typically negative when suspended in water. The negative charge attracts positive ions due to electrostatic forces which are distributed as shown in Figure 10.2.1. The inner layer (termed the stern layer) contains adsorbed ions and is typically about the thickness of a hydrated ion. The second diffuse layer contains a shear plane within which ions move with the particle. Outside the shear plane ions move independently of the particle as dictated by fluid and thermal motions.[2] The electrical potential difference between the shear plane and the bulk solution is termed the Zeta potential. Zeta potential is a measurable quantity and can be used qualitatively to predict the potential for coagulation.[1,2] As the Zeta potential approaches zero, coagulation should increase.[2]

However, the overall success of the coagulation/flocculation process is ultimately dependent upon the flocculating and settling characteristics of the particles. The rate at which coagulated particles coalesce is primarily related to the frequency of the collisions between the particles. Collisions occur as a result of heavier faster particles overtaking lighter slower particles. The collision frequency is proportional to the concentration of particles and the difference in settling velocities.[2] Since the total number of collisions increases with time, the degree of flocculation also generally increases with residence time in the reaction chamber.

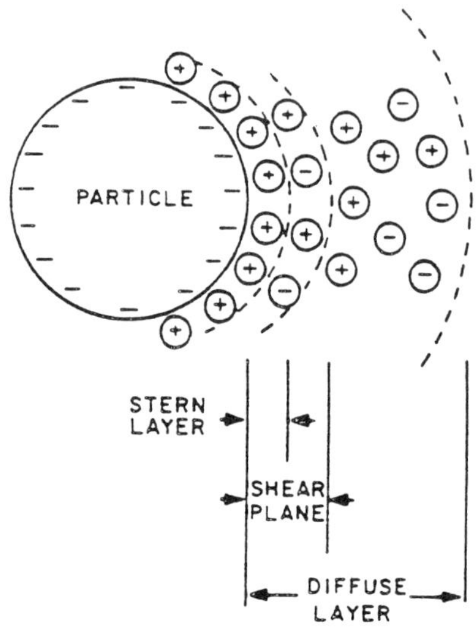

Figure 10.2.1. Double layer charge distribution.

Source: Reference 4.

The rate of flocculation cannot be predicted from collision frequency alone. The coalescence of particles depends upon many factors, such as the nature of the surface, the presence of charges, shape, and density. At present, there is no adequate theoretical model to predict the rate of flocculation in a suspension.

As with precipitation, most coagulation/flocculation processes operate under the same fundamental chemical principles and utilize similar types of equipment and process configurations.[5] For example, coagulation/flocculation processes typically entails the following three steps:

1. Addition of the coagulating/flocculating agent to the treated wastewater.

2. Rapid mixing to disperse the coagulating agent throughout the liquid.

3. Slow and gentle mixing to allow for contact between small particles and agglomeration into larger particles.

Coagulant Addition--Probably the most important parameters to be defined in the design of a coagulation/flocculation system are the type of and dosage of the coagulant, the pH and the mixing characteristics. The most common method of determining these parameters is through a jar test (described in references 1 and 6). The jar test is a laboratory scale test where the wastewater to be treated can be subjected to variable conditions of pH, coagulant type, dosage and mixing, flocculating and settling times. The effect of various coagulant aids can also be investigated in this test. The results of the test provide the following types of information:[4]

- Optimum pH value for efficient coagulation with different coagulants.
- Optimum coagulant dosages for effective flocculation.
- Effectiveness of coagulant aids.
- Most effective order of chemical addition.
- Correct mixing times.
- Flow settling characteristics.

- Quantity of sludge requiring disposal.
- Quality of clarified water to be expected from a particular treatment.

From these data the chemical requirements and unit sizes for coagulation, flocculation and sedimentation can be determined.

Rapid Mix--Rapid (flash) mixing residence times have been reported as 30 seconds to 5 minutes,[7] 2 to 5 minutes,[7] and 10 to 30 seconds.[1,8] Mixing characteristics are determined by the velocity gradient in the mixer (a measure of the shear intensity). Insufficient mixing will affect the performance of subsequent steps and overmixing can break up previously formed floc or the incoming wastewater solids.[1] Static mixers can also be employed although the mixing characteristics are a function of the flow which cannot be controlled. The velocity gradient, G, is usually chosen at about 300 ft/sec/ft.

Slow and Gentle Mixing--The slow and gentle mixing stage is usually carried out in a flocculator/clarifier. Clarification is defined as a quiescent flow condition with a hydraulic flow velocity sufficiently low to allow particles with some minimum settling velocity to separate from the waste overflow. The solid collect in the base of the chamber where a rake or suction device is used to remove the collected solids.

Sometimes, sludge recycle is practiced to gently mix the treated wastewater with a slurry of previously settled sludge solids. The recycle solids present a dense concentration of nucleation sites to promote particle growth.

Flocculator/clarifiers used for the removal of heavy metal contaminants come in three basic types:[9-14]

- Basic settling chambers, where the feed is distributed at one end and overflows at the other. This type of unit often requires a mixing zone to flocculate the particles before clarification. Units are available in rectangular or circular shapes with either flat or conical bottoms.

- Mixer-clarifiers where the incoming feed is mixed with the sludge maintained in the unit. This unit basically combines a flocculating chamber with sludge back-mixing and a settling chamber.

- Plate settlers (Lamella) where inclined plates reduce the distance particles must fall to be removed. These units are often well suited to application where space to house the equipment is limited since the units are mostly vertical rather than horizontal.

Mixers commonly used in flocculation/clarification units are typically either oscillating or rotary types. The oscillating types are most applicable to flocculation processes where very gentle flocculation is required. The rotary types consist of the paddle wheel and turbine designs. Typical design values for mixing and flocculation are shown in Table 10.2.1.

TABLE 10.2.1. MIXING AND FLOCCULATION DESIGN CRITERIA

	Detention time (minutes)	Velocity gradient (G) (m/s/m or sec^{-1})
Mixing	0.2 - 2	300 - 1,500
Flocculation	5 - 30	10 - 100
Fragile flocs (e.g., (e.g., biological flocs)	-	10 - 30
Medium strength flocs (e.g., flocs as encountered in turbidity removal)	-	20 - 50
High strength chemical flocs (e.g., flocs encountered in precipitation processes)	-	40 - 100

Source: Reference 15.

The major design parameters of a flocculator are:

- Residence time, t seconds;

- Velocity gradient, G, ft/sec/ft or sec^{-1}; and

- Ratio of floc to total volume of suspension, C, dimensionless.

Residence time is determined from the total flow rate and total flocculator volume. The velocity gradient for mechanically stirred units can be determined from:

$$G = \left(\frac{P}{\mu V}\right)^{1/2} \tag{1}$$

where: P = power requirement, ft-lb/sec

μ = fluid viscosity, lb_f-sec/ft^2

V = flocculator volume, cu ft

Power requirements are determined from:[4]

$$P = \frac{C_o A \rho v^3}{2} \tag{2}$$

where: A = paddle area, ft^2

ρ = fluid density

v = relative velocity of paddles in fluid, fps, usually about 0.7 to 0.8 of paddle tip speed

The velocity gradient can also be expressed as:[14]

$$G = (\text{power/viscosity x volume})^{1/2} = [(Nm/s)/(Ns/m^2)m^3]^{1/2} \tag{3}$$

where: N = force (N)

m = distance (m)

s = time (s)

Values of G from 20 to 90 s^{-1} are typical for flocculation units. Tapered flocculation employs high entrance values of G and lower values as the flow progresses to the exit. Values of G x t (where t = seconds of residence time) ranging from 30,000 to 150,000 are commonly employed for flocculation in domestic water treatment. Flocculator retention times of 5 to 30 minutes are typical. Experimentally derived values of G and G x t are advocated for industrial waste applications.[14]

10.2.1 Process Description

The coagulant/flocculants currently in commercial use are conveniently classified as inorganic, synthetic organic, and naturally occurring organic polymers. The following subsections are organized according to these three categories. Each subsection will highlight the unique aspects and typical uses of each coagulant/flocculant type.

10.2.1.1 Inorganic Flocculants--

Inorganic coagulants are used primarily for waste streams having dilute concentrations of constituents that become insoluble during neutralization/precipitation treatment. A major disadvantage of this technology is that it adds to the quantity of sludge generated by the precipitation process.

Many soluble salts can function as indifferent electrolytes, typically following the Schulz-Hardy rule for coagulation effectiveness, i.e., coagulation of sols is caused by the ions with charges opposite in sign to the charges on the sol particles; the flocculating power of bivalent ions is about 20 to 80 times greater than that of univalent ions, and the flocculating power of trivalent ions is many times greater than that of bivalent ions (see Table 10.2.2).[15,16]

Generally, inorganic coagulants destabilize colloidal particles in the following manner:

- Repression of the double layer.
- Charge neutralization.
- Entrapment by sweeping floc.

Repression of the double layer involves increasing the ionic strength of the solution. As ionic strength increases, the thickness of the layer is reduced, thus allowing particles to come in closer proximity to each other at which point VanderWaal forces may cause coagulation.[16] Repulsive forces can be reduced by charge neutralization. Destabilizing chemicals are added to the colloid within the stern layer so that the effective charge outside the shear layer is reduced.[8] In this case, overdosing can cause a charge reversal and restabilization. Entrapment requires the use of large doses of coagulants

TABLE 10.2.2. THE RATIO OF THE FLOCCULATION POWER OF SALTS WITH Me+, Me++ AND Me+++ IN SOLUTIONS

Sol	Salts	Ratio of the flocculation power
Ag	$NaCl_2$, $La(NO_3)_3$	1:60:10000
AS_2S_3	NaCl, $BaCl_2$, $Ce(NO_3)_3$	1:70:625
AS_2S_3	KNO_3, $UO_2(NO_3)_2$, $Ce(NO_3)_3$	1:80:625
Au	NaCl, BaCl, $Ce(NO_3)_3$	1:60:6660

Me = metal

Source: Reference 15.

which form gelatinous hydrolysis products. These products can effectively mesh the suspended matter. Because massive amounts of coagulants are used in this procedure, the volume of sludge created is greatly increased.

The three main classifications of inorganic coagulants are:

1. Aluminum derivatives.
2. Iron derivatives.
3. Lime.

Aluminum Derivatives--In the literature of coagulants the term alum refers to a commercial aluminum sulfate hydrate, $Al_2(SO_4)_3 \cdot 7H_2O$. It also is called papermakes' alum or filter alum, and is available either in the dry form or in solution. Dry alum is available in several grades, with a minimum aluminum content of 17 percent expressed as Al_2O_3.[17] Liquid alum is about a 49 wt percent solution of $Al_2(SO_4)_3 \cdot 14H_2O$, or about 8.3 wt percent aluminum as Al_2O_3. It can be stored indefinitely without deterioration.[17] Alum is the most widely used inorganic flocculant. Although alum may be considered as Al^{3+} for calculating the composition of the pure salt that ion does not exist in water environments. It forms complexes with water to give a compound such as $Al(H_2O)_6^{3+}$ and then loses protons by hydrolysis to assume a range of either positive or negative charges.[17]

The best range for alum coagulation is pH 5.5 to 8.0, however, actual removal efficiency depends to a large extent on competing ion and chelant concentrations. However, if the coagulation rate is too low, increasing the particle concentration through the use of synthetic organic polyelectrolytes can improve system performance.[18]

An alternative to aluminum sulfate is sodium aluminate which is commercially available either in dry form or in solution, with an excess of base present. It provides a strongly alkaline source of water-soluble aluminum, particularly useful when addition of sulfate ions is undesirable. Sometimes it is used in conjunction with alum for pH control. Another aluminum derivative is polyaluminum chloride (PAC) which is a partially hydrolyzed aluminum chloride solution with an aluminum content of

10 wt percent expressed as Al_2O_3. It is reported to provide faster and stronger flocs than alum in some applications but has yet to achieve widespread use.[17]

Iron Derivatives--Compared with aluminum, the hydrated ferric ion is more acidic,[17] it forms stronger complexes with simple anions, and its amorphous hydroxide is less acidic but the two show a gross similarity in hydrolysis reactions. Aging characteristics of the polynuclear products of the ferric ion are more dependent on the anions.[17] Minimum solubility of ferric hydroxide occurs in the pH range 6.8 to 8.4, where concentration of soluble Fe(III) species is about $10^{-11.5}$ M, but equilibration with polynuclear species in solution may be slow. Ferrous ions form analogous mononuclear species but comparable data on tendency to form polymers are not available. Minimum solubility of ferrous hydroxide occurs near pH 10.7, where concentration of soluble (Fe(II) species would be about $10^{-6.5}$ M, but a tendency for air oxidation to the ferric species complicates the system. Because of the color of iron compounds, they tend to be used in waste streams rather than in water supplies.

Liquid ferric chloride, a dark brown oily-appearing solution which is 35 to 45 wt percent $FeCl_3$, is the customary form for flocculant use. Ferric chloride also is available in solid form. Ferric sulfate is marketed as dry granules, $Fe_2(SO_4)_3 \cdot 7H_2O$. Ferrous sulfate, also known commercially as copperas, is generally available in dry form with the nominal composition $FeSO_4 \cdot 7H_2O$.[17]

Lime Derivatives--While lime is used primarily for pH control or chemical precipitation, it is also used as a co-flocculant. For a summary of properties, see hydroxide precipitation.

In general, inorganic coagulants are used sparingly in industrial waste treatment applications. Primary usage is in the precipitation/coagulation of soluble phosphates and trace metals at municipal POTW's. See Table 10.2.3 for summary of manufacturers of inorganic flocculants.

TABLE 10.2.3. MAIN PRODUCERS OF INORGANIC FLOCCULANTS IN THE UNITED STATES

Company	Products[a]					
Allied Chemical Corp.	a					f
American Cyanamid Company	a					
Associated Metals & Minerals Co.			c			
Burris Chemical, Inc.	a					
Catco, Inc.			c			
Cities Service Company, Inc.	a			d		
Conservation Chemical Co.			c			
The Cosmin Corp.					e	
Diamond Shamrock Corp.						f
The Dow Chemical Company			c			
E.I. duPont de Nemours & Co., Inc.	a		c			
Essex Chemical Corp.	a					
Filtrol Corp.	a					
Philip A. Hunt Chemical Corp.			c			
Imperial West Chemical Co.	a		c			
Nalco Chemical Co.		b				
NL Industries, Inc.					e	
Olin Corp.	a					
Pennwalt Corp.			c			
Pfizer, Inc.					e	
Philadelphia Quartz Co.						f
Quality Chemicals, Ltd.					e	
Reynolds Metals Co.		b				
Southern California Chemical Co.			c			
Stauffer Chemical Co.	a					
K. A. Steel Chemicals, Inc.			c			
Vinnings Chemical Co.		b				

[a]a = alum; b = sodium aluminate; c = ferric chloride; d = ferric sulfate
e = ferrous sulfate; f = sodium silicate.

Source: Reference 19.

10.2.1.2 Synthetic Organic Flocculants--

Synthetic organic polymers are used almost exclusively in the coagulation/flocculation of industrial heavy metal precipitates. Typically, synthetic organic coagulants/flocculants are water-soluble polymeric substances with average molecular weights ranging from about 10^3 to greater than 5×10^6. If some subunits of the polymer molecule are charged, it is termed a polyelectrolyte.[19] When the charge on the subunits is positive, the polymer is termed cationic; when the charge is negative, it is termed anionic. Polyelectrolytes containing both positive and negative charges in the same molecule are termed polyampholytes. Some water-soluble polymers contain little or no charged subunits (less than 1 percent). These are termed nonionic polymers.[19]

Polyelectrolytes operate through the mechanism of chemical bridging and physical enmeshment. The polymer is usually a long organic chain which contains many active sites with which particles can interact and adsorb. Bridging occurs where the polyelectrolyte acts as a bridge, joining colloidal particles together to form a larger particle. Destablization occurs by slowing down particle motion.[15,16]

The coagulant/flocculant most generally used in the agglomeration of metals-containing wastewaters is an anionic organic polyelectrolyte. This is because metallic precipitates and metal hydroxides in particular, possess a slight electrostatic positive charge resulting from charge density separation. The negatively charged reaction sites on the anionic polyelectrolyte attract and adsorb the slightly positive charged precipitate.[20] However, studies[21] have been conducted that show that anionic polyelectrolytes adsorb onto electronegative suspended particles as well. It is hypothesized that the attractive adsorption energies between the anionic polyelectrolytes and the electronegative particles are stronger than the repulsive electrical energies.

Synthetic organic polyelectrolytes are commercially available in the form of dry power, granules, beads, aqueous solutions, aqueous gels, and oil-in-water emulsions. High (M.W. $1\text{-}5 \times 10^6$) and very high (M.W. 5×10^6) molecular weight polymers such as anionic polyelectrolytes tend to be sold as dry products or as oil-in-water emulsions due to increases in viscosity. Generally, liquid polyelectrolyte systems are preferred because they require less floorspace, reduce the labor involved, and reduce the potential for side reactions because the concentrate can be diluted as used in automatic dispensing systems.[20]

Dosages for treatment of metals-containing wastes generally fall in the range of 0.5 to 2 mg/L with a mg/L being the most common.[22] The polyelectrolyte and wastewater are initially combined in the rapid-mix section of the clarifier. Usually, rapid-mix chambers provide a reactor volume equal to 5 to 30 seconds of the design flow rate. Excess polyelectrolyte dosing at this point could be detrimental in that it may waste chemicals and result in restabilization of the metallic precipitates.

The most commonly used commercial anionic polyelectrolytes are poly(acrylic acid-co-acrylamide) and hydrolized polyacrylamide.[19] Polyacrylamides are infinitely soluble in water but are limited in practical applications by viscosity. As a polyelectrolyte, polyacrylamide exhibits sensitivity to salts and variations in pH. For example, excess salt will cause an exponential decrease in viscosity with increasing salt concentration.[19] In addition, at alkaline and intermediate pH's, the flocculating power will increase but at pH's lower than 4.5 flocculating power is decreased. Bridging theory suggests that by increasing the number of NH_2 groups hydrolyzed to OH groups, the effectiveness of the polymer should increase because the polymer coil becomes more extended.[21] Experimental evidence (see Figure 10.2.2) shows that at pH 6.0, the chain is fully extended (20 percent hydrolysis) while at pH 4.5, the OH groups are unionized and hence increasing the hydrolysis level does not extend the length of the chain, thus decreasing effectiveness.

Table 10.2.4 lists typical properties of two low anionic charge (1 to 10 percent anionicity) polyacrylamides.[22] The liquid polyelectrolyte is an oil-in-water solution which is diluted to 3 percent concentration upon use. High shear action (above 475 rpm) is not recommended during make-up since it can cause degradation of the high molecular weight flocculant. Fifteen to 30 minutes of mix time in a flocculant make-up tank is recommended to insure complete dissolution and partially hydrolyze the polymer, prior to introduction into the rapid-mix tank.[22] The dry polyacrylamide is usually dissolved to a 0.1 to 0.5 percent solution with complete dissolution occurring

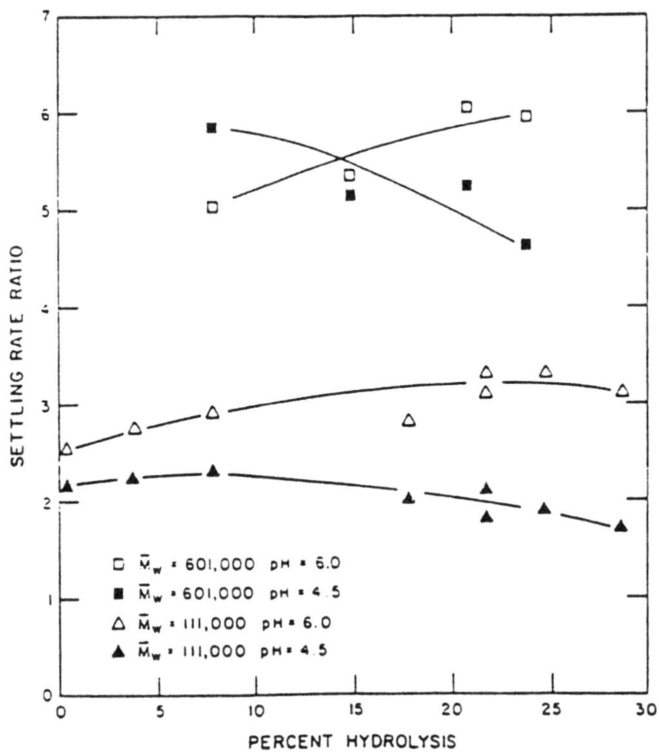

Figure 10.2.2. Settling rate ratio versus hydrolysis for linear polyacrylamide.

TABLE 10.2.4. PROPERTIES OF LIQUID AND DRY ANIONIC POLYACRYLAMIDES[a]

Typical Properties[a]

Appearance		Opaque white liquid		
Specific gravity at 25°C		1.0 ± .02		
Typical effective viscosity as is*, at 25°C, cps		300-600		
Typical viscosity,** as % solution, cps		0°C	20°C	40°C
	0.5	670	585	400
	1.0	1,295	1,130	790
	2.0	5,300	4,530	3,945
Freezing point		0°F (-15°C)		
Flash point, Tag Closed Cup		>200°F (93°C)		
Shelf life		9 months		

Environmental Properties

BOD, mg/L	870
COD, mg/L	7,060

Typical Properties[b]

Appearance			White powder
Degree of anionic charge			Low
Bulk density			43-45 lb/ft^3 (688-720 kg/m^3)
pH of 0.5% solution @ 25°C (77°F)			5.4
Viscosity***, cps			
% solution	0°C	20°C	40°C
0.1	50	40	30
0.5	300	250	200
1.0	1,400	1,200	1,000

Environmental Properties

BOD	approximately 0
COD	9,800 mg/L

*Viscosity at infinite shear speed (approximates the pumping situation).
**Brookfield.
***Brookfield.
[a]Magnifloc 1820A, American Cyanimid.
[b]Magnifloc 834A, American Cyanimid.
Note: Based on a 1 percent solution.
Source: Reference 22.

after 30 minutes. For best results, it is recommended that the solution be further diluted (100:1) with clean water prior to feeding into the rapid-mix tank. Stock solutions are usually stable for at least 2 weeks.[22]

10.2.1.3 Natural Organic Polymers--

Current flocculants derived from natural products include starch, starch derivatives, plant gums, seaweed extracts, cellulose derivatives, proteins, and tannins. Starch is the most widely used of these products, followed by guar gum.[23] Although price/kilogram for natural products tends to be low relative to synthetic flocculants, dosage requirements tend to be high. In addition, the composition of natural products tends to fluctuate, and they are more susceptible to microbiological attack which creates storage problems.[23] In recent years, the most promising of natural organic polymer flocculation technologies is a process utilizing insoluble starch xanthate (ISX).[24]

The ISX process was originally developed at the U.S. Department of Agriculture under a grant from the EPA. Production of ISX involved xanthating a relatively inexpensive chemically cross-linked, insoluble natural starch compound to form an anionic polymer capable of coagulating/flocculating heavy metals.[13]

The ISX process has been demonstrated to be capable of producing an effluent with very low residual metal concentrations (see Table 10.2.5).[25] The resulting ISX-metal sludge is said to dewater to 50 to 90 percent solids because it is nongelatinous. In addition, claims indicate that metal can be recovered from the ISX-metal sludge by acidification or incineration of the sludge.[26]

Two methods of ISX treatment have been applied on a commercial scale. The first method used in conjunction with commercial treatment, involves mixing an ISX slurry with neutralized/precipitated wastewaters in a reaction tank. The treatment is effective over a wide pH range, but for optimum coagulation/flocculation performance, this technology is typically operated at pH 9 in conjunction with a cationic polymer. In the coagulation reaction, ISX acts as an ion exchange liquid, bonding with heavy metal ions in exchange for sodium (or magnesium) ions as illustrated in the following reaction between ISX and nickel:

TABLE 10.2.5. HEAVY METAL REMOVAL EFFICIENCIES USING STARCH XANTHATE AS DETERMINED BY USDA

Metal	Concentration, mg/L	
	Influent	Effluent
Copper	31.8	0.007
Nickel	29.4	0.019
Cadmium	56.2	0.009
Lead	103.6	0.025
Trivalent Chromium	26.0	0.003
Silver	53.9	0.245
Zinc	32.7	0.046
Iron	27.9	not detectable
Manganese	27.5	1.630
Mercury	100.0	0.004

Source: Reference 25.

$$2ROC(=S)SNa + Ni^{+2} = (ROC(=S)S)_2Ni + 2Na^{+1} \qquad (4)$$

The second method involves using ISX as a filter precoat to polish effluent. In a typical operation, a sludge slurry would be pumped from a holding tank or clarifier to a precoated filter for dewatering of the sludge and removal of the metal ions remaining in solution. Table 10.2.6 contains removal efficiencies for a facility using ISX as a filter precoat.

Since ISX is susceptible to biological attack, it is typically shipped and stored under refrigeration (40°F). Shelf life is approximately 6 months and typical costs are $1.95 lb for a 25 lb container and $1.70/lb for a 250 lb container.[27]

Daily preparation of the ISX slurry would involve mixing predetermined amounts of ISX powder and water in a chemical feed tank. The slurry should be prepared at the ratio of approximately 2 pounds of ISX per gallon of water.[27]

ISX dosage is determined from laboratory testing. Calibration of the metering system involves monitoring the flow rate and adjusting the control system to deliver slurry in the required amount. The average capacity of ISX is in the range of 1.1 to 1.5 milliequivalents of metal ion per gram ISX. Thus, for a divalent nickel ion, one gram of ISX would remove 32 to 43 mg of nickel ions from solution.[25] Maintenance of this system involves periodic flushing of the lines to prevent build-up of ISX and restriction of the lines, and periodic checks of the metering system calibration.

In addition to storage and handling difficulties, disposal of process residuals or sludges is a major problem associated with the starch xanthate process. Laboratory test results indicate that heavy metal removal capacity is approximately 0.0011 moles per gram of starch.[28] Consequently, relatively large sludge volumes will be produced for the quantity of heavy metals removed. Conventional land disposal does not appear to be an environmentally acceptable alternative because the organic structure of the starch xanthate-metal sludge can decompose rapidly and release the metal to the environment. Incineration is being considered for possible metal recovery but off-gas scrubbing facilities would be necessary to insure that heavy metals are not emitted to the atmosphere. Additional costs associated with

TABLE 10.2.6. METAL REMOVAL RESULTS USING ISX AS A FILTER PRECOAT

Metal	Initial Concentration[a] (mg/L)	ISX Treated Concentration (mg/L)
Cr^{+3}	0.8	0.02
Cu^{+2}	7.0	0.02
Ni^{+2}	2.5	0.10

[a]Before ISX treatment.

Source: Reference 28.

these stack gas control facilities may be prohibitive. Also, the heavy metals collected in the scrubber liquor would again have to be removed before the liquor could be reused or discharged to a receiving stream.

10.2.1.4 Pretreatment and Post-Treatment Requirements--

Coagulation/flocculation is a well established technology, and in general, is very reliable. It is used primarily to treat aqueous metals-laden waste streams. The properties of the waste being treated which can affect performance include:

- Flow variations;
- Solids concentration variations;
- pH variations;
- Temperature variations;
- Cyanide content;
- Hexavalent chromium concentration; and
- Oil and grease concentration.

The effect of flow variations appears mainly in the sedimentation step. Temperature variations can also cause upsets in sedimentation by creating undesirable thermal currents. Changes in solids concentration and pH can affect the performance of the coagulation and flocculation process in systems where the agglomeration rate is a function of these parameters. Also, compounds in the wastewater that interfere with coagulation (such as sulfides and mercaptides) can result in reduced agglomeration effectiveness. To minimize these effects, equalization basins are generally recommended.[29] In addition to creating an influent of more consistent quality, sulfides or mercaptides can be oxidized to a less reactive or inert state. Also, in systems where pH influences the agglomeration rate, pH adjustment may be required. For a discussion of cyanide destruction, chromium reduction, and oil removal technologies refer to Sections 13.0, 10.3 and 10.1, respectively.

10.2.2 Process Performance

As previously indicated, most heavy metal coagulation/flocculation applications involve the use of an inorganic or polymeric reagent. The many disadvantages of naturally occurring organic polymers has currently limited their use to a few select applications. For example, insoluble starch xanthate which was once used at over 100 facilities is now utilized at less than 50 as a result of storage, application, and disposal problems.[27]

Process performance and costs for heavy metal coagulation/flocculation systems are very sensitive to coagulant dosage, type, and flow rate. Figures 10.2.3 and 10.2.4 illustrate the effect of iron dose and clarifier over flow rate on arsenic and selenium removal efficiencies.[30] An anionic polyelectrolyte was introduced into the feed line to the clarifier to assist in flocculation. These results show that arsenate removals exceeded 90 percent at clarifier overflow rates up to 1,200 gpd/ft^2. Selenium removal (56 to 89 percent) was limited by the fraction of selenate (which is not adsorbed by $Fe(OH)_3$) in the waste stream. Minimum iron and polymer doses for good performances were 14 mg/L and 0.15 mg/L (pH range 6.2 to 6.5), respectively.

Tables 10.2.7 through 10.2.9 demonstrate the various treatment options available for effectively removing such heavy metals as lead, zinc, cadmium, manganese, copper, and nickel.[31] In Table 10.2.7, lime (for precipitation and coagulation) was combined with either of two polymers, Magnifloc 1561/1820 or Percol 728 for flocculation. Both types of polyelectrolyte worked equally as well as a flocculant in removing lead, zinc, and cadmium. The optimum dosage for Magnifloc 1561/1820 was determined to be 1.5/0.5 mg/L, respectively, while 1.0 mg/L of Percol 728 was sufficient for greater than 99 percent removal. Table 10.2.8 shows the removal of cadmium, copper, iron, lead, manganese, nickel, and zinc using alum (35 mg/L) as a coagulant and an anionic polyelectrolyte (1 mg/L) as a flocculant. Sodium hydroxide is used to adjust the influent pH (7.3 to 8.9) to the 8.4 to 9.25 range for precipitation.

Table 10.2.9 presents performance data for a system using ferrous sulfate (Fe:Ni ratio = 0.7) as a coagulant to remove nickel from an aqueous waste stream.[32] Both an anionic polymer and a cationic polymer were evaluated.

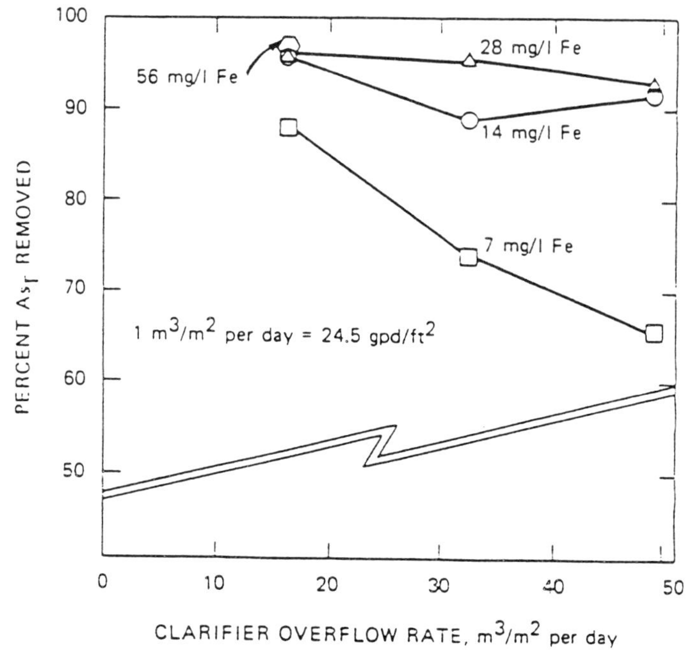

Figure 10.2.3. Effect of iron dose and clarifier overflow rate on arsenic removal efficiency.

Source: Reference 30.

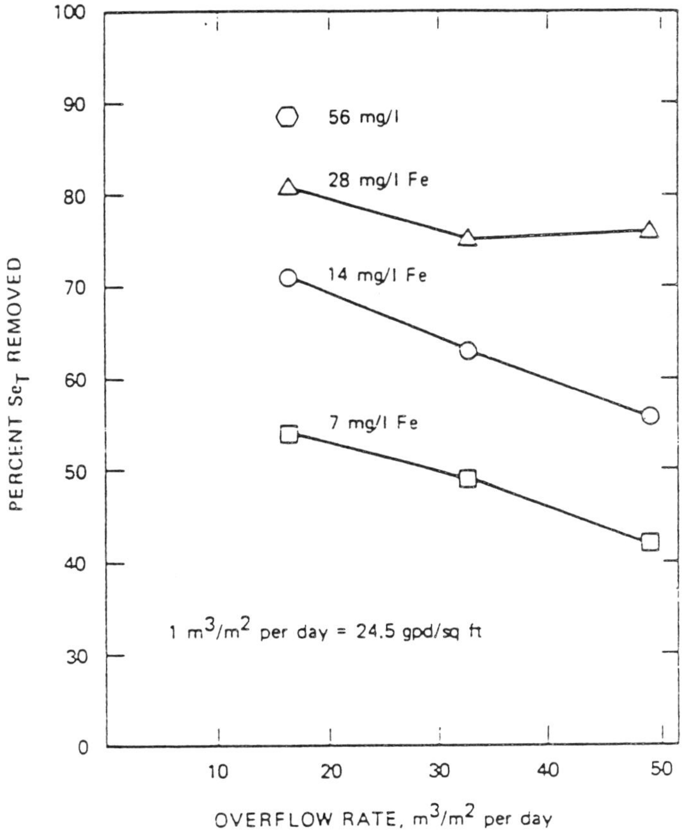

Figure 10.2.4. Effect of iron dose and clarifier overflow rate on selenium removal efficiency.

Source: Reference 30.

TABLE 10.2.7. METALS REMOVAL USING LIME AND ANIONIC POLYELECTROLYTES

Parameter	Clarifier production operation: Influent	Clarifier production operation: Effluent[b]	Dual media filter[a]: Effluent
pH, units	5.65-6.78	10.00	7.0
Suspended solids, mg/L	640	35	7.0
Calcium, mg/L	60	60	-
Lead, mg/L	59	0.77	0.25
Zinc, mg/L	72	1.26	0.41
Dissolved solids, mg/L	535	450	-
Cadmium, mg/L	0.45	0.01	-

[a]Typical filter effluent coal/sand.

[b]Typical clarified effluent lime/polymer treatment.

Source: Reference 31.

TABLE 10.2.8. METALS REMOVAL USING ALUM AND AN ANIONIC POLYELECTROLYTE

Parameter (mg/L)	Before new treatment plant	After new[a] treatment plant
Cadmium	0.036	0.029
Copper	0.084	0.02
Iron	12.0	0.25
Lead	1.5	0.09
Manganese	2.2	0.25
Nickel	LT 0.13	0.06
Total suspended solids	16 (avg.)	6.0
Zinc	8.2	0.29

[a]35 mg/L $Al_2(SO_4)_3$
1 mg/L anionic polyelectrolyte
PH = 8.4-9.25
Flow rate = 10 mgd

Source: Reference 31.

TABLE 10.2.9. RESIDUAL NICKEL CONCENTRATIONS FOR VARIOUS POLYMER ADDITIONS: Fe:Ni = 0.7, C_T = 0 mg/L

	Anionic polymer concentration (mg/L)			Cationic polymer concentration (mg/L)			No polymer (control)
	0.1	0.5	1.0	0.1	0.5	1.0	-
pH = 9							
Soluble	5.1	1.25	0.60	0.73	0.70	0.70	1.6
Total	5.6	1.30	0.70	0.91	0.80	0.85	1.8
pH = 10							
Soluble	0.22	0.12	0.10	0.12	0.15	0.12	0.12
Total	0.40	0.15	0.20	0.31	0.32	0.30	0.50

Source: Reference 32.

It was concluded that the addition of cationic and anionic polymers slightly enhanced settleability at both pH 9 and 10. Lime was used as the precipitation and neutralization reagent.

10.2.3 Process Costs

Table 10.2.10 contains the purchased equipment and installation costs, and annualized operating costs for a continuous coagulation/flocculation treatment system. The system consists of a continuous flocculation/clarification unit, sludge holding tank(s), and a filter press. The flocculation/clarification unit size is a function of the volumetric flow rate. The influent to the unit is assumed to contain 200 mg/L of heavy metals which have been previously precipitated with sodium hydroxide to form approximately 400 mg/L of suspended solids. The overflow from the clarification section is assumed to be solids-free, while the underflow is assumed to contain 6 percent solids. The coagulant alum is added (150 mg/L) along with the flocculant, Magnifloc 1820A (1 mg/L) in the flocculation tank prior to the clarifier. The sludge holding tanks (10 hours retention) and the filter press (8-hour cycle) have been sized to handle the solids content in the underflow. Capital and annualized operating costs are based on assumptions previously presented in Section 10.1.2.

A large percentage of total annual costs for the continuous coagulation/flocculation system developed for this section are a result of sludge disposal costs. Sludge production is increased roughly 20 percent by the addition of alum with an equivalent increase in sludge transportation and disposal costs. A 30 percent savings in reagent costs can be realized by using $FeSO_4$ ($145/ton) instead of $Al_2(SO_4)_3$ ($205/ton), but sludge generation will be equivalent, if all the aluminum is precipitated in the clarification section as an hydroxide.

Labor costs for this treatment technology are also a large percentage of the overall annual operating costs. This is due to the high operator skills required in making the coagulant/flocculant reagent additions. In addition, since the dosage requirements for coagulants such as alum, ferric chloride, and ferrous sulfite are nonstoichiometric, frequent jar tests are necessary to prevent underdosing or overdosing.

TABLE 10.2.10. CONTINUOUS COAGULANT/FLOCCULANT COST DATA[a]

	Flow rate (gph)		
	1,000	10,000	100,000
Purchased equipment and installation (PE&I) $			
Flocculator/clarifier	18,000	50,000	140,000
Sludge molding tank(s)	3,000	3,000	20,000
Filter Press	11,000	15,000	60,000
	32,000	68,800	220,000
Total capital investment (360% PEI)	115,200	244,800	792,000
Annual operating costs ($)			
Operating labor ($20/hr)	72,000	72,000	72,000
Maintenance (6% TCI)	6,900	14,700	47,500
General plant overhead (5.8% TCI)	6,700	14,200	45,900
Utilities (2% TCI)	2,300	4,900	15,800
Taxes and insurance (1% TCI)	1,200	2,500	7,900
Chemical costs: $Al_2(SO_{43}$ ($205/ton)	900	9,200	92,300
1820 A ($1.29/lb)	100	800	7,700
Sludge transportation ($0.25/ton-mile)	200	2,100	21,000
Sludge disposal ($200/ton)	11,300	112,600	1,126,100
Annualized capital (CFR = 0.177)	20,400	43,300	140,200
Total annual costs	122,000	276,300	1,576,400
Cost/1,000 gallon	17	4	2

[a]1987 Dollars.

10.2.4 Overall Process Status

Coagulation/flocculation is a well-developed process widely used for many industrial wastewaters containing suspended and colloidal solids. The equipment used is relatively simple, readily available, and can often be skid mounted in a modular design. In many cases, coagulation/flocculation can be added to existing process trains with only minor modifications. For high volume applications, the cost of this technology drops dramatically improving economic viability. In addition, the process is often improved by high ionic strength and is applicable to high influent metal loadings.

Disadvantages and primary environmental considerations result from a metals laden high-water-content sludge which must be treated (i.e., solidification, encapsulation, etc.) and then disposed. In addition, the process is also not readily applied to small intermittent flows and many of the coagulants used ($Al_2(SO_4)_3$, $FeCl_3$, etc.) form corrosive solutions. Finally, process efficiency is highly sensitive to initial contaminant concentration and the surface area of the primary floc formed in the rapid-mix chamber.

REFERENCES

1. Process Design Manual for Suspended Solids Removal, U.S. Environmental Protection Agency, EPA-625/1-75-003a. January 1975.

2. Sundstrom, D.W., and H.E. Klei. Wastewater Treatment. Prentice-Hall, Inc., Englewood Cliffs, NJ. 1979.

3. Clark, J.W. et al. Water Supply and Pollution Control. 3rd. Edition, Harper & Row. 1977.

4. Chillingworth, M.A. et al., Alliance Technologies Corporation. Industrial Waste Management Alternatives Assessment for the State of Illinois. February 1981.

5. Arthur D. Little, Inc. Physical, Chemical, and Biological Treatment Techniques for Industrial Wastes. U.S. EPA SW-148. November 1976.

6. Manual on Disposal of Refinery Wastes, Volume on Liquid Wastes, Chapter 9--Filtration, Flocculation and Flotation. American Petroleum Institute, Washington, DC. 1969.

7. Beychok, M.R. Aqueous Wastes from Petroleum and Petrochemical Plants. John Wiley & Sons, New York. 1967.

8. U.S. EPA. Treatability Manual. Volume III. Washington, DC. EPA-600/8-80-042c,d. July 1980.

9. Monk, R.D., and Willis, J.F. Designing Water Treatment Facilities. American Water Works Association Journal. February 1978.

10. Janson, C.F. et al. Alternatives for the Treatment of Heavy Metals. AICHE Paper No. 149c. November 1981.

11. Lanouette, K.H. Heavy Metals Removed. Chemical Engineering. October 17, 1977.

12. U.S. EPA. Reducing Water Pollution Costs in the Electroplating Industry. EPA-625/5-85-016. September 1985.

13. Cushnie, G.C., Centec Corporation. Navy Electroplating Pollution Control Technology Assessment Manual. Naval Civil Engineering Laboratory CR84.019. February 1984.

14. Cushnie, G.C. Removal of Metals from Wastewater: Neutralization and Precipitation. Pollution Technology Review, No. 107, Noyes Publications, Park Ridge, NJ. 1984.

15. Jurgensons, B., and M.E. Straumaris. Colloid Chemistry. 2nd. Edition, MacMillan Company, New York, NY. 1962.

16. Mysels, K.J. Introduction to Colloid Chemistry. Interscience Publishers, Inc., New York, NY. 1959.

17. Kirk-Othmer Encyclopedia of Chemical Technology. Vol. 24, 3rd. Edition, John Wiley & Sons, New York, NY. 1981.

18. Clifford, C., Subramonian, S., and T.J. Sorg. Removing Dissolved Inorganic Contaminants from Water. Environmental Science & Technology, Vol. 20, No. 11. 1986.

19. Kirk-Othmer Encyclopedia of Chemical Technology. Vol. 18, 3rd. Edition, John Wiley & Sons, New York, NY. 1986.

20. Capaccio, R.S., and R.J. Sarnelli. Flocculation and Clarification. Plating and Surface Finishing. October 1986.

21. Caskey, J.A., and R.J. Primus. The Effect of Anionic Polyacrylamide Molecular Conformation and Configuration on Flocculation Effectiveness. Environmental Progress. May 1986.

22. American Cyanamid Technical Brochure. Magnifloc 1839 A. 1986.

23. Kirk-Othmer Encyclopedia of Chemical Technology. Vol. 10, 3rd. Edition, John Wiley & Sons, New York, NY. 1981.

24. Wing, R.E. Dissolved Heavy Metal Removal by Insoluble Starch Xanthate (ISX). Environmental Progress. November 1983.

25. R.E. Wing. "Processes for Heavy Metal Removal from Plating Wastewaters." First Annual Conference on Advanced Pollution Control for the Metal Finishing Industry, U.S. Environmental Protection Agency, Washington, DC. EPA-600/8-78-010. May 1980.

26. R.E. Wing, et al. Removal of Heavy Metals from Industrial Wastewaters Using Insoluble Starch Xanthate. U.S. Environmental Protection Agency, Washington, DC. EPA-600/2-78-085. May 1978.

27. W. Stout. Stout's Supply telephone conversation with Stephen Palmer, Alliance Technologies Corporation. February 10, 1987.

28. U.S. EPA. Sources and Treatment of Wastewater in the Nonferrous Metals Industry. EPA-600/2-80-074. April 1980.

29. MITRE Corporation. Manual of Practice for Wastewater Neutralization and Precipitation. EPA-600/2-81-148. August 1981.

30. D.T. Merrill, et al. Field Evaluation of Arsenic and Selenium Removal of Iron Coprecipitation. Journal of the Water Pollution Control Federation. January 1986.

31. Osantowski, R., and J. Ruppersberger. Upgrading Foundry Wastewater Treatment. 39th. Industrial Waste Conference, Purdue University. 1982.

32. McFadden, F., Benefield, L., and Reed, R.B. Nickel Removal from Nickel Plating Wastewater Using Iron, Carbonate, and Polymers for Precipitation and Coprecipitation. 40th. Industrial Waste Conference, Purdue University. 1983.

10.3 CHEMICAL REDUCTION

Chemical reduction is a reaction in which one or more electrons are transferred to the chemical being reduced (reductant) from the chemical initiating the transfer (the reducing agent).[1] Chemical reduction can also be defined as a change in oxidation states where the oxidant (reducing agent) is an electron donor such as zinc in the reaction:[2]

$$Zn = Zn^{++} + 2e$$

The reductant is the substance which accepts electrons:

$$Cu^{++} + 2e = Cu$$

The overall reaction is called a reduction-oxidation (redox) reaction:

$$Zn + Cu^{++} = Cu + Zn^{++}$$

Redox processes are very common in aqueous systems since most organic and many inorganic reactions involve oxidation and reduction.[3] In reactions involving covalent bonds, the gain or loss of electrons by an element may not be clearly defined. The assignment of electrons to an atom is thus carried out according to rules. If two atoms share electrons in a covalent compound, the electrons are arbitrarily assigned to the atom that is more electronegative. If an electron pair is shared by two atoms of the same electronegativity, the electrons are split between them. After this division of charges has been made, the charge remaining on the atom is known as its oxidation number or state. The sum of oxidation numbers is equal to zero for molecules and is equal to the formal charge for ions.[2]

In principle, the equilibrium composition of a redox system can be determined from a thermodynamic analysis as in the case of acid-base reactions. Many inorganic redox reactions have fast reaction rates and chemical equilibrium is approached within typical process times. Redox reactions involving organic compounds, however, are often slow at ambient conditions.[2]

10.3.1 Process Description

Chemical reduction as a waste treatment process is an established and well-developed technology. The reduction of hexavalent chromium's valence state to decrease toxicity and encourage precipitation is presently used as a treatment technology in numerous electroplating facilities. Major advantages of chemical reduction when used to reduce hexavalent chromium is operation at ambient conditions, automatic controls, high reliability, and modular process equipment.[1] Process equipment typically requires a tank for pH adjustment and reduction, metering equipment, ORP (oxidation-reduction potential) and pH controls and instrumentation, mechanical agitation, adequate venting, and separate tanks for subsequent precipitation and sedimentation.[4] The retention time in the reduction tank is pH dependent but should be at least four times the theoretical time for complete reduction.

A number of chemicals are used as reducing agents. The most common include; sulfur dioxide, sodium metabisulfite, sodium bisulfite, ferrous sulfate, and sodium borohydride. Other reducing agents which can potentially be used for wastewater treatment are dithiocarbonate, hydrazine, aluminum, zinc, and formaldehyde.[5] The prevalent reducing agents are discussed in the following subsections.

Sulfur Dioxide--

For waste streams which contain chromates, gaseous sulfur dioxide is a widely used reducing agent. The net reaction involves chromic acid and sulfurous acid (produced through the reaction of sulfur dioxide and water) as follows:

$$3H_2SO_3 + 2H_2CrO_4 = Cr_2(SO_4)_3 + 5H_2O$$

Because the reaction proceeds rapidly at low pH, an acid (typically sulfuric) is added to maintain the pH between 2 and 3.[6] To prevent the release of sulfur dioxide during treatment, a pH of approximately 3 is recommended.[1,6] At pH levels above 5, the reaction rate slows drastically (see Figure 10.3.1).

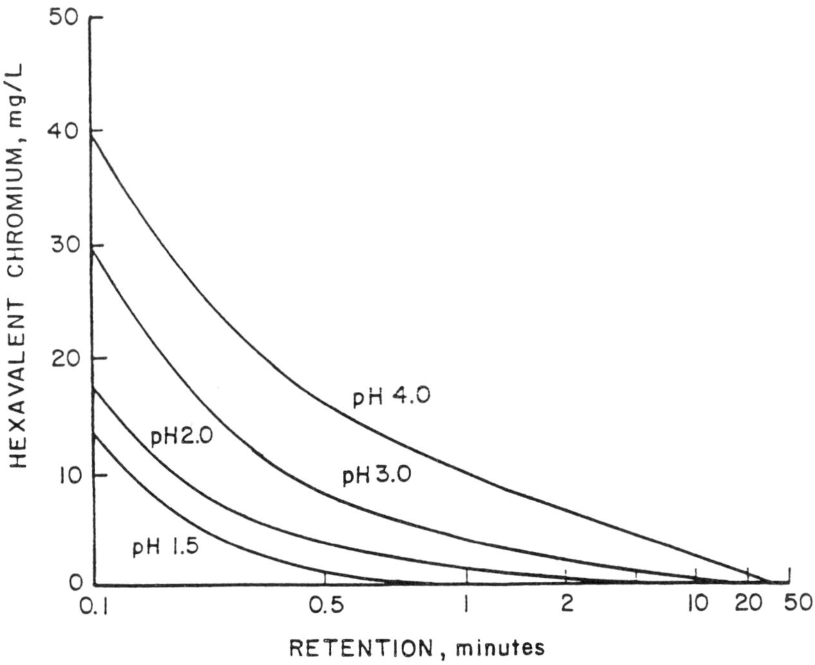

Figure 10.3.1. Effect of pH on chromium reduction rate.
Source: Reference 7.

Figure 10.3.2 shows a typical wastewater treatment process schematic for the reduction of chromates. The ORP control set point for this process varies by approximately 150 millivolts per change in pH unit, with SO_2 automatically metered to maintain ORP in the 250 to 300 range[6,7] (see Table 10.3.1). Consumption of SO_2 will normally average 50 to 100 percent of stoichiometric requirements. Dissolved oxygen or reducible organics will consume a significant portion of the reducing agent if the reaction vessel is open to air.

Sulfur dioxide as with all reduction processes can be employed either as a batch treatment or as a continuous process. Retention time is typically 30 to 45 minutes at a pH of 3, and reactor vessels should be sized accordingly. Theoretical chemical requirements per pound of chromium reduced are 2 lbs of SO_2 plus 35 mg for each liter of water being treated.[7] These relationships, however, should be confirmed by field tests (see Table 10.3.2 for summary of treatment levels).

Sodium Metabisulfite and Sodium Bisulfite--

Sodium Metabisulfite ($Na_2S_2O_5$) and Sodium Bisulfite ($NaHSO_3$) are soluble sulfite salts used as alternatives to gaseous SO_2 for the reduction of hexavalent chromium. These salts (see Section 10.1.3) for a description of physical properties) are available either as a dry powder flake (70 to 72 weight percent) or solution (44 to 60 weight percent).[8] The product is shipped either as flake in drums or as solutions in tank cars or tank trucks. Reagent is added either from storage in the case of liquid reagents or from rapid-mix tanks when using flakes. The reaction when using sodium bisulfite as a reducing agent is:

$$3NaHSO_3 + 3H_2SO_4 + 2H_2CrO_4 = Cr_2(SO_4)_3 + 3NaHSO_4 + 5H_2O$$

Sulfuric acid is added to depress the pH of the wastewaters to the optimum pH range of 2-3 (see Figure 10.3.1) as well as provide the required hydrogen for reaction completion. Table 10.3.3 lists a summary of treatment levels obtained by this technology.[6,9]

In this system, chromium bearing wastes are separated from the other metals waste streams and collected in a flow equalization chamber where flow and pH deviations are averaged. The equalization chamber is equipped with a

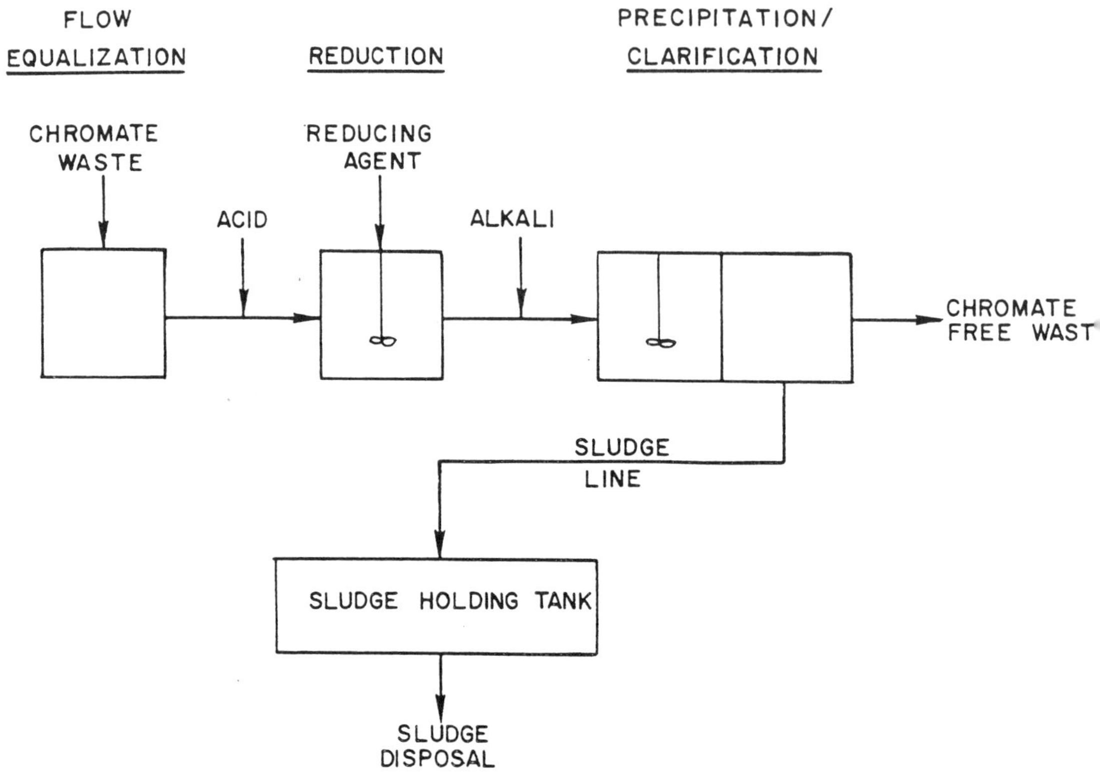

Figure 10.3.2. Continuous chromium reduction|precipitation system.

TABLE 10.3.1. RELATIONSHIP BETWEEN ORP AND HEXAVALENT CHROMIUM CONCENTRATION

ORP	Cr^{+6}
590	40 ppm
570	10 ppm
540	5 ppm
330	1 ppm
300	0

Source: Reference 7.

TABLE 10.3.2. SUMMARY OF TREATMENT LEVELS REPORTED FOR HEXAVALENT CHROMIUM WASTES

Reduction agent	$Chromium^{+6}$ Concentration (mg/L)	
	Initial	Final
Sulfur dioxide	--	0.3-1.3
	1,300	1.0
	--	0
	--	0.01
	--	0.05
	0.23-1.5	0.1

Source: Reference 6.

TABLE 10.3.3. SUMMARY OF TREATMENT LEVELS REPORTED FOR HEXAVALENT CHROMIUM

Reduction Agent	Chromium^{+6} Concentration (mg/L) Initial	Final
Bisulfite	140	0.7-1.0
Bisulfite	--	0.05-0.1
Bisulfite plus hydrazine	8-20.5	0.1
Metabisulfite	70	0.5
Metabisulfite	--	0.025-0.05
Metabisulfite	--	0.1
Metabisulfite	--	0.001-0.4

Source: Reference 7.

level controlled pump that delivers the wastewater to the reduction unit. Retention time is typically 30 to 60 minutes. Acid, usually sulfuric, is added at a point just prior to the reduction tank. Bisulfites are added directly to the reduction chamber by means of a metered feed system with pH and ORP controls. Retention times for the reaction tank are typically 10-60 minutes with theoretical reagent requirements of 1.5 lbs of $NaHSO_3$ and 1 lb of H_2SO_4 per pound of Cr reduced.[7] The trivalent chromium is removed by precipitation. Usually lime or caustic is added to increase the pH between 7.5 and 8.5 for minimum solubility of chromium hydroxide. Theoretical reagent requirements for precipitation would be 2.2 lbs $Ca(OH)_2$ or 2.5 lb NaOH or 3 lb Na_2CO_3[7] (see Section 10.1 for retention times and equipment specifications).

Treated wastewater is discharged to a mixer/clarifier where a flocculant may be added to improve hydroxide precipitate settling characteristics. The overflow from the clarifier is then discharged to the sewer system, while the solids in the underflow are collected in a holding tank for subsequent dewatering (see Section 10.2).

While this type of system is prevalent, many plants experience excess consumption of reducing agents. The major cause of excess sulfite consumption is hypothesized to be the dissolved oxygen present in the chromium wastewaters. For example, based on stoichiometry, one mole of oxygen will consume two moles of sulfite ion:

$$2\ SO_3^{-2} + O_2 = 2\ SO_4^{-2}$$

Oxygen molecules from the gas phase are transferred to the liquid phase in proportion to the difference between the existing concentration and the equilibrium concentration of gas in solution.[10] Since chromium reduction reaction vessels are usually open and the reaction is not instantaneous, oxygen diffusion into the chromium waste solution will continuously consume reducing agent. Therefore, to prevent excess reagent consumption due to dissolved oxygen and eliminate hydrogen sulfide odor problems, it is recommended that process vessels be enclosed and adequately vented.[10]

Another drawback to sulfite salt systems (as well as sulfur dioxide systems) is an excess consumption of acid and bases. Since historically chromium reduction has consisted of first a pH depression to reduce chromium to a trivalent state followed by a pH elevation to precipitate the chromium ions as hydroxides, acid and base reagent consumption adds significantly to the operation and maintenance of a chromium reduction system. This problem is compounded by the sodium sulfite salts which often form sodium hydroxide as a reaction byproduct, thus requiring an even greater excess of acid during the pH depression step. Therefore, chromium-bearing waste streams are typically segregated and treated separately to reduce reagent consumption. The reduced chromium-bearing stream can then be either precipitated/clarified separately or combined with other metal-bearing streams for further treatment.

Ferrous Sulfate--

Ferrous sulfate heptahydrate solids ($FeSO_4 \cdot 7H_2O$) are water soluble, blue-green crystals having a density of 1.898 g/cm^3 and a melting point of 64°C. Most ferrous sulfate is waste product derived from the pickling of steel surfaces in the steel industry. Supply exceeds the demand, and the major portion of the waste presents a serious disposal problem. Ferrous sulfate is available either in flakes or solution form. In moist air the flakes oxidize to basic iron (III) sulfate ($Fe_2(SO_4)_3$). Aqueous solutions are also subject to oxidation and are very sensitive to alkalis, temperature, and light.

In waste treatment applications, ferrous sulfate has been used in a variety of ways. Three methods reported in literature are acid reduction, alkaline reduction, and ferrite coprecipitation.

Acid reduction of hexavalent chromium with ferrous sulfate consists of adding ferrous sulfate heptahydrate to an acidic hexavalent chromium solution (pH 2-3). The ferrous ion (Fe^{+2}) will react with the hexavalent chromium, reducing the chromium and oxidizing the ferrous ion to basic iron (III) sulfate. The reaction occurs as follows:

$$2H_2CrO_4 + 6FeSO_4 + 7H_2O + 6H_2SO_4 = Cr_2(SO_4)_3 + 3Fe_2(SO_4)_3 + 15H_2O$$

In terms of reaction rate retention times, pH and chemical metering controls, acid ferrous sulfate reduction is similar to other sulfur-based reduction systems such as sodium metabisulfite and sodium hydrosulfite. The main advantage of this process is an abundant and inexpensive supply of ferrous sulfate. Disadvantages include excess acid and base requirements to adjust the wastewater pH to 2 for reduction and then back to 8.5 to 9 for precipitation.[6] Another disadvantage is that three moles of ferrous ions are required per mole of hexavalent chromium reduced. In addition the precipitation of the ferric ion (Fe^{+3}) as an hydroxide contributes greatly to the amount of sludge generated.[6] One study found that the use of ferrous sulfate rather than a soluble sulfite such as sodium hydrosulfite, for the reduction of hexavalent chromium results in a sludge product 31 times as great as the volume of sludge produced by the bisulfite process.[12]

Alkaline reduction of hexavalent chromium with ferrous sulfate was a process evaluated under a grant by Arizona State University.[13,14,15] The main advantages to this process are a rapid reduction of chromate at pH levels between 8 to 10 (eliminating the acid depression step) and a reduction of process equipment since the process can be accomplished in the same reactor as the neutralization/precipitation process. Disadvantages include sludge generation and a lack of control in chemical metering.

This process, like acid ferrous sulfate reduction is capable of reducing chromate concentrations to 0.05 mg/L. The process produces considerably more sludge and is consequently more expensive than the conventional process of pH reduction and the use of SO_2. However, for hexavalent chromium concentrations of 10 mg/L or less, ferrous sulfide reduction economics may be worth considering.[15]

Ferrite coprecipitation is a process similar to acid ferrous sulfate reduction for the conversion of soluble metal ions to insoluble metal hydroxides or ferrites. The process, which was developed in Japan, involves the mixing of ferrous sulfate heptahydrate with a heavy metal-bearing wastewater.[16] The ferrous ion will coexist with the heavy metal ions in solution. Alkali is added to neutralize the acidic solution and a dark green hydroxide is formed as follows:

$$XM^{++} + Fe^{++}(3-X) + 6(OH)^{-} = M_XFe_{(3-X)}(OH)_6$$

$$X = 1,2,3$$

In a variation on the traditional ferrous sulfate reduction process, oxidation with air is performed during which dissolution and complex formation occur yielding a black ferrite as follows:

$$M_xFe_{(3-x)}(OH)_6 + 1/2O_2 = M_xFe_{(3-x)}O_4 + 3H_2O$$

Tables 10.3.4 and 10.3.5 contain data from a test facility which uses a batch process to treat wastewater (4.5 gpm) and an installation which treat off-gas scrubber liquor from a municipal refuse incinerator (20 gpm). These data show that ferrite coprecipitation is an effective process for the removal of heavy metals.[16] However, little published data exist on the success of this process in the United States. It is reported to be labor intensive and like all iron precipitation technologies generates a voluminous sludge product.[16]

Sodium Borohydride--

Sodium borohydride ($NaBH_4$) is a mildly alkaline reducing agent available either as a 97 percent free-flowing powder or as a stabilized water solution of 12 percent sodium borohydride and 40 percent sodium hydroxide. The basic reduction reaction involves the donation of 8 electrons/molecule of SBH to an electron deficient metal cation. The net reaction is:

$$NaBH_4 + 4M^{++} + 2H_2O = 4M^\circ + NaBO_2 + 8H^+$$

Since one mole of sodium borohydride (SBH) can reduce four moles of divalent metal ion (or eight moles of monovalent), relatively low amounts of reagent usage can result in a substantial reduction of metallic contaminants.[17,18] Table 10.3.6 illustrates theoretical usage levels and the overall quantities of metals recovered. In practice, the metal/SBH ratio is lower since other reducible compounds (aldehydes, ketones, etc.) may react with borohydride, increasing reagent consumption. Typically, SBH requirements are 1.5 to 2 times the theoretical use level.[19]

Figure 10.3.3 illustrates a sodium borohydride treatment system, demonstrating the requirements for optimal metals reduction.[17] In this system, the pH is maintained between 6 and 7, although SBH can reduce under pH

TABLE 10.3.4. PERFORMANCE DATA FROM A FERRITE COPRECIPITATION TEST FACILITY (CONCENTRATION, mg/L)

Metal	Influent	Effluent
Mercury	7.4	0.001
Cadmium	240	0.008
Copper	10	0.010
Zinc	18	0.016
Chromium	10	0.010
Nickel	1,000	0.200
Manganese	12	0.007
Iron	600	0.06
Bismuth	240	0.100
Lead	475	0.010

Source: Reference 16.

TABLE 10.3.5. PERFORMANCE OF FERRITE COPRECIPITATION IN OSAKA UNIT (CONCENTRATION, mg/L)

Metal	Influent	Effluent
Mercury	6	0.005
Arsenic	0.7	0.01
Trivalent Chromium	25	0.01
Hexavalent Chromium	0.5	not detectable
Lead	480	0.05
Cadmium	15	0.01
Iron	3,500	0.04
Zinc	650	0.5
Copper	23	0.08
Manganese	60	0.5

Source: Reference 16.

TABLE 10.3.6. THEORETICAL SODIUM BOROHYDRIDE USE LEVELS AND QUANTITIES OF METALS RECOVERED

Metal	Oxidation State	Sodium Borohydride Theoretical Use Levels		Metal Recovery
		Powder (g SBH/kg metal)	SWS (mL SWS/kg metal)	(lb metal/lb SBH
Copper	Cu^{2}	143	850	7
Lead	Pb^{2+}	46	270	22
Nickel	Ni^{2+}	167	1000	6
Gold	Au^{3+}	16	430	14
Silver	Ag^{+}	72	260	23
Cadmium	Cd^{2+}	62	370	12
Mercury	Hg^{2+}	48	280	21
Palladium	Pd^{2+}	91	540	11
Platinum	Pt^{4+}	100	600	10
Cobalt	Co^{2+}	167	1000	6
Rhodium	Rh^{3+}	143	850	7
Iridium	Ir^{4+}	100	600	10

Treatment levels shown are for 97% active SBH powder and SWS", a stabilized water solution of 12% SBH and 40% NaOH (by weight).

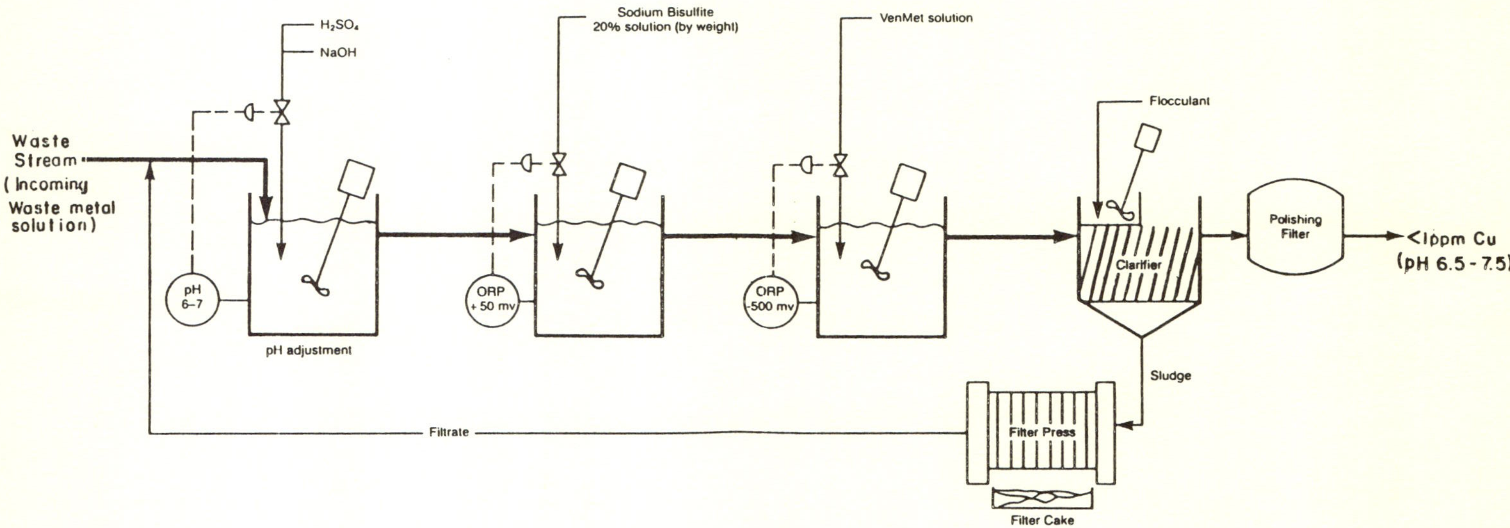

Figure 10.3.3. Sodium Borohydride recovery system.

conditions as low as 4.5 and as high as 11. Sodium bisulfite is added prior to sodium borohydride to lower the oxidation states of competing species, but does not totally reduce the metal cations present.[17]

In the second stage SBH solution is added. The stabilized water solution can be handled in a similar fashion to 50 percent sodium hydroxide (see Section 10.1.2). It is suitable for ORP control and can be metered from a storage tank or directly from a 55-gallon drum.[17] Some users further dilute the SBH solution 10:1 with deionized water prior to addition. Dilution allows for faster mixing and helps to prevent over-dosing which may impede downstream flocculation and settling. The solution can be used in processes with flow rates ranging from 5 gal/min up to 1,500 gal/min and metal concentrations from as low as 2 mg/L to as high as 20,000 mg/L.[19]

The sodium borohydride added in the second stage reacts with any residual bisulfite from stage one to form sodium dithionate:[17]

$$8NaHSO_3 + NaBH_4 = 4Na_2S_2O_4 + NaBO_2 + 6H_2O$$

The sodium dithionate further reduces any oxidizing agents left in the waste stream, partially reduces metal cations, and regenerates bisulfite which provides a mildly reducing environment.[17] Contact time between the SBH solution and wastewater can be as low as 5 minutes or as long as 60 depending on the metals concentration. the precipitated metal must be removed from the treated wastewater quickly (in less than 1-hour) because redissolution of the metals can occur.[5] The sludge produced is high in metal content, finely divided and of lower quantities than comparable technologies (68 percent less than lime on a dry weight basis).[12].

In this system a binary flocculation system is used to agglomerate the finely divided metallic fines. A cationic polymer (a polyamine) is added in stage 2 and an anionic polymer (hydrolyzed polyacrylamide) is added to a flash mix tank just prior to the clarifier.[17] This treatment method is capable of producing a high quality effluent without filtration, and when filtered with industrial filter media is capable of reducing many metallic contaminants to below detection limits.

The two main disadvantages of this process are high reagent cost and the introduction of boron to the effluent flow stream. The high cost of sodium borohydride solution ($2.40/lb vs. $.023/lb for hydrated lime) has limited this treatment technology to applications either low in competing reducible species or situations which require extremely low metallic effluent concentrations. In addition, SBH is unable to break cyanide complexes and it is necessary to first destroy the complex (via hypochlorite or chlorine oxidation) prior to treatment. Sodium borate, a by-product of the SBH reaction, introduces boron at 3 to 10 percent of the level of metals removed. Further treatment such as filtration or ion exchange may be necessary before discharge.

Pretreatment and Post-Treatment Requirements--

A pretreatment in itself, chemical reduction is typically applied to chromium bearing aqueous waste streams segregated from other process flowstreams. An exception to this is sodium borohydride reduction which is used to reduce a wide variety of metallic contaminants, although on a limited scale. The properties of the waste being treated which can affect performance include:

- Flow variations;
- pH variations;
- Presence of chelator/complexants;
- Competing nonpriority reducible species;
- Cyanide content; and
- Oil and grease concentration.

In facilities which experience a wide variation in flow rates, pH values, or pollutant concentrations of the wastewater, flow equalization as pretreatment is often used. A variety of process options exist (see Section 10.1), but all systems basically provide some sort of flow resistance, stream segregation, or influent concentration averaging to prevent waste treatment system overloading. In all methods of flow equalization, care must be exercised during the wastewater analysis to completely characterize any peak flows or concentrations. In addition, flexibility in system design should be provided for any future expansion, change in location, or deviation in flow rates.[20]

Oil and grease, cyanides, chelator/complexants; and nonpriority reducibles, are all factors which will increase reagent consumption and impede, if not prohibit, chemical reduction operations. Oil and grease removal is typically the first process step in any waste treatment train. A wide variety of treatment equipment and chemicals currently exists in both the literature and in industry. Cyanides also effect the feasibility of chemical reduction technologies by forming strong cyano-complexes or evolving toxic hydrogen cyanide gas at the acidic conditions required for many of the reduction technologies. Treatment of cyanide waste streams typically consists of segregation followed by oxidation (see Section 14). Chelator/complexants and nonpriority reducibles present a difficult problem when chemically reducing metallic contaminants such as hexavalent chromium. Since these compounds are often an integral part of the chromate waste stream, waste stream segregation is difficult if not impossible. Two established methods of pretreatment for the removal of chelator/complexants and nonpriority reducibles are pH depression and binary reduction systems. In the pH depression methods, the pH of the waste stream is lowered to approximately 2.0 through the use of acid. The low pH helps to break complexes and since it is already a part of the overall chromium reduction process eliminates the need for additional equipment. The second technology, binary reduction, uses a less expensive reductant such as hydrazine, dithiocarbonate, or sodium metabisulfite to "prereduce" waste streams containing excess chelators/complexes or oxidized compounds. The prereducer acts as a scavanger while the primary reductant works to reduce the metallic contaminants of concern.

Chemical Reduction in itself does not produce any residuals. However, to completely remove metallic species from the waste stream, chemical reduction is usually followed by precipitation, coagulation/flocculation, sedimentation, and sludge consolidation. The resulting toxic sludge must often then be treated (i.e., encapsulation) and land disposed. For a discussion of post-treatment techniques see Section 10.1.

10.3.2 Process Performance

Sulfur-based chromium reduction technologies have gained wide acceptance in industry for reducing waste stream hexavalent chromium concentrations. The most prevalent reduction reagents are sodium sulfite salts, sulfur dioxide, and sulfuric acid. However, iron salts such as ferrous sulfate have also shown potential for reducing chromium wastes. As demonstrated in Table 10.3.7 sulfur-based chromium reduction technologies operate under a wide range of influent conditions.[10,15,21-24] Variable flowrates (5 to 140 gpm), pH conditions (2 to 10 standard units), and hexavalent chromium concentrations (2.23 - 136 mg/L) were all successfully treated by the reduction technologies examined. The most prevalent method of treatment cited among the numerous examples in the literature is sulfuric acid adjustment to pH 2.0 followed by sodium bisulfite reduction and hydroxide precipitation.

The chromium reduction processes examined are very efficient in nature with complete reduction typically achieved in less than 1 hour. In addition, these technologies are able to successfully treat a wide range of chromate wastes as demonstrated in Table 10.3.8.

Since most of the reduction technologies examined utilize the same process equipment and are capable of reducing hexavalent chromium concentrations to less than 0.01 mg/L, system selection is usually based on economic considerations. Criteria such as reagent cost/lb Cr^{+6} reduced, excess reagent requirements, sludge generation, and pH adjustment costs will all influence the overall economics of the system selected.

Sodium borohydride ($NaBH_4$) has also shown promise in chromium and other heavy metals reduction.[17,26-28] Table 10.3.9 presents sodium borohydride performance data for a wide variety of waste streams and metallic contaminants. The success of sodium borohydride reduction is highly dependent on mixing, residence time, pH, nonpriority reducible concentrations, and reaction kinetics. Sodium borohydride treatment at facilities A-F removed to acceptable levels all metallic constituents of concern enabling the facilities to meet discharge standards. The design of the systems, as with sulfur-based technologies, is based on standard, automatically controlled (ORP) equipment, traditionally used in industrial wastewater treatment.

For facilities whose waste streams contained chelators and/or complexants (Facilities A, D, and F) pH adjustments and sodium bisulfite or ferric chloride were required to improve SBH reduction efficiencies. All metals except for nickel (50 percent or less removal efficiency) were removed effectively. All facilities reported improved sludge characteristics with the purity of the recovered metal limited only by the presence of other reducible species.

TABLE 10.3.7. PERFORMANCE DATA FOR SULFUR BASED REDUCTION SYSTEMS

Parameter	Facility					
	A[a]	B[b]	C[c]	D[d]	E[e]	F[f]
Wastestream	Nickel/ chromium rinse	Simulated chromium rinse	Chromium rinse	Chromium rinse	Simulated chromium rinse	Chromium rinse
Influent pH	7.1	6.0	2.5	2.2-3.0	7.0	NA
Influent flowrate (gpm)	12	35-140	10	90	5	70-87
Type of acid	Sulfuric	Sulfuric	--	Sulfuric	--	Sulfuric
Reduction pH	2.5	2.0	6.5-7.0	2.5	7.0-10.0	2.0
Reduction Reagent	Sodium bisulfite	Sodium bisulfite	Ferric chloride/ Sodium Sulfide	Sodium bisulfite Hydrazine	Ferrous suflate/ Sodium Sulfide	Sulfur dioxide
Retention Time (min)	60	10	240	NA	In-line mixing	NA
Precipitation Reagent	Caustic soda	Hydrated lime	--	Soda ash	Caustic soda	Lime
Precipitation pH	9.5-10.0	8.7	--	7.0-8.5	7.0-10.0	NA
Influent Cr^{+6} Concentration (mg/L)	2.23	6.60	136	8-21	5-50	NA
Effluent Cr^{+6} Concentration (mg/L)	0.01	0.01	0.01	.1	0.05-1.8	0.05
Effluent pH	7.5-8.0	8.7	6.5-7.0	7.1	7.0-9.0	6.0-8.5

NA = not available

[a]Source: Reference 21.

[b]Source: Reference 10.

[c]Source: Reference 22.

[d]Source: Reference 23.

[e]Source: Reference 15.

[f]Source: Reference 24.

TABLE 10.3.8. HAZARDOUS WASTES TREATED BY CHROMATE REDUCTION

	Waste code	Description	
Soluble	U032	Calcium chromate	
	D007	Chromium (hexavalent with chromic acid)	
	F019	Sludges from chemical conversion coatings	
Insoluble	K002	Production sludges from:	chromium yellow
	K003		molybdate orange
	K004		zinc yellow
	K005		chromium green oxide
	K006		chromium green oxide
	K008	Oven residues from chromium greenoxide	
	K086	Pigments and inks	
	F006	Insoluble chromates	

Source: Reference 25.

TABLE 10.3.9. SODIUM BOROHYDRIDE PERFORMANCE DATA

Parameter	Facility					
	A	B	C	D	E	F
Wastestream	Printed[a] Circuit Boards	Mercury[b] Cell Electrolytic	Tetraalkyl[b] Lead Manufacture	Lithographic[b] Film	Printed[c] Circuit Boards	Commercial[d] Wastetreatment Plant
Influent Flowrate (gpm)	NA	10-20	900-1,500	50-210	23	Batch (7,900 gal)
1st Stage						
pH	5.5	--	--	11.0	--	5.5
Retention Time (min)	20-40	--	--	15	--	15
Reagent	Sodium Bisulfite	--	--	Ferric chloride	--	Sodium Bisulfite
2nd Stage						
pH	8.0	NA	9.0	11.0	8.0-11	8.0
Retention Time (min)	20-30	15-30	15-20	30	30	30-45
Reagent	Sodium Borohydride (SBH)	SBH	SBH	SBH	SBH	SBH
Influent Metals Concentration (mg/L)						
Copper	20.0	--	--	--	786.0	237.0
Lead	--	--	5.35	--	0.57	0.32
Nickel	--	--	--	--	0.06	0.96
Zinc	--	--	--	--	3.86	5.10
Mercury	--	10-50	--	--	--	--
Silver	--	--	--	10-120	--	24.0
Cadmium	--	--	--	5-60	--	0.12
Effluent Metals Concentration (mg/L)						
Copper	1.0	--	--	--	1.49	0.47
Lead	--	--	0.1	--	0.10	0.14
Nickel	--	--	--	--	0.03	0.42
Zinc	--	--	--	--	0.03	0.08
Mercury	--	0.1-0.8	--	--	--	--
Silver	--	--	--	0.09	--	--
Cadmium	--	--	--	0.09	--	0.01

[a]Reference 17.

[b]Reference 26.

[c]Reference 27.

[d]Reference 28.

As with sulfur-based chemical reduction technologies, sodium borohydride implementation is influenced by overall process economics. Reagent costs are high, and several facilities (A, B, D, and F) reported impaired economic performance due to required pretreatments and/or excess reagent usage. All of these factors must be considered when evaluating sodium borohydride as either a primary or secondary treatment system for control of heavy or precious metal waste streams.

10.3.3 Process Costs

Figure 10.3.4 illustrates the basic process train developed for a continuous chromate reduction and precipitation system. the equipment for this system includes a flow equalization tank, a continuous chromate reduction tank, a precipitation reactor, a lamella type flocculator/clarifier, sludge holding tanks, and a plate and frame filter press. The continuous chromate reduction tank includes high level alarms, portable pH and oxidation-reduction potential (ORP) meters, a portable mixer, and storage tanks and feed pumps to add sodium metabisulfite and sulfuric acid. Reactor retention time is 30 minutes. The capital equipment cost data for the chromate reduction system is based on Figure 10.3.5.[29] The equipment specifications and cost assumptions for the remaining operations are based on assumptions previously presented in Section 10.1.[30]

The influent stream to the chromium waste treatment system is assumed to contain 200 mg/L of hexavalent chromium at pH of 6.0. Approximately 7.1 mg/L of sulfuric acid is required to depress the chromate reduction influent stream to a pH of 2.0. Sodium metabisulfite is added to the waste stream of a stoichiometric rate of 15:1 and a complete reaction is assumed to occur. Approximately 400 mg/L of hydrated lime is required to raise the pH and precipitate the trivalent chromium as an hydroxide. This reagent addition will result in the formation of 400 mg/L of chromium hydroxide sludge on a dry weight basis and a small quantity (1.4 mg/L) of calcium sulfate precipitate. In reality the quantity of sludge produced will be very much a function of the calcium added as hydrated lime and the quantity of sulfates present in the waste stream. In this hypothetical model the only sulfates present are those that were introduced by the sulfuric acid.

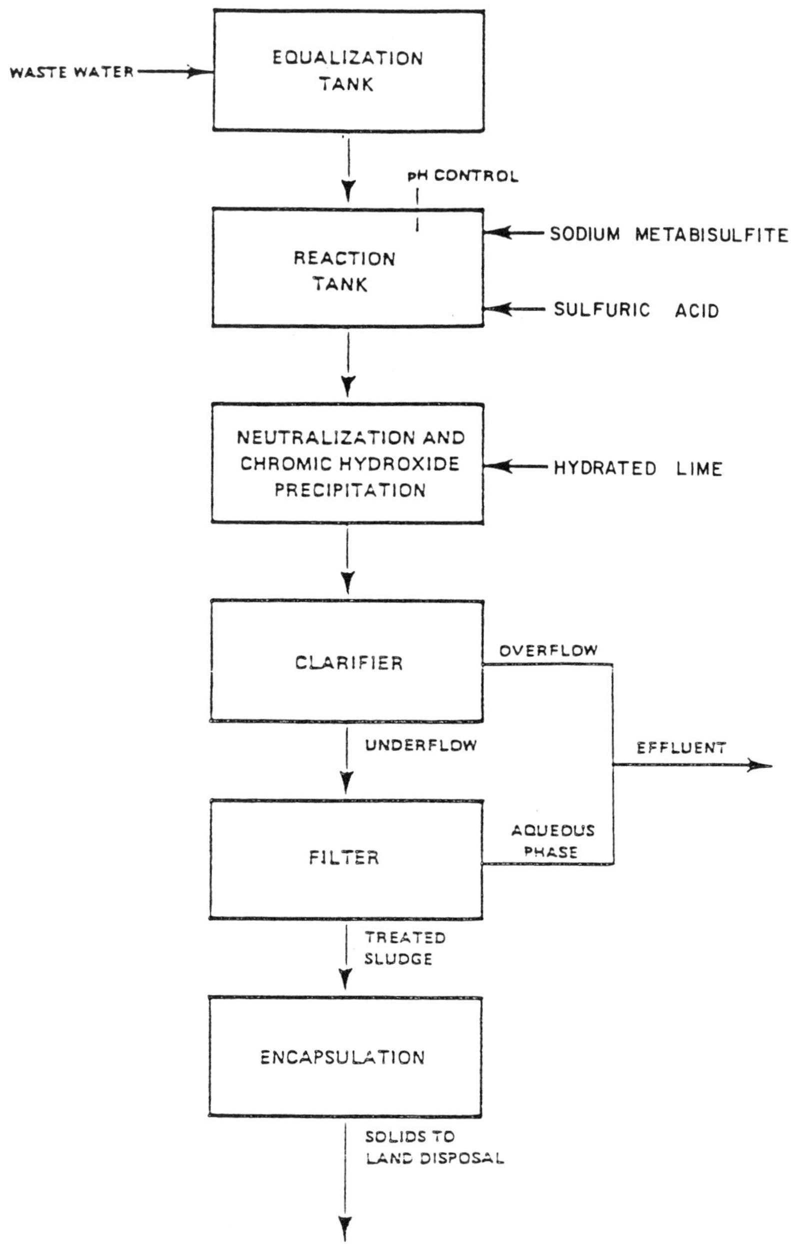

Figure 10.3.4. Chromate reduction system.

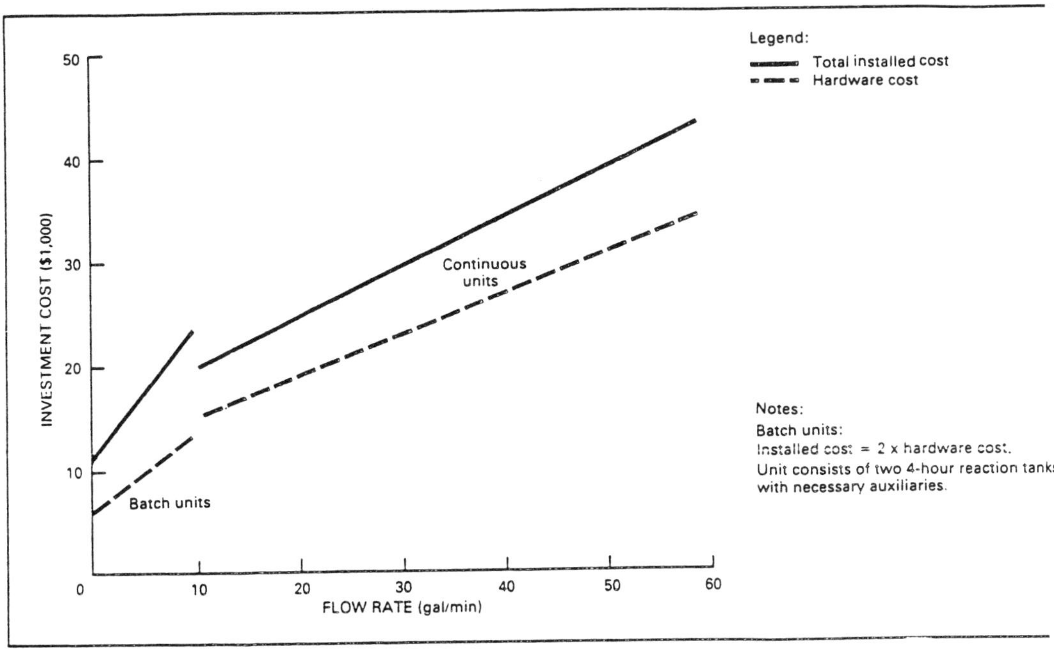

Figure 10.3.5. Investment cost for chromium reduction units.

Source: Reference 29.

Table 10.3.10 presents the annual costs for the continuous chromate reduction system illustrated in Figure 10.3.4. Flow rates are 1,000, 10,000, and 100,000 gal/hour. As with all waste treatment systems which rely on chemical precipitation as the primary method of contaminant removal, sludge disposal costs constitute a large percentage of the total annual costs. In addition the high cost of treatment chemicals may prohibit the use of this technology at high influent metals concentrations. As land disposal becomes increasingly expensive in anticipation of land disposal restrictions, waste treatment options such as chromium reduction which generate large quantities of potentially hazardous sludge will become less viable from both an economic and liability standpoint.

10.3.4 Overall Process Status

Chemical reduction of hexavalent chromium through sulfur-based reagents is a well established and fully-developed technology. Environmental considerations result primarily from residuals generated in the precipitation-sedimentation process following chromium reduction. In processes which use ferrous sulfate as the reducing agent, sludge generation can be significant. In addition, a potential hazard in reagent storage and handling is present for those facilities using gaseous sulfur dioxide. Table 10.3.11 contains a summary of the advantages and disadvantages of hexavalent chromium reduction.

Sodium borohydride, which has been applied on a limited basis as an alternative chemical reduction process in some chloralkali and metal finishing facilities, has a potential as a viable waste treatment option. Sludge production is less than comparable technologies and with the exception of nickel, metal removal efficiencies are sufficient to meet effluent limitation guidelines. The main limitations to this technology are high reagent costs, the introduction of boron into the effluent waste stream, and the evolution of hydrogen gas as part of the reduction process. None the less, as land disposal costs continue to increase, sodium borohydride's ability to produce a compact, high density, pure sludge product will enhance its selection as an alternate metals reduction process.

TABLE 10.3.10. ANNUAL COSTS FOR A CONTINUOUS CHROMATE REDUCTION SYSTEM[a]

	Flow rate (gph)		
	1000	10,000	100,000
Purchased Equipment and Installation (PE&I)			
Equalization Tank	17,000	29,000	50,000
Reduction Tank	15,000	66,000	114,000
Precipitation Reactor	24,000	40,000	69,000
Flocculator/Clarifier	18,000	50,000	140,000
Sludge Holding Tank(s)	3,000	6,000	48,000
Filter Press	10,000	25,000	100,000
	87,00	216,000	521,000
Total Capital Investment (360% PE&I)	313,200	777,600	1,875,600
Annual Operating Costs ($/Yr.)			
Operating Labor ($20/hr.)	72,000	72,000	72,000
Maintenance (6% TCI)	18,800	46,700	112,500
General Plant Overhead (5.8% TCI)	18,200	45,100	108,800
Utilities (2% TCI)	6,300	15,600	37,500
Taxes and Insurance (1% TCI)	3,100	7,800	18,800
Chemical Costs:			
Lime ($40/ton)	200	1,800	18,500
Sulfuric Acid ($72/ton)	2,200	21,600	216,100
Sodium Metabisulfite ($32/100lb)	1,500	15,200	152,200
Sludge Transportation ($0.25/ton-mile)	200	1,800	1,800
Sludge Disposal ($200/ton)	9,600	96,400	963,800
Annualized Capital (CFR-0.177)	55,400	137,600	332,000
Total Cost/year	187,500	461,600	2,034,000
Cost/1000 gallon	26	6	3

[a]1987 Dollars.

TABLE 10.3.11. ADVANTAGES AND DISADVANTAGES OF CHEMICAL REDUCTION OF HEXAVALENT CHROMIUM

Advantages

- Well proven technology with documented reduction efficiencies.
- Operates at ambient temperature and pressure lowering energy requirements.
- Process equipment is modular and widely available from a variety of manufacturers and suppliers.
- Is applicable to a wide range of chromium wastewaters from numerous industrial sources.

Disadvantages

- For high concentrations of influent chromium, the high cost of treatment chemicals may be prohibitive.
- Chemical interference by oxidizing agents in mixed waste streams may add substantially to reagent requirements.
- Sludge production is relatively high and in the case of ferrous sulfite can be significant.
- Storage and handling of gaseous sulfur dioxide is somewhat hazardous.

REFERENCES

1. U.S. EPA. Treatability Manual. Vol. III. Washington, DC. EPA-600/8-80-042c,d. July 1980.

2. Sundstrom, D.W., and H.E. Klei. Wastewater Treatment. Prentice-Hall, Inc., Englewood Cliffs, NJ. 1979.

3. Stumm, W., and J.J. Morgan. Aquatic Chemistry. John Wiley & Sons, New York, NY. 1970.

4. M.A. Chillingworth, et al., Alliance Technologies Corporation. Industrial Waste Management Alternatives Assessment for the State of Illinois. February 1981.

5. R.E. Wing. Complexed and Chelated Copper-Containing Rinsewaters. Plating and Surface Finishing. July 1986.

6. G.C. Cushnie, Centec Corporation. Navy Electroplating Pollution Control Technology Assessment Manual. Naval Civil Engineering Laboratory CR 84.019. February 1984.

7. K.H. Lanouette. Heavy Metals Removed. Chemical Engineering. October 17, 1977.

8. Kirk-Othmer Encyclopedia of Chemical Technology, Vol. 22, 3rd. Edition, John Wiley & Sons, New York, NY. 1981.

9. U.S. EPA. Proposed Development Document for Metal Finishing. EPA-440/1-82-091b. 1982.

10. Taylor, C.R., and S.R. Quasim. More Economical Treatment of Chromium-Bearing Wastewaters. Proceedings from the 37th. Industrial Waste Conference, Purdue University. 1982.

11. Kirk-Othmer Encyclopedia of Chemical Technology, Vol. 10, 3rd, Edition, John Wiley & Sons, New York, NY. 1981.

12. J.R. Aldrich, et al. Hazardous Sludge Reduction. 70th. AES Annual Technical Conference proceedings, Indianapolis, IN. June 1983.

13. Higgins, T.E., and V.E. Sater. Combined Removal of Cr, Cd, and Ni from Wastes. Environmental Progress. March 1984.

14. Higgins, T.E., and S.G. TerMaath. Treatment of Toxic Metal Wastewaters by Alkaline Ferrous Sulfate and Sodium Sulfide for Chromium Reductions, Precipitation, and Coagulation. 36th. Industrial Waste Conference, Purdue University. 1982.

15. Higgins, T.E., and B.R. Marshall. Combined Treatment of Hexavalent Chromium with Other Heavy Metals at Alkaline pH. 17th. Mid-Atlantic Industrial Waste Conference. June 1985.

16. U.S. EPA. Sources and Treatment of Wastewater in the Nonferrous Metals Industry. EPA-600/2-80-074. April 1980.

17. Lindsay, M.J., and M.E. Hackman, Morton Thiokol, Inc. Sodium Borohydride Reduces Hazardous Waste. Purdue Research Foundation, West Lafayette, IN. 1985.

18. J.A. Ulman. Control of Heavy Metal Discharge in the Printed Circuit Board Industry with Sodium Borohydride. AES SUR/FIN Annual Technical Conference and Exhibit. 1984.

19. Ven Met Brochure, Morton Thiokol, Inc., Ventron Division. 1984.

20. MITRE Corporation. Manual of Practice for Wastewater Neutralization and Precipitation. EPA-600/2-81-148. August 1981.

21. Rabosky, J.G., and T. Altares. Wastewater Treatment for a Small Chromium Plating Shop: A Case History. 38th. Industrial Waste Conference, Purdue University. 1983.

22. R.S. Talbot. Co-Precipitation of Heavy Metals with Soluble Sulfides Using Statistics for Process Control. 16th. Mid-Atlantic Industrial Waste Conference. 1984.

23. J.J. Martin. Chemical Treatment of Plating Waste for Removal of Heavy Metals. U.S. EPA-R2-73-044. May 1973.

24. Anonymous. Cleaning Up an Industrial Discharge. Environmental Science and Technology. August 1973.

25. D.W. Grosse. A Review of Alternate Treatment Processes for Metal-Bearing Hazardous Waste Streams. Journal of the Air Pollution Control Association. May 1986.

26. M.M. Cook, et al., Morthon Thiokol, Inc. Case Histories: Reviewing the Use of Sodium Borohydride for Control of Heavy Metal Discharge in Industrial Wastewaters. 34th. Industrial Waste Conference, Purdue University. 1979.

27. Palmer, S., and T.J. Nunno. Case Studies of Existing Treatment Applied to Hazardous Waste Banned from Landfill: Facility B Draft Final Report. Contract No. 68-03-3243.

28. Palmer, S., and T.J. Nunno. Case Studies of Existing Treatment Applied to Hazardous Waste Banned from Landfill: Facility A Draft Final Report. Contract No. 68-03-3243.

30. Versar Inc. Technical Assessment of Treatment Alternatives for Wastes-Containing Metals and/or Cyanides. Contract No. 68-03-3149, U.S. EPA/OSW. October 1984.

10.4 FLOTATION

Low density suspended material may often be separated from a liquid matrix by flotation. In this operation, fine air bubbles introduced into the solution result in the attachment of the bubbles to the particles.[1-3] The attached bubbles cause the particles to rise to the liquid surface, where they are removed by skimming. This process is referred to as dissolved or dispersed air/foam flotation.[4,5]

For materials that are dissolved and not suspended, other steps are needed to precipitate the contaminant prior to flotation. For example, a surfactant such as carboxylic acid, can be added. This process is referred to as ion flotation.[1,3,6] The "collector" reacts with the dissolved material to form an insoluble product and facilitates the attachment of bubbles to the particle surface. Improved cost-effectiveness can be achieved if the ion is precipitated first, and then floated with a subsequently smaller quantity of collector required. This is called precipitate flotation and is ideally carried out in a flotation column. Another technology, adsorbing colloid flotation, removes dissolved materials by adsorbing them onto colloidal particles which are then removed by flotation.

Since dissolved air flotation by itself is incapable of removing dissolved metallic contaminants, recent research efforts have centered on the technologies of ion, precipitate, and adsorbing colloid flotation. Therefore, the remainder of this section will be devoted to these technologies.

10.4.1 Process Description

The principal components of an air flotation system are a pressurizing pump, air supply, retention tank, and flotation unit, as shown in Figure 10.4.1. The system may also be operated with recycle as shown in Figure 10.4.2. In the recycle system, a portion of the clarified effluent is contacted with the dissolved air and pumped to the retention tank.[2] The aerated recycle stream is then mixed with fresh sludge at the entrance of the flotation unit. A recycle system avoids the shearing action of the pressurizing pump on the influent waste which impairs performance due to floc break-up. Table 10.4.1 summarizes typical operating parameters for this type of system.

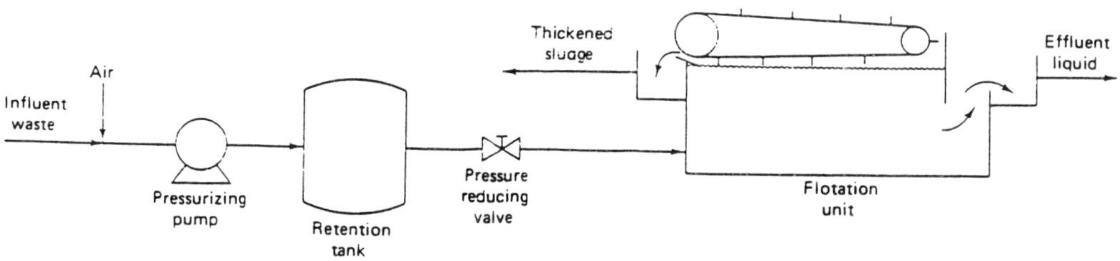

Figure 10.4.1. Schematic diagram of dissolved air flotation system without recycle.

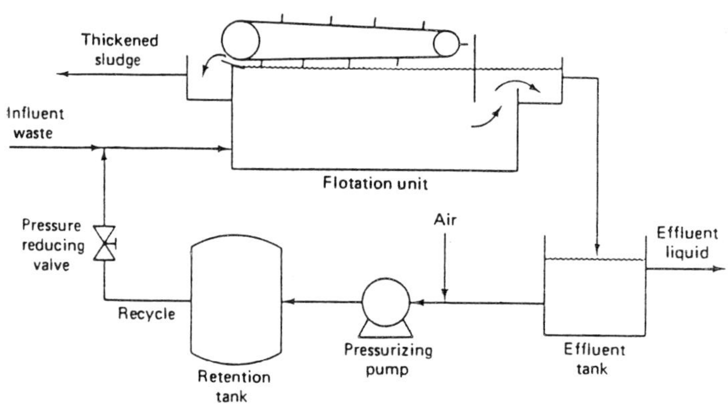

Figure 10.4.2. Schematic diagram of dissolved air flotation system with recycle.

Source: Reference 2.

TABLE 10.4.1. TYPICAL OPERATING PARAMETERS FOR DISSOLVED AIR FLOTATION UNITS

Parameter	Unit	Range
Air Pressure	psia	45 - 95
Air-to-Solids Ratio	mass/mass	1:10
Recycle Ratio	o/o	20 - 150
Overflow Rate	gpd/ft^2	700 - 2,500
Solids Loading	$lb/day\ ft^2$	25 - 100
Detection Time	min	20 - 60

Source: Reference 1, 2.

Central to any flotation technology, regardless of the separation mechanism used (i.e., ionic bonding, precipitation, adsorbtion), is the addition of a surfactant. Table 10.4.2 lists several commonly used flotation surfactants.

There are currently two physical models for the description of the attachment of contaminant particles to air-water interfaces in the presence of surfactant.[7,8] In the columbic model, ionic surfactant is adsorbed on the air-water interface resulting in a surface charge density on the bubbles (usually negative, because of the use of an anionic surfactant). The dissolved ion, precipitate, or absorbing floc is then given a surface charge by adjusting the pH or the concentration of other potential-determining ions. The adsorption mechanism involved is the electrostatic attraction between the ionic surfactant (collector) and the metallic contaminant of opposite charge.

In the contact angle model, surfactant ions adsorb to the primary layer of the metallic particle, presenting the ionic head of the surfactant to the solid and the long hydrophilic, hydrocarbon tail to the solution. The interfacial free energies (surface tensions) are now such that the contact angle of the air-water interface on the metallic particle is different from zero which permits the attachment of the particle to a bubble.

Some basic conclusions common to both models are:[8]

- Increasing ionic strength tends to decrease flotation efficiency.
- Increasing the length of the surfactant hydrocarbon tail decreases the bulk liquid concentration (moles/liter) of surfactant required to produce flotation.
- Increasing particle size increases flotation efficiency.
- Increasing temperature increases the concentration of surfactant required.

Recent advances in flotation equipment design have involved substituting vertical columns for the open, agitated vessels commonly used in conventional flotation systems.[4] Figure 10.4.3 illustrates a pilot plant developed by Thackston et. al. at Vanderbilt University, Nashville, TN.[8-12]

In this system, simulated wastewater is pumped from the storage tank through the pilot plant at the desired flow rate. The addition of the coagulant and NaOH, for pH control, occurs upstream of the main pump. The

TABLE 10.4.2. TYPICAL FLOTATION SURFACTANTS

Type	Formula[a]	Charge on the soluble ion
Sulfhydryl collectors[b]		
xanthate	ROCSSNa	anionic
dithiophosphate	$(RO)_2PSSNa$	anionic
monothiocarbamate	RHNCSOR	
thiol (mercaptan)	RSH	
dixanthogen	$(ROCSS)_2$	
thiocarbanilide	$(C_6H_5NH)_2CS$	
Colloidal electrolytes[c]		
fatty acids and their soaps	RCOOH, RCOONa	anionic
alkyl or aryl alkyl sulfonates	RSO_3Na	anionic
alkyl sulfate	$ROSO_3Na$	anoinic
primary amine salt	RNH_3Cl	cationic
secondary amine salt	R_2NH_2Cl	cationic
quaternary ammonium salt	$RN(CH_3)_3Cl$	cationic

[a] $R = CH_3(CH_2)n$

[b] For sulfides, $R = C_2 - C_5$.

[c] Generally straight chain C_{12} to C_{18}, or a benzene or naphthalene ring may be incorporated into the R group.

Source: Reference 3.

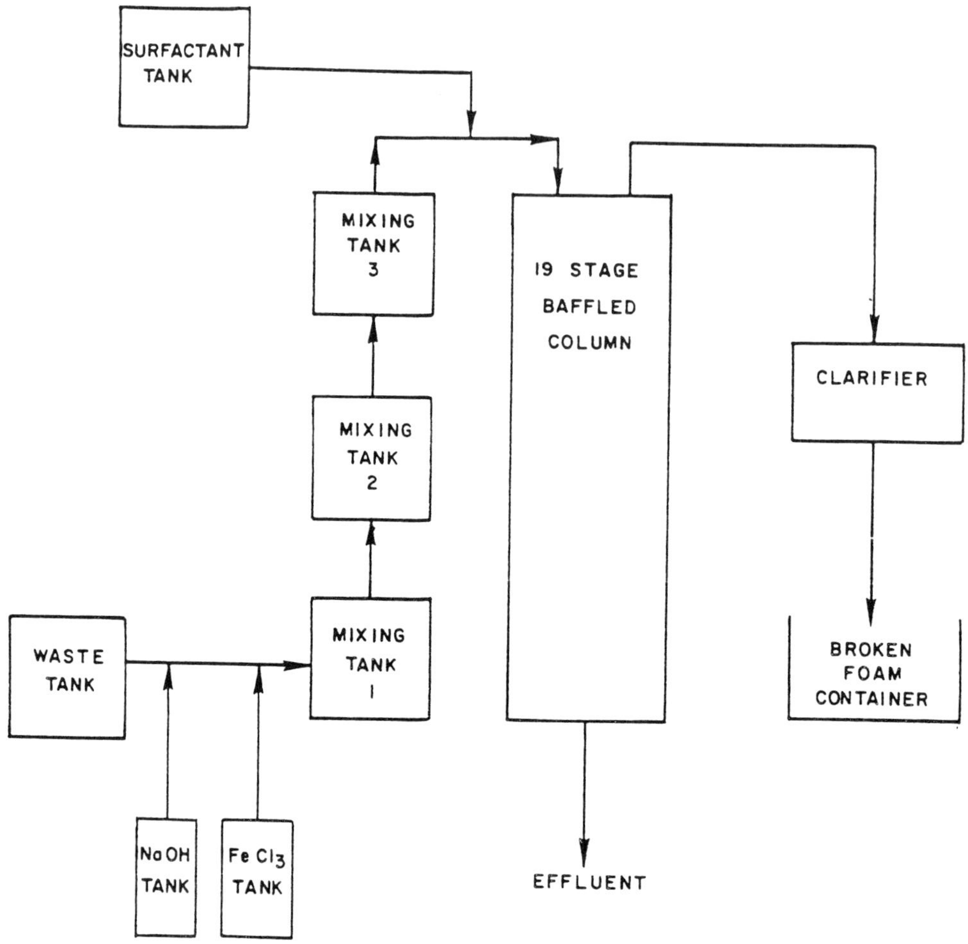

Figure 10.4.3. Schematic diagram of adsorbing foam flotation pilot plant.

coagulant is fed at the required rate by a variable feed pump. The NaOH solution flows by gravity through a solenoid valve connected to a pH controller set to produce the desired pH in the first mixing-flocculating chamber. After passing through the main pump, the wastewater enters a series of three mixing-flocculating chambers, after which NLS (sodium lauryl sulfate, an anionic surfactant) is injected at the required rate. The waste is then sent through the top of the flotation column to a spray nozzle. The wastewater then flows downward through the rising foam over an arrangement of 19 baffles installed to prevent foam overturning. The air is supplied through a fitted glass disk in the bottom of the column. The treated effluent leaves the column through the bottom and the foam is piped out of the top of the column to a rotating disk foam breaker. The effluent pH is monitored continuously.

Table 10.4.3 contains operating parameters for three colloidal adsorption foam flotation pilot plants.[8,11] All three studies were continuous flow applications which focused on lead removal from synthetic and industrial waste streams. Subsequent case studies have focused on removing Cu, Cd, Mg, Zn, Cr, and arsenate at both the bench and pilot scale.

Pretreatment requirements reported in the literature for chemical flotation include:

- Flow equalization;
- pH adjustment;
- Coagulation (absorbing colloid foam flotation); and
- Precipitation (precipitate flotation).

Some method of flow equalization (see Section 10.1) should be provided to average waste stream influent concentrations to prevent system overloading and maintain optimum performance characteristics. Other pretreatments such as pH adjustment (see Section 10.1), coagulation (see Section 10.3), and hydroxide precipitation (see Section 10.1) are all technology specific and have been extensively reported in the literature. Typical requirements would be a mixer-reactor for pH adjustment (equipped with appropriate reagent tanks and controls) followed by a precipitation or coagulation reactor depending on which flotation technology is utilized.

TABLE 10.4.3. OPTIMUM OPERATING PARAMETERS

Parameter, units	From Hanson	From Miller	From Thackston
pH	5.5-6.5	6.0-7.0	6.9-7.1
Fe(III), mg/L	150	100-150	90
NLS, mg/L	35-40	34-40	25
Hydraulic loading			
m^3/dm^2	118-176	148-178	326
gal/min/sq ft	2-3	2.5-3.0	5.5
Air supply			
$m^3/min/m^2$	0.2-0.3	0.4-0.5	0.2
cu ft/min/sq ft	0.66-0.98	1.31-1.64	0.7
Effluent Lead Concentrations (mg/L)	0.1	0.4	0.1

Source: Adapted from References 8, 11.

Oil and grease, cyanides, chelator/complexants, and competing ions for surfactant sites (i.e., carbonates) have not been considered in literature in any great detail. There is a real research need for work to determine whether or not these factors will make the waste acceptable for chemical flotation treatment. Most of the research to date has focused on low concentration (1,000 mg/L or less) simulated waste streams prepared by dissolving four or less metallic salts (alone or in combination) in tap water. If chemical flotation is to achieve wide acceptance in industry, more realistic (and consequently difficult to separate) waste streams will have to be investigated.

Residuals produced by chemical flotation consist primarily of a metals laden foam layer which is skimmed or drawn off the top of the reaction vessel/column. Post-treatment typically consists of sedimentation and sludge consolidation. The resulting hazardous sludge must often be treated (i.e., encapsulation) and then land disposed. In addition, air stripping of the foamate to recover surfactant may be desirable to reduce surfactant consumption (approximately 60 to 70 percent).

10.4.2 Process Performance

While most current research has focused on precipitate and absorbing colloid foam flotation techniques, Eastern European research has also encompassed the technology of ion flotation of dissolved metallic contaminants. For example, Skyleu et al. at I. I. Mechnivkov State University, Odessa, investigated the removal colloidal suspensions of metallic mercury (25 to 50 mg/L) using flotation apparatus.[13] The collectors were 0.01 to 0.1 percent aqueous solutions of potassium salts of pentadecoanoic, palmitic, heptadeconoic, and stearic acids. The best collector for mercury in ionic flotation treatment of these solutions was found to be potassium stearate, giving 98 percent extraction of a simulated waste stream containing 20 to 50 mg/L Hg and 10,000 to 15,000 mg/L NaCl and 78 percent extraction from a simulated waste stream containing 50 mg/L Hg, 1,000 mg/L NaCl, 3,000 mg/L NaOH, 500 mg/L Na_2CO_3, 10 mg/L CA^{+2}, and 500 mg/L SO_4^{-2}. However, as with most ionic flotation technologies, collector consumption was greater than stoichiometric requirements. Actual collector consumption reported by Skyleu et al. was 1.9 moles of potassium stearate/mole of mercury removed. The process was carried out at 25°C and required 10 to 12 minutes for completion.

In the United States, heavy metal removal through precipitate and absorbing colloid foam flotation treatment of industrial waste streams has received the most attention. In the early 1970's, Zeitlin's group at the University of Hawaii demonstrated the effectiveness of absorbing colloid flotation for removing zinc, copper, phosphate, and arsenate from seawater using dodecyl sulfate as a surfactant.[14,15] In the mid 1970's to the present, Wilson's group at Vanderbilt University investigated and refined Zeitlin's method using sodium lauryl sulfate (NLS) as a surfactant with iron and aluminum as coagulants.[16-19]

Using the apparatus shown in Figure 10.4.3, Wilson and Thackston, et al. investigated lead, copper, zinc, trivalent and hexavalent chromium removal using absorbing colloid flotation on actual and simulated industrial wastewaters. Table 10.4.4 describes optimum operating conditions as determined by pilot plant data for copper and zinc removal. Tables 10.4.5 and 10.4.6 illustrate foam flotation results for wastes containing a binary mixture of copper and zinc and a tertiary mixture of copper, divalent zinc, and trivalent chromium, respectively (influent metals concentrations 20 mg/L each). Tables 10.4.7 and 10.4.8 contain pilot plant data on hexavalent chromium and lead removal.

According to the researchers, for mixtures of metals, the optimum pH for metals removed is displaced to a higher value than those obtained for each individual metal. For example in the copper-zinc system, residual copper concentrations below 0.1 mg/L were obtained in pH range of almost one unit when using chemical doses of 100 mg/L Fe(III), 100 mg/L AL(III), and 70 mg/L NLS. Even when the chemical doses are reduced to 50 to 75 mg/L of Fe(III) and Al(III), and 40 to 50 mg/L of NLS, residual copper concentrations substantially below 0.2 mg/L were consistently achieved in the same pH range. Residual zinc concentrations below 1.0 mg/L were obtained in a pH range of 7.4 to 8, and values close to 0.5 mg/L were obtained even at the low coagulant-adsorbent and NLS dose concentrations indicated above for copper. When only Fe(III) was used as a coagulant-adsorbent, and no Al(III) was used, poor zinc removal was obtained. The presence of $Al(OH)_3$ exhibits some complementary beneficial effect on copper removal, and $Fe(OH)_3$ improves zinc removal, although the main effect is on copper.

TABLE 10.4.4. OPTIMUM OPERATING CONDITIONS AS DETERMINED IN THE 29- BY 244 CM PILOT PLANT

Parameter	Copper removal	Zinc removal
Effluent pH	6.9-7.3	7.5-7.8
Coagulant - adsorbent[a]	90-100 mg/L Fe(III)	100 mg/L Al(III)
NLS[b]	15-20 mg/L	30-40 mg/L
Hydraulic loading	7-14 $m^3/m^2/hr$ (3-6 $gal/ft^2/hr$)	7-14 $m^3/m^2/hr$ (3-6 $gal/ft^2/min$)
Air flow rate	12-14 N $m^3/m^2/hr$ (40-45 $ft^3/ft^2/hr$)	12-14 N $m^3/m^2/hr$ (40-45 $ft^3/ft^2/hr$)

[a]These values correspond to an initial metal concentration of 20 mg/L.

[b]These values refer to a floc concentration corresponding to the Fe(III) and Al(III) doses given in the table.

Source: Reference 10.

TABLE 10.4.5. FOAM FLOTATION OF WASTES CONTAINING Cu(I)+ Zn(II)[a]

pH	Fe(III) (mg/L)	Al(III) (mg/L)	NLS (mg/L)	Residual Copper (mg/L)	Residual Zinc (mg/L)
7.0	100	100	70	0.09	2.3
7.2	100	100	70	0.09	1.1
7.2	100	100	50	0.15	1.2
7.4	100	0	35	0.33	9.6
7.4	150	0	45	0.29	6.6
7.4	75	75	50	0.10	1.2
7.4	100	100	70	0.04	0.8
7.5	50	50	50	0.13	1.2
7.5	75	75	50	0.17	0.9
7.5	100	100	70	0.03	0.4
7.7	50	50	40	0.11	0.9
7.7	100	100	70	0.03	0.6
8.0	50	50	40	0.10	0.7

[a]All runs with initial copper and zinc at 20 mg/L each, influent flow rate at 6.9 $m^3/m^2/hr$, and air flow rate at 14 N $m^3/m^2/hr$.

Source: Reference 10.

TABLE 10.4.6. FOAM FLOTATION OF WASTES CONTAINING Cu(II)+ Zn(II)+ Cr(III)[a]

pH	Fe(III) (mg/L)	Al(III) (mg/L)	NLS (mg/L)	Residual Copper (mg/L)	Residual Zinc (mg/L)	Residual Chromium (mg/L)
7.0	100	100	70	0.10	2.00	0.12
7.2	100	100	70	0.07	0.90	0.13
7.2	115	75	50	0.12	1.10	0.13
7.2	150	100	85	0.07	0.60	0.12
7.3	75	75	50	0.42	2.40	0.60
7.3	100	100	70	0.08	0.56	0.20
7.3	115	75	70	0.08	0.56	0.20
7.3	115	75	50	0.14	0.75	0.33
7.3	150	100	85	0.06	0.40	0.12
7.4	115	75	70	0.07	0.16	0.20
7.4	150	100	85	0.11	0.30	0.21
7.6	100	100	70	0.22	0.20	0.26

[a]All runs made with initial copper, zince, and chromium at 20 mg/L each, influent flow rate at 6.9 $m^3/m^2/hr$, and air flow rate at 14 N $m^3/m^2/hr$.

Source: Reference 10.

TABLE 10.4.7. INFLUENCE OF Fe(II) DOSE AND pH ON CHROMIUM AND IRON REMOVAL[a]

pH controller lower set point	Steady-state effluent pH	Fe(II) dose (mg/L)	Effluent Cr (mg/L)	Effluent Fe (mg/L)
6.0	5.2	70	0.25	12
6.0	5.2	64	0.26	7
6.5	6.0	70	0.17	14
6.5	5.9	64	0.22	3
6.5	6.0	60.5	0.25	2
6.5	6.0	57.6	0.25	2
6.5	6.1	51.2	1.10	1

[a]Operating conditins: initial Cr(VI)=20, NLS=40 mg/L, H.L.=0.45 m^3/h (2 gal/min), H.L.R.=6.8 m^3/m^2 h (2.8 gal/min ft^2), air flow rate = 21.5 N m^3/m^2 (50 SCFH).

Source: Reference 9.

TABLE 10.4.8. EFFLUENT Pb(II) CONCENTRATION VERSUS pH

pH	Pb(II) concentration (mg/L)
5.5	>4.0
5.6	4.0
5.8	3.4
6.0	3.1
6.4	2.0
6.5	1.3
6.6	1.0
6.7	0.30
6.8	0.20
6.9	0.10
7.0	<0.10
7.1	<0.10
7.2	<0.10

Note - Q = 7.61 L/min (2 gal/min[a])
Hydraulic loading rate = 163 $m^3/d\ m^2$(2.77 gal/min/sq ft)
Fe(III) = 90 mg/L
NLS = 25 mg/L
Air flow = 850 std L/min (30 SCFH)

[a]Same trend occurs at 15 L/min (4 gal/min).

Source: Reference 11.

In the copper-zinc-chromium system good results were obtained over a reasonably wide pH range at 100 to 115 mg/L Fe(III) and 75 to 100 mg/L Al(III), and adjusting the NLS dose as a function of the total floc concentration in the system. Zinc removal was more effective at pH values higher than 7.3 in the range studied. Copper and chromium were better removed at pH values between 7 and 7.3, although very good results were achieved throughout the experimental range. Similar low effluent concentrations were obtained for the hexavalent chromium (0.22 mg/L) and lead (<0.1 mg/L) single component systems when iron was used as the coagulant. However, Wilson et al., did not determine the effect of greater than 100 mg/L influent metal concentrations, metallic precipitates, and chelation agents, on filtration removal efficiencies.

Bench-scale experimentation was performed by Brooks et al. in 1984 to examine the effect of these factors on the potential for separation and concentration of strategic metals, such as chromium, copper, zinc, and nickel.[20] Specific waste systems selected for the evaluation tests were electrochemical machining solids from high nickel-alloy processing and electroplating wastes, as well as brass industry pickling waste sludges.

The experiments which simulated waste metal hydroxides using alkali precipitation from the salts of the individual metals (see Table 10.4.9) indicate that without coagulants (i.e., iron or alum), only nickel provided any measure of efficient flotation. Similarly, when chelation agents were used in conjunction with NLS (see Table 10.4.10) only the nickel-dimethyl glyoxime system obtained a high selectivity in flotation separation.

Clearly, while flotation technologies offer promise as an efficient method for removal of low concentrations (<100 mg/L) of soluble metals in wastewaters, further research remains to be performed.

10.4.3 Process Costs

Realistic costs for the foam flotation process can not be developed at this time due to the lack of commercial-scale testing. It is expected that the primary cost would be for the flotation column. Operating costs would be expected to be equivalent with that of lime precipitation, and savings would be realized in reduced disposal costs and reduced purchase costs for recovered chemicals.[10,12]

TABLE 10.4.9. FLOTATION OF METAL HYDROXIDES WITH SODIUM LAURYL SULFATE (NLS) at 200 ppm

Metal as hydroxide	Metal concentration (ppm)	NLS/Metal (Wt/Ratio)	pH	Flotation Performance
Nickel	1,000	0.2	9.5	Good
Zinc	1,000	0.2	8.0	Partial
Copper	1,000	0.2	9.5	Partial
Iron	200	1.0	9.5	Poor
Chromiun	200	1.0	9.5	Poor

TABLE 10.4.10. METAL HYDROXIDE FLOTATION WITH CHELATION AGENTS COMBINED WITH SODIUM LAURYL SULFATE (200 ppm)

Hydroxide (1,000 ppm metal)	Chelation Agent	Weight ratios		pH	Flotation performance
		Surfactant solids	Chelate solids		
Nickel	Dimethyl Glyoxime	0.2	0.3	8	Good
Zinc	Zincon	0.2	0.36	8	Partial
Copper	Neocuproine	0.2	0.4	8	Partial
Iron	Bathophenanthroline	1.0	2.5	9.5	Poor
Chromium	Diphenylcarbazide	0.8	1.7	9.5	Poor
Chromium	Aliquat 336	1.0	1.0	9	Poor

Source: Reference 20.

Preliminary cost estimates for absorbing foam flotation metal treatment systems have been prepared by the developer (Wilson et al.) based on pilot-scaling testing. Capital costs (1983) for a 50,000 gpd plant treating 60 mg/L of heavy metal was hypothesized to be approximately $20,000. However, until a full-scale commercial unit is actually in place, the economic feasibility of absorbing foam flotation has yet to be adequately determined.

10.4.4 Process Status

Chemical flotation, while currently at the bench and pilot-plant level of development, shows promise for reducing low concentrations (~100 mg/L) of effluent metals to acceptable levels. The process operates at ambient temperature and pressure and is well suited to automatic and computer control. Its ability to treat both singular and mixed metals waste streams has been well demonstrated by researchers at the University of Hawaii and Vanderbilt University. However, further research is necessary before this technology is applicable on a wider scale.

Environmental impacts result primarily from the production of potentially hazardous sludge. The sludge product is generated during the foam breaking process in the supernatant clarifier. At this point in time, little information is available in the literature on the quality and composition of the sludge product produced. Research dissertations outlining experimental results have focused primarily on metals removal. However, foam flotation is reported to generate less sludge than comparable precipitation processes, approximately 2 to 3 percent of the influent volume. The demonstrated performance, and possible lower sludge generation rates for this waste treatment technology warrant further research efforts.

REFERENCES

1. U.S. EPA. Treatability Manual, Vol. III. EPA-600/18-80-042. July 1980.

2. Sundstrom, D.W., and H.E. Klei. Wastewater Treatment. Prentice-Hall, Inc., Englewood Cliffs, NY. 1979.

3. Kirk-Othmer Encyclopedia of Chemical Technology, Vol. 10, 3rd. Edition, John Wiley & Sons, New York, NY. 1981.

4. G. Parkinson. Improved Flotation Routes Get Separation Tryouts. Chemical Engineering. March 31, 1986.

5. T.D. Reynolds. Unit Operations and Processes in Environmental Engineering. PWS Publishers, Boston, MA. 1982.

6. R.R. Klimpel, Dow Chemical Company. Use of Chemical Reagents in Flotation. Chemical Engineering. September 3, 1984.

7. Rubin, A.J., and W.L. Lapp. Foam Separation of Lead (11) with Sodium Lauryl Sulfate. Analytical Chemistry. July 1969.

8. E.L. Thackston, et al. Lead Removal with Adsorbing Colloid Flotation. Journal of the Water Pollution Control Federation. February 1980.

9. Huang, S., and D.J. Wilson. Hexavalent Chromium Removal in a Foam Flotation Pilot Plant. Separation Science and Technology, 19. 1984.

10. G. McIntyre, et al. Inexpensive Heavy Metal Removal by Foam Flotation. 36th. Industrial Waste Conference, Purdue University. 1982.

11. Slapik, M.A., Thackston, E.L., and D.J. Wilson. Improvements in Foam Flotation for Lead Removal. Journal of the Water Pollution Control Federation. March 1982.

12. G. McIntyre, et al. Inexpensive Heavy Metal Removal by Foam Flotation. Journal of the Water Pollution Control Federation. September 1983.

13. L.D. Skrylev, et al. Flotation Separation of Colloidally Dissolved Metallic Mercury Collected with the Aid of Potassium Salts of Saturated Fatty Acids. Plenum Publishing Corporation. 1985.

14. Kim, Y.S., and M. Zeitlin. The Separation of Zinc and Copper From Seawater by Adsorption Colloid Flotation. Separation Science 7. 1972.

15. Chainc, F.E., and M. Zeitlin. The Separation of Phosphate and Arsenate from Seawater by Adsorption Colloid Flotation. Separation Science 9(1). 1974.

16. Ferguson, B.B., Hinkle, C., and D.J. Wilson. Foam Separation of Lead (11) and Cadmium (11) from Wastewater. Separation Science 9(2). 1974.

17. Wilson, J.W., and D.J. Wilson. Electrical Aspects of Adsorbing Colloid Flotation. Separation Science 9(5). 1974.

18. Robertson, R.P., Wilson, D.J., and C.S. Wilson. The Adsorbing Colloid Flotation of Lead (11) and Zinc (11) by Hydroxides. Separation Science 11(6). 1976.

19. Currin, B.L., Potter, F.S., and D.J. Wilson. Surfactant Recovery in Adsorbing Colloid Flotation. Separation Science 13(4). 1978.

20. C.S. Brooks. Recycle Metals, Metal Recovery From Waste Sludges. 39th. Industrial Waste Conference, Purdue University. 1984.

11. Biological Treatment for Metal-Containing Wastes

Biological treatment is a separation process rather than a destruction technology for metal-containing wastes. Biological separation mechanisms include sulfide precipitation, adsorption, and bioflocculation. The type of biological treatment technology vary considerably. Those that are considered in this section are activated sludge, anaerobic digestion, and algal treatment. Common parameters for the design and operation of these technologies are outlined and information on recently developed biological organisms is presented.

High concentrations of heavy metals are toxic to most microorganisms and often cause serious upsets in biological systems. Thus, influent heavy metal concentrations which can be tolerated and removed is the major criterion on which these technologies are evaluated in this section. In addition, factors such as type of influent, its strength, and the extent of system acclimation are also used to evaluate the viability of biological treatment as a technology for the removal of heavy metals from wastes.

11.1 PROCESS DESCRIPTION

As previously stated, several mechanisms can affect the removal of heavy metals during biological treatment include sulfide precipitation, adsorption, and bioflocculation. The first mechanism, hydrogen sulfide precipitation (see Section 10.1), is initiated by the pH dependent generation of hydrogen sulfide by bacteria.[1] Soluble metal ions react with the hydrogen sulfide and are precipitated as insoluble metallic sulfides. The second mechanism, adsorption of cationic metallic ions, may result from the anionic nature of certain cellular material, clay particles, and industrial wastes constituents.[2]

Also, the organic part of organo-metallic complexes may be adsorbed through the cell walls of the biological organisms, thus trapping the metals.[3,4] The third mechanism, bioflocculation,[5,6] is related to the synthesis of insoluble extracellular polymer strands. These extracellular polymers can act as non-specific sorbers for metal ions.

Typically, the removal of heavy metals in a biological system and the type of mechanism which dominates are dependent on the species of heavy metal present (see Table 11.1.1).[7] The distribution of a particular heavy metal among various chemical forms, however, largely depends upon the physical and chemical properties of the environment established by the treatment process itself.[7] Upon introduction into the biological treatment system, species of heavy metal make adjustments toward a new equilibrium state defined by chemical environment parameters such as pH, oxidation reduction potential (ORP), the presence of complexing agents, and concentrations of precipitant ligands.[8] At this point, adsorption to solid phases[8] or biomass,[9] and intracellular storage can occur.[10,12]

It has been found that the microbial removal of heavy metals consists of initial rapid uptake followed by slow, but consistent long-term uptake.[13] The rate of uptake is greatly affected by solution pH.[14] Sludge age, as well as the extent of acclimation, can also affect the extent of metal removal in an activated sludge system.

The following is a brief description of the three main technologies used in biological treatment of heavy metals. More details can be found in standard texts and the references cited herein.

11.1.1 Activated Sludge

The activated sludge process uses biological populations in a completely mixed, oxygen-rich environment to treat wastes. Dissolved oxygen and mixing are provided by mechanical aerators or fine-bubble air diffusers. A settling tank is then used to remove the biological floc, part of which is mixed with incoming waste in the aeration tank.[15]

A variation to the activated sludge process is the use of high purity oxygen instead of air for aerobic treatment. Oxygen can be supplied from onsite gas generators with liquid oxygen storage as back-up. In addition to

TABLE 11.1.1. POSSIBLE SPECIES OF HEAVY METALS IN BIOLOGICAL TREATMENT

Soluble

- Ionic
- Organic complexes
- Inorganic complexes

Co-precipitates in metal oxides

Precipitates

Adsorbates

- Physical
- Chemisorption
- Clay lattice

Biological residues

Source: Reference 7.

oxygen use, the aeration tank is covered which helps to eliminate odors and maintain temperatures in cold-weather periods.

There are many design variations to the conventional activated sludge process besides the use of high purity oxygen. These include: multiple units with series and/or parallel flow patterns; a tapered distribution of air along the tank length; stepwise addition of raw waste; reaeration of the recycled sludge before mixing with the raw influent; and extended aeration (e.g., 24 hours or longer) used for small wastewater flows.[15]

Activated sludge systems lack stability because of the microbial growth pattern within the tank. A high rate of growth exists at the influent but decreases along the length of the tank. This problem is greatly amplified during flow surges. Variation of pH, temperature, and the presence of toxic waste constituents can all contribute to instability. Extended aeration and the use of high purity oxygen can help eliminate the effects of shock loadings.[16-20] An extended aeration system is depicted in Figure 11.1.1.

Table 11.1.2 presents typical ranges in values for activated sludge system design parameters.[20]. Additional design factors to be considered include sludge settling and accumulation rates and air requirements. These factors will vary depending upon the type of waste to be handled. Other factors which may limit the viability of the activated sludge process include climate/temperature conditions, available land area, and variations in flow rate and organic loading.

11.1.2 Anaerobic Digestion

Anaerobic digestion is a process commonly used to convert raw wastewater sludge into inoffensive forms by decreasing its organic content. The process biologically reduces the amount of volatile suspended solids that must be handled by subsequent dewatering and disposal operations rendering the organic material nonputrescible. In addition, its major gaseous end product, methane, can be harnessed to supply plant energy needs and the digested sludge can be used as a soil conditioner.[15]

The biodegredation mechanism proceeds in two discreet steps. First, facultative organisms called "acid formers" degrade the complex organics of wastewater sludge to volatile organic acids, primarily acetic acid. In the

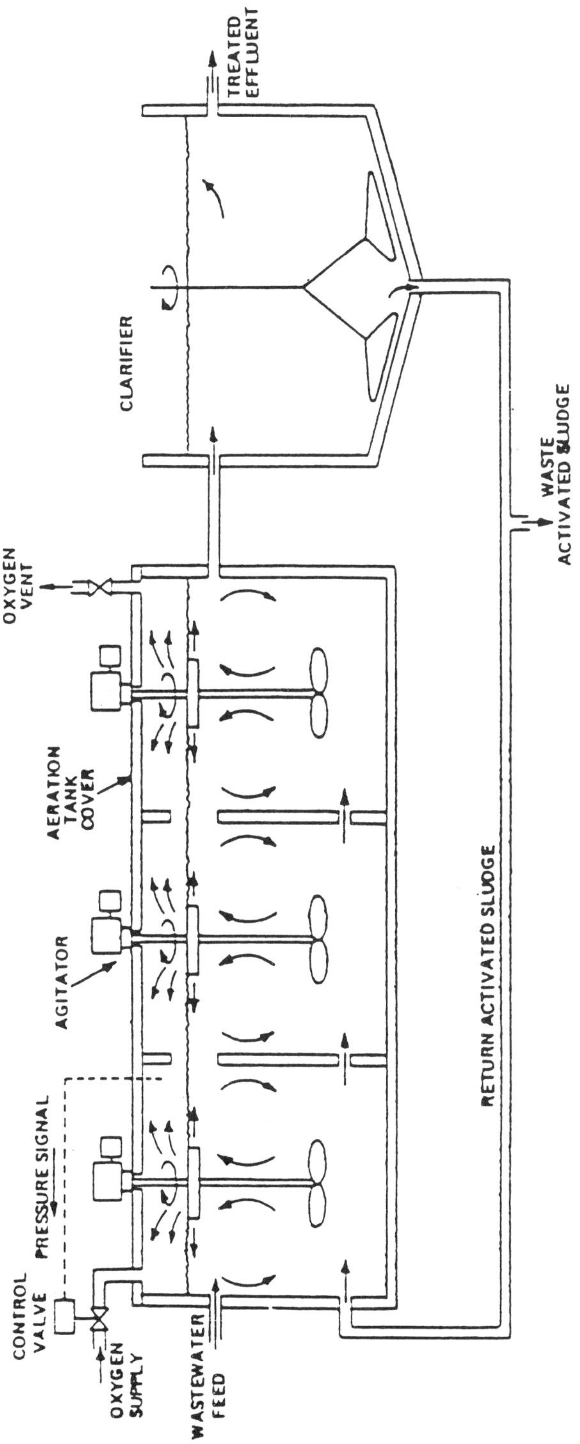

Figure 11.1.1. Schematic diagram, three-stage Unox system.

Source: Reference 20.

TABLE 11.1.2. TYPICAL ACTIVATED SLUDGE DESIGN PARAMETERS[a]

Process modification	Flow regime	Sludge retention time (days)	Food to microorganism ratio-#BOD_5/MLVSS/day	Aerator loading #BDO_5/1,000 ft^3 tank volume	Mixed liquor suspended solids (mg/l)	Detention time (hr)	Recirculation ratio
Conventional	Plug	5-15	0.2 - 0.4	30-40	1500-3000	4-8	0.25-0.75
Complete mix	Complete mix	3-10	0.2 - 0.6	50-120	3000-6000	4-6	0.25-1.0
Step aeration	Plug	5-15	0.2 - 0.4	40-60	2000-3500	3-6	0.25-0.75
Contact stabilization	Plug	5-15	0.2 - 0.6	30-75	1000-4000[b] 4000-10000[c]	0.5-1.5[c]	0.5-1.5
Extended aeration	Complete mix	20-30	0.05-0.15	10-15	2000-6000	24	0.5-2.0
Pure oxygen systems	Complete mix reactors in series	8-20	0.25-1.0	100-250	4000-8000	1-4	0.25-0.5

[a]Values given are for organic removals only; no nitrification.

[b]Contact unit.

[c]Stabilization tank.

Reference 20.

second step, these volatile acids are fermented to methane and carbon dioxide by a group of strict anaerobes called "methane bacteria."

Two main anaerobic digestive processes are used: the standard rate and the high rate systems. Schematics of these processes, as well as their operating criteria are provided in Figure 11.1.2. A modification of these systems, the two-stage process, has also been successfully used (see Figure 11.1.3). A brief description of each of these systems follows.

In a standard-rate system, the tank is not mixed and, in some cases, is not heated. Sludge is added at the top and withdrawn at the bottom. During progression from top to bottom of the digestion tank, the sludge is compressed and gradually dewatered. Stratification develops within this plug-flow system due to a lack of mixing. As a result, much of the digester volume is wasted, and many operational problems result. Acidification sometimes takes place in the top and middle layers while methane fermentation is confined to the lower layers. This can lead to areas of low and high pH in the system, which restrict optimum biological activity. Also, chemicals added for pH control are not dispersed throughout the tank, and their effectiveness is therefore limited. Grease breakdown is poor because the grease tends to float to the top of the digester while the methane bacteria are confined to the lower levels. Methane bacteria are removed with the digested sludge and are not recycled to the top, where they are required.[21]

The high-rate system differs from the low-rate system in that the contents are well mixed, either continuously or intermittently, and the digester is heated. This procedure avoids most of the difficulties inherent in low-rate systems. Consequently, this system demonstrates improved operation at lower retention times and higher organic loadings.[21]

The two-stage process evolved as an attempt to provide additional gas production and a separate settling and thickening process in the secondary digester. The process can be used successfully when the feed consists of primary sludge or combinations of primary sludge and limited amounts of secondary sludge. With the advent of wastewater treatment systems that are more efficient than simple sedimentation, large quantities of activated and sometimes advanced waste treatment (AWT) sludges are produced. When placed in a two-stage anaerobic digestion process, this additional sludge can cause high operating costs and poor plant efficiencies since the additional solids do not readily settle after digestion.[21]

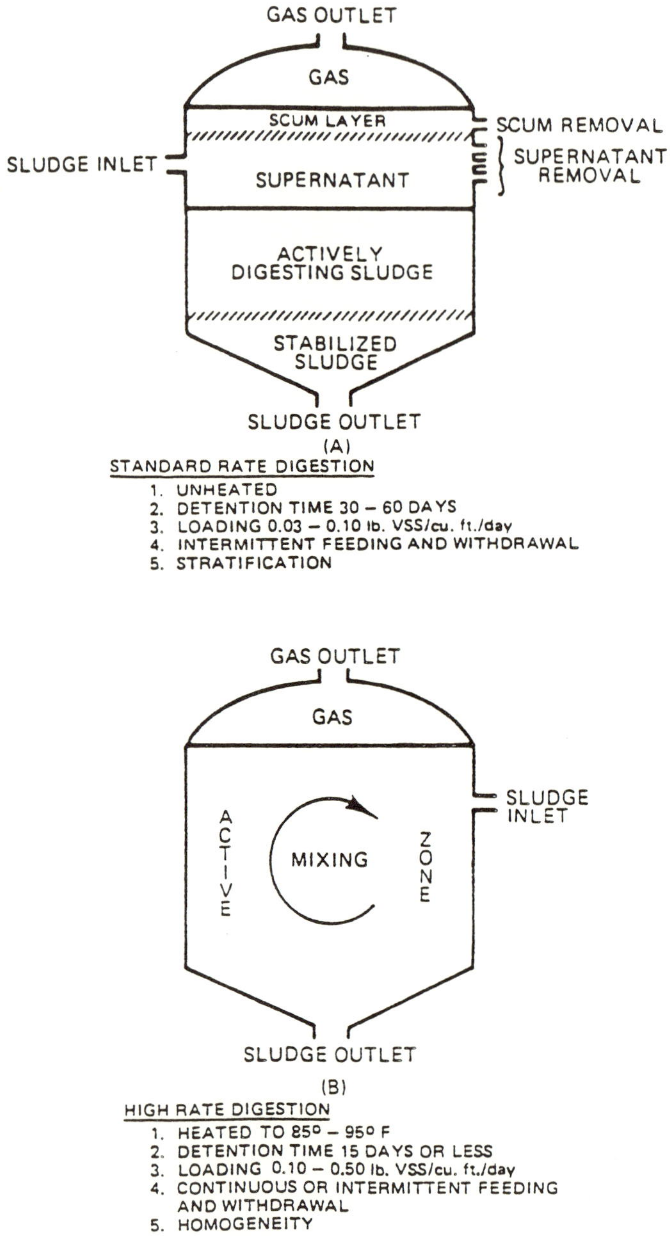

Figure 11.1.2. Standard rate and high rate digestion.

Source: Reference 21.

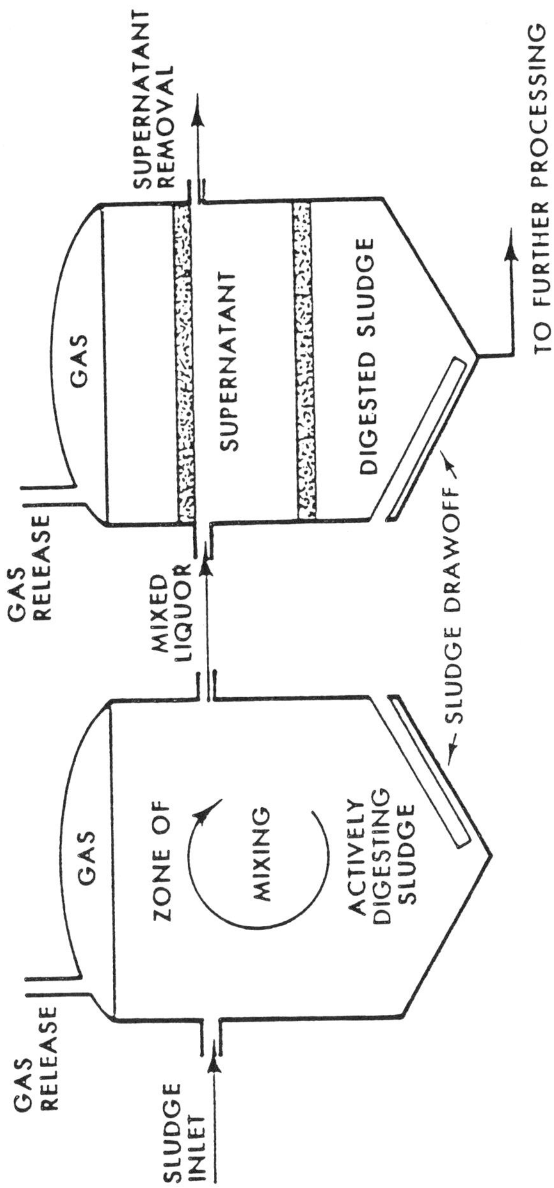

Figure 11.1.3. Two-stage anaerobic digestion.

Source: Reference 21.

Anaerobic digestion is suitable for nontoxic, organic-containing sludges resulting predominately from primary settling. It is in widespread use, accounting for 60 to 70 percent of biological treatment applied for primary and secondary sludge in plants having a capacity of 1 mgd or more. The two-stage system, while having roughly twice the capital cost of single-stage digestion, is gaining in popularity. This is attributable to the increased gas production, clearer supernatant liquor, and lower heat requirements (due to smaller tank size) associated with its use. Anaerobic digestion is very sensitive to process upsets due to the difficulty the bacteria have in adjusting to environmental changes. However, despite its operating sensitivity, anaerobic digestion is widely used due to the production of methane.

Since a biological mechanism is involved, the applicability of this treatment process to the digestion of any given industrial sludge can only be determined by specific pilot plant studies. Chemical factors are of greatest importance to industrial sludge treatment. Close pH control is required because methane bacteria are extremely sensitive to slight changes in pH. The usual pH range required is 6.6 to 7.4, and the pretreatment of incoming sludge to a pH of 7.0 is desirable.[15]

The optimum temperature for sludge digestion is related to the temperature response of the methane bacteria. The rate of bacterial growth and, therefore, the rate of process stabilization increase and decrease with temperature within certain limits. Systems operated at high temperatures cost more to heat, but may be justified by increased efficiencies. Essentially all digesters in the United States operate between 80°F and 110°F. More important than selection of a particular temperature is maintaining it at a constant level. A temperature change of 1 or 2°C is sufficient to disturb the dynamic balance between the acid and methane formers. This will lead to an upset because the acid formers will respond much more rapidly to changes in temperature than will the methane bacteria.[15]

Knowledge of the specific nutritional requirements of methane bacteria is limited. Domestic wastewater appears to contain all of the nutrients required by these organisms. However, due to the uncertainty of the precise nutritional mix required, difficulty may be encountered when treating wastewater of industrial origin.

11.1.3 Algal Biodegradation Technology

Recent research indicated that algae may be used to remove metal ions from wastewater or possibly concentrate valuable metals from dilute solutions.[22,23] Filip et al.[23] found that when algae grown in a sewage lagoon were mixed with heavy metal solutions and subsequently dewatered by an intermittent sand filter, 98 percent of the copper in solution and 100 percent of the cadmium had been removed. Kerfoot and Jacobs[24] reported rapid uptake of cadmium by algae used in the first stage of a tertiary treatment system.

Typically, algae is contacted with the influent metals-containing wastewater in an aerated lagoon. The lagoon is usually a lined, flat-bottom pond enclosed by earthen dikes.[15] Oxygen transfer between the air and water is accomplished through algae photosynthesis, although platform-mounted mechanical aerators can be used to enhance transfer. Influent wastewater enters near the center of the lagoon and effluent discharges at the windward side.

Advantages of this type of system relative to previously mentioned biological processes include lower capital and operating costs. In addition, operational flexibility is increased since the effluent flow can be regulated. Disadvantages include extensive physical space requirements, poor industrial waste treatment capacity, and seasonal performance variations. Table 11.1.3 presents the major design parameters and typical values for algae lagoon processes for aerobic and facultative systems, with and without supplemental mechanical aeration.

11.1.4 Pretreatment and Post-Treatment Requirements--

Industrial influents to biological waste treatment plants are often characterized by periodic changes in waste volume, strength, and composition, all of which can have a detrimental impact on maintenence of desirable conditions. Flow equalization can be used to lessen the chance for system upset by dampening changes in waste quantities and qualities. Similarly, concentrated sludge discharges can be mixed with the feed to maintain constant solids concentration.

TABLE 11.1.3. EMPIRICAL DESIGN CRITERIA FOR WASTE STABILIZATION LAGOONS

Parameter	Aerobic	Facultative	Aerated	
			Aerobic	Facultative
Flow regime			Completely mixed	Mixed surf. layer
Lagoon size (acres multiples)	10 acre multiples	2-10 multiples	2-10 multiples	2-10 multiples
Operation	Series or parallel	Series or parallel	Series or parallel	Series or parallel
Hydraulic retention time (days)	10-40	7-30	3-20	7-20
Depth (ft)	3-4	3-6	6-20	3-8
Hydraulic loading (in./day)	3-5	0.5-1.5	-	-
BOD_5 loading:				
(lb/day/acre)	60-120	15-50	20-400	-
(lb BOD/day/1,000 ft^3)	-	-	-	-
Optimum temperature (°C)	20	20	20	20
Temperature range (°C)	0-40	0-50	0-40	0-50
BOD_5 removal efficiency (%)	80-95	80-95	80-95	80-95
Algal concentration (mg/l)	80-200	40-160	-	-
Coliform removal %	>99	>99	-	-
Algae	(0.4-1.2)$(BOD_5)_i$	(0.2-0.8)$(BOD_5)_i$	(0.02-0.1)$(BOD_5)_i$	(0.02-0.1)$(BOD_5)_i$
Microorganisms	(0.2-0.5)$(BOD_5)_i$	(0.2-0.5)$(BOD_5)_i$	(0.2-0.5)$(BOD_5)_i$	(0.2-0.5)$(BOD_5)_i$
Other	Low	(0.1-0.4)$(SS)_i$	(1.1-1.4)$(SS)_i$	(0.1-0.4)$(SS)_i$
Effluent BOD_5 (mg/l):				
Soluble BOD_5	(0.02-0.1)$(BOD_5)_i$	(0.02-0.1)$(BOD_5)_i$	(0.02-0.1)$(BOD_5)_i$	(0.02-0.1)$(BOD_5)_i$
Insoluble BOD_5	(0.3-1.2)$(SS)_e$	(0.3-1.0)$(SS)_e$	(0.5-0.8)$(SS)_e$	(0.3-0.8)$(SS)_e$
Typical effluent quality, (mg/l):				
BOD_5	15-40	15-40	20-70	20-70
SS	25-50	25-50	-	-
pH	6.5-10.5	6.5-9.0	6.5-8.5	6.5-8.5
Oxygen source	Algae	Algae	Aerators	Aerators
Aerator design goals	-	-	Aeration plus mixing	Aeration

i = influent, e = effluent.

Source: Reference 15.

Where the equalization basins are not on-line with the continuous waste flow, special overflow weir or sensor-actuated flow gates must be provided for temporary diversion of flows to the basin structure. Where the basin is used for protecting the on-line biological treatment processes, methods must be provided for anticipating qualitative changes so that the appropriate waste volume can be diverted before the normal treatment scheme can be restored.

The acidity or alkalinity (pH) of the waste stream introduced to the bioreactor must be maintained within a specific range to preserve microbial populations. To accomplish this, different industrial waste streams may be combined for treatment based on their neutralizing effects, or chemicals may be purchased and added to the influent wastes. In the former situation, lower operational costs are realized (no chemicals must be purchased), but maintenence of pH is dependent on the consistency of each component stream. The stockpiling of chemicals such as sulphuric acid, caustic soda or lime will increase chemical costs, but will also provide the capability for responding to variations in waste stream characteristics or flows by adjusting chemical additions. Neutralization of highly concentrated waste streams may be most effectively achieved before they are mixed with other, more dilute waste streams.

Other pretreatment techniques may also be practiced in order to enhance the biodegradation of problematic waste streams. The use of cooling towers should be considered as an effective means of enhancing biodegradation. If inlet wastes are not cooled to at least 40°C to 45°C, they may adversely affect the microorganisms in the bioreactor. Some technologies, such as solvent extraction, are best applied to individual process waste streams before they are combined with other industrial waste streams prior to treatment. Other techniques applicable to single or mixed waste flows include: reverse osmosis, chemical precipitation, evaporation, ion exchange, distillation, resin adsorption, and gravity separation. Powdered activated carbon has been shown to be effective in adsorbing and attenuating compounds thereby limiting the toxic effects of concentrated wastes.

Post-treatment of biological residues containing heavy metals are often restricted by the presence of these metals. For example, high metal levels can result in air pollution, ash disposal, and mechanical operating problems during sludge incineration (see Section 12). The presence of cadmium and other metals at excessive levels can prevent the sludge from being disposed via land application.[25] Depending on the level of heavy metals accumulated within the sludge product, the most likely method of disposal would be solidification followed by landfilling. Note that these sludges are unlikely to contain metals at concentrations which would prohibit them from being land disposed under the 1984 HSWA regulations. Alternatively, extraction for metals removal or incineration may be the most viable post-treatment methods.

11.2 PROCESS PERFORMANCE

Numerous research studies have been conducted to investigate the toxic effects of heavy metals on conventional biological treatment processes. The following is a brief summary, by metal type, of the adaption of biological systems to heavy metals removal. This is followed by a discussion of the effects of synergism and recent developments concerning the use of novel organisms for biological treatment. While the available literature emphasize activated sludge treatment, anaeorbic and algal systems have been increasingly explored in recent years and are also discussed.[26]

11.2.1 Zinc

The percent removal of zinc in activated sludge treatment is normally very good compared to other metals. Typical values range from as low as 22 percent to as high as 68 percent, averaging better than 50 percent. Both soluble and insoluble zinc is removed mainly in the aeration basin absorbed by the microbial floc. Pilot plant studies on activated sludge treatment[27] showed zinc removals ranging from 74 to 95 percent at concentrations ranging from 2.5 to 20 mg/L. BOD removals for these zinc concentrations were only slightly affected. It is not expected that municipal plants achieve these removals, but it is interesting to note that the pilot studies have no supernatant recycle.

Anaerobic digestor operation has been found to be tolerant of zinc influent levels up to 20 g/L.[27] Zinc-cyanide complexes can cause digestor problems if the digestor is not previously acclimated to low cyanide levels. An algal system using *Chlorella pyrevoidosa* was reported to tolerate up to 10 mg/L zinc over a 24-hour period. However, removal efficiency was poor (29 percent) and 103 mg/L of *Chlorella* was required.[28]

11.2.2 Copper

In municipal treatment plants, copper is often encountered as copper sulphate or copper cyanide complex since these are common plating wastes. A field survey of treatment plants[29] showed influent copper values ranging from 0.2 to 6.8 mg/L. While copper removals of only 37.2 percent were obtained at a full-scale facility (the Ukima Treatment facility in Japan)[30], results in Table 11.3.1 from pilot plant studies[31] show 50 to 75 percent removals are obtainable when copper is provided as copper cyanide. Of interest in this study are the low Cu residuals obtained at 1.2 mg/L Cu feed levels. It was also found that 1 mg/L Cu actually increased sludge settleability leaving BOD and COD removals essentially unchanged (Table 11.3.2). In another study, an algal system using Nostoc muscorum on influent copper concentration of 10 mg/L reported a 48 percent reduction over a 24-hour period.

These results tend to support the theories of Wood et al.[1] concerning the growth of filamentous organisms which can enhance metal removal through adsorption. Sludge feed doses of copper sulfate were largely absorbed by the activated sludge and then released slowly, minimizing toxic effects on the microorganisms.

11.2.3 Nickel

Nickel removals by activated sludge were found by McDermott, et al.,[32] to be roughly 30 percent for influent concentrations ranging from 2.5 to 10 mg/L. The BOD removal efficiency was reduced only 5 percent at these concentrations, however, a sludge dose at 200 mg/L was found to have significant adverse effects on system operation. Anaerobic digestion was found to reduce soluble Ni content of sludge to a constant level of 8 to 14 mg/L, regardless of the initial feed levels. However, these results were contrary to the findings of other investigators[33].

Another experiment tested six strains of algae and Eugleve sp. for their ability to bioaccumulate nickel.[4] The researchers found that, at pH 8.0, algae could tolerate up to 7.8 mg/L of nickel before cell lysis would occur.

TABLE 11.2.1. FATE OF COPPER FED AS COPPER CYANIDE COMPLEX IN ACTIVATED SLUDGE TREATMENT

Type of check sample	Location of check sample	Copper in sewage feed (mg/L) 0.4	1.2	2.5	5
Copper fed found in outlet (%)	Primary sludge	--	12.5	10.7	7
	Excess activated sludge	--	43.3	25.6	23
	Final effluent	43	25.1	43.3	50
	Unaccounted for	--	20.0	20.0	20
Efficiency of copper removal (%)		57	75	57	50
Soluble copper in primary effluent (mg/L)	Total	0.22	0.19	--	--
	Reactive	--	--	--	2.56
Soluble copper in effluent (mg/L)	Total	0.12	0.10	--	--
	Reactive	--	--	0.67	0.92

Source: Reference 31.

TABLE 11.2.2. EFFICIENCY OF ACTIVATED SLUDGE TREATMENT OF SEWAGE-CONTAINING COPPER FED CONTINUOUSLY

Copper (mg/L)	BOD removal, average (%)	COD removal, average (%)
0	95	85
0.1	95	85
1.2	93	84
2.5	91	85
5.0	89	76
10.0	88	69

Source: Reference 31.

11.2.4 Lead

Lead removals obtained by Brown, et al.,[34] were on the order of 55 percent for secondary plants. Lead removal was found to be enhanced by longer settling times and larger floc size in the activated sludge process. Lead removals using Anabaena Flos-Aquae investigated by Sloan et al.[28], were on the order of 65 percent when initial lead concentrations were 4.0 mg/L.

11.2.5 Cadmium

Cadmium removals are normally not very high in biological treatment, partly due to the low concentrations usually encountered. Low removals of cadmium can be attributed to its high solubility, ability to form other complexes at the pH of sewage, and competition from other metal ions in the influent. However, when an influent concentration of 2.0 mg/L of cadmium was contacted with the algae Anabaena Flos-Aquae, a removal efficiency of 70 percent was reported over a 24-hour period.[28]

11.2.6 Chromium

Chromium removals by operating activated sludge plants are generally around 40 to 60 percent.[34,35] Hexavalent chromium is normally reduced to the trivalent state before it is removed by microbial flocs. However, in a pilot-plant study by Moore, et al.,[2] the prior reduction of Cr^{+6} by means of biological reduction showed approximately 92 percent removal of CR^{+6} at feed levels of 46.5 mg/L. The biological reductor is a complete mix reducing basin, with chromate serving as the principle source of oxygen. The BOD removals of the activated sludge process in conjunction with the biological reductor were around 94 percent, indicating little decrease in efficiency due to presence of the metal. Gas production in the digestor was also not affected by chromium in the sludge, although these results were contrary to those found by other investigators[36].

11.2.7 Synergistic Effects

Barth, et al.,[29] wrote a summary report on the effect of heavy metals on various biological units. Using the data compiled from an activated sludge pilot plant, they concluded that the aerobic biological treatment system could tolerate a total heavy metal concentration of up to 10 mg/L (Cr, Cu, Ni, and Zn), either singly or in combination, with only about a 5 percent reduction of the BOD removal efficiency.[29,37] They further concluded that a small dose of metal could noticeably reduce the treatment efficiency, but this effect diminished with larger doses. Nitrifying microorganisms were also reported to be sensitive to heavy metals. The investigators reported that 5 mg/L of copper in influent sewage was the highest dose that could still allow satisfactory anaerobic sludge digestion. Finally, they reported that limited metals could have beneficial effects; e.g., reducing the degree of sludge bulking problems in aerobic systems. These conclusions have been confirmed in other studies including Dawson and Jenkins,[36] Jenkins,[38] and Tarvin.[39]

11.2.8 Recent Developments

In the past 5 or 6 years, researchers have been working on the development of specific organisms designed primarily for the bioaccumulation of heavy metals. The following is a brief discussion of some of the more promising technologies.

Researchers at the Hebrew University in Rehovot, Israel, have developed a new method for removing metals from wastewater using water ferns.[40] Azolla, a water fern found in Asia, East Africa and Central America, can be used to remove metals such as copper, zinc, chromium, cadmium, nickel, silver, titanium and uranium from industrial waste. It can be grown in settling ponds and, when harvested and dried, used as filtering material in paint and metals-processing plants.

In the United States, a process developed by Kerr-McGee of Oklahoma City, Oklahoma, was used to biologically remove selenium from wastewater.[40] The researchers found that selenium can be removed from uranium-mine wastewater by anaerobic *Clostridium* bacteria. At a scale of 100 gal/d, selenium concentrations have been lowered from 1.6 mg/L to below 0.5 mg/L. The organism is proving to be more effective in this application than ion exchange and reverse osmosis which were also tried.

At New Mexico State University in Las Cruces, Dennis Darnell and his research group are using green algae, <u>Chlorella vulgaris</u>, to recover metals from waste streams at a cost that is a mere 1 percent to 2 percent of the cost of exchange resins which are currently used. The researchers immobilized the algae on silica gel and have run as many as 20 cycles with no decrease in effectiveness. Acidity and salt content of solutions can reportedly be adjusted to retrieve metals selectively. Chromium, silver, mercury, platinum, and other metals have been removed in trials, but gold was found to be the most tightly bound by the algae. More than 90 percent of gold was removed from a test solution, although the inlet concentration was not specified in the reference.[40]

Researchers at Austria's Institute of Microbiology at the University of Innsbruck have discovered that certain fungi and bacteria are able to "filter" silver from dumped waste material and store it in their cells. This discovery could lead to the recovery of precious metals from industrial sewage.[40]

In Sweden, Rolf O. Hallberg has discovered a process for removing heavy metals from wastewaters containing sulfate ions by means of sulfate-reducing bacteria.[41] The bacteria can be any of the known sulfate reducers, including <u>Desulfovibrio</u> and <u>Desulfotomaculum</u>. The bacteria reduce the sulfate to sulfides, producing hydrogen sulfide gas, leaving the heavy metals to precipitate out as sulfides. Two vessels are used, one for culturing the bacteria in a nutrient and the wastewater, and the other for precipitation. Holding time in the culture vessel may be 10 to 40 hours. Aqueous solution of hydrogen sulfide produced in the culturing vessel is fed continuously into the precipitation vessel along with the remainder of the wastewater. The resulting precipitate is flocky and settles easily.

An example presented in a patent for the process used simulated wastewater with a sulfate ion content of about 600 mg/L, 10 mg/l copper, 600 mg/L zinc, and 500 mg/L iron. Unfiltered water from the precipitation vessel contained up to 0.1 mg/L copper, 0.1 mg/L zinc, 10 mg/l hydrogen sulfide, and 10 to 50 mg/L iron. Iron content could be decreased to zero by adjusting pH in the precipitation vessel. Another possible process variation would be to aerate the output to oxidize residual hydrogen sulfide to sulfate, and iron to Fe^{+2} and Fe^{+3} for reuse.[41]

A strain of Pseudomonas fluorescens that reduced chromate ions to a precipitatable form has been found by Lawrence H. Bopp.[41] The strain, designated LB300, reduces Cr^{+6} to Cr^{+3}, which precipitates and is thus removed from the wastewater. The organism can be used to detoxify chromate in a contaminated sewage digester in which microflfora have been killed by chromate-bearing sewage. After detoxification with LB300, normal microflfora growth can be reestablished.

LB300 is resistant to potassium chromate concentrations as high as 2,000 ppm in a minimal salts medium. That is high enough to include most industrial effluent of chromate, such as chrome plating wastes or wastes from chromate ore processing. LB300 can be used under aerobic or anaerobic conditions at temperatures between -4° and +35°C.[41]

11.3 PROCESS COSTS

Widespread use of aerobic biological treatment systems has led to well developed cost estimation procedures basing capital, operational, materials, and labor costs on system capacity. Estimated treatment system outlays can be determined using Table 11.4.1 and Figures 11.4.1 and 11.4.2, although these do not include the additional costs of seed chemostatic organisms to be used in bioaugmented processes. More complete and up-to-date cost information can be found in the EPA publication " Estimating Water Treatment Costs".[42] Although the breadth of this document prevents its inclusion in this section, the data presented here do show the relative costs and scaling factors used for various cost elements.[43,44] All costs have been updated to 1987 dollars, using the Chemical Engineering Index.[45]

Standardized cost data for anaerobic treatment systems were not found. An example of a modern anaerobic system is the "Celrobic" process developed by Celanese.[44] In 1983, a 1.08 million gallon/day waste stream with an influent COD of 3.3 g/L, incurred outlays of $8,100,000 in capital costs and $400,000 in annual operating costs (1982 dollars). This plant was expected to produce 220 million cubic feet of methane gas annually which considerably reduced its net annual operating costs.

TABLE 11.3.1. ESTIMATED CAPITAL COST FOR WASTEWATER TREATMENT UNITS

Treatment unit	Parameter	Model cost (1987 dollars)
Raw wastewater pumping	Capacity (mgd)	$C = 6.5 \times 10^3 (mgd)^{1.0}$
Screening, grit removal and flow measurement	Capacity (mgd)	$C = 67.5 \times 10^4 (mgd)^{0.62}$
Equalization	Volume (mg)	$C = 18.0 \times 10^4 (mg)^{0.52}$
Primary sedimentation or secondary clarification	Surface area (in 1,000 ft^2)	$C = 7.0 \times 10^4 (A)^{0.88}$
Aeration-basin	Volume (in 1,000 ft^3)	$C = 10.5 \times 10^3 (V)^{0.79}$
Aeration-diffused air system	Blower capacity (in 1,000 cfm)	$C = 22.5 \times 10^4 (cap)^{0.72}$
Aeration-surface	Capacity (horsepower)	$C = 2.5 \times 10^3 (hp)^{0.89}$
Trickling filter	Media volume (in 1,000 ft^3)	$C = 8.5 \times 10^3 (V)^{0.84}$
Recirculation pumping	Capacity (mgd)	$C = 6.3 \times 10^4 (mgd)^{0.70}$
Sludge digesters and buildings	Sludge volume (in 1,000 ft^3)	$C = 3.5 \times 10^4 (V)^{0.64}$
Lagoon	Volume (mg)	$C = 18.0 \times 10^4 (V)^{0.71}$
Vacuum filtration	Filter area (ft^2)	$C = 14.8 \times 10^3 (A)^{0.67}$
Centrifugation	Capacity (gpm)	$C = 8.3 \times 10^4 (gpm)^{0.54}$
Incineration	Dry solids capacity (lb/hr)	$C = 3.5 \times 10^4 (cap)^{0.56}$

Source: References 43 and 44.

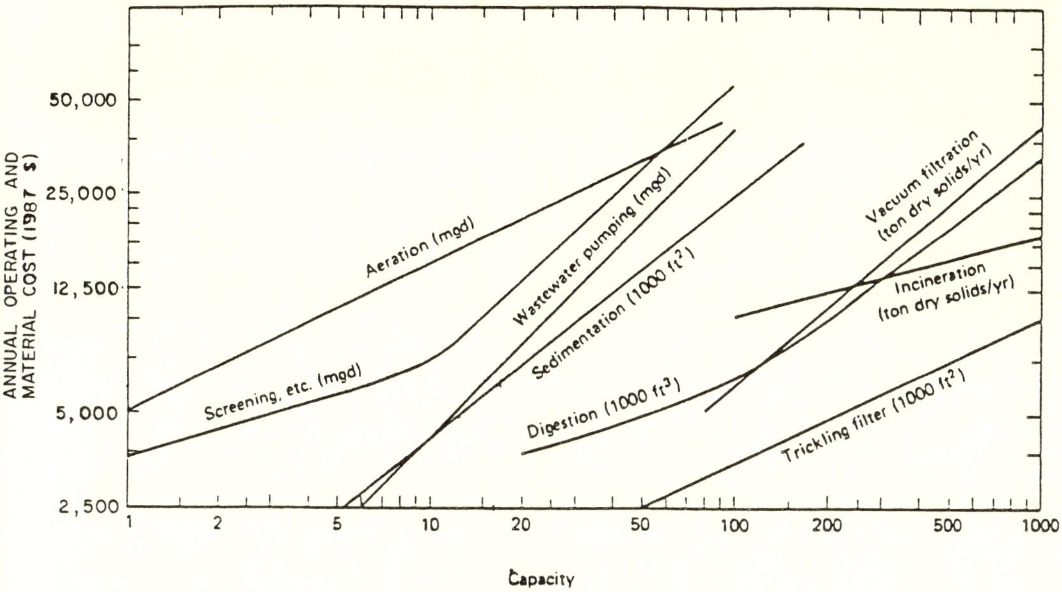

Figure 11.3.1. Estimated annual operating and materials costs as a function of wastewater treatment facility capacity.

Source: References 43 and 44.

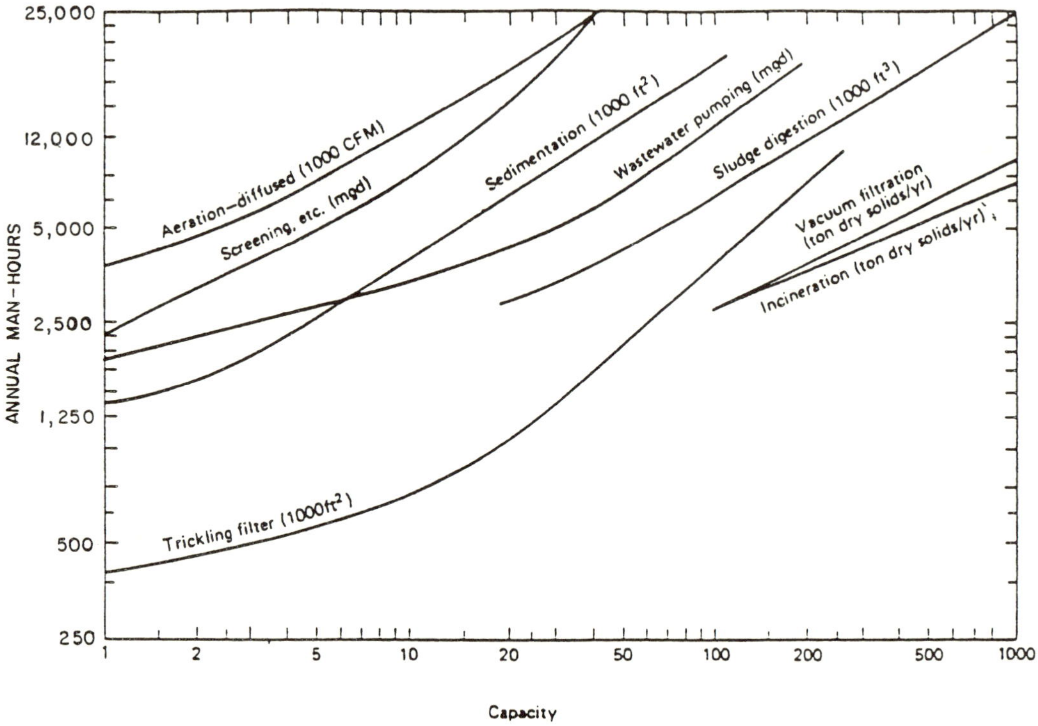

Figure 11.3.2. Estimated annual man-hours needed for wastewater treatment facility operation.

Source: References 43 and 44.

Actual treatment costs will depend on specific characteristics of the waste stream. Pertinent data needed for cost estimation are: waste stream volumetric rate, organic compound constituents and concentrations, other waste characteristics such as influent BOD, COD, or level of toxins, treatment design, and overall treatment objectives.

11.4 OVERALL STATUS OF BIOLOGICAL TREATMENT

A large number of companies exist that specialize in the design and construction of biological treatment systems. Aerobic systems are the most readily available, and their design and operation are complex, but manageable. The total number of facilities using some sort of aerobic biological treatment is over 2,000.[46] Conversely, the number of companies offering expertise in bioaugmentation and anaerobic treatment is relatively small, but this segment is expected to grow rapidly.

Biological treatment of metals using conventional equipment and acclimated strains is typically only capable of treating combined heavy metal influents of 10 mg/L.[37] While improvements in process tolerance for inorganic priority pollutants is encouraging, most advancements are still in the developmental stage and have yet to be widely applied.[41]

REFERENCES

1. Wood, D.K., et al. Trace Elements in Biological Treatment with Specific Reference to the Activated Sludge Process. Proc. 29th Industrial Waste Conference, Part II, Purdue University. 648. May 1974.

2. Knauss, H.J., and S.W. Porter. The Adsorption of Inorganic Ions by Chlorella Pyrenoidosa. Plant Physiology, 29(3). May 1954.

3. Wood, J.M. Proc. Nobel Conference Inorganic Biochemistry, Chemical Service, 21, 155-160. 1983.

4. Wang, H.K., and J.M. Wood. Bioaccumulation of Nickel by Algae. Environmental Science Technology, Vol. 18, No. 2. 1984.

5. Dugan, P.R., and H.M. Dickrum. Removal of Mineral Ions from Water by Microbially Produced Polymers. Proc. 27th Industrial Waste Conference, Purdue University, Lafayette, Indiana. 1019. May 1972.

6. Unz, R.F. Microcultures Studies of Activated Sludge Bacterial Zoogloeas. Abstr. Ann. Amer. Soc. Microbiol. 22(32). 1972.

7. Hayes, T.D. et al., Department of Agricultural Engineering, Cornell University. Heavy Metal Removal from Sludges Using Combined Biological/Chemical Treatment. 34th Industrial Waste Conference, Purdue University. 1979.

8. Gould, M.S., and E.J. Genetelli. Heavy Metal Complexation Behavior in Anaerobically Digested Sludges. Water Research, 12, 505. 1978.

9. Eckenfelder, W.W. Water Quality Engineering for Practicing Engineers. Barnes and Noble, New York, NY. 1970.

10. Doyle, J.J., Marshall, R.T., and W.H. Pfander. Effects of Cadmium on the Growth and Uptake of Cadmium by Microorganisms. Applied Microbiology, 29(4), 562. 1975.

11. Hayes, T.D., and T.L. Theis. The Distribution of Heavy Metals in Anaerobic Digestion. Journal Water Pollution Control Federation, 50(1), 61. 1978.

12. Patrick, F.M., and M. Loutit. Passage of Metals in Effluents, Through Bacteria to Higher Organisms. Water Research, 10, 333. 1976.

13. Chang, S.H., et al. Effects of CD (II) and Cu (II) on a BioFilm System. Journal of Environmental Engineering, Vol. 112, No. 1. February 1986.

14. Sicrp, F., and F. Fransemier. Copper and Biological Sewage Treatment. Vom Wasser, 7. 1933.

15. Chillingworth et al., Alliance Technologies Corporation. Industrial Waste Management Alternatives Assessment for the State of Illinois. Volume IV. February 1981.

16. Benjes, H.H., Jr. Handbook of Biological Wastewater Treatment. Garland STPM Press, New York. 1980.

17. Clark, Viessman, Hammer. Water Supply and Pollution Control. Harper & Row. 1977.

18. Gurnham, C.F. Industrial Wastewater Control. Academic Press, New York. 1965.

19. Shreve, R.N., and J.A. Bunk, Jr. Chemical Process Industries. McGraw-Hill, New York. 1977.

20. Hinrichs, D.J. Inspector's Guide for Evaluation of Municipal Wastewater Treatment Plants. EPA-430/9-79-010, U.S. Environmental Protection Agency. April 1979.

21. U.S. EPA Process Design Manual for Sludge Treatment and Disposal. EPA-600/1-74-006. October 1974.

22. Hasset, J.M., Jennett, J.C., and J.E. Smith. Heavy Metal Accumulation by Algae. In: Contaminants and Sediments, R.A. Baker Editor, Ann Arbor, Ann Arbor Science Publishers, Inc. 1980.

23. Filip, D.S. Peters, T., Adams, V.D., and E.J. Middlebrooks. Residual Heavy Metal Removal by an Algae-Intermittent Sand Filtration System. Water Resources. 13:305. 1979.

24. Kerfoot, W.B., and S.A. Jacobs. Cadmium Accrual in Combined Wastewater Treatment Aquaculture System. Environmental Science & Technology. 10(7):662. 1976.

25. Patterson, J.W., and Hao, S.S. Heavy Metals Interactions in the Anaerobic Digestion System. 34th Industrial Waste Treatment Conference, Purdue University. 1979.

26. Russell, H.H. et al. Impact of Priority Pollutants on Publicly Owned Treated Works Processes: A Literature Review. 37th Industrial Waste Treatment Conference, Purdue University. 1982.

27. McDermott, G.N., et al. Zinc in Relation to Activated Sludge and Anearobic Digestion Processes. Proc. 17th Industrial Waste Conference, Lafayette, Indiana, May 1962. Engineering Ext. Series 112, Engineering Bulletin, Purdue University, 47(2):461. 1963.

28. Sloan, F.J. et al., Clemson University. Removal of Metal Ions from Wastewater by Algae. 38th Industrial Waste Conference, Purdue University. 1983.

29. Barth, E.F., English, J.N., Salotto, B.V., Jackson, B.N., and M.B. Ettinger. Field Survey of Four Municipal Wastewater Treatment Plants Receiving Metallic Wastes. W.P.C.F. Journal 37:1101. August 1965.

30. Kondo, J. Some Problems on the Joint Treatment of Industrial Wastes and Sewage in the Ukima Treatment Plant. Water Research 7: 375-384. 1973.

31. McDermott, G.N., et al. Effects of Copper on Anerobic Biological Treatment. W.P.C.F. Journal, 35(2): 227. February 1963.

32. McDermott, G.N., et al. Nickel in Relation to Activated Sludge and Anaerobic Processes. W.P.C.F. Journal, 37(2): 163. February 1965.

33. Lawrence, A.W., and P.L. McCarty. W.P.C.F. Journal, 37(3): 392. March 1965.

34. Brown, H.G., Hensley, C.P., McKinney, G.L., and J.L. Robinson. Efficiency of Heavy Metals Removal in Municipal Sewage Treatment Plants. Environmental Letters, 5(2): 103-114. 1973.

35. Moore, W.A., et al. Effects of Chromium on the Activated Sludge Process. W.P.C.F. Journal, 33: 54. January 1961.

36. Dawson, P.S., and S.H. Jenkins. The Oxygen Requirements of Activated Sludge Determined by Manometric Methods. Sewage and Industrial Wastes, 22: 490. 1950.

37. Barth, E.F., Salotto, B.V., McDermott, G.N., English, J.N., and M.B. Ettinger. Effects of a Mixture of Heavy Metals on Sewage Treatment Processes. Proc. 18th Industrial Wastes Conference, Purdue University. p. 616. May 1963.

38. Jenkins, S.H. Trade Waste Treatment. The Institute of Sewage Purification, 28: 1371. 1956.

39. Tarvin, D. Metal Plating Wastes and Sewage Treatment. Sewage and Industrial Wastes, 28: 1371. 1956.

40. Greene, R. Biotechnology and Pollution Control. Chemical Engineering. March 4, 1985.

41. Technical Insights, Inc. New Methods for Degrading/Detoxifing Chemical Wastes. Emerging Technologies, No. 18. 1986.

42. U.S. EPA. Estimating Water Treatment Costs. U.S. Environmental Protection Agency, Municipal Environmental Research Laboratory, Cincinnati, Ohio. EPA-600/2-79-162(a,b). August 1979.

43. Black & Veatch. Estimating Costs and Manpower Requirements for Conventional Wastewater Treatment Facilities. U.S. Government Printing Office, Washington, D.C. 1971.

44. Sundstrom, D.W., and H.E. Klei. Wastewater Treatment. Prentice-Hall, Inc., Englewood Cliffs, N.J. pp. 439. 1979.

45. Chemical Engineering. Chemical Engineering Plant Cost Index. April 27, 1987.

46. U.S. EPA. Background Document for Solvents to Support 40 CFR Part 268 Land Disposal Restrictions. Volume II. January 1986.

47. U.S. EPA. Selected Biodegradation Techniques for Treatment and/or Ultimate Disposal of Organic Materials. EPA-600/2-79-006. March 1979.

12. Thermal Destruction/Recovery Processes

Processes included here may be used to treat hazardous wastes containing many of the toxic heavy metals. Processes such as incineration and the "pyrometallurgical" processes such as calcination or smelting, and recovery processes such as evaporation and crystallization, may be used to separate the metal compounds from the other waste constituents. The application of these processes will allow either their recovery (e.g., by crystallization) or concentration and ultimate disposal through techniques such as encapsulation of incineration ash that are most appropriate for concentrated wastes. Since the metal compounds cannot be "broken down" in the same manner as, for example, an organic compound may be broken down through pyrolysis to simpler compounds or oxidized to form CO_2 and water, the usefulness of the thermal destruction processes as a means of concentrating metal wastes strongly depends upon the nature of the other waste constituents. To a lesser extent, the applicability of a thermal technology may also depend upon the volatility of the metals, and the physical form of the waste.

In this chapter, the available technologies to be discussed are as follows:

- Incineration;
- Calcination/Smelting (Pyrometallurgy);
- Evaporation; and
- Crystallization.

The last two processes are recovery processes which are more physical than thermal in nature. Evaporation depends upon removal of a volatile, nonmetallic component of the waste, usually water, to concentrate metallic salts. Crystallization, often used in conjunction with evaporation, involves cooling a solution to reduce the solubility of the metal salts and bring about precipitation.

While many hazardous wastes containing heavy metals may be good candidates for thermal destruction processes, as classified above, there are some which should definitely never be handled in this manner. Such wastes include:

- Organoarsenic compounds, which when combusted yield arsenic III oxide, which has a fairly low boiling point. To effectively capture this material from combustion vent gases, the gases need to be cooled to near ambient temperatures before dry or wet collection of particulate arsenic III oxide can be satisfactorily conducted.

- Selenium compounds; the selenium dioxide generated is a low boiling compound and which like arsenic oxide will be difficult to contain.

- Wastes containing chlorides and chromates, which during incineration may generate chromyl chloride (CrO_2Cl_2), which has boiling point of only 117°C. For this substance, the same problems as those discussed for organoarsenic compounds are encountered.

- Explosives; explosive mixtures should not be handled by incineration for obvious reasons.[1]

In general, there are several disadvantages inherent in using thermal destruction processes to handle metal-bearing wastes, including the following:

- All thermal destruction processes will create air emissions, in many cases including heavy metal particles or vapors;

- Thermal destruction processes will often form chemical by-products such as hydrochloric acid which may be damaging to the systems themselves, or which, like dioxins, may present serious environmental hazards which are as significant as those posed by the toxic heavy metal wastes;

- Thermal destruction processes require the generation or removal of heat energy through fuel burning or consumption of electrical power and therefore may not be economically attractive.

In this chapter, the amenability of metal wastes to the specific processes will be reviewed, focusing on specific performance test results to develop an understanding of which physical/chemical parameters dictate the acceptability of each process.

12.1 INCINERATION

Incineration is the process of applying sufficient heat energy and oxyger to cause the oxidation and/or pyrolysis of compounds such that they are broker down to form more "basic" chemical species such as water, CO_2, HCl, elemental metals or metal oxides, etc. Incineration processes are often considered an attractive management alternative for hazardous wastes because they possess many advantages over other technologies, including the following:

- Thermal destruction by incineration provides the ultimate disposal of hazardous wastes, minimizing future liability from land disposal;
- Toxic components of hazardous wastes can be converted to harmless or less harmful compounds;
- The volume of waste material may be reduced significantly by incineration; and
- Resource recovery (i.e., heat value recovery) is possible through combustion.

Unfortunately, because metals are not destroyed by incineration and have no heat recovery value there is little incentive to treat most metal bearing wastes by incineration. An exception might be an organometallic compound containing waste such as tetraethyl lead or a cyanide complex which is highly toxic and not readily treated by more conventional methods.

Several problems must be faced when incinerating metal wastes. A primar consideration is the extent to which air emissions of toxic heavy metal particles or vapors will be generated. Certain metals and their oxide such a mercury, lead, selenium, and arsenic are volatile, particularly at the elevated temperatures of incineration. A significant percentage of the input of these relatively volatile metals will be emitted as a vapor or as fine particulates which are difficult to control. A second problem involved in th incineration of such wastes is the generation of an incinerator ash or sludge containing metals or metal oxides, which will require safe disposal. Third, wastes containing high concentrations of noncombustible materials require greater energy input via auxiliary fuel combustion, thus increasing processing costs significantly. Finally, such wastes may be difficult to handle in certain incineration systems. Liquid injection incinerators may not be used, for example, should the solids content of the waste be such that the injectors will be come clogged.

Although numerous studies of incinerator performance have been conducted in which organic wastes containing metals were burned, the available data are limited in content relative to the effect of metals on combustion. Based on the available data, it does not appear as though the presence of metals in small concentrations will hinder the destruction of organics. The data do show, however, that certain metal species may present more of a concern relative to potential air emissions than do others.

Incineration facilities permitted to operate by EPA under RCRA are required to achieve at least a three tiered environmental standard:[2]

1. They must achieve a destruction and removal efficiency (DRE) of 99.99 percent for each principal organic hazardous constituent (POHC);

2. They must achieve a 99 percent HCl scrubbing efficiency or emit less than 4 lbs/hr of hydrogen chloride; and

3. They must not emit particulate matter in excess of 0.08 grains/dscf, (0.18 grams/dscm) corrected to 7 percent oxygen.

The HCl and particulate matter standards will be exceeded by most uncontrolled incinerators burning even relatively clean wastes. For example, the HCl standard of 4 lb/hr will be exceeded by units larger than 3.8 x 10^6 Btu/hr burning a 19,000 Btu/lb waste containing 2 percent chlorine. The ash content corresponding to the 0.18 g/dscm particulate emission standard would be about 0.3 percent assuming incineration of a similar high Btu fuel at 7 percent oxygen.[3] Thus, in order to comply with the particulate emission standard, control devices will be required to reduce air emissions when burning a waste containing higher concentrations of metals For example, a control device of about 90 percent efficiency will be needed to achieve the standard for a waste containing 3 percent metal. The exact value will depend upon the heating value and composition of combustibles in the waste (and auxiliary fuel) and the fraction of the metal input emitted with the flue gas.

According to Oppelt,[4] more than half of the incinerators operating in 1981 used no air pollution control system at all. These uncontrolled incinerators would not be suitable for the incineration of metal bearing wastes. Standards for emissions of toxic air pollutants such as toxic metals may also limit incineration of metal bearing wastes.

12.1.1 Process Description

Hazardous waste incineration technologies range from those with widespread commercial application and many years of proven effective performance, to those currently in development. As many as 67 companies may be involved in the design and development of hazardous waste incinerators, with more expected as limitations on land disposal of hazardous wastes increase.[5]

As mentioned previously, there are several incineration technologies which have become established commercially as the primary options available for the incineration of hazardous wastes. These technologies have been demonstrated extensively for a wide range of hazardous wastes. They comprise about 80 percent (by number) of the U.S. market.[6,7,8] They include:

- Liquid injection incinerators;
- Rotary kilns;
- Fluidized-bed incinerators;
- Fixed hearth incinerators, particularly the starved air or pyrolysis type units; and
- Multiple hearth incinerators.

Liquid injection (64 percent), rotary kiln (12.3 percent), and fixed hearth incinerators (18.5 percent) are the most widely used for the disposal of hazardous wastes. A detailed discussion of the design and operation of these systems can be found in Reference 3, or in the open literature.

The relative ease with which hazardous wastes containing toxic heavy metals or metal compounds may be incinerated has been studied within the context of a general study of hazardous waste incinerability conducted by EPA.[9] A summary of the "incinerability" ratings developed by EPA for such wastes is presented in Table 12.1.1. As shown, almost all of the metal-bearing hazardous wastes are perceived to be "poor" candidates for incineration. The results of a study analyzing "incineration risk" conducted for EPA in 1981 by ICF,[10] however, showed that 35 of a total of 139 hazardous waste streams currently incinerated (25 percent) contain toxic metals or metal salts. The metal-containing wastes do not, however, account for a significant percentage of the total volume of hazardous waste incinerated. In fact, the study showed that approximately 90 percent of the volume of wastes incinerated are characterized as D001 and D002 wastes, about which little is known. It is not expected that such waste would contain an appreciable amount (if any) of heavy metal.

The primary characteristic of metal-bearing hazardous wastes which might limit incinerability is the concentration of the metal itself. Other limiting factors relate to the characteristics of the other waste constituents, e.g., organic solvents. A detailed discussion of those components is presented in Reference 3. Most commercial incineration facilities surveyed by Alliance indicated metals concentration limits in the 1 to 500 ppm range.[11-20] Most are limited to such low feed concentrations by air emission regulations or effluent guidelines. Thus, although a higher metals concentration may not necessarily render a waste less combustible, the air emissions and/or ash or effluents generated may preclude the incineration of a particular waste.

As a result of this limitation, many commercial facilities blend metal-bearing wastes with other compatible waste streams in order to achieve a proper concentration level. Blending may also serve to enhance the combustibility of the waste stream (i.e., raise the heat value). No other form of pretreatment appears to be used by the commercial incineration industry.

TABLE 12.1.1. METAL WASTE INCINERABILITY

Waste code	Description	Rating[a]
D004	Arsenic	
D005	Barium	
D006	Cadmium	
D007	Chromium	
D008	Lead	
D009	Mercury	
D010	Selenium	
D011	Silver	
F006	Wastewater treatment sludges from electroplating	Poor
F007	Spent bath solutions from electroplating	Poor
F008	Planting bath sludge from electroplating	Poor
F009	Spent stripping and cleaning solutions from electroplating	Poor
F010	Sludge from metal treating	Poor
F011	Cleaning solutions from metal treating	Poor
F012	WW trtmt. sludge from metal treating	Poor
F019	Conversion coating sludge from metal treating	Poor
K002	Chromium pigment production sludges	Poor
K003	Chromium pigment production sludges	Poor
K004	Chromium pigment production sludges	Poor
K005	Chromium pigment production sludges	Poor
K006	Chromium pigment production sludges	Poor
K007	Chromium pigment production sludges	Poor
K008	Chromium pigment production sludges	Poor
K021	Spent antimony catalyst	Poor
K031	Cacodylic acid production by-products	Poor
K046	Sludge from lead detonator production	Poor
K053	Chromium trimmings from leather tanning	Poor
K054	Chromium trimmings from leather tanning	Poor
K055	Buffing dust from leather tanning	Poor
K056	Screenings from leather tanning	Poor
K057	WW trtmt sludges from leather tannning	Poor
K058	WW trtmt sludges from leather tanning	Poor
K059	WW trtmt sludges from leather tanning	Poor
K060	Lime sludge from coking operations	Poor
K061	Furnace dusts	Poor
K062	Spent pickle liquor	Poor
K063	Lime treatment sludge from steel finishing	Poor
K065	Surface impoundment solids from primary lead smelting	Poor
K066	WW trtmt sludge from primary zinc production	Poor
K067	Electrolytic anode sludge from primary zinc production	Poor
K068	Cadmium plant leach residue from lead smelting	Poor
K069	Lead smelting dusts	Poor

(continued)

TABLE 12.1.1 (continued)

Waste code	Description	Rating[a]
K071	Muds from mercury chloroalkali cell	
K084	Organarsenic production wask	
K086	Slude from ink and pigment manufacturing	
K087	Tars, sludges from coking operations	
K100	Lead processing leachate	
K101	Organoarsenic production waste	
K102	Organoarsenic production waste	
K106	Mercury chloroalkali cell sludge	
P006	Aluminum phosphide	Poor
P010	Arsenic acid	Poor
P011	Arsenic pentoxide	Poor
P012	Arsenic trioxide	Poor
P015	Beryllium dust	Poor
P036	Dichlorophenylarsine	Poor
P038	Diethylarsine	Poor
P065	Mercuric fulminate	Poor
P073	Nickel carbonyl	Poor
P074	Nickel carbide	Poor
P087	Osmium tetroxide	Poor
P092	Phenyl mercuric acetate	Poor
P103	Selenourea	Poor
P107	Strontium sulfide	Poor
P110	Tetraethyl lead	Low
P113	Thallic oxide	Poor
P114	Thallous selenite	Poor
P115	Thallous sulfate	Poor
P119	Ammonium vanadate	Poor
P120	Vanadium pentoxide	Poor
P122	Zinc phosphide	Poor
U032	Calcium chromate	Poor
U136	Cacodylic acid	Poor
U139	Iron dextran	Poor
U144	Lead acetate	Poor
U145	Lead phosphate	Poor
U151	Mercury	Poor
U204	Selenous acid	Poor
U205	Selenium disulfide	Poor
U214	Thallium acetate	Poor
U215	Thallium carbonate	Poor
U216	Thallium chloride	Poor
U217	Thallium nitrate	Poor

[a] Low = Low potential
Poor = Poor potential
Blank = No information provided

12.1.2 Performance of Incinerators Burning Metal-Bearing Hazardous Wastes

Numerous studies have been conducted to assess the effectiveness of incinerators in destroying various hazardous wastes. These studies, and the accompanying available performance data have, however, focused upon the extent to which organic waste constituents are destroyed (i.e., the destruction and removal efficiency, or DRE), and thus in the majority of cases the wastes tested are organic waste streams. While those wastes may contain metals in very low concentrations, few if any could be considered equivalent to the metal-bearing wastes generally considered here. The available data which show the effect of toxic heavy metals on hazardous waste incinerator performance are, therefore, quite limited.

Although the concentrations of metals in the wastes tested are not generally significant,[4] some valuable conclusions may be drawn from an evaluation of metals analysis data derived from several of the available performance tests. Tables 12.1.2 and 12.1.3 present summaries of data obtained from studies conducted by MRI[21] and GCA Corporation.[22] These data indicate the fate of metals introduced with the waste feeds, showing their resultant concentrations in stack emissions and effluents (which include incinerator ash, and control system effluents such as wet scrubber sludges). The data, while obviously not conclusive, suggest that many metals are retained as bottom ash and that the amount of metal in the fly ash which is not caught by the air pollution control systems and is thus emitted from the incinerator stack may be significant (i.e., as high as 10 percent or more). These phenomenon appears to be related to both the concentration of metal in the waste feed and the volatility of the metal species. The ratio of emissions to input were higher for lead, for example, than other metals.

While not shown in the table, it was concluded that in all cases, the organics under study were destroyed well beyond the required limits. On the basis of these studies, therefore, it appears that incineration can be an effective means of managing certain metal-bearing wastes, particularly in its ability to significantly reduce the overall waste volume to be handled, to convert much of the other toxic components of such wastes to harmless or less harmful compounds, and to render a metal-bearing waste more amenable to land disposal.

TABLE 12.1.2. SUMMARY OF METALS ANALYSIS DATA - MASS INPUT AND OUTPUT RATES (G/MIN) FOR TOXIC HEAVY METALS STUDY

Facility	Stream	Metals							
		Arsenic	Cadmium	Chromium	Mercury	Nickel	Lead	Selenium	Thallium
Ross[b]	Input		5.04	20.1	0.33	19.7	71.7		0.65
	Effluent from absorber		23.4	14.4	56.7	6.08	248.0		113.4
	Emissions from stack		0.071	0.187	0.004	0.032	6.93		0.011
Plant B	Input	0.90	0.11	0.84	0.50	0.16	2.1	5.2	0.9
	Control device ash	0.03	0.04	0.1	0.06	0.03	0.9	5.1	0.05
	Emissions	0.035	0.069	0.47	0.037	0.32	1.5	4.5	0.06
American Cyanamid[d]	Input[c]	0.024	0.002	0.555	0.009	0.377	0.002	0.022	0.011
	Emissions		0.005	0.188	0.064	0.188	0.016		0.04
DuPont[b,d]	Input		0.007[e]	0.016[e]		0.155	0.117		
	Emissions[b]	0.049	0.007	0.007[e]	0.01	0.032	0.146	0.17	0.009
TWI[d,f]	Input	0.100	0.141	0.234		0.050	2.51		
	Emissions	0.020	0.044	0.012	0.020	0.004	0.810	0.500	0.040

[a]Where no data are shown, no data were available.

[b]Average for three test runs.

[c]Calculated from concentration data.

[d]No effluent data given.

[e]One test run only.

[f]Average for four test runs.

Source: Reference 21.

TABLE 12.1.3. CONCENTRATIONS OF METALS IN COMBUSTIBLE WASTE FEED AND COMPARISON OF INPUT RATES TO EMISSION RATES

	Average concentration in combustible waste (ug/g)	Average feed rate (mg/min)	Average emission rate (mg/min)	Ratio of emission to input from combustible waste
Arsenic	19.30	88.90	40.10	0.45
Barium	121.00	558.00	56.20	0.10
Beryllium	4.67	21.50	1.31	0.06
Cadmium	4.06	18.70	23.70	1.30
Chromium	166.00	765.00	34.20	0.05
Iron	20,800.00	95,900.00	5,370.00	0.06
Lead	458.00	2,110.00	2,340.00	1.10
Mercury	0.52	2.40	0.02	0.01
Selenium	0.50	2.30	0.82	---

Source: Reference 22.

12.1.3. Costs

The overall costs associated with hazardous waste incineration, whether onsite or at commercial incineration facilities, are high relative to other hazardous waste treatment or disposal methods. Incineration facilities require large capital costs due to the size and complexity of the systems involved, and the requirements associated with the handling of hazardous wastes and their combustion products. Operating costs are high due to the large energy input required, and also due to high raw materials and environmental control costs. Incineration costs are difficult to specify in general, because in each situation the number of factors impacting costs is large. These factors may be classified fundamentally as follows:

- Waste characteristics;
- Facility design characteristics;
- Operational characteristics.

A detailed discussion of these factors is presented in Reference 3.

Costs for commercial incineration of metal-bearing hazardous wastes were obtained by Alliance within a survey conducted for commercial incinerators.[11-20] In general, it can be stated that incineration costs are higher and in certain cases much higher depending upon the type of metals involved and the metal concentrations involved (see Table 12.1.4). It is useful to note that the costs of incinerating such wastes at a commercial facility employing a cement kiln are significantly lower than those for standard incinerators.

12.1.4 Overall Process Status

As discussed in Reference 3, there are a number of companies actively involved in the development, manufacture, and installation of hazardous waste incineration systems. There are also numerous commercial facilities which operate hazardous waste incinerators capable of handling wastes containing toxic heavy metals, up to certain limits. A telephone survey of several such

TABLE 12.1.4. SUMMARY OF INFORMATION OBTAINED IN ALLIANCE SURVEY OF COMMERCIAL HAZARDOUS WASTE INCINERATORS

Type of incinerator	State	Types of metal wastes handled	Limitations	Costs
Fluidized Bed	Colorado	Organic wastes (e.g., refinery wastes), Paint sludges, waste oil. Blend all wastes.	Based on metal concentrations. Two examples: Lead = 500-600 ppm Chromium = 1000 ppm Also somewhat based on volatility of metals (e.g., metal hydrides are more volatile, and they may not choose to handle them due to potential air emission). Will burn wastes containing cyanides up to 2000 ppm CN.	Could not specify. Metal concentration will be directly proportional to cost increase factor.
Liquid Injection/ Rotary Kiln	Illinois	Any except those containing cyanides. Blend all wastes.	Only limitations on lead and mercury, as shown: Lead = if concentration is over 100 ppm, container size is limited to 50 lbs/ container. Mercury = if conc. is over 100 ppm, must limit to 300 mg/container.	Typical cost is \$85/5 lb for lead-bearing sludges. Costs are proportional to metal concentration
Liquid Injection	New Jersey	Organic wastes (no inorganic) No cyanides Blend all wastes.	Based on metal concentration. Examples: Chromium = 100-200 ppm Lead = 100-200 ppm Mercury = 5 ng/kg Cadmium = 5 mg/kg	Could not specify.
Cement Kiln	Ohio	Any. No cyanides. Blend all wastes.	Limitations exist on wastes containing Pb, Bn, Zn, Cr (most prevalent, also have limits on others). General limit to these species is 4000 ppm Can handle up to 10,000 ppm and reduce by blending. Only bulk wastes burred	Based on metal concentraton. Typical costs as follows: Base = \$20/gal 4-600 ppm = \$25-35/gal 6-10,000 ppm = \$35-45/gal

Source: References 11-20.

facilities nationwide was conducted to determine the prevalence of incineration of metal-bearing wastes and the associated incineration costs.[11-20] The results from four facilities in this survey are presented in Table 12.1.4.

Overall, it may be concluded that incineration appears to be a limited and potentially costly alternative for the treatment of hazardous wastes containing heavy metals. The wastes which may be handled in this manner are limited to organic wastes (including organometallic compounds such as cyanides and tetraethyl lead) which contain metals in fairly low concentrations. Most commercial incineration facilities will handle such wastes, but will charge a premium based on metals concentration.

12.2 PYROMETALLURGICAL PROCESSES

12.2.1 Process Descriptions

Most of the pyrometallurgical processes identified for metal waste treatment are classified as "calcination" or "smelting" operations. Calcination processes are generally those which form metal oxides, while smelting produces pure metal. The relationship of these processes within the overall realm of the pyrometallurgical processes may be depicted as shown in Figure 12.2.1. The first step shown in the figure eliminates volatiles from the waste matrix. The oxidation step is similar to incineration, where combustible materials such as organics will be eliminated. From this point on, the various available pyrometallurgical technologies are numerous, and, in some cases, quite different from one another.

Calcination is essentially comprised of Steps 1 and 2 of the flow process depicted in Figure 12.2.1. The product or products of calcination are metal oxides. Metal oxides may be separated for use by further chemical processing or may be disposed of through encapsulation. Generally, if the purpose of treating the metal waste stream is to render it more amenable to land disposal, calcination is conducted. Smelting essentially involves Steps 3 and 4 of the flow diagram. The product of a smelter is the pure metal. The feed to a smelter may be the metal oxides formed through calcination of metal

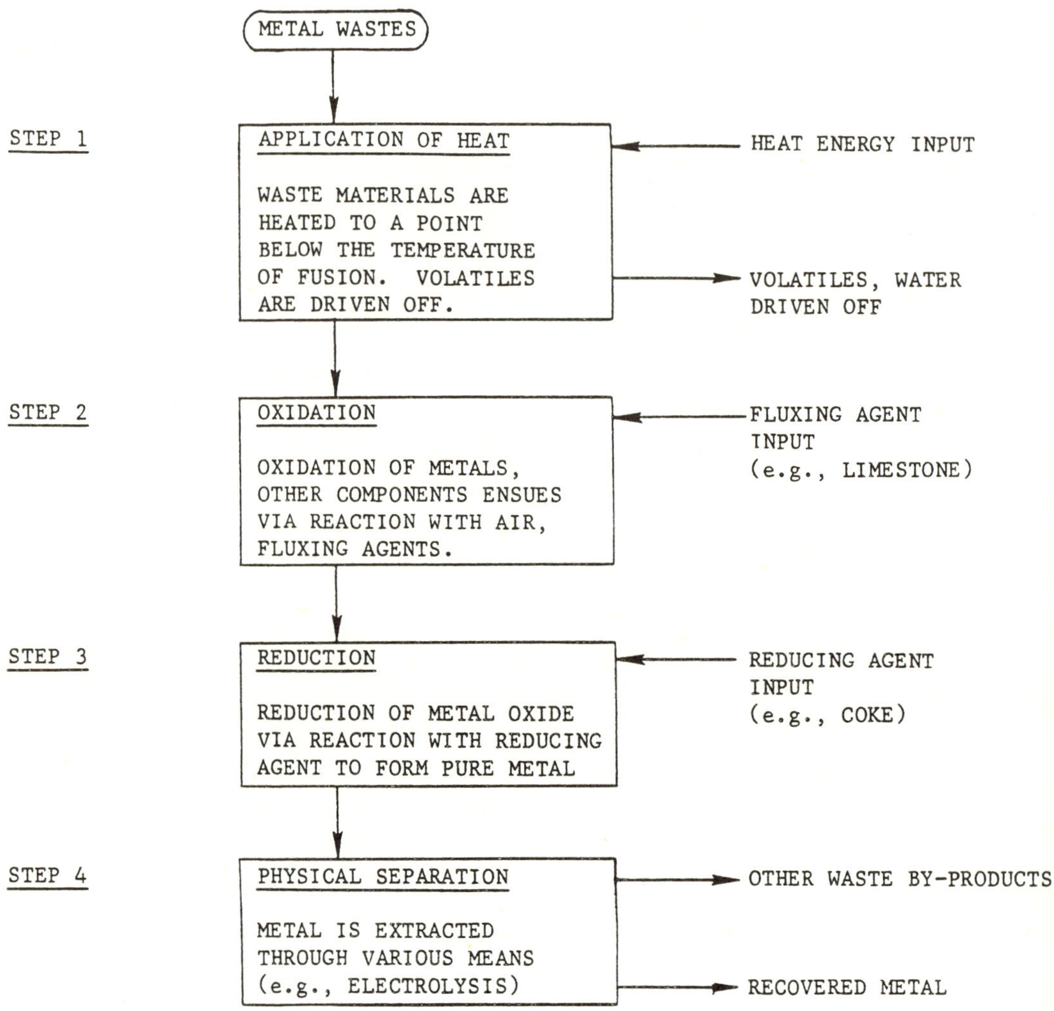

Figure 12.2.1. Flow diagram for pyrometallurgical processes.

wastes, or wastes containing metal oxides. In many smelting operations, coke is employed as a reducing agent. The most common smelting operations involve recovery of iron, lead, or copper.

Usually, the waste matrix cannot be reduced to the metal in a single operation and a preparation process is needed to modify the physical or chemical properties of the raw material. Furthermore, most pyrometallurgical reductions do not yield a pure metal and an additional step, refining, is needed to achieve the chemical purity that is specified for the commercial use of the metal.[24]

Drying and calcination are usually carried out in various types of kilns such as rotary kilns, shaft furnaces, and rotary hearths. Smelting operations are conducted in blast or reverbatory furnaces as described in reports and texts dealing with metal processing.[23-25] Many nonferrous metals can be extracted by reduction smelting: copper, tin, nickel, cobalt, silver, antimony, bismuth, and others. Blast furnaces are sometimes used for the smelting of copper or tin, but reverberatory furnaces are more common for most metals.

One of the newer pyrometallurgical processes to be developed is one which employs the ultra-high temperatures of a plasma arc furnace. Waste dusts from furnace operations may be fed to a plasma burner operating at temperatures as high as 5000°C. The high heat will pyrolyze (break apart) the molecules of the waste mixture. Recovery may then be effected through selectively precipitating metals at their appropriate condensation point. This is a proposed method for handling solid, metallic wastes, particularly those in which a variety of metals are contained, such as dusts from specialty steelmaking furnaces. A flow diagram of one such system, the PLASMADUST process developed by SKF Industries, is shown in Figure 12.2.2.

Plasma Energy Corporation (Raleigh, N.C.) is testing and demonstrating a plasma ladle-reheating system for maintaining or increasing the temperature of molten steel drawn from basic oxygen or electric arc furnaces. The plasma maintains the necessary temperature for vertical and horizontal continuous casting. The company is testing prototype plasma systems for recovering precious metals from automobile catalysts and electronic scrap, for making fused quartz and superalloys, and for destroying PCBs. It also is conducting refuse conversion tests for the Canadian government.

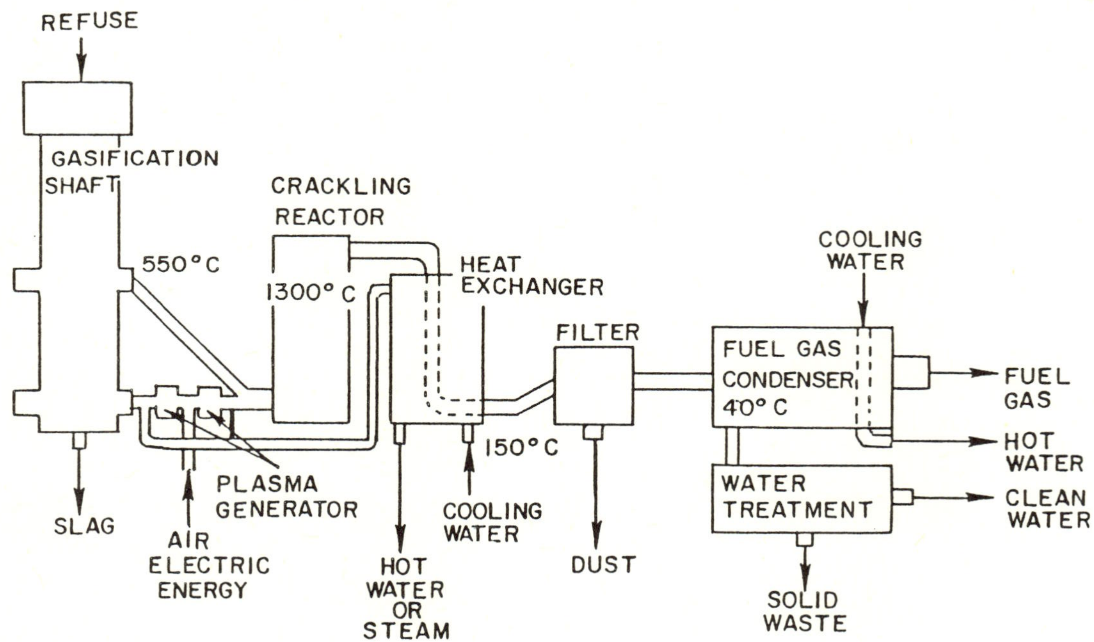

Figure 12.2.2. Pilot waste destruction system.

Source: Reference 28.

12.2.2 Process Performance

Because the potential for recovery of metals and other valuable waste constituents often constitutes the primary incentive for selection of a pyrometallurgical process, it is most meaningful to evaluate their performance on that basis. While the amount of available performance data for such systems are limited, several studies of pyrometallurgical process performance were reported in the literature. The results of three such studies[26-28] have been summarized below. Overall results are as follows:

- The three test studies represent both a variety of metal-bearing wastes and technologies. The different metal-bearing hazardous wastes tested include electroplating bath sludges, metal manufacturing sludges, and furnace dusts from specialty steelmaking operations. The results presented may be skewed somewhat, however, due to the fact that only a limited number of the metals were represented (i.e., only the recovery of chromium, nickel, and lead were shown). Both bench-scale and full-scale tests were conducted.

- In all cases, the percentage of metal recovered was high, ranging from approximately 70 percent to 99 percent. In most cases, this represented recovery as essentially pure metal. In certain cases, recovery was as a metal oxide, which is generally a less valuable material. Several studies were referenced indicating high recovery of other waste constituents, primarily, acids from treatment of plating wastes.

- Strong dependency was exhibited between metal recovery yields and operating parameters.

Summary of Performance Test Results: Recovery of Chromium and Nickel from Speciality Steelmaking, Other Wastes--

A series of tests were conducted by the U.S. Bureau of Mines[26] to assess the recovery of chromium and nickel, and other metals, from specialty steelmaking dusts (from processes such as stainless steelmaking). The smelting process was also tested for the recovery of other metal-bearing wastes such as sludges from electroplating.

Four bench-scale test series were conducted and the results obtained were as follows:

- In the first series, waste feed containing 15 percent AOD dust, 20 percent EF dust, 20 percent mill scale, and 40 percent grindings dust were smelted at different temperatures (ranging from 2850° to 3050°F), for different times (20 to 30 minutes), and with different chromium oxide reductants. The results showed high level recovery of chromium and nickel (i.e., greater than 85 percent) in all cases. The results also indicated that the recovery of chromium and nickel is directly related to temperature, time, and amount of reductant (as well as to the type of reductant).

- In the second series, feed composition was varied between a "low mill scale" grade and a "high mill scale" grade, which employed twice as much mill scale. Pelletizing tests indicated an optimal value of 35 percent mill scale, although pellets containing up to 55 percent mill scale could be produced. Smelting results showed high metal recovery (from 82 to approximately 100 percent) for both grades. The recovery was greater, however, for the "low mill scale" grade.

- In the third series, smelting of pellets made from another specialty steelmaking dust was conducted. This material contained several other types of metals in addition to those found in the previous tests. The results showed that metal recovery was still high for this material, although not as high as in previous tests (i.e., 60 to 90 percent as opposed to 80 to 99 percent).

- In the fourth series, the smelting process was applied to several chromium and nickel-bearing sludges. Data from these tests are limited. Recovery of 95 percent of the nickel in one electroplating sludge was reported, however.

Seven test runs were conducted on five full-scale process equipments. Recoveries similar to those found in the bench-scale tests were observed for these wastes.

Summary of Performance Test Results: Recovery of Chromium From Plating Wastes--

This test program[27] involved evaluation of conditions for recovery of chromium from plating wastes using a roasting (calcination) system. Soda ash (Na_2CO_3) and limestone (CaO) were used as fluxing agents (i.e., to supply oxygen for calcination reaction). In this process, chromium oxide reacts with the soda ash to form sodium chromate. Sodium chromate can be recovered through acid leaching. A total of four test series were conducted, varying soda ash input, lime output, temperature, and reaction time. Recoveries of over 90 percent were obtained under optimal conditions.

Summary of Performance Test Results: Pilot Testing of the PLASMADUST Process--

SKF Steel Engineering has a plasma pilot plant at Hofors, Sweden, where many different kinds of waste oxides have been tested with very promising results. The pilot plant used 1.5 MW plasma generator for heat supply. Pneumatic charging systems are used for the raw materials, coal, and the slag formers. The plant is equipped with a commercially used splash-type zinc condenser and a venturi scrubber.

All tests in the pilot plant have shown that the valuable metals in typical waste oxides from steelmaking and other secondary materials can be recovered with a high yield (96 to 99 percent) in the PLASMADUST process. They also report that the process is stable and simple to control.[28]

12.2.3 Costs

The economics of using pyrometallurgical processes to treat metal-bearing hazardous wastes represent the most significant potential drawback to their attractiveness. The systems involved are highly capital-intensive, as large-scale furnaces and attendant systems are used. Typically, capital costs for an installed system can range from $0.5 to 7 million for a full-scale pyrometallurgical system.[29] Compounding the high capital cost considerations is the fact that such systems are limited by design to fairly specific waste streams, and thus may not be attractive to a generator of a wide variety of metal-bearing wastes. The operating costs of such systems may also be quite high, considering the high demand for energy and raw materials (i.e., fluxing and reducing agents). There are also associated costs including the cost of disposal of by-products.

The cost of pyrometallurgical systems may be offset, however, by the economic benefits accrued through recovery of metals from specific wastes. Of the heavy metals under consideration in this document, the most valuable are nickel, selenium, and thallium, while the least valuable are cadmium and lead. A listing of prices available for several of the metals is presented in Table 12.2.1.

Cost data for the pyrometallurgical systems are limited. Furnace manufacturers were contacted, but none could adequately define a cost range for such a system without knowledge of the specifics which would affect

TABLE 12.2.1. PRICES OF METALS, METAL OXIDES

Substance	Price ($/lb)
Arsenious Trioxide	0.42
Cadmium Metal	1.20
Chromium Oxide	1.90
Lead Metal	0.28
Lead Dioxide	0.66
Mercury Metal (precipitate)	7.89
Mercury Oxide	7.00
Nickel Metal	3.45
Nickel Oxide	2.60
Selenium	13.00
Thallium	35.00

Source: Chemical Marketing Reporter, March 1987.

process design and operation. One commercial waste processor known to be involved in this type of treatment was contacted, but was unable to give specifics on costs without clear definition of the nature of wastes to be handled. The operating costs of such a facility were estimated by Higley, et al., at $78/ton for the specialty steelmaking dusts tested in their study. In that study, the value of the recovered materials was also estimated to be $280/ton; thus, a resultant economic gain of $0.10/lb could be realized.[26] An economic evaluation of a similar thermal system was also presented in Reference 30. Economic data from this evaluation is presented in Table 12.2.2. This system was designed to recover acids, but as described realizes some value from the recovery of iron oxide.

Overall, the key element in evaluating the economic attractiveness of pyrometallurgical systems is the value which may be derived from recovery of metals. However, systems which can not produce reusable materials may be attractive in terms of providing good volumetric reduction of wastes, but may not be viable economically.

12.2.4 Process Status

Usage of pyrometallurgical processes for treatment of metal-bearing hazardous wastes is not well established. Commercial waste processing by pyrometallurgical processes is not extensive, based on a survey conducted of waste processors. Due to the potential economic benefits associated with metals recovery, some furnace dusts wastes are now being recovered. A number of examples of facilities where pyrometallurgical systems have been implemented were described by Franklin Associates.[23]

12.3 EVAPORATION

12.3.1 Process Description

Evaporation is a common unit operation used in the chemical process industry to separate materials on the basis of their relative volatilities. In the metal finishing and electroplating industry evaporation, it used to concentrate and recover plating solutions, chromic acid, nitric

TABLE 12.2.2. ECONOMIC EVALUATION OF HYDROCHLORIC ACID REGENERATION USING THERMAL DECOMPOSITION

Item	Cost basis	System size (gpd)		
		10,000	100,000	200,000
Capital Costs	TIC*	$3,907,000	$14,974,000	$23,487,000
Operating Costs				
Labor	1.8% of TIC	69,000	266,000	427,000
Maintenance	3% of TIC	120,000	460,000	720,000
Fuel	12,000 Btu/gal/waste	59,000	225,000	363,000
Electricity	0.10 kwh/gal waste	11,000	40,000	65,000
Water	1 gal/gal waste	6,000	24,000	39,000
	Total	$265,000	$1,015,000	$1,614,000
Cost Savings				
Acid value	50% of PMV**	457,000	4,570,000	9,133,000
Iron oxide value	$100/ton	187,500	1,875,000	3,750,000
Treatment and disposal costs	Caustic soda	12,000	120,000	240,000
	Total	$656,500	$6,565,000	$13,123,000
Net Annual Savings	Savings - operating	$391,500	$5,550,000	$11,509,000
Payback Period (years)	Capital-Net savings	10.0	2.7	2.0

*TIC = Total installed cost.

**PMV = Present market value.

Source: Reference 30.

acid/hydrofluoril acid pickling liquors, and metal cyanides from spent baths and rinsewaters.[31] It is also commonly used as a post-treatment following reverse osmosis to concentrate metal solutions to the levels needed for replenishment of the plating bath. Fractionation of volatile components in a distillation column is a procedure that is seldom required when treating these metal bearing wastes. The usual purpose is to achieve satisfactory concentration levels by evaporation of water from the aqueous solution. An extensive discussion of evaporation/distillation processes can be found in Reference 3.

12.3.2 Process Performance

The available data describing the performance of systems used for the recovery/treatment of metal-bearing hazardous waste are limited to studies of the treatment of pickle liquors, where acids are also recovered. Despite the widespread usage of evaporation/distillation in the recovery of volatile metals, no detailed studies of performance were described in the literature. Numerous references, however, indicated, for example, that recovery of as much as 95 percent of pure mercury is regularly achieved through the distillation process.[24,32]

The results of several studies of the distillation of pickle liquor wastes were discussed extensively in Reference 30. Summaries of these tests are presented in Tables 12.3.1 to 12.3.3, and the results are briefly summarized below:

- High percentage levels of acids and metals (metal salts) were recovered;
- Low temperature operation was maintained (thus reducing energy demand/cost);
- Environmental impacts were negligible;
- Waste volume was reduced significantly.

TABLE 12.3.1. SUMMARY OF PERFORMANCE TEST: SUPERIOR PLATING INC.

Parameter	Result
Purpose of Test: Recovery of cadmium, sodium cyanide from a cadmium cyanide plating solution	
Heat Pump Capacity	300,000 BTU/hr
Heat Pump Exit Temperature	125°F
Evaporator Capacity	200 - 250 Btu/lb water distilled
Evaporation Temperature	110°F
Chiller Exit Temperature	95°F
Freon Condensor Exit Temperature	140°F
Coefficient of Performance (COP)	4.35
Recovered Cadmium Conc.	2.14 oz./gal.
Recovered Sodium Cyanide Conc.	15.3 oz./gal.

Source: Reference 30.

TABLE 12.3.2. SUMMARY OF PERFORMANCE TEST: A SINGLE-STAGE EVAPORATOR AT A CHINESE STEEL PLANT

Item	Volume (liters)	Concentration (g/l)					
		H^+	F^-	NO_3^-	Fe^{+2}	Ni^{+2}	Cr^{+3}
Purpose of Test: Recovery of acid and metal from pickling liquor wastes							
Spent Acid	1140	2.58	2.67	1.70	21.50	3.58	4.27
Residual Liquid	350	13.47	0.88	0.23	80.05	12.30	13.7
Regenerated Acid	858		3.23	2.12			
Sulfuric Acid	150	32.40					
Recovery Ratio (%)		92.90	93.90				

Source: Reference 30.

TABLE 12.3.3. SUMMARY OF PERFORMANCE TEST: TESTING OF HIGH VACUUM VAPOR COMPRESSION EVAPORATION AT THE CHARLESTON NAVY YARD

Parameter	Result
Purpose of Test: Recovery of chromic acid from a hard chromium plating line rinse	
Compressor Efficiency	
Coefficient of Performance (COP)	10.3
Adiabatic Efficiency	25 %
Capacity	25 gph @ 700 rpm speed 40 gph @ 1170 rpm speed
Total Chrome Recovered (70 gals x 54,900 mg/l)/7484	32.1 lb (513.5 oz.)
Dragout Rate (32.1 lb/320 hrs/month) x (1 gal/2 lb Cr+6)	0.05 gal/hr
Rinse Ratio (Ratio of plating bath concentration to final rinse concentration using 3 countercurrent rinse tanks)	20,000
Rinse Flow Rate	27 gph per 1 gph dragout
Evaporator Capacity (Required Rinse Rate) 27 gph x 0.05 gph = 1.35	1.35 gph
Recovered Process Water	
Quantity	8.75 gph (33,600 gpy)
Conductance	10 mmho
Operating Temperatures	95 - 122°F
Electrical Requirements	9 kw

Source: Reference 30.

12.3.3 Process Costs

Capital costs for an evaporative recovery system will vary with the waste type, waste quantity, process flow rates, type of heat exchanger employed, and system size. Operating costs generally include 1-2 hours labor for system maintenance and operation (labor requirements will be reduced if the system is operated continuously), electrical and fuel energy requirements for heat supply, taxes and insurance, and depreciation costs. Approximately 10 lbs of low pressure steam (15 psig) is required for every gallon of liquid evaporated.[33] Typical capital equipment costs are shown in Table 12.3.4.

Evaporation/distillation processes require large amounts of heat energy, which can make the process quite costly. However, efficient use of energy systems can lower these costs significantly. Waste heat from other industrial processes (diesel generators, incinerators, boilers, and furnaces) within the plant can be recovered for use in the evaporation/distillation system. The use of multi-effect evaporators and vapor compression systems can also improve thermal efficiencies. Cost savings will be realized in reduced neutralization costs, reduced sludge disposal costs, and reduced purchase requirements for fresh bath makeup solutions.

12.3.4 Process Status

Evaporation/distillation is one of the oldest recovery techniques, and is widely used in industry. Over 600 metal waste recovery units are currently in operation in the United States.[30,33] They are most commonly used in metal finishing and electroplating industries to recover plating solutions, chromic acid and other concentrated acids, and metal cyanides. In addition, water recovered from the evaporation process is of high purity and can be reused in process waters. The percentage of these units used in various plating applications is presented in Table 12.3.5. These systems are most effective in recovering acids, bases, and metals from rinsewaters. Systems can be designed cost-effectively with capacities ranging from 20 gph to 300 gph.[33] These system are cost-competitive with conventional neutralization and disposal technologies. Greater cost savings are realized with larger operations.

TABLE 12.3.4. TYPICAL CAPITAL EQUIPMENT COSTS FOR VARIOUS EVAPORATION SYSTEM CAPACITIES

Evaporator capacity (gph)	Capital costs ($)
20	25,000
40	33,800
55	39,199
120	44,129
300	115,000

Source: Reference 30.

TABLE 12.3.5. PERCENTAGE BREAKDOWN BY PLATING TYPE OF EVAPORATION UNITS CURRENTLY IN OPERATION

Plating chemical	Percent of units
Chromium	50
Chromium Etch	10
Nickel	20
Cyanide	10
Other	10

Source: Reference 33.

12.4 CRYSTALLIZATION

Crystallization is a recovery technique in which metal contaminants in spent corrosive solutions are precipitated through temperature reduction and then are removed by settling or centrifugation. The applicability of crystallization as a treatment alternative for metal-bearing hazardous wastes is limited to liquid waste with appropriate solubility characteristics. As such, crystallization is most applicable to spent acid wastes from pickling, plating, etching, or other types of metal finishing operations, such as caustic soda etching of aluminum.[30]

12.4.1 Process Description

The general process employed in crystallization is simple, focusing primarily on controlled cooling. A typical crystallization process is depicted in Figure 12.4.1. In this process, pickle solution is pumped directly to the crystallizer, which is essentially an insulated tank. Cooling of the solution is effected and crystallization of metal salts occurs. The crystals settle to the bottom and in some system, flow out of the crystallizer. The crystallization process often is conducted over long periods of time, e.g., 8 to 16 hours, and involves temperature reductions of 30 to 100°F. Crystallization may be done on a batch or continuous basis. Eventually all crystals are removed from solution and settle out. Acid is then recovered by filtration, or centrifugation, or some other physical separation operation, and the metal salt crystals are collected for disposal, or in some cases, for further treatment for metals recovery.

The critical operating parameters involved in crystallization processes include the solubility of the metal salts in solution, waste composition, process time, and temperature. The process is more efficient when concentrations and crystallization temperatures are high. Freezing point characteristics are also significant consideration, as some waste constituents (most typically water) may begin to freeze at or before the applicable crystallization temperature. To counter freezing, pretreatment is often conducted. In particular, dewatering of wastes may be done through thermal

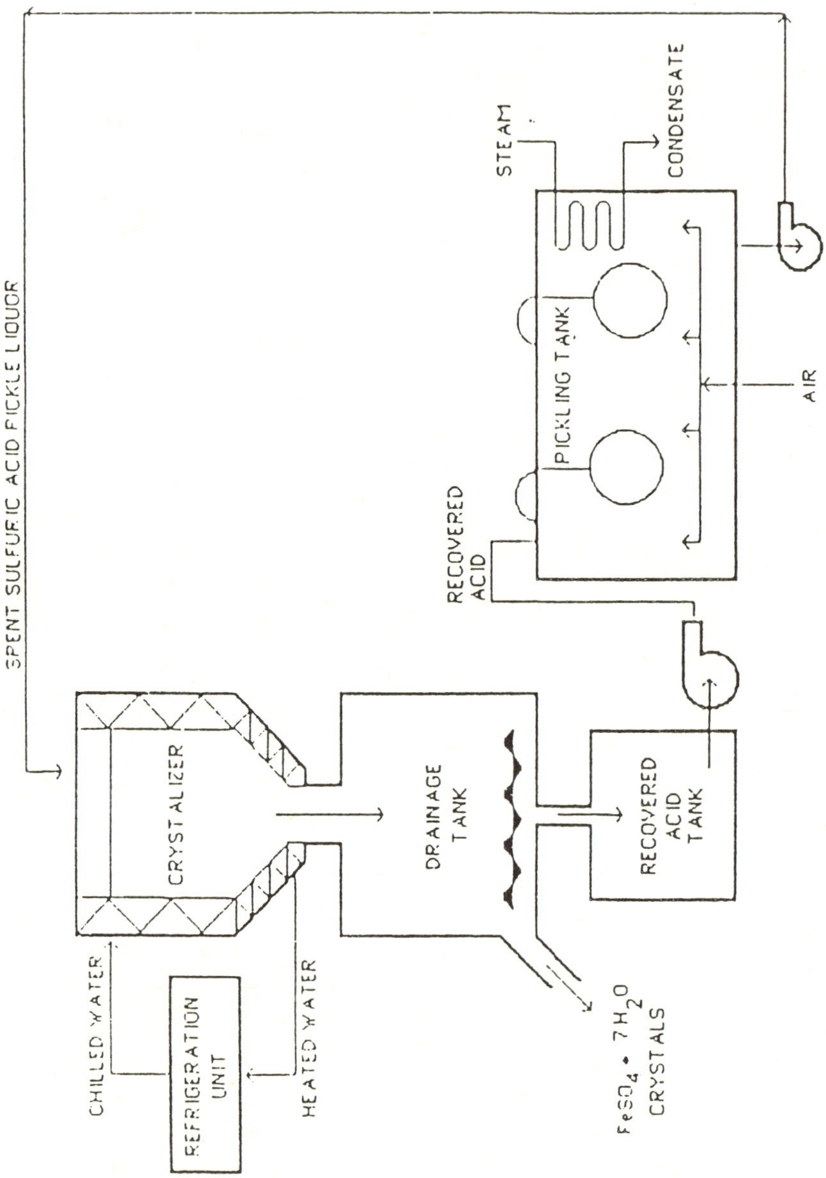

Figure 12.4.1. Flow diagram of crystallization system for recovery of sulfuric acid pickling liquor.

Source: Reference 34.

evaporation, countercurrent rinsing, chemical treatment, or air agitation. Processes in which evaporation is conducted prior to crystallization are thought of as two-stage systems.

12.4.2 Process Performance

The performance of a crystallization system is typically evaluated on the basis of percentage acid recovery, percentage metals removal, overall product quality (purity), processing time, and overall economics (i.e., recovery value vs. operating costs). Typical performance data for crystallization are summarized in Tables 12.4.1 and 12.4.2. In general, metal recoveries are in the 50 to 90 percent range.

12.4.3 Process Costs

The costs of such systems are moderate compared to other thermal-based recovery processes, primarily due to the great simplicity of crystallization systems. Capital costs may typically include construction of tank-type evaporation and crystallization units, refrigeration system, and connections. Operating costs are primarily based on disposal, energy, and maintenance. Cost-effectiveness depends strongly on the value of the acid or other substances recovered. A typical economic profile is presented in Table 12.4.3.

12.4.4 Process Status

Crystallization systems have been applied on a commercial scale, primarily by generators of large volumes of spent solutions (e.g., iron and steel plants). There are several different commercially applied processes for recovery of sulfuric acid from spent pickle liquor. All processes, however, rely upon the basic principles of crystallization of iron salts (mainly ferrous sulfate) from the spent pickle liquor and the addition of enough fresh sulfuric acid to return the pickling solution to its original acid strength. These commercial acid recovery systems allow the free sulfuric acid remaining in the spent pickling solution to be reused. The processes differ in the methods used to crystallize the ferrous sulfate.

TABLE 12.4.1. TYPICAL OPERATING PARAMETERS AND RESULTS FOR SULFURIC ACID RECOVERY SYSTEM USING CRYSTALLIZATION

Parameter	Result
Optimum iron content in the waste feed	10 to 14%
Iron removal efficiency	80 to 85%
Acidity losses in recovered acid	2 to 3%
Average cycle time	6 hrs

Source: Reference 35.

TABLE 12.4.2. TYPICAL PERFORMANCE OF A TWO-STAGE CRYSTALLIZATION SYSTEM FOR THE RECOVERY OF NITRIC-HYDROFLUORIC ACID

	Concentration, weight-percent (lbs/hr)					
Parameter	Fe	Cr	Ni	NO_3	F	Water
Feed to evaporator	3.4 (26.5)	1.1 (8.6)	1.6 (12.5)	12.0 (93.6)	6.0 (46.8)	75.9 (592)
Feed to crystallizer	6.5 (26.5)	2.1 (8.6)	3.1 (12.5)	22.1 (89.9)	10.1 (41.2)	56.1 (228.3)
Condensed vapor	-	-	-	1 (3.7)	1.5 (5.6)	97.5 (363.7)
Residue from crystallizer	25 (20.0)	4.6 (3.7)	0.8 (0.6)	6.0 (4.8)	30.8 (24.6)	32.9 (26.3)
Filtrate from crystallizer	2.0 (6.5)	1.5 (4.9)	3.5 (11.9)	26.0 (15.1)	5.1 (16.6)	61.7 (202.0)
Total concentration recovered	0.9 (6.5)	0.7 (4.9)	1.7 (11.9)	12.7 (88.8)	3.2 (22.2)	80.8 (565.7
Total required additions	-	-	-	(43)	(32)	(261.1

Source: Reference 36.

TABLE 12.4.3. ECONOMIC EVALUATION OF ACID RECOVERY SYSTEM USING CRYSTALLIZATION TECHNIQUE

Item	Small unit ($)	Medium unit ($)	Large unit ($)
Flow rate (gal/day)	2,400	16,000	30,000
CAPITAL COSTS			
Equipment	175,000	460,000	850,000
Tank (2 tanks @ $1.25/gal)	5,000	40,000	75,000
Installation (10% of investment)	17,500	46,000	85,000
Total capital costs:	197,500	546,000	1,010,000
OPERATING COSTS			
Maintenance (6% of investment)	10,500	27,600	51,000
Taxes & insurance (0.5% of investment)	875	2,300	4,250
Utilities (@ $0.02/KW-h)	8,000	10,000	12,000
Depreciation (10% of investment)	17,500	46,000	85,000
Total operating costs:	36,875	85,900	152,250
COST SAVINGS			
Neutralization savings	22,653	139,400	261,375
Disposal savings	51,025	314,000	588,750
Process water savings	2,633	16,200	30,375
Acid makeup savings	16,250	100,000	187,500
Total cost savings:	92,560	569,600	1,068,000
NET SAVINGS: (Gross savings-Operating costs)	55,685	483,700	915,750
PAYBACK PERIOD (Capital costs/Net savings)	3.59 yrs (43 months)	1.16 yrs (14 months)	1.14 yrs (14 months)

Source: References 34, 35, 37.

Copper recovery from sulfuric acid-hydrogen peroxide pickle liquors is being used more and more in the U.S. and Europe. One of the main advantages of peroxide is the ability to regenerate spent liquors and to recover copper electrolytically or by crystallization of copper sulfate. Copper recovery regeneration of the sulfuric acid has been accomplished for many years with simple sulfuric acid pickling solutions. This method is suitable also for peroxide pickles and low peroxide concentration.[23]

REFERENCES

1. Versar, Inc. Technical Assessment of Treatment Alternatives for Wastes Containing Metals and/or Cyanides. Draft Final Report. Versar, Inc., Springfield, VA. Prepared for U.S. Environmental Protection Agency, Office of Solid Waste, Washington, D.C. EPA Contract No. 68-03-3149. October 31, 1987.

2. Federal Register 1982, 47 27516-35.

3. Breton, M. *et. al*. Technical Resource Document, Treatment Technologies for Solvent-Containing Wastes. Final Report. GCA Technology Division, Inc., Bedford, MA. Prepared for U.S. Environmental Protection Agency, Hazardous Waste Engineering Research Laboratory, Cincinnati, OH. EPA Contract No. 68-03-3243. August, 1986.

4. Oppelt, E. T. Hazardous Waste Destruction, Environmental Science and Technology, Vol. 20, No. 4. 1986.

5. U.S. EPA . National Survey of Hazardous Waste Generators and Treatment, Storage and Disposal Facilities Regulated under RCRA in 1981, U.S. Government Printing Office Order No. 055000-00239-8. U.S. Environmental Protection Agency, Office of Solid Waste, Washington, D.C. 1984.

6. MITRE Corp. A Profile of Existing Hazardous Waste Incineration Facilities and Manufactures in the United States, PB-84-157072, MITRE Corporation, McLean, VA. Prepared for U.S. Environmental Protection Agency, Office of Solid Waste, Washington, D.C., 1984.

7. Incineration and Treatment of Hazardous Waste: Proceedings of the 8th Annual Research Symposium. EPA-600/9-83-003. Article cited: Frankle, J., N. Sanders, and G. Volgel, "Profile of the Hazardous Waste Incinerator Manufacturing Industry";

8. MITRE Corporation. Survey of Hazardous Waste Incinerator Manufacturers, 1981. MITRE Corporation, METREK Division, McLean, VA. 1982.

9. Advanced Environmental Control Technology Research Center. Research Planning Task Group Study - Thermal Destruction. EPA-600/2-84-025. Prepared for U.S. Environmental Protection Agency, Industrial Research Laboratory, Cincinnati, OH. January 1984.

10. ICF Incorporated. RCRA Risk/Cost Policy Model - Phase III Report. Prepared for U.S. Environmental Protection Agency, Office of Solid Waste, Washington, D.C. 1984.

11. Young, C. Telephone conversation with M. Kravett, Alliance Technologies Corporation. Waste-Tech Services, Inc., March 1987.

12. Mullen, D. Telephone conversation with M. Kravett, Alliance Technologies Corporation. SCA Chemical Services, March 1987.

13. Anderson, R. Telephone conversation with M. Kravett, Alliance Technologies Corporation. IT Corporation, March 1987.

14. Warren, P. Telephone conversation with M. Kravett, Alliance Technologies Corporation. Stablex Corporation, March 1987.

15. Garcia, G. Telephone conversation with M. Kravett, Alliance Technologies Corporation. TWI, Inc., March 1987.

16. Bell, R. Telephone conversation with M. Kravett, Alliance Technologies Corporation. SYSTECH Corp., March 1987.

17. Klotzbach, K. Telephone conversation with M. Kravett, Alliance Technologies Corporation. Rollins Environmental Services Inc., March 1987.

18. Frost, D. Telephone conversation with M. Kravett, Alliance Technologies Corporation. Rollins Environmental Services, Inc., March 1987.

19. Bush, R. Telephone conversation with M. Kravett, Alliance Technologies Corporation. IT Corporation, March 1987.

20. Cooper, D. Telephone conversation with M. Kravett, Alliance Technologies Corporation. Rollins Environmental Services, Inc., March 1987.

21. Trenholm, A., Gorham, P., and G. Sungclaus. Performance Evaluation of Full Scale Hazardous Waste Incinerators. EPA-600/2-84-181. Midwest Research Institute, Kansas City, MO. Prepared for U.S. Environmental Protection Agency, Office of Research and Development, Cincinnati, OH. November 1984.

22. Hall, R.R., et al. Union Chemical Trial Burn Sampling and Analysis. Draft Final Report. GCA Technology Division, Bedford, MA. GCA Report No. GCA-TR-84-22-G. Prepared for Union Chemical Co., Union, ME. February 1984.

23. Franklin Associates, LTD. Industrial Resource Recovery Practices: Metals Smelting and Refining (SIC 33). Final Report. Franklin Associates, LTD, Prairie Village, KS. Prepared for: U.S. Environmental Protection Agency, Office of Solid Waste, Washington, D.C. EPA Contract No. 68-01-6000. January, 1983.

24. Duby, P. Extractive Metallurgy. In: Kirk-Othermer Encyclopedia of Chemical Technology, Third Edition, Volume 9. John Wiley and Sons, New York, NY. 1980. pp. 739-767.

25. Shamsuddin, M. Metal Recovery from Scrap and Waste. Journal of Metals. February 1986. pp. 24-31.

26. Higley, L.W., Crosby, R.L., and L.A. Neaumeir. In-Plant Recovery of Stainless and Other Specialty Steelmaking Wastes. Report No. 8724. U.S. Department of the Interior. Bureau of Mines, Washington, D.C. 1982.

27. Bolto, B.A., Kotowski, M., and L. Pawlowski. Recovery of Chromium from Plating Wastes. In: ASTM Special Publication No. 851, "Hazardous and Industrial Waste Management and Testing: Third Symposium, March 7-10, 1983, Philadelphia, PA". American Society of Testing Methods, Philadelphia, PA. 1984.

28. Herlitz, H., and S. Erikson. Metal Recovery from Hazardous Baghouse Dust using Plasma Technology. SKF Steel Engineering, Inc., Sweden. In: Proceedings, Second Conference on Management of Municipal, Hazardous, and Coal Wastes. DOE/METC-84-34. Miami University. September 1984.

29. Alben, P. Telephone conversation with M. Kravett, Alliance Technologies Corp. World Resources Corp. March 1987.

30. Wilk, L. et. al. Technical Resource Document: Treatment Technologies for Corrosives-Containing Wastes, Volume II. Final Report. Alliance Technologies Corporation, Bedford, MA. Prepared for U.S. Environmental Protection Agency, Hazardous Waste Engineering Research Laboratory, Cincinnati, OH. EPA Contract No. 68-03-3243. October, 1986.

31. Stephenson, J.B., Hogan, J.C., and R. S. Kaplan. Recycling and Metal Recovery Technology for Stainless Steel Pickling Liquors. U.S. Department of the Interior, Bureau of Mines. In: Environmental Progress, Volume 3, No. 1. February 1984. pp. 50-52.

32. Jones, H.R. Mercury Pollution Control. Pollution Control Review No. 1. Noyes Data Corporation, Park Ridge, NJ. 1977.

33. Constantine, D. Corning Process Systems, Corning, NY. Technical Data Sheet No. RT-1: Rinse Theory. May 12, 1980.

34. Camp, Dresser, and McKee, Inc. Technical Assessment of Treatment Alternatives for Wastes Containing Corrosives. Camp, Dresser, and McKee, Inc., Boston, MA. Prepared for: U.S. Environmental Protection Agency, Office of Solid Waste, Washington D.C. EPA Contract No. 68-01-6403. September 1984.

35. Luhrs, R. Telephone conversation with L. Wilk, GCA Technology Division, Inc. Acid Recovery Systems, Inc. September 1986.

36. Krepler, A. Apparatus for Recovery Nitric Acid and Hydrofluoric Acid from Solutions. U.S. Patent No. 4, 252, 602. Assigned to Ruthner Industrieanlagen-Aktiengasellschaft, Vienna, Austria. February 24, 1981.

37. Crown Technology, Inc. Product Literature: Crown Acid Recovery Systems. Received in communication with L. Wilk, GCA Technology Division, Inc. July 1986.

13. Physical Removal Processes for Cyanides

The two treatment processes discussed in this section are based on physical methods of separation and do not result in destruction of the contaminants in the waste feed stream. The processes discussed are:

13.1 Ion Exchange

13.2 Flotation/Foam Separation

Both of these processes are used to some extent for the treatment of cyanide wastes, but differ in their applicability to various types of wastes and their need for pretreatment and post-treatment procedures. However, since successful physical treatment merely concentrates the free and complexed cyanides into smaller volume residuals, some sort of secondary treatment will be required prior to disposal.

13.1 ION EXCHANGE

Ion exchange has successfully removed metal cyanide complexes from plating, coke plant, and gold mill effluents.[1] However, backwashing of the cyanide complexes from the strong base anion resins has often proved difficult, resulting in a continual loss of capacity through repeated cycles.[2] This problem has been apparently overcome by the use of weak base anion resins or using a three-bed ion exchange installation consisting of a strong acid cation, weak base anion, and a strong base anion resins.[3] Laboratory experiments and pilot testing programs have demonstrated the removal of ferrocyanide from synthetic solutions and industrial wastewaters to below 1 mg/L of total cyanide.[4]

13.1.1 Process Description

Ion exchange is the process of removing undesirable anions and cations from a wastewater by bringing the wastewater in contact with a resin that exchanges the ions in the wastewater with a set of substitute ions.[1,5] There are three principal operating modes for ion exchange systems: concurrent fixed-bed, countercurrent fixed-bed, and continuous countercurrent. Figure 13.1.1 illustrates the three principal operational modes, while Table 13.1.1 contains a comparison summary.

Fixed-bed ion exchange operations require a cylindrical ion exchange bed, tanks for solution storage, and pumps.[1,5] The choice of materials is governed by the chemical environment. Continuous ion exchange systems are much more complex, requiring solids handling equipment and more intricate control systems. Table 13.1.2 gives design parameters and a range of typical design values for ion exchange.

In selective cyanide removal through ion exchange, free cyanide is often first complexed with iron and then contacted with a basic anion exchanger which is highly selective for ferrocyanide. The ion exchange column removes cyanides as follows:[6]

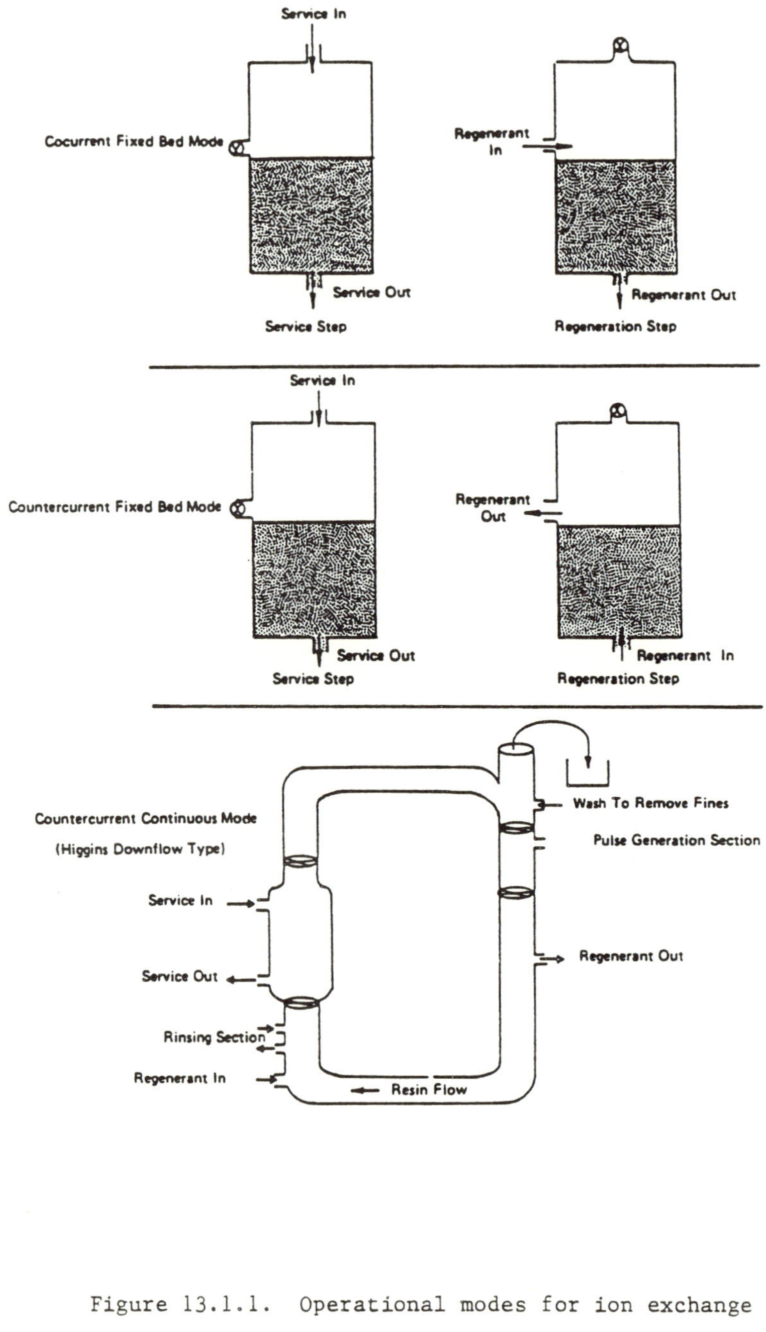

Figure 13.1.1. Operational modes for ion exchange

Source: Reference 1.

TABLE 13.1.1. COMPARISON OF ION EXCHANGE OPERATING MODES

Criteria	Concurrent fixed bed	Countercurrent fixed bed	Countercurrent continuous
Capacity for high feed flow and concentration	Least	Middle	Highest
Effluent quality	Fluctuates with bed exhaustion	High, minor fluctuations	High
Regenerant and rinse requirements	Highest	Somewhat less than concurrent	Least, yields concentrated regeneration waste
Equipment complexity	Simplest; can use manual operation	More complex; automatic controls for regeneration	Most complex; completely automated
Equipment for continuous operation	Multiple beds, single regeneration equipment	Multiple beds, single regenertion equipment	Provides continuous service
Relative costs (per unit volume:			
Investment	Least	Middle	Highest
Operating	Highest chemicals and labor; highest resin inventory	Less chemicals, water, and labor than concurrent	Least chemicals and labor; lowest resin inventory

Source: Reference 1.

TABLE 13.1.2. ION EXCHANGE DESIGN CRITERIA

	Units	Design criteria
Ion exchange operation		
Bed height	m (ft)	1.2 to 1.8 (4 to 6)
Wastewater loading rate	bed volume/hour	7.5 to 20
Pressure drop	cm of water/m (in. of water/ft)	11 (8.4)
Cycle time	bed volumes* bed volumes**	100 to 150 200 to 250
Regeneration		
Solution flow rate	bed volumes/hour liter/sec/m^2 (gal/min/ft^2)	4 to 10 3 to 7 (4 to 8)
Total solution volume	percent of treated wastewater (or 10 bed volumes)	2.5 to 5
Cycle time backwash	hours liter/sec/m^2 (gal/min/ft^2)	1 to 3 5 (8)

*For one 1.8 m (6 ft) bed.
**For two 1.8 m (6 ft) beds.

Source: Reference 1.

$$x\ [\text{Resin-N-R}_3^+ + Cl^-] + Fe(CN)_6^{x-} = [(\text{Resin-N-R}_3^+)_x + Fe(CN)_6^{x-}] + xCl^-$$

where x = 4 in ferrocyanide and x = 3 in ferricyanide. Once the resin is exhausted, it can be regenerated with aqueous sodium chloride as follows:

$$[(\text{Resin-N-R}_3^+)_x + (Fe(CN)_6^{x-})] + NaCl = x[\text{Resin-N-R}_3^+ + Cl^-]$$
$$+ Na_x\ Fe(CN)_6^{x-}$$

Table 13.1.3 contains a list of suitable anion and chelator exchangers. The properties and performance specifications for anionic exchange resins are usually fully detailed by the resin manufacturer.[7] For a more detailed description of ion exchange equipment, materials, and process configurations see Section 8.0.

Ions exchange units are only able to treat contaminants in solution. High concentrations of suspended solids which can foul the resin bed are typically removed through some form of filtering.[8] Examples of filters which may be employed include activated carbon, deep bed, diatomaceous earth precoat, and resins. The filters eventually become clogged with particulates, and are replaced when overall cycle time decreases to unacceptable levels. The filter size and replacement frequency will depend on the quantity of suspended particulates passed through the filter per unit time.

For large volume systems, which require more frequent changing of the filter cartridges, it may be more cost-effective to use a multimedia sand filter with a backwashing system to regenerate the filter. Although initial capital costs for this type of filtration system are higher, savings in filter replacement costs may be realized.[10]

The use of weak acid and weak base exchangers for treating corrosive wastes will require additional pretreatment. The exchange capacity of weak acid exchangers is limited below pH 6.0, and weak base exchangers are not effective above pH 7.[11] Therefore, a pH adjustment system must be incorporated prior to feeding the waste stream through weak ion exchangers.

TABLE 13.1.3. CHELATING AND ANION EXCHANGERS USED

Ion exchanger	Produced by	Matrix structure[a]	Functional group
Amberlite IRA-45	Rohm Haas, U.S.A.	S + DVB	$-NH_2$, -NHR, $-NR_2$
Amberlite IRA-93		S + DVB (M)	$-NR_2$
Amberlite IRA-94S		S + DVB (M)	$-NR_2$
Amberlite IRC-50		M + DVB	carboxyl
Amberlite IRC-84		A + DVB	carboxyl
Amberlite IRC-718		S + DVB	iminodiacetic
Amberlyst A-21		-	weak base
Chelex 100	Bio-Rad, U.S.A.	S + DVB	iminodiacetic
Diaion CR-10	Mitsubishi Chemical	S + DVB	iminodiacetic
Diaion CR-20	Industries, Japan	S + DVB	polyamine
Diaion CR-40		-	polyethylenimine
Diaion WA-11		A + DVB	weak base
Diaion WA-21		S + DVB	weak base
Duolite A-6	Dia Prosim, France	F	$-NR_2$
Duolite A-7		F	$-NH_2$, -NHR, $-NR_2$
Duolite S-30		F	polyphenol
Duolite S-37		F	-NHR, $-NR_2$
Duolite ES-63		S + DVB	phosphonic
Duolite A-368		S + DVB	$-NR_2$
Duolite ES-346		-	amidoxime
Duolite ES-465		S + DVB	thiol
Duolite ES-466		S + DVB	iminodiacetic
Duolite ES-467		S + DVB	aminophosphonic
Imac A-20S	Imacti, Holland	S + DVB	$-NR_2$
Imac GT-73		S + DVB (M)	weak acid type
Imac TMR		S + DVB	-SH, $-SO_3H$

(continued)

TABLE 13.1.3 (continued)

Ion exchanger	Produced by	Matrix structure[a]	Functional group
Kastel A-101	Montedison, Italy	S + DVB	$-NR_2$
Lewatit MP-62		S + DVB (M)	weak base
Lewatit TP-207		S + DVB (M)	iminodiacetic
Merck II	Merck, F.R.G.	S + DVB	$-NH_2$, $-NHR$, $-NR_2$
Relite 4MS	Resindion, Italy	S + DVB	$-NR_2$
Relite MGl		A + DVB	moderate base
Thiol resin	Chemical Industry Works A.E., Poland	-	-SH
Wofatit AD-41	Veb. Chemiekombinat	S + DVB (M)	$-NR_2$
Wofatit MC-50	Bitterfeld, G.D.R.	S + DVB	iminodiacetic andacetic
Zerolit HXiP	Permutit, England	S + DVB	$-NR_2$, $-NR_3OH$
Zerolit 216		F	phenylcarboxylic
Zerolit S-1006		S + DVB	EDTA type

[a]S + DVB = Copolymer of styrene with divinylbenzene;
S + DVB (M) = copolymer of styrene with divinylbenzene of macroporous matrix structure;
M + DVB = copolymer of methacrylic acid with divinylbenzene;
A + DVB = copolymer of polyacrylic acid with divinylbenzene;
F = polycondensation exchanger of phenolformaldehyde matrix structure;
- = no data available.

Source: Reference 7.

Other pretreatment requirements include flow equalization for waste streams experiencing flow or pollutant concentration surges and oil separation to prevent resin fouling.

Waste streams from the ion exchange process include: spent regenerant solution, wash streams, and solids from the filtering system. Typically, since both the spent regenerant solution and the wash stream contain cyanides these streams will require treatment and disposal, although in some cases the recovered cyanide can be reused or marketed. Solids from the filtering system can generally be land disposed without further treatment.[12] The quantities of wastes generated will depend on the types and concentrations of contaminants present in the solution being treated.

13.1.2 Process Performance

Numerous studies have been conducted by various researchers into the effectiveness of ion exchange for treating metal/cyanide-bearing waste streams. In 1979, Trachtenberg and Murphy described studies on iron cyanide removal from leachate from a storage dump containing discarded linings from aluminum reduction cells.[13] Data from the full-scale treatment system showed an average reduction from 48 mg/L of ferrocyanide to 0.5 mg/L (99 percent removal) at a hydraulic loading of 0.13 mL/min/mL resin. However, no information was provided pertaining to wastewater volumes treated before regeneration or exchange capacities.

Bessent et al. reported on the use of ion exchange for the treatment of coke plant wastewaters.[14] Pilot-scale, glass columns, 6 in. x 6 ft high with metal headers, were used to simulate the filtration and ion exchange systems. Initially, sand was used as the media in the filter column. This was replaced, following filtration problems, with a media consisting of sand, anthracite coal, garnet and granite. In the case of the ion exchange system, early pilot study runs utilized only one resin column. Later pilot study runs utilized two ion exchange columns operated in an alternating series mode. Backwash, regenerant and rinse facilities consisted of various sizes of tanks and containers applicable to the specific operation being performed. Figure 13.1.2 presents a schematic of the cyanide removal pilot plant.

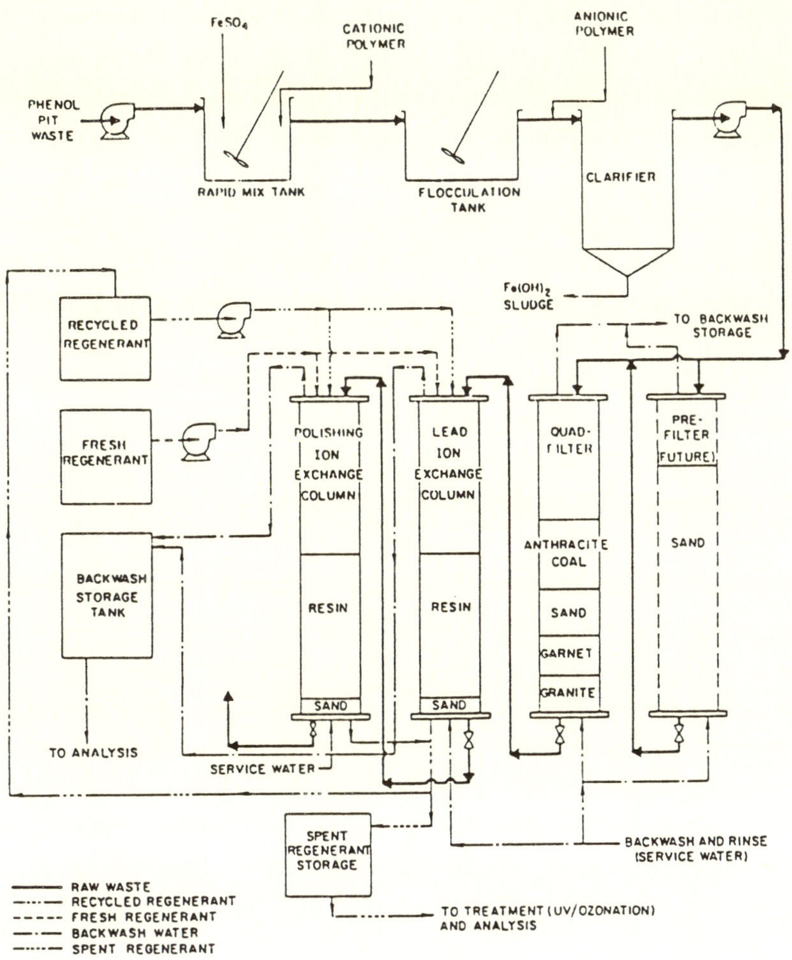

Figure 13.1.2. Schematic of pilot-plant cyanide removal system.

Source: Reference 14.

The coking effluents were first treated with ferrous sulfate to convert the free cyanide to ferrocyanide. The ferrocyanide was then removed by an ion exchange process employing a strongly basic ion exchange resin, Amberlite IRA-958. Based on the pilot-scale studies, they concluded that for a full-scale application, the average resin loading would be 11 mg CN_{Fe}/mL resin at a nominal concentration breakthrough of 10 mg/L of iron ferro-cyanide. Breakthrough would occur after 200 bed volumes if the influent cyanide concentration was 80 mg/L and the hydraulic loading rate was 0.13 mL/min/mL resin.

Table 13.1.4 presents a summary of the test conditions at the pilot plant while Table 13.1.5 presents a summary of the test results. Total cyanide in the polishing column effluent was consistently below 10 mg/L and in most cases, the free cyanide concentration was less than 2 mg/L. Runs 6 and 7 closely simulated the equipment configuration, hydraulic loading, operational mode, and performance recommended for a full-scale system. In runs 6 and 7, a filter (sand, anthracite, garnet, and granite) followed by two exchange columns in series was evaluated. The filter replaced a sand filter in an attempt to: (a) assess performance of an alternative filtration media, and (b) provide greater protection for the ion exchange resin from solids with maximum run times between backwashes.

Resin capacity to cyanide breakthrough points were calculated for various ion exchange runs from cyanide breakthrough data. Average resin capacity was determined to be 17.6 kg Cn/M^3. This approaches the lower limit of published capacity for Amberlite IRA 958. It should be noted that the runs with the 32 bed volume (BV)/hr feed rate showed the highest resin capacity of approximately 20.98 kg CN/M^3. Final hydraulic loading requirements will depend on desired throughput.

In 1985, Vachon investigated the removal of iron cyanide from synthetic and actual gold mill effluents, using the strongly basic anion exchanger Amberlite IRA-958.[15] As indicated in Table 13.1.6, ion exchange was effective in removing iron cyanide to concentrations of less than 3 mg/L. The general trend observed was that exchange capacity increased under conditions of increasing cyanide concentration, lower hydraulic loading, and increasing

TABLE 13.1.4. PILOT STUDY RUN SUMMARY - RUNS 1 THROUGH 7

Purpose	Run 1[a]	Run 2[b]	Run 3[c]	Run 4[d]	Run 5[e]	Run 6[f]	Run 7[g]
Flow	8 BV/hr 0.6 gpm	16 BV/hr 1.2 gpm	8 BV/hr 0.6 gpm	32 BV/hr 2.4 gpm	32 BV/hr 2.4 gpm	16 BV/hr 1.2 gpm	16 BV/hr 1.2 gpm
Run length	400 BV	400 BV	400 BV	400 BV	400 BV	400 BV	400 BV
$FeSO_4$ dose	500 mg/L	300 mg/L	300 mg/L	300 mg/L	300 mg/L	300 mg/L	300 mg/L
Anionic polymer dose	None	2 mg/L	2 mg/L	3 mg/L	3 mg/L	3 mg/L	3 mg/L
Cationic polymer dose	None	None	None	15 mg/L	15 mg/L	15 mg/L	15 mg/L
Type of filtration	Sand	Sand	Sand	Sand	Sand	Quad-media[h]	Quad-media[h]
Ion exchange Resin col. 1	Lead/polishing $0.59 ft^3$ resin bed	Lead/polishing $0.59 ft^3$ resin bed	Lead/polishing $0.59 ft^3$ resin bed	Lead $0.59 ft^3$ resin bed	Polishing $0.59 ft^3$ resin bed	Lead $0.59 ft^3$ resin bed	Polishing $0.59 ft^3$ resin bed
Ion exchange Resin col. 2	None	None	None	Polishing $0.59 ft^3$ resin bed	Lead $0.59 ft^3$ resin bed	Polishing $0.59 ft^3$ resin bed	Lead $0.59 ft^3$ resin bed
Regeneration mode	4 BV fresh NaCL (reg.)	2 BV recycled (reg.) 2 BV fresh (reg.)	2 BV recycled (reg.) 2 BV fresh (reg.)	2 BV recycled (reg.) 2 BV fresh (reg.)	2 BV recycled (reg.) 5 BV fresh (reg.)	4 BV fresh (reg.)	None

[a] Development of cyanide breakthrough curve to determine initial cyanide loading to the column.

[b] Determine effect of increasing hydraulic loading relative to run length and cyanide removal.

[c] Verify breakthrough curve of run 1; determine resin deterioration.

[d] Observe 2-column operation; determine effect of high hydraulic loading on the columns relative to cyanide removal and run length on a virgin resin.

[e] Determine performance of exposed resin columns operated at a high hydraulic loading; duplicate run 4 cyanide breakthrough curve.

[f] Evaluate multi-media filtration system and ion exchange system operated at its high recommended hydraulic loading.

[g] Duplicate cyanide breakthrough curve of run 6; determine evidence of any resin deterioration.

[h] Sand, anthracite, garnet, and granite.

Source: Reference 14.

TABLE 13.1.5. COMPOSITE SAMPLE SUMMARY

Run No.	Site No.	Total cyanide (mg/L)	Free cyanide (mg/L)	Ammonia (mg/L)	Grease & oil (mg/L)	Phenol (mg/L)	Thiocynates (mg/L)	Total suspended solids (mg/L)	Total iron (mg/L)	Chemical oxygen demand (mg/L)
$FeSO_4$ study	1	36.0	20.0	-	-	-	-	41.0	4.9	-
$FeSO_4$ study	1	65.0	48.0	-	136	450	430	-	4.1	3,200
1	1	53.0	42.0	1,300	38	480	-	24.0	-	3,100
	2	64.0	-	-	47	-	186	110.0	-	-
	3	63.5	0	-	16	-	164	14.0	13.0	-
	4	1.0	0[b]	1,006	38	326	170	11.0	12.0	2,400
1	1	72.5	-	1,024	28	475	28	35.0	-	2,117
	2	80.5	-	-	77	-	255	196.0	55.0	-
	3	70.5	0[b]	-	38	-	192	226.0	33.0	-
	4	12.5	0[b]	922	31	364	275	240.0	45.0	2,195
2	1	74.0	-	-	63	-	-	42.0	6.1	4,185
	3	55.1	35.0	-	57	-	-	8.0	5.8	-
	4	31.5	0[b]	-	24	-	-	48.0	14.0	5,952
3	1	67.0	-	-	43	-	-	35.0	3.6	2,880
	3	70.0	-	-	21	-	240	12.0	7.5	-
	4	9.0	-	-	7	325	190	3.0	5.3	2,680
4	1	34.0	-	-	47	-	-	18.0	5.3	5,214
	2	68.0	-	-	26	-	-	127.0	6.0	-
	3	61.9	2	-	20.3	0	310	24.0	17.0	-
	4	2.0	3.6	1,015	12	1,020	230	21.0	0.2	4,547
5	1	38.0	-	-	-	-	-	-	6.1	3,919
	2	-	-	-	-	-	-	25.0	-	-
	3	54.0	-	-	-	-	-	56.0	-	-
	4	2.5	-	-	-	812	310	-	1.9	3,298
6	1	-	-	-	64	-	-	39.0	5.4	1,680
	2	-	-	-	22	-	-	31.0	10.2	-
	3	62.0	-	-	13	-	-	40	11.6	-
	4	2.0	-	-	-	-	-	2	1.03	1,416
7	1	61.0	19.0	-	-	-	-	26	6.0	-
	2	-	-	-	-	-	-	36	-	-
	4	-	-	1,722	-	142	-	2	-	-

[a]Site 1 = Raw phenol pit waste; Site 2 = Clarifier effluent; Site 3 = Filtration effluent/ion exchange influent; Site 4 = Final ion exchange effluent.

[b]Number actually is negative due to interferences; reported as zero.

Source: Reference 14.

TABLE 13.1.6. RESULTS OF CONTINUOUS FLOW TESTS ON GOLD MILL EFFLUENTS USING A STRONG BASE ANION EXCHANGER

Test	Feed solution	Initial $[CN_T]$ (mg/L)	Hydraulic loading (mL/min/mL resin)	Breakthrough point $[CN_{Fe}]$ (mg/L)	$[CN_T]$ (mg/L)	No. of bed volumes	Effective exchange capacity (mg/mL resin) CN_T	CN_W	CN_{Fe}	Fe	CU	Zn
1	Ferrocyanide	110	0.41	1.8	1.8	280	30		30	11		
2	Ferricyanide	290	0.41	3.0	3.0	120	34		34	13		
3	Ferro/Ferri cyanides	210	0.31	2.6	2.6	120	25		25	8.2		
4	Iron, copper, zinc cyanides	470	0.30	0	3	50	24	11	13	4.7	6.1	8.0
5	Raw barren bleed	378	0.30	0.5	64	75	26	20	5.7	2.0	14	1.1
6A	Raw barren bleed	365	0.40	0.1	62	90	27	20	6.8	2.6	17	
7	Raw barren bleed	370	0.28	2.2	31.6	70	26	20	6.0	1.7	13	2.6
8	Raw barren bleed	365	0.09	0	109	100	26	18	7.6	2.9	15	
9	Raw barren bleed	378	1.7	10	245	80	-19	-14	-5	-2	-9	-3
10	Treated barren bleed	54	0.30	3.3	3.3	280	14		14	4.5		
11	Tailings Pond decant	15	0.37	0.4	0.6	1,480	21	-1	20	10		
12	Field test	20	0.38	3.0	16.5	1,840	26	6.7	19	7.8	4.3	

Source: Reference 15.

pH. However, for feed solutions containing copper and zinc cyanides (tests 4 through 9), the number of bed volumes treated prior to breakthrough was significantly lowered due to competition for available exchange capacity.

Limited testing was done using cyanide stream from the cyanide leaching process to evaluate the resin exchange capacity after multicycling. The data presented in Table 13.1.7 indicates a 25 percent loss of resin capacity for total cyanide after the first regeneration with subsequent losses of about 1 percent/cycle. This phenomena has been reported by numerous investigators and presents a continuing problem in using strong base anion exchangers to remove cyanides.

Earlier, Union Carbide investigated several anion exchange resins (16 to 50 mesh) in the chloride form for effectiveness in treating zinc and zinc cyanide electroplating wastes.[16] The anion exchangers evaluated were:

- Dowex 1 - A strongly basic anion exchanger;
- Dowex NC-20771 - A weakly basic anion exchanger;
- Amberlite IRA-93 - A weakly basic anion exchanger; and
- Amberlite XE-275 - A macroreticular, weakly basic anion exchange resin possessing tertiary amine functionality in a cross-linked acylic matrix.

Removal of zinc and zinc cyanide (Table 13.1.8) from the electroplating waste solutions was greater than 97 percent of equilibrium pH for all four of the anion exchange resins tested. However, as indicated in Table 13.1.9, relatively large concentrations of stripping solution (NaOH) were required to regenerate the column except in the case of Amberlite XE-275.

13.1.3 Process Costs

Process costs for ion exchange processes have been provided in detail in Section 8.2.

13.1.4 Overall Process Status

Typically, ion exchange for cyanides removal has been applied as a polishing step to sorb any ferricyanide or other complexed cyanide residuals

TABLE 13.1.7. RESULTS OF MULTICYCLE ION EXCHANGE STUDIES ON RAW BARREN BLEED

Test	Cycle	Initial [CN_T] (mg/L)	Hydraulic loading (mL/min/mL resin)	Breakthrough point			Effective exchange capacity (mg/mL resin)				
				[CN_{Fe}] (mg/L)	[CN_T] (mg/L)	No. of bed volumes	CN_T	CN_W	CN_{Fe}	Fe	Cu
6A	1	365	0.40	0	62	90	27	20	6.8	2.6	17
6B	2	365	0.40	1	134	90	21	14	6.8	2.5	12
6C	3	365	0.40	12	159	90	19	13	5.8	2.3	11
6D	4	365	0.40	8	144	90	20	14	6.1	2.3	12
6E	5	365	0.40	11	161	90	18	13	5.9	2.2	9.2
6F	6	365	0.40	5	152	90	19	13	6.4	2.2	11
6G	7	365	0.40	11	159	90	19	13	5.9	2.2	9.5

Source: Reference 15.

TABLE 13.1.8. SORPTION OF ZINC CYANIDE AND CYANIDE FROM AN INDUSTRIAL ELECTROPLATING WASTE SOLUTION BY VARIOUS ANION EXCHANGERS

Anion exchanger	Equilibrium pH	Percent sorbed	
		Zn	Cn
XE-275	11.8	7.0	8.0
	10.4	48.0	53.0
	9.9	72	97
	9.8	87	97
	9.4	94	97
Dowex-1	11.8	97	97
	9.9	98	97
	8.9 - 7.5	99	97
IRA-93	8.4 - 7.3	91.0	97
NC-20771	8.4 - 7.6	87	97

Source: Reference 16.

TABLE 13.1.9. STRIPPING OF ZINC CYANIDE FROM VARIOUS ANION EXCHANGERS AS A FUNCTION OF NaOH CONCENTRATION

	Zinc stripped (percent)			
NaOH, M	XE-275	Dowex-1	IRA-93	NC-20771
10	91.8	97.4	97.0	96.5
8	92.1	96.9	97.7	97.0
6	98.0	96.3	98.3	97.7
4	99.2	93.8	98.2	97.6
2	98.8	88.8	97.1	97.9
1	98.7	74.2	92.5	94.7
0.5	97.9	49.7	79.6	89.0
0.1	96.7	8.9	40.0	45.5

Source: Reference 16.

from oxidation processes such as alkaline chlorination. The environmental impact from this technology is that it concentrates cyanides in the regeneration step, creating a secondary stream that needs to be treated.

The advantages of this technology are that it has been demonstrated at both the bench-scale and pilot-scale. The equipment is compact, versatile, and is generally applicable to many different waste treatment situations. Limitations include the high cost of regenerative chemicals and the waste streams originating from the regeneration process are relatively high in pollutant concentration. In addition, if more than 25 mg/L of suspended solids and/or more than 20 mg/L of oil exists in the influent, filtration is required as pretreatment. Also, the stream to be treated should not contain any materials that cannot be removed by the backwash operation. Some organic compounds, particularly aromatics, will be irreversibly absorbed by the resins, and this will result in decreased capacity.

REFERENCES

1. U.S. EPA. Treatability Manual, Volume III. EPA-600/8-80-042. July 1980.

2. Kunz, R.E., Casey, J., and J. Huff. A Review of Cyanide in Refinery Wastewaters. 3rd. Annual Conference on Treatment and Disposal of Industrial Wastewaters and Residues. 1978.

3. J. Ciancia. New Waste Treatment Technology in the Metal Finishing Industry. Plating 60 (10). 1973.

4. Avery, N.L., and W. Fries. Selective Removal of Cyanide from Industrial Waste Effluents with Ion-Exchange Resins. Industrial Engineering Chemical Products Research and Development 14 (2). 1975.

5. Wilk, L., Palmer, S., and M. Breton, Alliance Technologies Corporation. Treatment Technologies for Corrosive-Containing Wastes. Contract No. 68-02-3997. October 1986.

6. Avery, N.L., and W.H. Waitz. Ion-Exchange Treatment Process for Selective Removal of Cyanide. Amber-Hi-Lites, Rohm and Haas Technical Brochure. 1977.

7. Z. Hubicki. Purification of Nickel Sulfate Using Chelating Ion Exchangers. Hydrometallurgy, 16. 1986.

8. T.J. Reynolds. Unit Operations in Environmental Engineering. Rheinhold Publishers, New York, NY. 1982.

9. Robertson, W.M., James, C.E., and J.Y. Huang. Recovery and Reuse of Waste Nitric Acid From an Aluminum Etch Process. 35th. Industrial Waste Conference, Purdue University. 1981.

10. C. Fontana, Eco-Tech, Ltd. Personal communication with L. Wilk, Alliance Technologies Corporation. August 21, 1986.

11. U.S. EPA. Control and Treatment Technology for the Metal Finishing Industry-Ion Exchange. EPA-625/8-81-007. June 1981.

12. C. Fontana, Eco-Tech, Ltd. Personal communication with L. Wilk, Alliance Technologies Corporation. August 26, 1986.

13. Trachtenberg, J.J., and M.A. Murphy. Removal of Iron Cyanide Complexes from Wastewater Utilizing an Ion Exchange Process. Light Metal. 1979.

14. R.A. Bessent, et al. Removal of Cyanides from Coke Plant Wastewaters by Selective Ion Exchange Results of Pilot Testing Program. 34th. Industrial Waste Conference, Purdue University. 1979.

15. D.T. Vachon. Removal of Iron Cyanide From Gold Mill Effluents by Ion Exchange. Water Science Technology, Vol. 17. 1985.

16. F.L. Moore. Oak Ridge National Laboratory. An Improved Ion Exchange Resin Method for Removal and Recovery of Zinc Cyanide and Cyanide from Electroplating Wastes. Journal of Environmental Science, (7). 1976.

13.2 FLOTATION/FOAM SEPARATION

Flotation/foam separation is the separation of finely divided solid particles from a bulk solution by attachment to fine air bubbles introduced into the solution. The bubbles contact the suspended solids and bring them to the liquid surface where they are retained as a foam.[1,2,3]

The mechanism of bubble attachment is accomplished through the addition of a suitable surfactant called a collector.[3,4] The principles and physical models used to describe the attachment of contaminant particles to air-water interfaces in the presence of a surfactant are well understood and have been previously described in Section 10.4. With respect to cyanide removal through flotation/foam separation, iron salts are introduced to the wastewater stream to complex free cyanide and reduce its toxicity. When precipitated with excess iron, the iron-cyanide complexes can be removed by flotation using a cationic surfactant (see Table 13.2.1 for list of commonly used flotation surfactants).[3]

A disadvantage of this process is that flotation like ion exchange is physical separation technology.[2,5] Therefore, use of this technology will result in a low volume, but highly concentrated toxic by-product wastestream. This wastestream will require either some sort of secondary oxidative treatment (ozone and ultraviolet radiation, wet air oxidation, etc.) or solidification/encapsulation prior to land disposal.

Currently, flotation/foam separation of cyanide bearing wastestreams is still in a preliminary stage of development. Research into possible applications has been ongoing for over 15 years, but no large scale commercial applications have been reported in the literature. Therefore, when considering flotation for possible industrial utilization, it is important to note that further research will be needed to determine its applicability to specific waste streams.

13.2.1 Process Description

The general process equipment used for the flotation of complexed cyanides is similar to equipment utilized in the flotation of complexed metals.[2,5-7] Figure 13.2.1 illustrates a simple flotation system used to

TABLE 13.2.1. TYPICAL FLOTATION SURFACTANTS

Type	Formula[a]	Charge on the soluble ion
Sulfhydryl collectors[b]		
xanthate	ROCSSNa	anionic
dithiophosphate	$(RO)_2PSSNa$	anionic
monothiocarbamate	RHNCSOR	
thiol (mercaptan)	RSH	
dixanthogen	$(ROCSS)_2$	
thiocarbanilide	$(C_6H_5NH)_2CS$	
Colloidal electrolythes[c]		
fatty acids and their soaps	RCOOH, RCOONa	anionic
alkyl or aryl alkyl sulfonates	RSO_3Na	anionic
alkyl sulfate	$ROSO_3Na$	anionic
primary amine salt	RNH_3Cl	anionic
secondary amine salt	R_2NH_2Cl	anionic
quaternary ammonium salt	$RN(CH_3)_3Cl$	anionic

[a] $R = CH_3(CH_2)n$

[b] For sulfides, $R = C_2 - C_5$.

[c] Generall straight chain C_{12} to C_{18}, or a benzene or naphthalene ring may be incorporated into the R group.

Source: Reference 3.

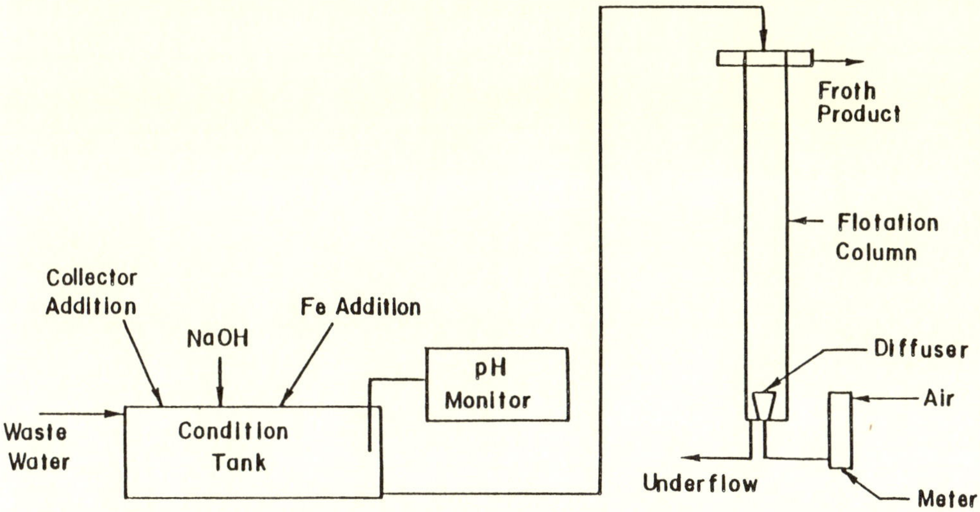

Figure 13.2.1. Flotation system.

Source: Reference 8.

treat cyanide wastestreams. In this system, the solutions to be treated are conditioned over a given period of time in an agitated conditioning cell.[8] The resulting slurry is then introduced to the top of a flotation column. Air is introduced through a sintered glass diffuser at the bottom of the column and the froth product is removed at the top of the column as overflow.

Important parameters which affect cyanide removal and which must be experimentally determined prior to full-scale application include: type of surfactant, conditioning tank retention time, flotation column retention time, air flow rate, feed concentration, and feed pH.[9] The type and dosage of surfactant added are important since at low surfactant dosages the recovery is impaired because there is not enough surfactant present to react with all the influent ferrocyanide. At high surfactant dosages the feed becomes emulsified and restabilized thereby limiting separation efficiencies. Condition tank retention time will vary with the influent feed and type of conditioning chemicals used, but generally recovery will increase rapidly with increasing conditioning time until a steady state is reached (10-30 minutes).[9]

Residuals produced by chemical flotation consist primarily of the cyanide-laden foam which is skimmed or drawn off the top of the reaction vessel/column.[8,9] Post-treatment typically consists of sedimentation and sludge consolidation. The resulting hazardous sludge or by-product wastestream must often be treated (e.g., oxidation encapsulation) and then discharged to the sewer or land disposed depending on the post-treatment method utilized.[10]

13.2.2 Process Performance

One of the first experimental investigations into flotation of cyanide bearing wastestreams was performed by Battelle Laboratories in 1971.[11] The experimental apparatus for the study consisted of a specially designed bench scale glass flotation cell. In the first series of experiments several anionic collectors were screened for flotation effectiveness. The compounds selected consisted of primary, tertiary, or quaternary-ammonium compounds while the complexing material consisted of either 10.8 ppm cadmium or 5.64 ppm nickel in an aqueous stream.

The results of the first series of tests is shown in Table 13.2.2. It was found that the nickel cyanide complexes could be removed much more effectively than cadmium cyanide complexes. Subsequent experiments examined the effect of a quaternary-ammonium compound collector (tetradecylamine) on iron cyanide solutions at various feed pHs and cyanide concentrations.[11] The results presented in Table 13.2.3 are nonconclusive, but do show some general trends. For example, high extractions were obtained only when the solution was prepared by adding ferrous iron to a basic cyanide solution. In slightly acid solution, the complex between ferrous or ferric ion and cyanide did not occur and low extractions were obtained.

Later investigators such as Clarke and Wilson, Bucsh and Lower, and Szarawara built on this earlier work by further studying the utility of flotation for treatment of cyanide bearing wastestreams. Clarke and Wilson reported that adsorbing colloid flotation could remove 92 percent of available free cyanide at an optimum pH of 5.5.[12] In this technique, the iron cyanide precipitate was adsorbed onto ferric hydroxide flox using sodium lauryl sulfate as a collector. Bucsh and Lower used ion flotation to concentrate ferrocyanide (pH 4-10) using Aliquot 336 as a surfactant.[13] Removals of approximately 70 percent were achieved for both ferri and ferrocyanide while free cyanide had only a 28 percent removal. Szarawara reported that upon addition of ferrous iron to cyanide solutions, the cyanide concentration would be at a minimum between pH 8 and 9 as a result of the formation of the complex $Fe(CN)_6^{-4}$ followed by the precipitation of $Fe_2Fe(CN)_6$.[14]

In 1983, researchers at Michigan Technological University performed an evaluation of the removal of cyanide from coke making waters by precipitate flotation of iron cyanides.[8] Parameters investigated included type of surfactant, surfactant dosage, interference of salt concentration, pH, and reaction time. Table 13.2.4 lists the type of flotation reagents used, while Table 13.2.5 reports the effect of surfactant type for ARMAC 12D and Sodium Lauryl Sulfate (NLS). As shown, much higher recoveries were obtained with the amine collector then with the sulfate collector. Subsequent tests indicate that the most efficient collector was the primary amine, dodecylamine acetate (pH 8.0). The quaternary amine, aliquat A-336 was effective up to pH 6.0, but recoveries dropped sharply at higher pH values. The investigators ultimately concluded that recovery efficiencies of 95 to 99 percent could be expected

TABLE 13.2.2 EXPERIMENTAL DATA ON VARIOUS COLLECTORS FOR FLOTATION OF CADMIUM CYANIDE AND NICKEL CYANIDE COMPLEXES

Collector Used	Cadmium cyanide runs				Nickel cyanide runs			
	Amount used cc[a]	ppm CN Initial[b]	ppm CN Final	Indicated percent removal	Amount used cc[a]	ppm CN Initial[c]	ppm CN Final	Indicated percent removal
(1) Dodecylamine HCl	4.0	10.0	5.25	47				
(2) Tetradecylamine HCl	2.0	10.0	6.25	37	2.0	10.0	0.75	93
(3) Hexadecylamine HCl	2.0	10.0	7.50	25				
(4) N,N-Dimethyldodecylamine HCl	4.0	10.0	6.75	32	3.0	10.0	0.75	93
(5) Decyltrimethylammonium bromide	4.0	10.0	4.25	57	2.0	10.0	0.50	95
(6) Ethylhexadecyldimethylammonium bromide	0.5	10.0	7.75[d]	22	0.5	10.0	0.50[d]	95
(7) Hexadecylpyridinium chloride	0.5	10.0	7.50[d]	25	1.0	10.0	1.00[d]	90

(a) Solutions were made by dissolving collector in isopropanol and adjusting to 20 gal/L. The amine collectors also were neutralized with HCl to a pH or 7.

(b) Initial solutions also contained 10.8 ppm cadmium as cadmium chloride.

(c) Initial solutions also contained 5.64 ppm nickel as nickel sulfate.

(d) Excessive foaming occurred during these runs causing loss of some solution.

Source: Reference 11.

TABLE 13.2.3. FLOTATION DATA ON IRON CYANIDE SOLUTIONS

Expt. No.	Initial Solution ppm CN	ppm Fe	pH[a]	pH During Flotation[a]	Final Solution Analyses,[b] ppm CN	Apparent Percent Extraction
16A	10.0	3.58 ferrous	Basic	4.0	1.25	87.5
16B	10.0	3.58 ferrous	Acid	4.3	7.75	22.5
16C	10.0	3.58 ferric	Basic	4.0	6.25	37.5
17A	10.0	3.58 ferrous	Basic	8.4	2.35	76.5
17B	10.0	3.58 ferrous	Basic	6.5	1.95	80.5
17C	10.0	3.58 ferrous	Basic	5.1	2.10	79.0
17D	10.0	3.58 ferrous	Basic	4.0	2.65	73.5
18A	10.0	3.58 ferrous	Basic	≈4	2.70	73.0
18B	10.0	5.37 ferrous	Basic	≈4	1.05	89.5
18C	10.0	7.16 ferrous	Basic	≈4	1.25	87.5
18D	10.0	3.58 ferrous 3.58 ferric	Acid	≈4	9.50	5.0

(a) Adjustments of pH were made by adding dilute HCl or NaOH.

(b) Solutions were floated by adding 0.5 cc of tetradecylamine collector and aerating for 10 minutes.

Source: Reference 11.

TABLE 13.2.4 FLOTATION REAGENTS

Reagent name	Manufacturer	M.W.	Formula
ARMAC 12D Dodecylamine Ethyl acetate	Armour Industrial Chemical Company	245	$Ch_3(CH_3)_{11}NH_3OCOCH_3$
Sodium Lauryl Sulfate	Aldrich Chemical Company	288	$CH_3(CH_2)_{11}OSO_3Na$
Aliquat 336 Tricaprylyl Methyl Ammonium Chloride	General Mills Chemical Division	X = 442	R_3NCH_3Cl $R = C_8 - C_{10}$ carbon
4-Methyl 2-Pentanol	Unknown	1-2	$(CH_3)_2CHCH_2CHOHCH_3$

TABLE 13.2.5 EFFECT OF TYPE OF COLLECTOR

Run	Surfactant	Level mg/L (mM)	pH	Fr	Rf
29	NLS	46.2(0.16)	4	.45	.28
30	12D	156(0.64)	4	.90	.88
31	NLS	23.1(0.08)	4	.42	.28
32	12D	31.3(0.13)	4	.87	.85
33	NLS	46.2(0.16)	6	.91	.89
34	12D	156(0.64)	6	.94	.80
36	NLS	23.1(0.08)	6	.84	.80
37	12D	31.3(0.13)	6	.90	.89

100 mg/L Total CN
215 mg/L Total Fe II
Fr = fraction removed

$$R_f = \text{removal factor} = 1 - \frac{\text{cyanide in the underflow}}{\text{cyanide in the feed}}$$

Source: Reference 8.

from cyanide solutions containing little or no competing ions. For actual cyanide wastewaters such as coke plant effluents, maximum efficiencies were reduced to 91 percent.

13.2.3 Process Cost

Presently, little cost data has been reported on flotation/foam separation as a treatment process for the physical removal of cyanides from wastestreams. While capital and operating costs associated with this technology are expected to be low, no precise costs have been developed. A primary cost (and environmental liability) anticipated from the use of flotation/foam seperation is for the secondary treatment and final disposal of the iron-cyanide flotation sludge. Secondary treatment will consist either of a destruction technology (e.g., wet air oxidation or UV ozonation) which can successfully treat iron cyanides or a solidification/encapsulation technology which will immobilize the cyanide pollutants contained in the flotation sludge. The inclusion of these secondary treatment costs are expected to add significantly to the overall treatment costs.

13.2.4 Overall Process Status

Flotation/foam separation of cyanide bearing wastewaters has not yet been tested on a pilot-scale at an actual commercial facility. Most of the research that has been performed to date with flotation has focused on equipment development and process parameter definition. Although preliminary research has demonstrated the technical feasibility of the process, pilot-scale testing is needed to determine if sufficient cyanide recoveries can be achieved. Flotation could prove to be a cost-effective alternative to conventional treatment practices because of its minimal operating requirements.

As with all physical separation processes, any process which will concentrate the cyanide waste material should be followed by a process which will detoxify or immobilize the concentrate. Since the concentrate of the process considered here contains precipitated ferri- and ferrocyanide which are not amenable to conventional oxidation technologies such as alkaline chlorination, alternate technologies such as wet air oxidation or UV/ozonation may be more appropriate. In addition, solidification or encapsulation of the residuals may be required prior to land disposal.

REFERENCES

1. U.S. EPA. Treatability Manual. Volume III. EPA-600/8-80-042. July 1980.

2. Sundstrom, D.W., and H.E. Klei. Wastewater Treatment. Prentice-Hall, Inc., Englewood Cliffs, NJ. 1979.

3. Kirk-Othmer Encyclopedia of Chemical Technology. Vol. 10, 3rd. Edition, John Wiley & Sons, New York, NY. 1981.

4. R.R. Klimpel, Dow Chemical Company. Use of Chemical Reagents in Flotation. Chemical Engineering. September 3, 1982.

5. T.D. Reynolds. Unit Operations and Processes in Environmental Engineering. PWS Publishers, Boston, MA. 1982.

6. E.L. Thackston, et al. Lead Removal with Adsorbing Colloid Flotation. Journal of the Water Pollution Control Federation. February 1980.

7. Huang, S., and D.J. Wilson. Hexavalent Chromium Removal in a Foam Flotation Pilot Plant. Separation Science and Technology, 19. 1984.

8. Lower, G.W., and D.J. Spottiswood, Michigan Technological University. Cyanide Removal from Coke Making and Blast Furnace Wastewaters. EPA-600/2-83-066. August 1983.

9. Busch, R.O., Spottiswood, D.J., and G.W. Lower. Ion Precipitate Flotation of Iron-Cyanide Complexes. Journal of the Water Pollution Control Federation. December 1980.

10. Currin, B.L., Potter, F.J., and D.J. Wilson. Surfactant Recovery in Adsorbing Colloid Flotation. Separation Science 13(4). 1978.

11. Battelle Laboratories. An Investigation of Techniques for Removal of Cyanide from Electroplating Wastes. Water Pollution Control Research Series. 12010 EIE. November 1971.

12. Clark, A.N., and D.J. Wilson. The Removal of Mettallo-Cyanide Complexes by Foam Flotation. International Conference on Management and Control of Heavy Metals in the Environment. September 1979.

13. Bucsh, R.O., and G.W. Lower. Cyanide Removal from Coke Making and Blast Furnace Waste Waters. EPA-600/9-81-017. March 1981.

14. J. Szarawara, et al. Studies of the Equilibria of Complex Formation of Cyanides with Ferrous Sulphate. Chemia Stosowana, IVI, 79. 1972.

14. Chemical Destruction of Cyanides

The cyanide destruction processes discussed in this section are based on chemical methods of separation and destruction of cyanide contaminants in the waste feed stream. These unit processes are:

14.1 Alkaline Chlorination

14.2 Ozonation

14.3 Wet Air Oxidation

14.4 Sulfur-Based Technologies

14.5 Miscellaneous Processes

The cyanide waste streams treated by these processes are produced by several industries including ore extraction (cyanide leaching), photographic processing, synthetic organic and inorganic compound manufacturing, and metal finishing. The most significant source of hazardous cyanide waste is the metal finishing industry. Aqueous solutions with free cyanide, ionic cyanides, and highly soluble metal cyanide complexes are of major environmental concern. Aqueous cyanide waste solutions from the metal finishing industry include contaminated rinse water and spent process solutions.

14.1 ALKALINE CHLORINATION

14.1.1 Process Description

Alkaline chlorination of dilute cyanide waste streams is a waste treatment technology which has been in commercial use for over 25 years.[1] The process is suitable for destroying free dissolved hydrogen cyanide and for oxidizing all simple and most complex inorganic cyanides in aqueous media. The process is operated at ambient temperature; with good pH and oxidation-reduction potential (ORP) control, the effluent typically contains less than 1.0 ppm cyanide.[1,2]

The destruction reaction is an oxidation process in which one or more electrons are transferred from the chemical being oxidized (cyanide) to the chemical initiating the transfer (oxidizing agent).[3,4] Chlorine in elemental form or hypochlorite salt are the two most common oxidizing agents used in industrial cyanide oxidation systems.

The mechanism of cyanide destruction by alkaline chlorination is shown by the following equations:

$$Cl_2\ (g) + NaCN = CNCl + NaCl$$

$$CNCl + 2NaOH = NaCNO + NaCl + H_2O$$

$$3Cl_2\ (g) + 2\ NaCNO + 6\ NaOH = 2NaHCO_3 + N_2\ (g) + 6\ NaCl + 2H_2O$$

In this reaction chlorine gas (Cl_2) plus sodium hydroxide (NaOH) are used to oxidize cyanides to cyanates (CNO^-) and ultimately to carbon dioxide and nitrogen. The formation of cyanide chloride (CNCl) is essentially instantaneous. Sodium hypochlorite (NaOCl) is often used in place of chlorine gas due to the danger and higher equipment costs involved with chlorine usage. The stoichiometry is the same in terms of equivalents of chlorine added, but alkali additions and unit reagent costs (sodium hypochlorite is approximately twice as expensive as chlorine gas) will vary with the oxidizing agent used.

Alkaline chlorination treatment of cyanide solutions can be conducted in one or two stages.[1,2,5] In the more commonly used two stage process, solution pH is initially raised to a pH of 10 of higher. Hydrolysis of the cyanogen chloride complex is rapid and the reaction is typically 80-90 percent complete within two minutes. In the second stage, the pH of the solution is reduced to the 8.0-8.5 range for rapid oxidation of cyanate. Retention time in the second stage is generally 30 minutes to 1 hour in order to ensure complete cyanide destruction.[6] Alternatively, an intermediate pH between 8.5 to 10.0 can be maintained in a single tank for simultaneous completion of both stages. In the single stage system close pH control is essential, and retention time will depend upon the selection of pH and the amount of hypochlorite present. Figure 14.1.1 illustrates a conventional two-stage cyanide oxidation system. The system features separate pH controlled addition of caustic and ORP controlled addition of chlorine to each stage if necessary. Table 14.1.1 presents treatment levels for cyanide wastewaters using both single and two-stage chlorination processes.

Reagent requirements for the theoretical oxidation of cyanide to cyanate are 2.7 lbs. of chlorine and 3.1 lbs of caustic per pound of cyanide.[6] Overall reagent requirements for the complete destruction of cyanides are 6.8 lbs of chlorine and 7.3 lbs of caustic per pound of cyanide. Practical experience, however, has demonstrated that typically 8 lbs of chlorine or more (a 10 percent excess) are required to completely destroy cyanide and meet effluent guidelines.[6] The excess chlorine is used to account for side reactions (organics and reduced metals) and ensure rapid and complete hydrolysis of cyanogen chloride.

The rate equation for the hydrolysis of cyanogen chloride to cyanate is:[7]

$$-d\,[CNCl]/dt = k_1\,[CNCl]\,[OH^-]$$

As indicated by the presence of the hydroxyl group $[OH^-]$, the rate equation shows cyanogen hydrolysis to be pH dependent. The greater the concentration of hydroxyl ions, the more rapid the reaction rate. The reaction has been found experimentally to be most rapid above pH 10, a region of high alkalinity (i.e., excess of hydroxyl ions).

Hydrolysis of cyanogen chloride is greatly accelerated by the presence of hypochlorite, which apparently has a catalytic effect. Competition between CNCl and CNO^- for excess hypochlorite may result in incomplete cyanogen hydrolysis at low pH values. However, at high pH values (greater than pH 10.0) cyanogen hydrolysis is complete before significant CNO^- oxidation occurs.[7]

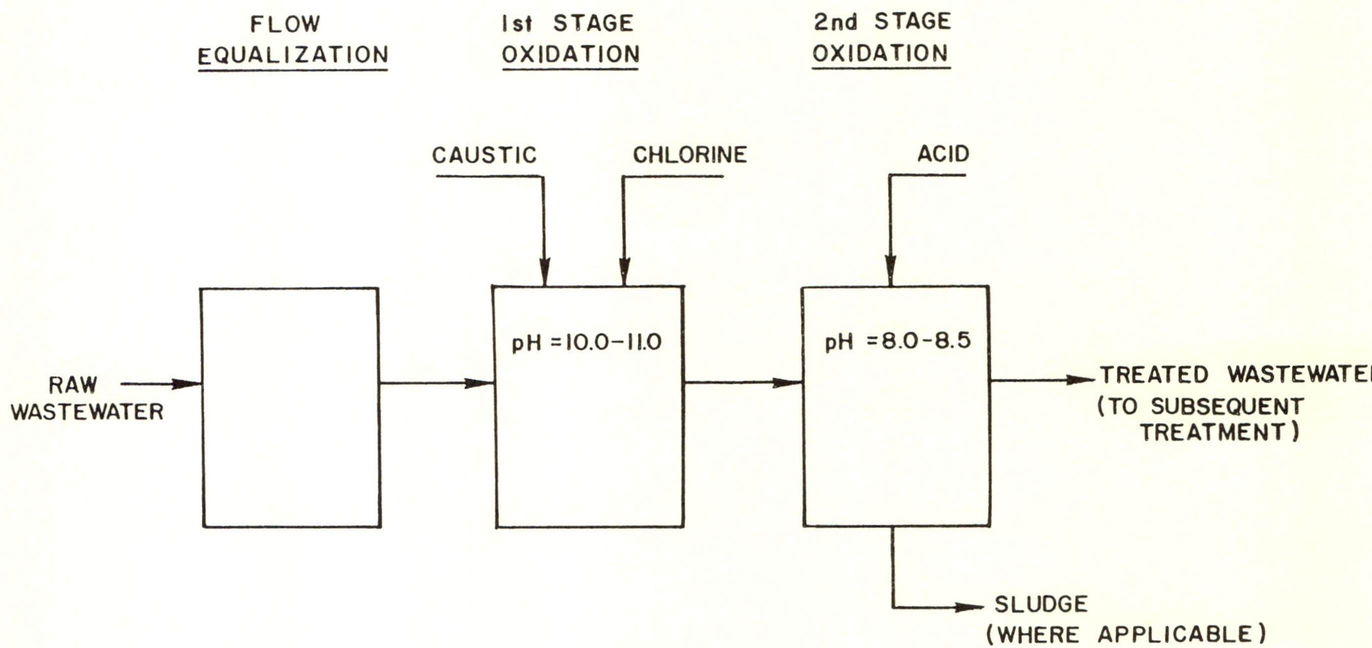

Figure 14.1.1. Treatment flow schematic for 2-stage oxidation process.

TABLE 14.1.1. TREATMENT LEVELS FOR CYANIDE WASTEWATERS

Treatment process	Cyanide Concentration mg/L)		Percent Removal
	Initial	Final	
Alkaline chlorination[a]	---	1.7	---
Alkaline chlorination[a]	---	0.1	---
Alkaline chlorination[b]	---	0.4	---
Alkaline chlorination[b]	700	0.0	100
Alkaline chlorination	32.5	0.0	100
Alkaline chlorination	5.1	0.1	98

[a]Single-stage chlorination.

[b]Two-stage chlorination.

Source: Reference 2.

General processing equipment and construction materials for cyanide oxidation units are identical to those of precipitation, reduction, and coagulation/flocculation processes. A fully engineered two-stage cyanide oxidation system would consist of the following components:[1,2,8]

- 2 treatment tanks
- 3 reagent storage tanks (caustic, chlorine, acid)
- 5 agitators
- 6 pumps
- 2 pH controller/probes
- 2 ORP controller/probes
- piping and valves
- electrical fit-up

The two treatment tanks can be fabricated from a wide range of construction materials, but most industrial systems use fiber reinforced plastic (FRP).[9,10] Most vessels are of a flat-bottomed configuration, equipped with air tight covers or air ducts to minimize exposure to any volatile, toxic reaction products which might be evolved. Each stage should be designed to provide approximately 1-hour retention volume.

Agitation serves the purpose of equalizing the concentration profile within the reaction vessel as the influent is dispersed in the reaction tanks. Vessels with large stagnant areas provide little mixing between reactants and causes large disturbances when concentrated materials are released into the system. For accurate process control, Hoyle has suggested that agitator capacity should be measured as a ratio of the system dead time (the interval between the addition of a reagent and the first observable process change) to the retention time (volume of the vessel divided by the flow through the vessel).[12] A ratio of dead time to retention time of 0.05 approaches an optimum value. Typically agitation is provided overhead in line with the vertical axis. In addition, mechanical agitation should be provided in the reagent storage/slurry tanks to maintain reagent homogeneity.

Pumps and piping are required for all aspects of fluid transfer within the cyanide oxidation system. Pumps are necessary to transport the cyanide waste to the first stage, pump it to the second stage, and then displace the treated fluid from the second tank to whatever post-treatment processes may be appropriate. In addition, a separate chemical metering pump is required to transfer reagent from each of the reagent storage tanks (in smaller system it is sometimes possible to meter directly from a 55-gallon drum) to the treatment system. The many different factors influencing the final choice of pump type and size for fluids are discussed in detail by Peters and Timmerhaus in Reference 13.

At the heart of the alkaline chlorination cyanide destruction system are the pH and ORP control systems. The pH control systems for batch precipitation processes can be quite simple with only on-off control provided via solenoid or air activated valves. Control system designs for continuous flow cyanide oxidation system are more complicated because the wastewater feeds often fluctuate in both flow and concentration. Systems currently available include: proportional, cascade, feedforward, or feedback pH control. Each system has distinct advantages and disadvantages which have been reviewed in the literature.[12,14,15] Both pH and ORP control systems consist of a probe (to take the reading), monitor (to compare the reading set point and make the appropriate adjustment), and a recorder to visually display the resultant data. In addition, there is typically a control panel with an indicator, starters and controls for metering pumps, all relays, high/low alarms, switches, and mixer motor starters.

Table 14.1.2 summarizes typical operating parameters for a two-stage alkaline chlorination systems. Improper chlorination of cyanide ion, hydrogen cyanide, or thiocyanate ion, particularly under conditions below pH 10, will result in increased evolution of cyanogen chloride, a gas which is considered to be at least as hazardous as hydrogen cyanide. Cyanide in combination with nickel, cobalt, silver, or gold is oxidized slowly, but is still treatable if sufficient time is provided.[2]

A pretreatment in itself, alkaline chlorination is usually applied to cyanide bearing aqueous waste streams segregated from other process flowstreams. Segregation is essential to prevent the formation of difficult to treat wastes or the evolution of toxic gases.

TABLE 14.1.2. TYPICAL OPERATING PARAMETERS OF A TWO-STAGE ALKALINE CHLORINATION CYANIDE DESTRUCTION UNIT

Parameter	Unit	Range
Influent		
Cyanide concentration	mg/L	1 - 1,000[a]
Influent		
Flowrate pressure	gpm	10 - 350
Pressure	atm	1
Temperature	°C	20-22
Agitation	turnover/minute	1
First-Stage		
pH	--	9.5 - 11
ORP	Mv	350 - 400
Chlorine	lb/lb CN	2.7 - 3.0
Caustic	lb/lb CN	3.1 - 3.4
Retention time	Min	30 - 60
Second-Stage		
pH	--	8.0 - 8.5
ORP	Mv	600
Chlorine	lb/lb CN	4.1 - 4.5
Caustic	lb/lb CN	4.2 - 4.6
Retention time	min	30 - 60
Effluent		
Cyanide	mg/L	<0.1

[a]Initial cyanide concentrations of up to 5,000 mg/L are possible, but require batch treatment. Optimum influent cyanide concentrations for continuous systems are <100 mg/L.

Source: Adapted from References 1, 2, 6, 9.

For example, alkaline chlorination cannot effectively oxidize stable iron and nickel cyanide complexes. As most cyanide discharge limits are based on total cyanide levels, provisions should be made to ensure that cyanide solutions do not mix with iron and nickel compounds. Similarly acid-bearing waste streams should be segregated from the cyanide bearing wastestream to prevent pH depression and the evolution of toxic hydrogen cyanide gas (HCN). Following successful cyanide destruction, the treated cyanide wastestream may then be combined with other waste streams for subsequent treatment (i.e., metals precipitation, coagulation, filtration, etc.).

Other properties of the waste being treated that can affect alkaline chlorination performance include:

- Flow variations
- pH variations
- Presence of chelators/complexants
- Competing nonpriority oxidizable species
- Oil and grease concentration

In facilities which experience a wide variation in flow rates, pH values, or pollutant concentrations of the wastewater, flow equalization as pretreatment is often used.[1,9] A variety of process options exist (see Section 10.1) but all systems basically provide some sort of flow resistance, stream segregation, or influent concentration averaging to prevent wastetreatment system overloading. In all methods of flow equalization, care must be exercised during the wastewater analysis to completely characterize any peak flows or concentrations. In addition, flexibility in system design should be provided for any future expansion, change in location, or deviation in flow rates.

Oil and grease, chelator/complexants; and nonpriority oxidizables, are all factors which will increase reagent consumption and impede if not prohibit chemical oxidation operations.[1] Oil and grease removal is typically the first process step in any waste treatment train. The removal of chelator/complexants and nonpriority oxidizable compounds present more difficult problems since many of these compounds are often an integral part of the cyanide wastestream. The presence of organic compounds and reduced metals can increase chlorine or sodium hypochlorite consumption by as much as 25 to 100 percent over stoichiometric requirements.[2] In addition the presence of cupric cyanide can cause precipitation during the chlorination process. This results in a sludge containing cyanide complexes that may require separate post-treatment. Other inorganic salts which cannot be effectively treated by this process and may require segregation and/or pretreatment include ferro and ferric-cyanides, nickel cyanide, and zinc cyanide.

Residuals generated in the alkaline chlorination process occur from the use of caustic with chlorine gas. Smaller quantities of residual product will result from alkaline chlorinations using hypochlorites. The sludge product consists primarily of insoluble hydroxide compounds generated during the hydrolysis of cyanogen chloride in the first-stage reactor. Therefore some provision for sludge removal or batch clean-out should be provided. However, alkaline chlorination post-treatment is more likely to consist of such unit processes as precipitation, coagulation/flocculation/ sedimentation, and sludge consolidation. The resulting toxic sludge must often then be treated (i.e., encapsulation) and land disposed.

14.1.2 Process Performance

Alkaline chlorination with chlorine or hypochlorites has become the most widely accepted conventional method of cyanide destruction. The stoichiometry and rate factors in cyanide destruction by alkaline chlorination have been researched and thoroughly reported in the literature. Use of this method however becomes increasingly difficult as cyanide and stable inorganic salt-cyanide complex concentrations increases.

Table 14.1.3 summarizes effluent cyanide concentrations for 15 metal finishing plants reviewed in the literature.[8] Total cyanide influent concentrations ranged from 0.045 to 1,680 mg/L, with a median of 77.4 mg/L. As can be seen, alkaline chlorination was successful in reducing 65 percent of the total cyanide waste streams to a final effluent concentration of less than 0.10 mg/L. However, two of the facilities were unable to detoxify total cyanide concentrations to less than 1.0 mg/L. If was postulated that inefficient operation, the presence of stable inorganic complexes (i.e. iron,

TABLE 14.1.3. EFFLUENT CYANIDE PERFORMANCE DATA USING ALKALINE CHLORINATION

Plant ID	Total Cyanide[a] mean effluent concentration (mg/L)	Amenable Cyanide[b] mean effluent concentration (mg/L)
1	0.04	---
2	0.15	---
3	0.09	---
4	2.20	---
5	0.09	---
6	0.10	---
7	1.21	---
8	0.05	---
9	0.001	---
10	0.13	
11	0.46	0.09
12	0.04	---
13	----	0.004
14	0.01	---
15	0.06	0.007

[a]Average daily total cyanide influent concentrations ranged from 0.045 - 1,680 mg/L with a median concentration of 77.4 mg/L.

[b]Amenable cyanide influent concentrations ranged from 0-1,560 mg/L with a 7.63 mg/L median concentration.

Source: Reference 8.

nickel, zinc cyanide), or excessive influent total cyanide concentrations (1,000 mg/L or greater) were responsible for the poor removal efficiencies experienced at these plants.

Table 14.1.4 presents detailed alkaline chlorination operation and performance data for five more facilities. Facilities A, B, and D are batch processes used in lieu of continuous alkaline chlorination.[16,17,19] Facility A uses its system to collect and batch treat spent cyanide baths and floor spills. Therefore, equipment usage is intermittent and process conditions are variable. Facility D operated a batch pilot plant with limited throughput to determine treatment feasibility. Facility B is a commercial wastetreatment plant which accepts and treats large volumes of concentrated cyanides with lime and sodium hypochlorite. The process is limited by two factors. First the initial content of cyanide (CN^-) must not exceed 2,000 mg/L in order for the process to achieve a final cyanide concentration 0.5 mg/L. Secondly the total amount of Cl_2 used should not surpass 6,000 mg/L in order to limit the levels of cyanogen chloride formed during the process. If either process parameter is exceeded, a dilution operation is performed.

Facilities C and E are continuous alkaline chlorination operations.[18,20] Facility E treats cyanide contaminated ore leaching wastewater generated during gold milling operations (see Figure 14.1.2). The gold in the ore is mainly locked in fine grained arsenopyrite (FeAsS), but also contains copper and zinc. The capacity of the treatment plant was 2.5 to 16.8 gpm with tanks constructed of protected (lined) mild steel and of plastic. Plastic piping and rubber hose were used for ease of changing flow patterns. Process operations consisted of oxidation of reduced species (cyanides and arsenites), alkaline precipitation of metallic hydroxides, ferric sulfate precipitation of pentavalent arsenic, and liquid-solid separation. The levels achieved are as follows:

Cyanide	1.0 mg/L
Arsenic	0.2 mg/L
Copper	0.3 mg/L
Zinc	0.2 mg/L
Iron	3.0 mg/L

TABLE 14.1.4. ALKALINE CHLORINATION PERFORMANCE DATA

Parameter	A[a]	B[b]	C[c]	D[d]	E[e]
Wastestream	Metal Finishing	Commercial wastetreat- Facility	Coke and coke by-products	Gold mill barren bleed	Gold mill effluent
Influent Flowrate (gpm)	batch (1000 gal)	batch	167	batch	2.5-16.8
1st stage					
pH	---	11	---	12	11.2-11.8
Retention	---	250-350	---	---	---
Time (min)	---	60	---	90	82
Reagent	---	Lime/ Sodium hypochlorite	---	Lime/ Sodium hypo-chlorite	Lime/ Sodium hypo-chlorite
2nd Stage					
pH	NA	8.5	9.0-9.5	8.5	7.5-10.4
ORP(MV)	NA	NA	120-180	---	---
Retention Time (Min)	NA	5-60	90	60	100
Reagent	Sodium hypochlorite Sodium hydroxide (NaOH)	Waste acid	Chlorine NaOH	Sulfuric acid	NA
Influent Total Cyanide Concentration (mg/L)	0.5-6.8	2,000	83-104	63	300
Effluent total Cyanide Concentration	0.1	0.5	4.7	0.4	0.07

[a]Reference 16.
[b]Reference 17.
[c]Reference 18.
[d]Reference 19.
[e]Reference 20.

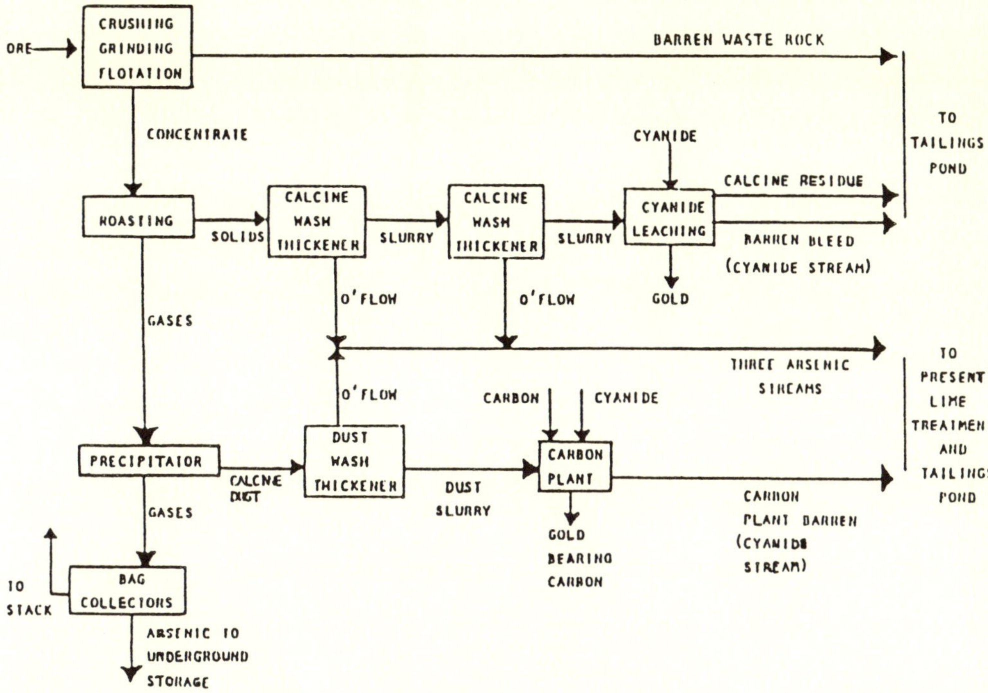

Figure 14.1.2. Gold processing flow diagram.

Source: Reference 20.

Facility C was one of the few facilities to report an inability to achieve effluent limits with alkaline chlorination. High influent ammonia and thiocyanate concentrations were felt to have reacted with some of the excess chlorine. In addition, Facility C reported difficulty in maintaining efficient automatic ORP control. Subsequent test results indicate that chlorine dosage rates of less than 2,000 mg/L should be sufficient to oxidize the cyanide to permitted levels while 2500 mg/L was sufficient to oxidize thiocyanates to below detection limits. Once the chlorination system is effectively automated, it is anticipated that effluent guidelines for cyanide will be met.

While most research on alkaline chlorination has focused on stoichiometry, rate factors, and destruction efficiencies, little work has been performed on sludge generation and handling characteristics. Researchers at the University of Tennessee and Illinois have investigated sludge and supernatant quality following cadmium cyanide destruction and precipitation.[7] The first objective in the investigation was to examine the alkaline chlorination of cadmium cyanide solutions in the pH region of carbonate precipitation. Previous research and field data have shown that carbonate precipitation results in reduced metal solubilities and improved sludge characteristics (see Section 10.1.3).[21,22] As shown previously, an equivalent level of carbonate is produced from the destruction of the cyanide radical.

The second objective was to investigate the effects of two forms of hypochlorite on cadmium solubility and solid phase characteristics. The two forms of hypochlorite investigated were sodium (NaOCl) and calcium ($Ca(OCl)_2$) hypochlorite. Previous work has indicated that sludge produced from calcium hypochlorite oxidation dewater more effectively than the more gelatinous sodium hypochlorite oxidation sludges. This is primarily due to the coprecipitation of calcium carbonate and metallic carbonate which due to calcium granular nature results in distinctly different filterability characteristics.

In the pH region between 7 and 10 for the calcium hypochlorite system both cadmium carbonate and calcium were formed as separate crystals. However, in the optimum range for cyanate oxidation (pH 8.5-9.0) twice as much calcium carbonate as cadmium carbonate was precipitated (on a molar basis). This resulted in a dry weight sludge product of only 22-30 percent cadmium. In contrast sodium hypochlorite cyanate oxidation in the pH range of 8.5-9.0 resulted in 70 to 100 percent of the precipitate formed consisting of cadmium carbonate. This represents a dry weight sludge yield of approximately 65 percent cadmium. Therefore while calcium-based hypochlorite systems may produce a precipitate which filters more readily and to a higher solids content, sodium-based hypochlorite systems theoretically yield a sludge which is more amenable to metals recovery.

14.1.3 Process Costs

Figure 14.1.3 illustrates the process flow schematic developed for the continuous alkaline chlorination system costs contained in this section. The influent waste water stream is assumed to contain 50 mg/L of cyanide ion and 200 mg/L of heavy metal ions. Three flow rates were costed 1,000, 10,000 and 100,000 gallons per hour. These systems were assumed to operate 24 hours per day, 300 days per year. Complete reaction in the cyanide chlorination tanks is assumed to occur and the heavy metals are rendered insoluble in the precipitation reactor.

Cost data and design and operating cost assumptions for the equalization tank, precipitation reactor, flocculator/clarifier, sludge holding tanks, and filter press have been presented previously in Section 10.1.[5,13,23,24] The capital costs for the alkaline chlorination unit has been adapted from Figure 14.1.4. The unit uses sodium hydroxide for pH adjustment and sodium hypochlorite as the oxidizing agent. The operations are conducted in two series-connected reaction tanks in which reagent demand in each stage is determined by measuring pH and ORP. The reaction time in each stage is assumed to be 60 minutes to ensure complete cyanide destruction. The cost for the system also includes storage and feed systems for the treatment reagents.

Table 14.1.5 contains the capital and operating costs for the continuous alkaline chlorination system developed for this section. It is immediately apparent that at the higher flow rates chemical and sludge disposal costs can constitute up to 60 percent of the total annual costs. In addition, the presence of other oxidizable species, stable complexes, or higher influent cyanide concentrations could render this process economically nonviable.

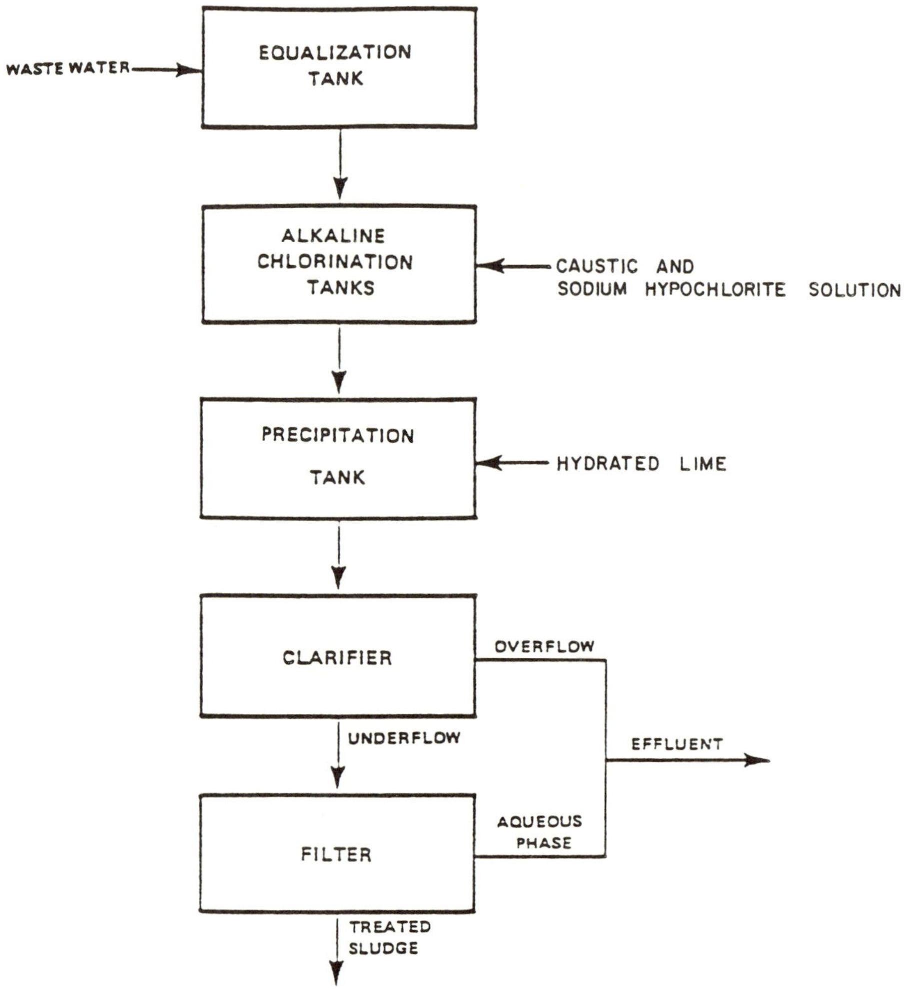

Figure 14.1.3. Alkaline chlorination process.

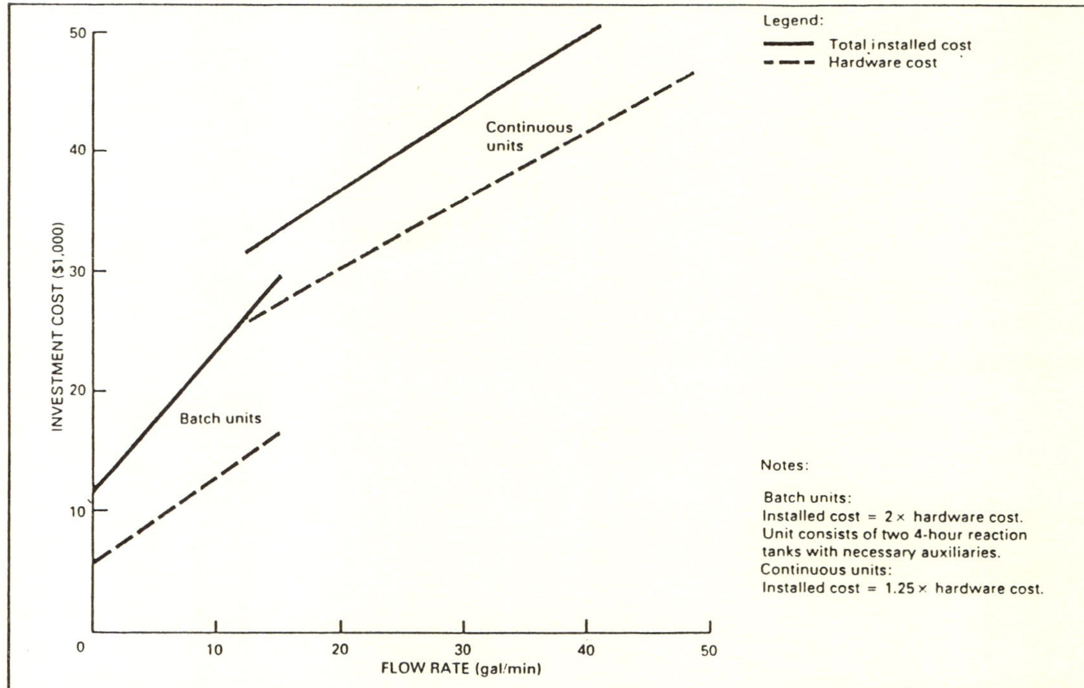

Figure 14.1.4. Investment cost for cyanide oxidation units.

TABLE 14.1.5. CONTINUOUS ALKALINE CHLORINATION TREATMENT COSTS[a]

	Flow rate (gph)		
	1,000	10,000	100,000
Purchased Equipment and Installation (PE&I)			
Equilization Tank	17,000	29,000	50,000
Cyanide Oxidation Units	28,000	81,000	267,000
Precipitation Reactor	24,000	40,000	160,000
Flocculator/Clarifier	18,000	50,000	203,000
Sludge Holding Tank(s)	3,000	6,000	48,000
Filter Press	10,000	25,000	100,000
	100,000	231,000	828,000
Total Capital Investment (360% PE&I)	360,000	831,600	2,980,800
Annual Operating Costs ($/Yr.)			
Operating Labor ($20/hr.)	72,000	72,000	72,000
Maintenance (6% TCI)	21,600	49,900	178,800
General Plant Overhead (5.8% TCI)	20,900	48,200	172,900
Utilities (2% TCI)	7,200	16,600	59,600
Taxes and Insurance (1% TCI)	3,600	8,300	29,800
Chemical Costs:			
NaOH ($175/ton)	1,900	19,200	191,700
NaOCl ($0.38/gal)	7,800	78,200	781,700
Lime ($40/ton)	500	5,300	53,000
Sludge Transportation ($0.25/ton-mile)	200	2,300	22,500
Sludge Disposal ($200/ton)	12,000	120,000	1,200,000
Annualized Capital (CFR-0.177)	63,700	147,200	527,600
Total Cost/year	211,400	567,200	3,289,600
Cost/1000 gallon	29	8	5

[a]1987 dollars.

14.1.4 Status of Technology

Alkaline chlorination systems have generally proven reliable if well maintained and equipped with well-designed ORP control. The treatment technology cannot oxidize stable cyanide complexes such as ferrocyanides and has difficulty treating nickel cyanides. The most widespread application of cyanide oxidation through alkaline chlorination is in facilities using cyanides in electroplating operations.

The evolution of toxic hydrogen cyanide gas may be a problem if pH levels are lowered excessively. In cases where alkaline chlorination is used to treat dissolved complex cyanides and dissolved cyanides of heavy metals, sludges of metal hydroxides and carbonates are generated. These sludges can be recovered by filtration and treated by chemical fixation/solidification.

Many of the chemicals used in this process have potential for hazardous and or toxic effects if catastrophically released during shipment, storage, or handling. Liquid sodium hydroxide (greater than 40 percent) and concentrated sulfuric acid are extremely corrosive.[9,11] Chlorine gas and hypochlorite salts are powerful oxidizers and must be segregated to avoid reaction with other chemicals. For a summary of the advantages and disadvantages of alkaline chlorination see Table 14.1.6.

TABLE 14.1.6. ADVANTAGES AND DISADVANTAGES OF ALKALINE CHLORINATION

Advantages

- Proven technology with documented cyanide destruction efficiencies.
- Operates at standard operation temperatures and pressures and is well suited to automatic control.
- Modular design allows for plant expansion and can be used in different configurations.
- When treating dissolved HCN, calcium, potassium, or sodium cyanide no sludges are generated.

Disadvantages

- Need for careful pH and ORP control.
- Possible chemical interference in the treatment of mixed wastes (i.e., large oxidation chemical excesses required for complete reactions).
- Process is not selective and therefore restricted to specific product wastestreams.
- Potential hazard of shipping, storing, and handling of chlorine gas, hypochlorite salts, sodium hydroxide, and concentrated sulfuric acid.
- Unable to treat ferro and ferricyanides and has difficulty treating nickel cyanide.
- Potential for creating toxic residue which will require post-treatment (i.e., fixation/solidification/ encapsulation).

Source: Adapted from References 1, 2, 6, 7, and 11.

REFERENCES

1. U.S. EPA. Treatability Manual, Volume III. EPA-600/8-80-042. July 1980.

2. Cushnie, G. C. CENTEC Corporation, Navy Electroplating Pollution Control Technology Assessment Manual. CR 84.019. February 1984.

3. Sundstrom, D. W., H. E. Klei. Wastewater Treatment. Prentice-Hall. Englewood Cliffs, N.J. 1979.

4. Chillingworth, M. A. Alliance Technologies Corporation, Industrial Waste Management Alternatives and their Associated Technologies/Processes. GCA-TR-80-80-G. February 1981.

5. U.S. EPA. Reducing Water Pollution Costs in the Electroplating Industry. EPA-625/5-85-016. September 1985.

6. Lanoutte, K. H. Heavy Metals Removal. Chemical Engineering. October 17, 1977.

7. Butcher, B. T., R. A. Minear, and R. B. Robinson. University of Tennessee, Destruction of Cadmium Cyanide Waste by Alkaline Chlorination Treatment. 17th Mid-Atlantic Industrial Waste Conference. 1984.

8. U.S. EPA. Development Document for Effluent Limitations Guidelines and Standards for the Metal Finishing Point Source Category (Proposed). EPA-440/1-82-091-b. August 1982.

9. Mitre Corporation. Manual of Practice for Wastewater Neutralization and Precipitation. EPA-600/2-81-148. August 1981.

10. Cushnie, G. C. Removal of metals from wastewater: Neutralization and Precipitation. Pollution Technology Review, No. 107. Noyes Publication, Park Ridge, NJ. 1984.

11. Kirk-Othmer Encyclopedia of Chemical Technology. Vol. 14, 3rd Edition. John Wiley and Sons, New York, NY. 1981.

12. Hoyle, D. L. Designing for pH control. Chemical Engineering. November 8, 1976.

13. Peters, M. S., and K. D. Timmerhaus. Plant Design and Economics for Chemical Engineers. McGraw-Hill Book Company, New York, NY. 1980.

14. Hoffman, F. How to select a ;pH control system for neutralizing waste acids. Chemical Engineering. October 30, 1972.

15. Jungels, P. R., and E. T. Eoytowicz. Practical pH Control. Industrial Water Engineering. February/March 1972.

16. Martin, J. J. Chemical Treatment of Plating Waste for Removal of Heavy Metals. EPA-R2-73-044. May 1973.

17. Sund, S. Physical/Chemical Processing Options. Hazardous Waste and Hazardous Materials. Vol. 3, No. 2. 1986.

18. Zwikl, J. R., N. S. Buchko, and D. R. Junkins. Physical/Chemical Treatment of Coke Plant Wastewaters.

19. Schmidt, J. W. L. Simovic. and E. E. Shannon. Development Studies for Suitable Technologies for the Removal of Cyanide and Heavy Metals from Gold Milling Effluents. 36th Industrial Waste Conference, Purdue University. 1981.

20. Erkku, H., and L. S. Price. Treatment of Gold Milling Effluents. 36th Industrial Waste Conference, Purdue University. 1981.

21. HSU, D. Y., et al. Soda Ash Improves Lead Removal in Lime Precipitation Process. 34th Industrial Waste Conference, Purdue University. 1978.

22. Mabbett, Cappacio and Associates. Industrial Wastewater Pretreatment Study: Preliminary Engineering Design Report. January 1982.

23. Versar Inc. Technical Assessment of Treatment Alternatives for Wastes Containing Metals and/or Cyanides. Contract No. 68-03-3149. U.S. EPA/OSW. October 1984.

24. U.S. EPA. Economics of Wastewater Treatment Alternatives for the Electroplating Industry: Environmental Pollution Control Alternatives. EPA 625/5-79-016. 1979.

14.2 OZONATION

Chemical oxidation has the potential for removing from wastewaters organic materials which are resistant to other treatment methods, e.g., refractory materials which are toxic to biological systems. Ozone (O_3) is one of the strongest oxidants available, as shown in Table 14.2.1, which lists the oxidation potential and relative oxidation power of a number of oxidizing agents. Ozone, as an oxidant, is sufficiently strong to break many carbon-carbon bonds and even to cleave aromatic ring systems.

Ozone has been used for years in Europe to purify, deodorize, and disinfect drinking water.[1] More recently, it has been used in the waste treatment area to oxidize cyanide wastewaters. Cost and mass transfer considerations restrict usage of ozone to the treatment of wastewaters with 1 percent or lower contaminant concentration levels.[2,3] Since oxidation by ozone occurs nonselectively, it is also generally used only for aqueous wastes which contain a high proportion of hazardous constituents versus nonhazardous oxidizable compounds, thus focusing ozone usage on contaminants of concern. Ozonation may be particularly useful as a final treatment for waste streams which are dilute in oxidizable contaminants, but which do not quite meet standards.

14.2.1 Process Description

Ozone is generated on site by the use of corona discharge technology. Electrons within the corona discharge spilt the oxygen-oxygen double bonds upon impact with oxygen molecules. The two oxygen atoms formed from the molecule react with other oxygen molecules to form the gas ozone, at equilibrium concentration levels of roughly 2 percent in air and 3 percent in oxygen (maximum values of 4 and 8 percent, respectively). Ozone must be produced onsite (ozone decomposes in a matter of hours to simple, molecular oxygen[4]) and ozonation is restricted to treatment of streams with low quantities of oxidizable materials. Using a rule of thumb, two parts of ozone are required per part of contaminant. A large commercial ozone generator producing 500 lb/day of ozone could treat 1 million gallon/day of wastewater containing 30 ppm of oxidizable matter, or equivalently, 3,000 gallons/day of

TABLE 14.2.1. RELATIVE OXIDATION POWER OF OXIDIZING SPECIES

Species	Oxidation potentail, volts	Relative oxidation power[a]
Flourine	3.06	2.25
Hydroxyl radical	2.80	2.05
Atomic oxygen	2.42	1.78
Ozone	2.07	1.52
Hydrogen peroxide	1.77	1.30
Perhydroxyl radicals	1.70	1.25
Permanganate	1.70	1.25
Hypochlorous acid	1.49	1.10
Chlorine	1.36	1.00

[a]Based on chlorine as reference (= 1.00).

Source: References 1 and 2.

wastewater containing 1 percent of oxidizable matter.[2] Extensive information related to the generation of ozone and its application to the treatment of industrial wastewaters can be found in References 5 through 9.

While direct ozonation of industrial wastewater is possible and is practiced commercially, other technologies have been combined with ozonation to enhance the efficiency and rate of the oxidation reactions. These technologies, which supply additional energy to the reactants, involve the use of ultraviolet light or ultrasonics.

Cyanides are decomposed by ozone according to the general rate expression:

$$\frac{-d\,[O_3]}{dt} = k\,[CN_t]^{0.63 \pm 0.04}\,[O_3] \tag{1}$$

where $[O_3]$ and $[CN_t]$ are the concentrations of ozone and the total cyanide (including thiocyanates) and k is the reaction rate constant. Pilot plant and bench-scale data indicates that the reaction is first order with respect to ozone, and fractional order with respect to the cyanide ion.[10] This fractional order of the cyanide ion indicates the ozone-cyanide reaction is not a simple, bimolecular reaction but involves the formation and reaction of free radicals. Reactions of OH^- and HO_2^- with O_3 can initiate the radical chain reactions.[10] Therefore, pH considerations, as indicated by the following rate relation for the decomposition of ozone, are important in determining the overall rate equations:

$$\frac{-d\,[O_3]}{dt} = k\,[OH^-]^{0.55}\,[O_3]^2 \tag{2}$$

However, it should be noted that the limiting factor in ozone rate equations is the mass transfer of ozone gas to the liquid phase. Pilot plant data will be required to determine mass transfer characteristics. Research at Drexel University has focused on these rate relations in an efforts to generate fundamental kinetic and mechanistic data for the reactions of ozone with cyanide by distinguishing between mass transfer of ozone and the oxidation and decomposition reactions of ozone.[11,12] Figure 14.2.1 illustrates the profiles obtained for total cyanide, cyanate, and ozone residuals using an ozone bubble column and pHs of 11.2, 7.0, and 2.5. The results show that reaction rate increases with increasing pH and demonstrates a varying dependence on cyanide concentration at different pH values.

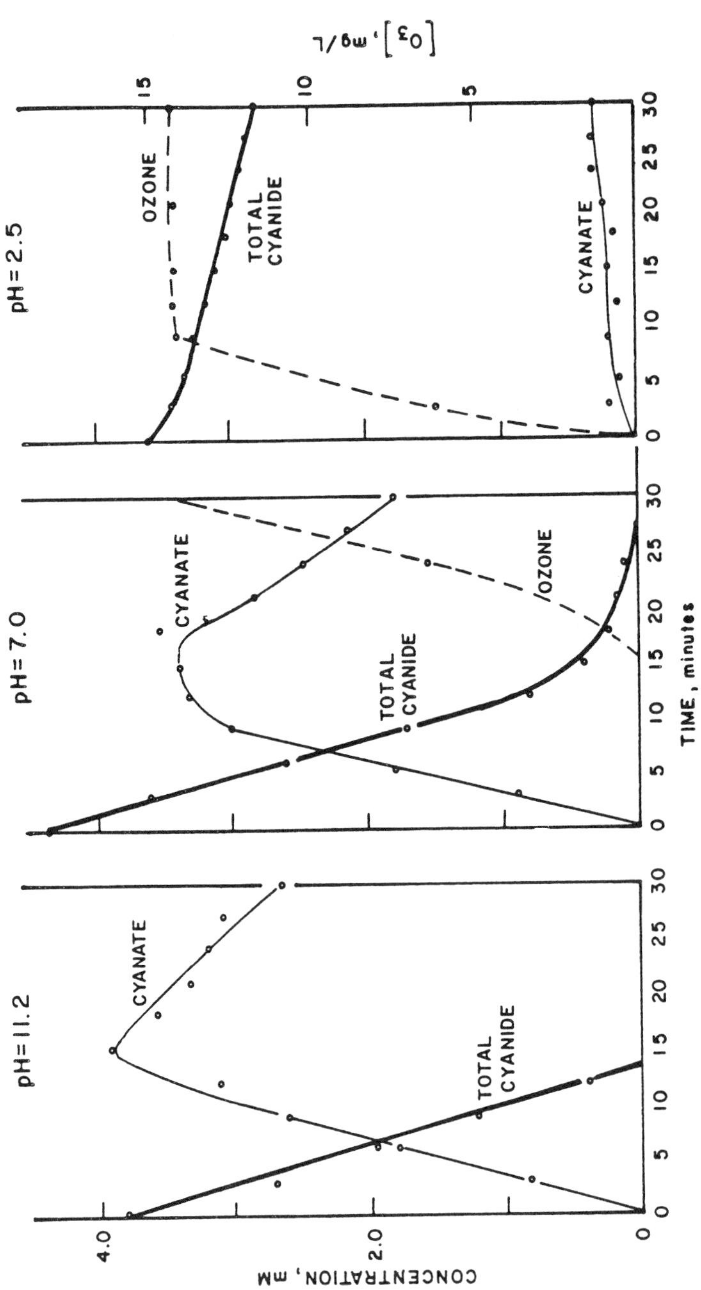

Figure 14.2.1. Profiles of total cyanide, cyanate, and ozone residual in the bubble column for pH 11.2, 7.0, and 2.5.

Upon oxidation of each mole of cyanide, 1.2 $\pm$ 0.2 mol of ozone is consumed and 1 mole of cyanate is produced as the reaction product. At pH 11.2, the removal rate of cyanide is mass transfer limited because of its very high oxidation rate with ozone, as indicated by the zero-order behavior of the cyanide profile. Cyanate appears in the solution at a rate which is equal to the rate of removal of cyanide. After cyanide is oxidized completely, cyanate starts to react with ozone at a much slower rate. During the course of the experiment at this pH, ozone does not appear in solution because of its rapid consumption by the oxidation and the decomposition reactions. It is postulated that if the ozone and cyanate were allowed to react further, the cyanate would be completely decomposed into harmless constituents.[11]

The removal rate of cyanide at pH 7.0 is equal to the rate at pH 11.2 and is mass transfer limited for the first 12 min. of ozonation. However, after the total cyanide concentration is reduced to about 0.8 mM, the oxidation reaction becomes the rate-limiting step. Ozone appears in solution as soon as the system becomes reaction rate limited and accumulates until reaching a plateau at about 14 mg/L. Cyanate is produced at an equal to cyanide rate oxidation; however, oxidation of cyanate starts while cyanide still exists in solution.

At pH 2.5, volatilization of HCN contributes more to the removal of cyanide than its oxidation by ozone, as demonstrated by independent experimentations with pure oxygen. Nevertheless, oxidation of cyanide produces equal moles of cyanate. Due to slow oxidation and decomposition of ozone at this pH, ozone appears in solution instantaneously and stabilizes at a saturation value of about 14 mg/L.[11,12]

To effectively bring about the reaction of ozone with reactive contaminants, it is important that mass transfer of ozone and its reactants through the gas-liquid interface be maximized. Also, to increase ozone solubility in water, temperatures should be maintained as low as possible and pressures as high as possible. Under conditions leading to maximum reactivity rates, costs may also increase due to less efficient use of ozone. Decisions will have to be made on a case-by-case basis to establish the most effective operating conditions.

Several commercial designs are available for the conduct of gas/liquid reactions which bring reactants into contact as effectively as possible (see Table 14.2.2 for a list of some commercial equipment vendors).[13] The types

TABLE 14.2.2. MAJOR U.S. MANUFACTURERS OF OZONE GENERATING EQUIPMENT

Manufacturer	Address	Equipment	Models-capacities lb O_3/day-air feed	Cooling method	Typical O_3 concentration in air, percent
Crane Cochrane	P.O. Box 191 King of Prussia, PA 19406 (215) 265-5050	Concentric tubes SS/glass/aluminum Series C - cabinet Series P - skid mounted	Series C, 1-18 lb/day Series P, 18-122 lb/day	Water on outer electrode	1
Emery Industries, Inc., Ozone Technology Group	4900 Estee Avenue Cincinnati, OH 45232 (513) 482-2100	Concentric tubes SS/glass/nichrome Skid mounted	Series 9270, 1-23 lb/day Series 9260, 21-400 lb/day	Water on outer electrode	1
Ozone Research & Equipment Corporation	3840 North 40th Avenue Phoenix, AZ 85019 (602) 272-2681	Concentric tubes SS/glass/SS Series V, B & D cabinet Series H skid mounted	Series B & V, 1/4-2 lb/day Series D & H 4-250 lb/day	Water on outer electrode	1
PCI Ozone Corporation	One Fairfield Crescent West Caldwell, NJ 07006 (201) 575-7052	Concentric tubes SS/glass/silver Series G - cabinet Series B - skid mounted	Series G, 1-28 lb/day Series B, 35-1400 lb/day	Water on inner electrode Oil on outer electrode	2
Welsbach Ozone Systems Corporation	3340 Stokely Street Philadelphia, PA 19129 (215) 226-6900	Concentric tubes SS/glass/SS Series CLP & GLP Both skid mounted	Series CLP, 24-127 lb/day Series GLP, 170-322 lb/day	Water on outer electrode	1
Infilco Degremont, Inc.	Koger Executive Center Box K-7 Richmond, VA 23288 (804) 285-9961	Concentric tubes SS/glass/aluminum Skid mounted	No model designations, 10-600 lb/day	Water on outer electrode	1
Union Carbide Linde Division Environmental Systems	P.O. Box 44 Tonawanda, NY 14150 (716) 877-1600	Parallel ceramic coated steel Lowther plates	No model designations, 1-1200 lb/day	Air on outside both electrodes	1
U.S. Ozonair Corp.	464 Cabot Road S. San Francisco, CA 94080 (415) 952-1420	Concentric tubes Titanium/ceramic/aluminum	Series HF, 5-570 lb/day	Water on inner electrode Air on outer electrode	2

Source: Reference 13.

of reactor designs available range from mechanically agitated reactors to more complex spray, packed, and tray type towers. Their advantages and limitations are discussed in detail in many standard texts and publications (for example, see References 2 through 5).

The process of UV/ozone treatment operates in the following manner. The influent to the system is mixed with ozone and then enters a reaction chamber where it flows past numerous ultraviolet lamps as it travels through the chamber (see Figure 14.2.2). Flow patterns and configurations in the UV exposure chamber are designed to maximize exposure of the total volume of ozone-bearing wastewater to the high energy UV radiation. Although the nature of the effect appears to be influenced by the characteristics of the waste, the UV radiation enhances oxidation by direct dissociation of the contaminant molecule or through excitation of the various species within the waste stream. In industrial systems, the system is generally equipped with recycle capacity. Gases from the reactor are passed through a thermo catalytic unit, destroying any volatiles, replenished with ozone, and then recycled back into the reactor. The system has no gas emissions.

Another alternative process involves the coupling of ultrasonic energy with ozonation. It has been shown that significant increases in the rate of oxidation can be obtained by the use of ultrasonic energy as apposed to ozone alone. Experimental details were not available in Reference 3, although different oxidation pathways were reported operating in the presence or absence of ultrasonics. Regardless of the reaction mechanisms, there appears to be no doubt that the combination of ozonation with either UV or ultrasonic excitation leads to increased oxidation rates. Typical design data for one 40,000 gal/day UV/ozone treatment process are shown in Table 14.2.3.[14]

In addition to reactor design, contactor system optimization and UV radiation utilization, two key factors in ozone equipment selection and design are power consumption and ozone generator cooling.[13] Typically, the major operating cost for ozone manufacturing is the cost of electric power. Power consumption figures in facilities using air as the source of ozone range from 6 to 8 kWh/lb O_3 for the ozone generator alone, and 10 to 13 kWh/lb O_3 total consumption including air handling and preparation. Using oxygen as feed gas reduces these ranges to 3 to 4 kWh/lb O_3 for ozone generation and 7 to 12 kWh/lb O_3 total consumption (depending on the source of oxygen).[13] However, when pure oxygen is used as the ozone manufacturing reagent, chemical costs will also have to be included.

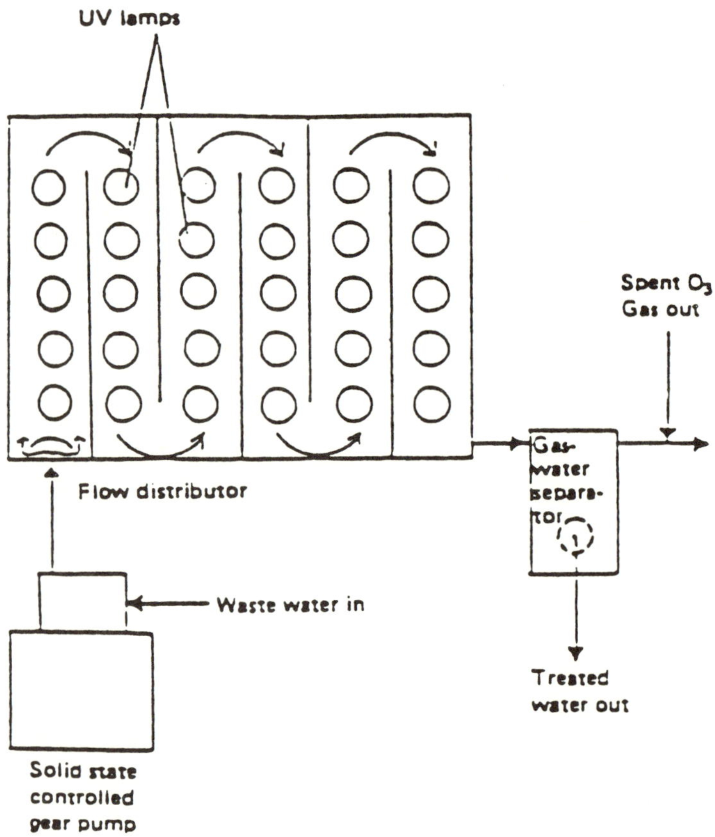

Figure 14.2.2. Schematic of top view of ULTROX pilot plant by General Electric (ozone sparging system omitted) (Edwards, B. H., 1983).

Source: Reference 4.

TABLE 14.2.3. DESIGN DATA FOR A 40,000 GPD (151,400 L/DAY) ULTROX PLANT

Reactor	
Dimensions Meters: (LxWxH)	2.5 x 4.9 x 1.5
Wet volume, liters	14,951
UV lamps:	
Number of 65 watt lamps	378
Total power, KW	25
Ozone generator	
Dimensions Meters: (LxWxH)	1.7 x 1.8 x 1.2
gms ozone/minute	5.3
kg ozone/day	7.7
Total power, kW	7.0
Total energy required (KWH/day)	768

Source: Reference 14.

Ozone generator cooling costs arise since ozone generators must be continuously cooled to maintain optimum efficiency and to avoid deterioration of the dielectric. Generators are usually air- or water-cooled. All manufacturers but one use water to cool their medium and large size generators. Typical generator cooling fluid requirements are 100,000 ft^3 air/lb O_3 or 500 gal water/lb O_3 for systems using air as the ozone source.

Due to the nonselective nature of the ozonation reactions it is important that the concentration levels of nonhazardous, but oxidizable, contaminants in the feed stream be reduced as much as possible prior to treatment. The strong electrophilic nature of ozone imparts to it the ability to react with a wide variety of organic functional groups, including aliphatic and aromatic carbon-carbon double and triple bonds, alcohols, organometallic functional groups, and some carbon-chlorine bonds. It is important to recognize that many functional groups can be present which compete for the ozone reactant and can add significantly to the cost of the treatment.

The waste to be treated should also be relatively free of suspended solids, since a high concentration of suspended solids can foul the equipment normally used to bring about contact between ozone and the aqueous phase contaminants. When ozonation is combined with UV radiation or ultrasonics, a high concentration of suspended solids also can impede the passage of UV radiation or attenuate the energy supplied by ultrasonics to enhance the oxidation rate. Other pretreatments include flow equalization, neutralization and oil and grease removal.

Post-treatment of industrial wastewaters that have been contacted with ozone will involve elimination of residual ozone, usually be passing the effluent through a thermocatalytic unit. Some by-product residuals may be formed in the feed water and some contaminants, if present, will not undergo reaction. Compounds considered unreactive include many chlorinated aliphatic compounds. If these compounds are present in the waste, technologies other than ozonation should be considered.

14.2.2 Process Performance

Although there has been a great deal of research into the ozonation of cyanide in the last 30 years, only a few commercial plants have been installed. Data are limited and additional studies are needed to establish the utility of ozone treatment.

Two applications of ozone for the oxidation of cyanides that have been reported in the literature are those at San Diego Plating, San Diego California, and Sealector Corporation, Mamaroneck, New York. The San Diego Plating System was installed by Ozodyne Corporation to treat wastewaters from an automobile recycling operation. The Sealector system was installed by PCI Ozone Corporation under funding from EPA's R&D branch in Cincinnati.[15,16]

San Diego Plating's ozone system consists of 300 gallon reactor, a vacuum precoat filter, and a solids collection unit. Prior to the ozone reactor, ozone gas under negative pressure is drawn into the waste stream to be treated. The wastewater containing dissolved ozone and ozone gas in then formed into fine particulates to enhance mass transfer using a spinning dial type aspirator. The treated wastewater is then pumped to the filter where the solids are dewatered while the filtered effluent is discharged to the sewer.

Table 14.2.4 summaries sampling results for the ozone system at San Diego Plating. Cyanide effluent concentrations were consistently reduced to very low levels. In addition, oxidation of the metal hydroxide solids reduced the degree of hydration and improved the dewatering characteristics (74 percent solids versus 20-30 percent for conventional precipitation solids).

The treatment system at Sealector was similar except that the ozone reaction tank consisted of two separate compartments. One tank was used to treat the wastewater with ozone while the second tank recovered unreacted ozone from the off-gas and recycled it back to the incoming cyanide waste stream. However, constant equipment failure, operational problems, and process unreliability resulted in the ozone system being replaced with more conventional alkaline chlorination technology.[17] San Diego Plating has also taken its ozone system out of cyanide oxidation operations and replaced it with a batch chlorine unit.[18] The main drawback of the commercial systems discussed above systems was high capital investment and operational costs. Figure 14.2.3 which compare conventional waste treatment costs with those of ozone oxidation at San Diego Plating. Figure 14.2.3 indicates that ozone oxidation appears to be more expensive than conventional treatment systems over the range shown.[16]

TABLE 14.2.4. SUMMARY OF SAMPLING RESULTS - SAN DIEGO PLATING

Parameter	Influent Range	Influent Average	Effluent Range	Effluent Average	Average and removal
Cyanide	3.75 - 0.05	1.02	0.87 - <0.02	0.08	>92.5
Total chrome	6.62 - 0.82	1.41	1.55 - 0.05	0.40	>71.6
Copper	33.0 - 5.05	9.45	1.32 - 0.04	0.05	99.5
Nickel	60.0 - 10.2	20.32	0.37 - <0.10	0.13	>99.4
TSS	559 - 35	135	93 - <1	11.6	>91.5
pH	12.2 - 3.4	6.4[a]	12.4 - 5.8	8.4[a]	--

[a]Median.

Average solids content of sludge = 74 percent.

Influent and effluent values, except pH, in mg/L.

Source: Reference 16.

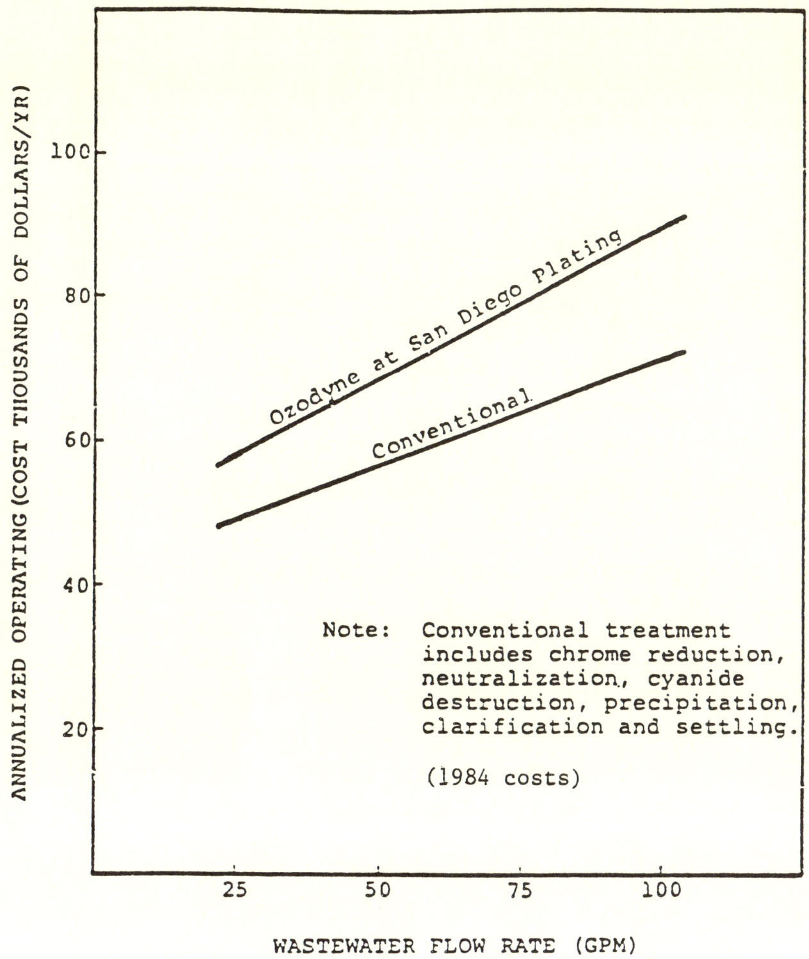

Figure 14.2.3. Annualized operating costs for conventional treatment. Ozodyne treatment system at San Diego Plating.

Source: Reference 16.

14.2.3 Process Costs

Table 14.2.5 lists the costs for a 40,000 gpd UV/Ozone plant for which design data were shown in Table 14.2.3. Cost estimates were based on wastewater containing 50 ppm PCB, designed to achieve an effluent concentration of 1 ppm. Costs were considered to be competitive with activated carbon.[14,19] The unit cost for treatment of the waste is greatly affected by whether or not the cost for a monitoring system is included. The cost of PCB destroyed is in excess of $10/pound. PCB data were used for costing purposes because of its availability. However, the costs will increase substantially if ozonation is to be used as treatment for a waste containing 1 percent organic contaminants. This is 200 times the concentration used to develop the costs in Table 14.2.5. Assuming capital equipment costs follow a simple "sixth-tenths" factor scaling relationship,[20] the costs of the reactor and generator would be about $3,000,000 (or 24 times the costs shown in Table 14.2.5) for treatment of this higher concentration. Scale factors would be variable for the operating and maintenance cost items listed in Table 14.2.5. However, the net result of scale-up to handle the more concentrated waste would drastically increase the cost/1,000 gallons treated, but would also result in far lower costs when calculated on the basis of the amount of contaminant destroyed. Costs of roughly $10/pound of contaminant destroyed would be reduced to an estimated $1/pound, assuming comparable efficiencies. However, destruction efficiencies may be adversely affected at higher concentrations due to mass-transfer and other considerations. Thus, the cost benefits per pound of contaminant destroyed, as stated above, may not be fully achievable. An optimal tradeoff must be made on the basis of pilot-scale or full scale test results.

14.2.4 Overall Status of Process

Availability--

Ozonation equipment is available commercially from several manufacturers within the United States. The Chemical Engineering Equipment Buyers' Guide published by McGraw Hill lists nine manufacturers of ozone generators and 10 manufacturers of ozonators. The latter classification includes firms that usually provide the ozone generator, the reactor, and auxiliaries such as the

TABLE 14.2.5. EQUIPMENT PLUS OPERATING AND MAINTENANCE COSTS: 40,000 GPD UV/OZONE PLANT

Reactor	$ 94,500
Generator	30,000
	$124,500
O & M costs/day	
Ozone generator power	$ 4.25
UV lamp power	15.00
Maintenance (Lamp replacement)	27.00
Equipment amortization (10 years at 10 percent)	41.90
Monitoring labor	85.71
Total/day:	$ 173.86
Cost per 1,000 gallons (3,785 liters) with monitoring labor	$ 4.35
Cost per 1,000 gallons without monitoring labor	$ 2.20

Source: Reference 14.

catalytic unit for destruction of ozone from the treated stream. The status of UV/ozonation is far less advanced. Processes such as the Ultrox process[4,14,19] have been concerned with highly refractory compounds such as PCBs. Equipment specifically designed and available for UV/ozonation of industrial wastewaters, is not available as a standard commercial item.

Application--

Ozonation appears best suited for treatment of very dilute waste streams, similar to those streams treated by the ozone based water disinfection processes now used in Europe. It does not appear to be cost competitive or technically viable for most industrial waste streams where organic concentration levels are 1 percent or higher. However, it may be viable for certain specific wastes with high levels of a contaminant of special concern and high reactivity.

Environmental Impact--

Assuming adequate destruction of a contaminant by ozonation, the principal environmental impact would appear to be associated with ozone in the effluent vapor and liquid streams. However, thermal decomposition of ozone is effective and is used commercially to destroy ozone prior to discharge. Unreacted contaminants or partially oxidized residuals in the aqueous effluent may be a problem necessitating further treatment by other technologies. Presence of many such residuals will generally result in selection of a more suitable alternative technology.

Advantages and Limitations--

There are several factors which suggest that ozonation may be a viable technology for treating certain dilute aqueous waste streams:[1,4]

- Capital and operating costs are not excessive when compared to incineration provided oxidizable contaminant concentration levels are less than 1 percent.
- The system is readily adaptable to the onsite treatment of hazardous waste because the ozone can and must be generated onsite.

- Ozonation can be used as a final treatment for certain wastes since effluent discharge standards can be met.
- It can be used as a preliminary treatment for certain wastes (e.g., preceding biological treatment).

However, there are limitations which often will preclude use of ozonation as a treatment technology. These include:

- Ozone is a nonselective oxidant; the waste stream should contain primarily the contaminants of interest.
- Certain compounds because of their structure are not amenable to ozonation, e.g., chlorinated aliphatics.
- Ozone systems are generally restricted to 1 percent or lower levels of toxic compounds. The system is not amenable to bulky wastes.
- Toxic intermediates may persist in the waste stream effluent.
- Ozone decomposes rapidly with increasing temperature, therefore, excess heat must be removed rapidly.
- Ozone oxidation is currently not as cost effective or reliable as alkaline chlorination.

REFERENCES

1. Prengle, H. W., Jr. Evolution of the Ozone/UV Process for Wastewater Treatment. Paper presented at Seminar on Wastewater Treatment and Disinfection with Ozone. Cincinnati, Ohio. September 15, 1977. International Ozone Association, Vienna, VA.

2. Harris, J. C. Ozonation. In: Unit Operations for Treatment of Hazardous Industrial Wastes. Noyes Data Corporation, Park Ridge, NJ. 1978.

3. Rice, R. G. Ozone for the Treatment of Hazardous Materials. In: Water-1980; AICHE Symposium Series 209, Vol. 77. 1981.

4. Edwards, B. H., J. N. Paullin, and K. Coghlan-Jordan. Emerging Technologies for the Destruction of Hazardous Waste - Ultraviolet/Ozone Destruction. In: Land Disposal: Hazardous Waste. U.S. EPA 600/9-81-025. March 1981.

5. Ebon Research Systems, Washington, D.C. In: Emerging Technologies for the Control of Hazardous Waste. U.S. EPA 600/2-82-011. 1982.

6. Rice, R. G., and M. E. Browning. Ozone for Industrial Water and Wastewater Treatment, an Annotated bibliography. EPA-600/2-80-142, U.S. EPA RSNERL, Ada, OK. May 1980.

7. Rice, R. G., and M. E. Browning. Ozone for Industrial Water and Wastewater Treatment, A Literature Survey. EPA-600/2-80-060. U.S. EPA RSKERL, Ada, OK. April 1980.

8. International Ozone Institute, Inc., Vienna, VA. First International Symposium on Ozone for Water and Wastewater Treatment. 1975.

9. International Ozone Institute, Inc., Vienna, VA. Second International Symposium on Ozone Technology. 1976.

10. Gurol, M. D., W. M. Bremen, and T. E. Holden. Drexel University Oxidation of Cyanides in Industrial Wastewaters by Ozone. Environmental Progress. February 1985.

11. Gurol, M. D., and W. M. Bremen. Drexel University Kinetics and Mechanism of Ozonation of Free Cyanide Species in Water. Environmental Science and Technology. Vol. 19, No. 9. 1985.

12. Gurol, M. D., and P. C. Singer. Drexel University Kinetics of Ozone Decomposition: A Dynamic Approach. Environmental Science and Technology, Vol. 16, No. 7. 1982.

13. Derrick, D. G., and S. R. Perrich. Guide to Ozone Equipment Selection. Pollution Engineering. November 1979.

14. Arisman, R. K., and R. C. Musick. Experience in Operations of a UV-Ozone Ultrox Pilot Plant for Destroying PCBs in Industrial Waste Effluent. Paper presented at the 35th Annual Purdue Industrial Waste Conference. May 1980.

15. Militello, P. Centec Corporation. Assessment of Emerging Technologies for Metal Finishing Pollution Control: Three Case studies. EPA-600/2-81-153. August 1981.

16. Cushnie, G. C. Centec Corporation, Navy Electroplating Pollution Control Technology Assessment Manual. CR 84.019. February 1984.

17. Sacco, S. Sealector Corporation. Telephone conversation with Stephen Palmer. February 26, 1987.

18. Needham, B. San Diego Plating. Telephone conversation with Stephen Palmer. February 26, 1987.

19. Zeff, J. D. Westgate Research Corporation. Ultrox Process Treatment of Organic Wastewater. Third Annual Conference on Treatment and Disposal of Industrial Wastewaters and Residues. 1978.

20. Peters, M. S., and K. D. Timmerhaus. Plant Design and Economics for Chemical Engineers. McGraw-Hill, New York, NY. 1980.

14.3 WET AIR OXIDATION

Wet air oxidation (WAO) is the oxidation of dissolved or suspended contaminants in aqueous waste streams at elevated temperatures and pressures. It is generally considered applicable for the treatment of certain organic-containing media that are too toxic to treat biologically and yet too dilute to incinerate economically.[1,2] A leading manufacturer of commercial available WAO equipment reports that WAO takes place at temperatures of 175 to 320°C (347 to 608°F) and pressures of 2,169 to 20,708 kPa (300 to 3,000 psig).[1] Although the process is operated at subcritical conditions (i.e., below 374°C and 218 atmospheres), the high temperatures and the high solubility of oxygen in the aqueous phase greatly enhances the reaction rates over those experienced at lower temperatures and pressures. In practice, the three variables of pressure, temperature and time are controlled to achieve the desired reductions in contaminant levels.

In addition to serving as the source of oxygen for the process, the aqueous phase also moderates the reaction rates by providing a medium for heat transfer and heat dissipation through vaporization. The reactions are exothermic and proceed without the need for auxilliary fuel at feed chemical oxygen demand (COD) concentrations of 20 to 30 grams per liter.[3]

14.3.1 Process Description

A schematic of a continuous WAO system is shown in Figure 14.3.1.[4] The Zimmerman WAO System,[5] as shown in the figure, has been developed by Zimpro, Inc. Rothschild, Wisconsin. It represents an established technology for the treatment of municipal sludges and certain industrial wastes. While industrial applications of cyanide destruction through wet air oxidation have been few, a wet-air oxidation unit developed by Zimpro Corporation was placed into operation in 1983 for commercial off site treatment.[3,6-8] During test runs the unit effectively treated cyanide wastes; a destruction efficiency of 99.7 percent of the influent cyanide was achieved with cyanide concentrations of 25,000 mg/L reduced to 82 mg/L.

In the WAO process shown in Figure 14.3.1, the waste stream containing oxidizable contaminants is pumped to the reactor using a positive displacement, high pressure pump. The feed stream is preheated by heat

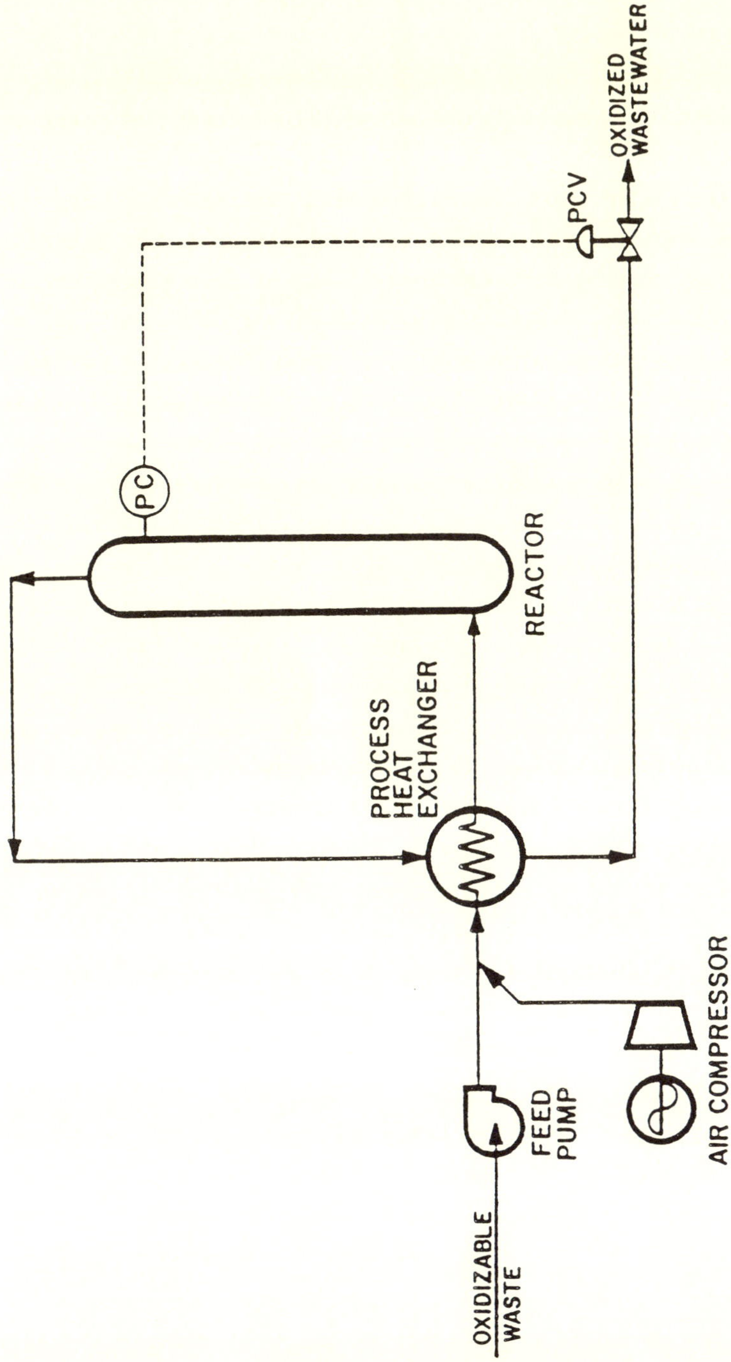

Figure 14.3.1. Wet air oxidation general flow diagram.

Source: Reference 4.

exchange with the hot, treated effluent stream. Steam is added as required to increase the temperature within the reactor to a level necessary to support the oxidation reactions in the unit. As oxidation proceeds, heat of combustion is liberated. At feed COD concentrations of roughly 2 percent the heat of combustion will generally be sufficient to bring about a temperature rise and some vaporization of volatile components. Depending upon the temperature of the effluent following heat exchange with the feed stream, energy recovery may be possible or final cooling may be required. Following energy removal, the oxidized effluent, consisting mainly of water, carbon dioxide, and nitrogen, is reduced in pressure through a specially designed automatic control valve. The effluent liquor is either suitable for final discharge (contaminant reduction achieves treatment standards) or is now readily biodegradable and can be piped to a biotreatment unit for further reduction of contaminant levels. Similarly, noncondensible gases can either be released to the atmosphere or passed through a secondary control device (e.g., carbon adsorption unit) if additional treatment is required to reduce air contaminant emissions to acceptable levels.[8]

The continuous reactor can reportedly take two forms: a tower reactor[1] or a reactor consisting of a cascade of completely stirred tank reactors (CSTRs).[10] The bubble tower reactor available commercially from Zimpro is a vertical reactor in which air is passed through the feed. The reactor is sized, based on feed rate, to provide the holding time required for the reactions to proceed to design levels. The stirred tank cascade reactor consists of a series of horizontal reactor chambers contained within a horizontal cylinder. The wastewater cascades from one chamber to the next, and then is released for discharge or post-treatment. Air is generally injected into each of the CSTRs.

Although operation of a WAO system is possible, by definition, under all subcritical conditions; i.e., below 374°C and 218 atm (3220 psig), commercially available equipment is designed to operate at temperatures ranging from 175 to 320°C and at pressures of 300 to 3,000 psig.[1]

Of all variables affecting WAO, temperature has the greatest effect on reaction rates. In most cases, about 150°C (300°F) is the lower limit for appreciable reaction. About 250°C (482°F) is needed for 80 percent reduction of COD, and at least 300°C (572°F) is needed for 95 percent reduction of COD within practical reaction times. Destruction rates for specific constituents may be greater or less than that shown for COD reductions.[2]

Initial reaction rates and rates during the first 30 minutes are relatively fast. After about 60 minutes, rates become so slow that generally little increase in percent oxidation is gained in extended reaction times.[2]

An increase in reaction temperature will lead to increased oxidation but generally will require an increase in system pressure to maintain the liquid phase and promote wet oxidation. A drawback to increasing the temperature and pressure of the reaction is the greater stress placed on the equipment and its components, e.g., the increased potential for corrosion problems.

In addition, increased temperatures and pressures increase both capital and operating costs as well as greatly decreasing liquid phase equilibrium oxygen concentration. Decreased equilibrium oxygen concentrations decrease gas mass transfer rates which thereby restricts overall reaction rates.

As noted by Zimmerman, et al., the object of WAO is to intimately mix the right portion of air with the feed, so that under the required pressure, combustion will occur at a speed and temperature which will effectively reduce the organic waste to desired levels. Pressures should be maintained at a level that will provide an oxygen rich liquid phase so that oxidation is maintained.[5] Charts and curves are provided in this reference[5] to aid in the determination of waste heating valve, stoichiometric oxygen requirement, and the distribution of water between the liquid and vapor phases at given temperatures and pressures. More information can be obtained from the manufacturer.

A model has been developed to gain insight into the key system parameters using a common industrial waste stream and fixed temperature, residence time, and COD reduction. The model was used to estimate costs for the system.[11] Its value, as a predictive tool, along with that of supplementary kinetic studies[12] of batch wet oxidation, is limited by the sparsity of experimental data concerning reaction products and their phase distributions at the elevated temperatures and pressures encountered during WAO.

Very little discussion is found in the literature concerning the physical form of wastes treatable by WAO. However, WAO equipment and designs have been used successfully to treat a number of municipal and industrial sludges.[13] According to a representative of the leading manufacturer of WAO systems, wastes containing up to 15 percent COD (roughly equivalent to 7 to 8 percent organics) are now being treated successfully in commercial equipment.[14]

Treatment of solid bearing wastes is dependent upon selection of suitable pump designs and control devices. WAO units used for activated carbon regeneration now operate at the 5 to 6 percent solids range.[14] Treatment of higher solid levels is not precluded by fundamental process or design limitations. Column design must also be consistent with the need to avoid settling within the column under operating flow conditions. Thus, pretreatment to remove high density solids (e.g., metals by precipitation) and accomplish size reduction (e.g., filtration, gravity setting) would be required for some slurries. It should be noted that the WAO unit operated by the California facility does not accept slurries or sludges for treatment. This may be a result of design factors precluding their introduction into the system.[15]

Under typical WAO operating conditions it is likely that both contaminant residuals and low molecular weight process by-product residuals may be present. While it is entirely possible that imposition of more stringent operating conditions will serve to reduce these residuals to acceptable levels, the manufacturers and users of commercial WAO system stress that the major applications involve the pretreatment of waste, usually for subsequent biological treatment.

Even under conditions that are favorable for wet oxidation, it is also likely that certain contaminants, particularly some of the more volatile components, will partition between the vapor phase and the liquid phase. Empirical testing will be necessary to establish vapor and liquid phase residuals and some post-treatment of both streams may be necessary. Existing post-treatment methods for the liquid generally involves bacteriological treatment. Although the results of post-treatment schemes for vapors from the WAO system have not been found in the literature, a two-stage water scrubber/activated carbon adsorption system has been used to treat WAO vapor emissions.[3] Presumably carbon adsorption or scrubbing systems could be routinely employed if necessary.

14.3.2 Process Performance

Tables 14.3.1 and 14.3.2 summarize cyanide wet air oxidation demonstration test results conducted in 1983 at a commercial waste treatment facility in California.[16] The WAO unit is currently still in operation at this site. Table 14.3.3 contains cyanide oxidation data for the WAO treatment of spent caustic scrubbing liquor from a natural gas based ethylene plant.[17]

TABLE 14.3.1. RESULTS OF WET AIR OXIDATION UNIT - OXIDATION OF CYANIDE WASTE

Sample	Influent	Effluent
COD, g/L	32.2	9.3
COD Reduction, %	-	71.1
Cyanide, mg/L	28,630	0.82
Cyanide Reduction, %	-	99.99
pH	12.9	9.1
Zinc, mg/L	15,700	3,500
Nickel, mg/L	120	15
Copper, mg/L	1,900	536
Log Book No.	2082-66-1	2082-66-2

Source: Reference 16.

TABLE 14.3.2. WET AIR OXIDATION DEMONSTRATION OF CYANIDE WASTEWATER

		Raw Influent	Oxidized Effluent
A.	Oxidation Conditions		
	Oxidation Temperature	495°F (257°C)	
	*Nominal Residence Time	80 minutes	
	Waste Flowrate	7.5 gpm	
	Reactor Pressure	1200 psig	
	*Nominal Residence Time = Reactor Volume Divided by Waste Flowrate		
B.	Oxidation Results		
	COD, g/L	37.4	4.2
	COD Reduction, %	--	88.8
	Total Cyanide, mg/L	25,390	82
	Total Cyanide Reduction, %	--	99.7
C.	Off-Gas Grab Sample Analysis		
	Carbon Dioxide	1.5%	
	Oxygen	8.5%	
	Nitrogen	82.8%	
	Carbon Monoxide	Not detected	
	Methane	9.0 ppm	
	Total Hydrocarbons	61.1 ppm as methane	

Source: Reference 16.

TABLE 14.3.3. TREATMENT OF SPENT CAUSTIC SCRUBBING LIQUOR FROM NATURAL GAS BASED ETHYLENE PLANT, 608°F, 3000 PSIG (320°C, 210 Kg/cm^2)

	Influent liquor	Effluent	% Reduction
pH	12.7	12.5	--
COD, g/L	21.0	1.0	95.2
BOD_5, mg/L	--	650	--
Sulfide Sulfur, mg/L	5.8	<0.001	>99.99
Cyanide, mg/L	110	0.035	99.97

Source: Reference 17.

As shown in the tables, the treatment of cyanide bearing waste streams at high temperatures achieves almost complete cyanide destruction in addition to high sulfides and COD removal. As indicated by the data, cyanide destructions of 99.7 percent are typical and, in some cases, total cyanide reductions as high as 99.995 have been observed.

At present two commercial Zimpro wet oxidation units, one in Japan and the other Europe, are engaged in treating spent caustic scrubbing liquors from petrochemical plants. Another unit, presently under construction in the U.S., will be using pure oxygen as the oxygen source. In addition, five WAO units are currently in service in Japan for the treatment of cyanide-bearing wastewaters from acrylonitrile production plants.[18]

14.3.3 Process Costs

Treatment costs for wet air oxidation systems will be affected by a number of parameters including the amount of oxidation occurring, the hydraulic flow, the design operating conditions necessary to meet the treatment objectives, and the materials of construction. These factors account for the band of capital costs shown in Figure 14.3.2. The figure was taken from Reference 2 and updated to reflect changes in the 1982 to 1986 Chemical Engineering (CE) plant cost index. The costs do not include any costs associated with pretreatment of the feed or post-treatment of the vapor phase component of the treated liquor. However, post-treatment costs were included in another capital cost estimate of $2.45 million (adjusted to 1986 using the CE plant cost index) for a 20 gpm plant.[4] This estimate is within the capital cost band shown in Figure 14.3.2.

Operating costs for the wet oxidation unit area shown in Figure 14.3.3. These data were also derived from data given in Reference 2 with adjustment made for the costs of labor and cooling water. As noted in Reference 2, power accounts for the largest element of cost. This power cost is primarily the result of air compressor operation. Additional power for supplying energy for the oxidation of very dilute wastewaters would be at most 500 Btu/gallon. The associated costs for this energy would be less than one (1) cent/gallon.

The use of pure oxygen instead of compressed air will help to lower power costs, particularly with respect to handling and consumption. However, increased reagent costs may more than offset decreased operating costs and decisions should be made on a case by case basis.

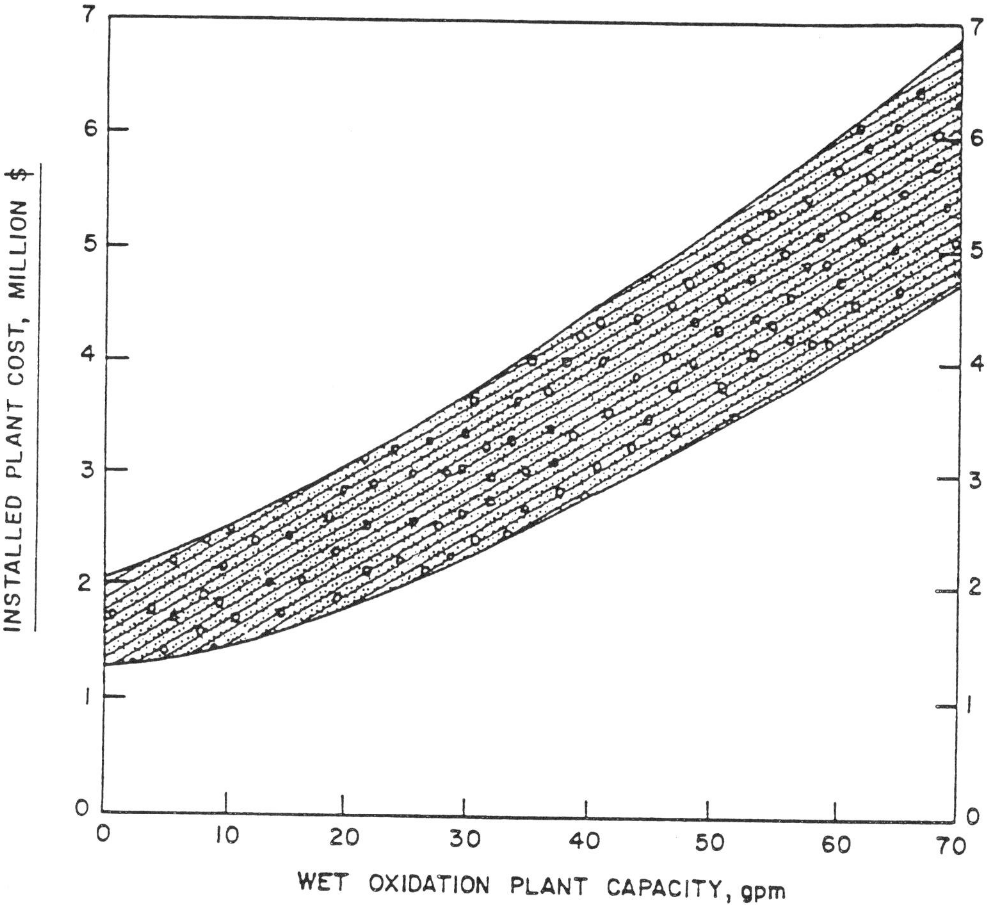

Figure 14.3.2. Installed plant costs versus capacity.

Source: Reference 2.

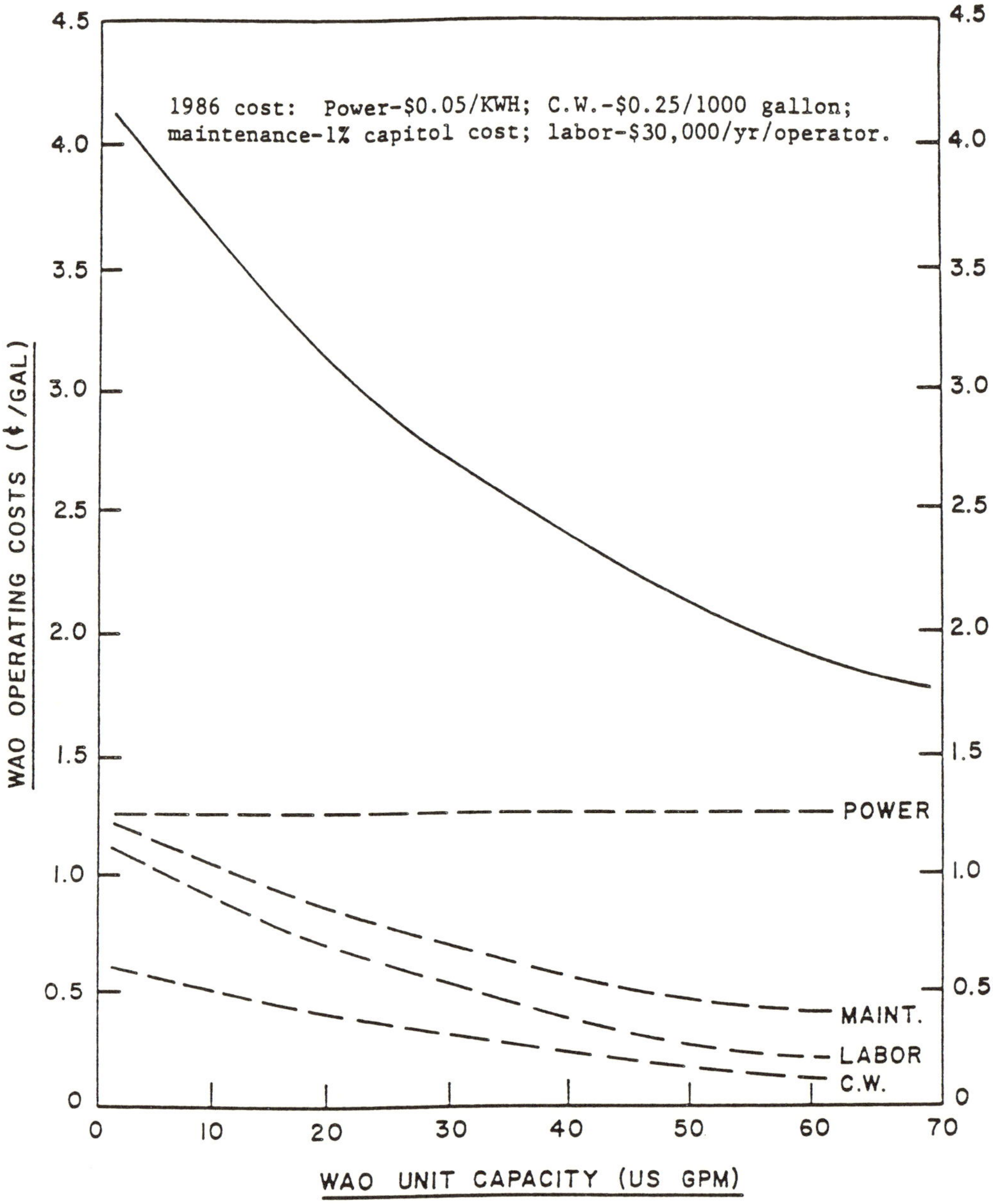

Figure 14.3.3. Unit operating costs versus unit flow rate.

Source: Reference 2.

Total costs, capital plus operating, on a per unit of feed basis, requires assumptions on life cycle, depreciation, taxes, and current interest rates for the capital cost. One avenue for financing that has been used commercially, common lease terms, are 5 years and 20 percent value at end of term.* Table 14.3.4 illustrates the effect on total costs per unit of feed.

TABLE 14.3.4. WAO COSTS VERSUS FLOW

Hydraulic flow (gpm)	Cost elements per gallon, cents		
	Operating	Capital	Total
2.0	23	31	54
10	6	7	13
20	3	5	8
40	2-3	4-5	6-8

At Casmalia Resources, the prices, (April, 1985) for treatment of wastes are computed based on the oxygen demand of the material. Prices range from a minimum of $120 per ton to a maximum of $700 per ton versus $15 per ton for the land disposal of low risk wastes.[15]

14.3.4 Status of Technology

The WAO process is available commercially, and reportedly well over 150 units are now operating in the field treating municipal and various industrial sludges.[14] The process is used predominately as a pretreatment step to enhance biodegradability. Only a few units are now being used to treat industrial cyanide wastes. These include the unit in California and six other units currently operating in Japan and Europe.

*Assume lease charges of $17/1,000 per month based on total installed cost.

The oxidation of specific contaminants in waste streams by the wet oxidation process is not highly predictable. Equipment manufacturers rely largely on the result of bench-scale results to tailor the design of full-scale WAO continuous units for specific wastes. Full-scale data confirm the results of WAO performance data obtained in bench and pilot-scale studies.[1]

As noted, the process is thermally self-sustaining when the amount of oxygen uptake is in the 15-20 g/liter range. Below this range, some energy input will be required to initiate and sustain reaction. However, the energy requirement will be appreciable less than that required for incineration.

The environmental impacts of WAO will hinge upon the residuals remaining after treatment. Wet scrubbing and carbon adsorption cleanup systems have been used to treat the HCl formed as a product of chlorinated organic oxidation and to remove volatile organics from the waste off gases. Residuals in the liquid phase may also require post treatment if, for example, 100 percent conversion to CO_2 and H_2O is not realized when treating hydrocarbon contaminants. The available data do suggest that some form of post treatment of both liquid and vapor phases will be required to meet EPA treatment standards.

There are several advantages associated with the use of WAO as noted by the developer and stated in Reference 2.

1. The process is thermally self-sustaining when the amount of oxygen uptake is in the 15-20 gram/liter range.

2. The process is well suited for wastes that are too dilute to incinerate economically, yet too toxic to treat biologically.

3. Condensed phase processing requires less equipment volume than gas phase processing.

4. The products of WAO stay in the liquid phase. Off-gases from a WAO system are free of NO_x, SO_2, and particulate. Water scrubbing and, if need be, carbon adsorption or fume incineration are used to reduce hydrocarbon emissions or odors.

5. WAO also has application for inorganic compounds combined with organics. The oxidation cleans up the mixture for further removal of the inorganics. WAO can detoxify most of the EPA priority pollutants. Toxic removal parameters are in the order of 99+ percent using short-term, acute, static toxicity measurements.

Limitations of the WAO process relate to the sensitivity of destruction efficiency associated with the chemical nature of the contaminant, the possible influence of metals and other contaminants on performance, the unfavorable economics associated with low and high concentration levels, and the presence of residuals in both the vapor and liquid phases which may require additional treatment. Costly materials of construction and design features may also be required for certain wastes which will form corrosive reaction products or require extreme temperature/pressure conditions to achieve destruction to acceptable treatment standard levels. In particular, chlorinated aromatic compounds are more resistant to degradation and can result in the production of HCl byproduct.

REFERENCES

1. Dietrich, M. J., T. L. Randall, and P. J. Canney. Wet Air Oxidation of Hazardous Organics in Wastewater, Environmental Progress, Vol. 4, No. 3. August 1985.

2. Freeman, H. Innovative Thermal Hazardous Treatment Processes, U.S. EPA, Hazardous Waste Engineering Research Laboratory, Cincinnati, Ohio. 1985.

3. California Air Resources Board. Air Pollution Impacts of Hazardous Waste Incineration: A California Perspective. December 1983.

4. Wilhelmi, A. R., and P. V. Knopp. Wet Air Oxidation - An Alternative to Incineration, Chemical Engineering Progress. August 1979.

5. Zimmerman, F. J., and D. G. Diddams. The Zimmerman Process and its Applications in the Pulp and Paper Industry, TAPPI Vol. 43, No. 8. August 1960.

6. Copa, W., J. Jeimbuch, and P. Schaeffer. Full Scale Demonstration of Wet Air Oxidation as a Hazardous Waste Treatment Technology. In: Incineration and Treatment of Hazardous Waste, Proceedings of the Ninth Annual Research Symposium, U.S. EPA 600/9-84-015. July 1984.

7. Copa, W., M. J. Dietrich, P. J. Cannery, and T. L. Randall. Demonstration of Wet Air Oxidation of Hazardous Waste. In Proceedings of Tenth Annual Research Symposium, U.S. EPA 600/9-84-022. September 1984.

8. U.S. Environmental Protection Agency, Background Document for Solvents to Support 40 CFR Part 268, Land Disposal Restrictions, Volume II. January 1986.

9. Radimsky, J., et al. California Department of Health Services, Recycling and/or Treatment Capacity for Hazardous Waste Containing Cyanides. Staff Report. March 1983.

10. Baillod, C. R., and R. A. Lamporter. Applications of Wet Oxidation to Industrial Waste Treatment. Presented at 1984 AICHE National Meeting, Philadelphia, PA. August 19-22, 1984.

11. Baillod, C. R., R. A. Lamporter, and B. A. Barna. Wet Oxidation for Industrial Waste Treatment, Chemical Engineering Progress. March 1985.

12. Baillod, C. R., B. M. Faith, and D. Masi. Fate of Specific Pollutants During Wet Oxidation and Ozonation, Environmental Progress. March 1985.

13. Randall, T. R. Wet Oxidation of Toxic and Hazardous Compounds. Zimpro, Inc. Technical Bulletin 1-610. 1981.

14. Telephone conversation with A. Wilhelmi on April 3. 1986.

15. Metcalf & Eddy, Inc. Hazardous Waste Treatment Storage and Disposal Facility - Site Evaluation Report, Casmalia Resources, Casmalia, California, Publication NS J-1074. April 8, 1985.

16. McBride, J. L., and J. A. Heimbuch. Casmalia Resources. Hazardous Waste Treatment Using Wet Air Oxidation of Casmalia Resources. 1983.

17. Treatment of Spent Caustic Liquors by Wet Oxidation. Zimpro Inc. Technical Bulletin 3260-T.

18. Wilhelmi, A. R. Zimpro Inc. Telephone conversation with Stephen Palmer of Alliance Technologies Corporation. March 3, 1987.

14.4 SULFUR-BASED CYANIDE TREATMENT TECHNOLOGIES

In recent years, sulfur-based cyanide treatment technologies have been the focus of an increasing number of research efforts and commercial applications. Sulfur-based cyanide treatment technologies have shown potential for removing cyanide from aqueous waste streams and are not subject to many of the limitations associated with more conventional cyanide treatment technologies. For example, alkaline chlorination, the most common treatment procedure for cyanide wastes, has the potential for generating hazardous and/or toxic by-products[1] (i.e., chlorinated aliphatic hydrocarbons) while ozonation and wet air oxidation require large capital investments.[2,3]

The three sulfur-based cyanide treatment technologies which have shown the most promise are polysulfide treatment, the INCO process, and ferrous sulfate treatment. The use of polysulfides for treating cyanide waste streams was first reported in 1940.[4] Polysulfide solutions have been recently adapted to scrub hydrogen cyanide from fluid catalytic cracking and coking gases, treat concentrated cyanide electroplating solutions, and remove cyanides present in coal gasification wastewaters.[4] The INCO process and ferrous sulfate treatment have also shown promise in treating a wide variety of cyanide wastewaters such as ore leaching and electroplating effluents.

14.4.1 Process Description

14.4.1.1 Polysulfide Treatment--

Polysulfides species are formed when neutral sulfur atoms combine with monosulfide species. They can be represented by the chemical formulas, H_2S_x, HS_x, where x = 2 - 5. Equilibrium calculations show that the tetrasulfide and pentasulfide species should be the predominant polysulfide forms in neutral and slightly alkaline solutions, but recent experimental work has detected only the pentasulfide species.[5]

In the cyanide-polysulfide reaction it has been postulated that 1 mole cyanide reacts with 1 mole of polysulfide to produce 1 mole of less toxic thiocyanate.

$$CN^- + S_xS^{-2} = SCN^- + S_{x-1}S^{-2}$$

However, it should be noted that in sufficient quantities, thiocyanates can cause toxic inhibition to biological treatment systems.

During the reaction, one polysulfide sulfur atom, poly-S°, is reduced from oxidation state 0 to $^-1$, while the cyanide carbon atom is oxidized from oxidation state $^+2$ to $^+3$.[5]

The rate equation for the change of total free cyanide ([HCN] + [CN^-]) per unit time in the presence of polysulfide is as follows:[6]

$$-d[CN_T]/dt = K\ [CN_T]\ [S_x - S^{-2}]$$

where:

$[CN_T]$ = [HCN] + [CN^-], moles/L
$[S_x - S^{-2}]$ = polysulfide, moles/L
K = reaction rate constant (liter/mole/min)
t = time, minutes

The role of the hydroxyl ion (OH^-) in the cyanide-polysulfide reaction system is unknown at this point in time. The hydroxyl ion may initiate or impede the oxidation process via a free radical chain mechanism. The following rate relations for the cyanide-polysulfide reaction system were derived by researchers at Carnegie-Mellon University to determine the effect of pH.[5]

pH = 8.2 $\quad {}^-d[CN_T]/dt = 1.41\ [CN^-]^{1.04}\ [\text{poly-S°}]^{0.85}$

pH = 10.0 $\quad {}^-d[CN_T]/dt = 0.27\ [CN^-]^{0.51}\ [\text{poly-S°}]^{0.87}$

pH = 12.0 $\quad {}^-d[CN_T]/dt = 0.14\ [CN^-[^{0.49}\ [\text{poly-S°}]^{0.78}$

The initial rate kinetic data shows that the reaction is mixed order and that both reaction order and reaction rate change are heavily influenced by pH.

Three common forms of polysulfides are sodium polysulfide, ammonium polysulfide, and calcium polysulfide (limesulfur). Sodium and ammonium polysulfide are manufactured according to the following stoichiometric equations:[4]

$$2NaOH + H_2S + (X-1)\ S° = Na_2S_x + 2H_2O$$

$$2NH_4OH + H_2S + (X-1)\ S° = (NH_4)_2\ S_x + 2H_2O$$

Limesulfur is commercially available as a commonly used pesticide and fungicide.[7]

Equipment needs are similar to those described for other chemical precipitation processes. Storage tanks, reaction vessels, agitation, materials handling, and process control equipment are standard process items.

14.4.1.2 INCO Process--

In 1982 Inco Metals Company announced the development of a technology for the destruction of cyanide in gold mill waste streams. The process involves the selective oxidation to cyanate of both free and complexed cyanide species using a mixture of SO_2 and air at controlled pH in the presence of copper as a catalyst. Metals are precipitated from solution as hydroxides. The process also removes iron cyanide, not by oxidation, but by precipitation as an insoluble copper or zinc ferrocyanide.[10-12]

The oxidation cyanide occurs according to a simplified reaction as follows:

$$CN^- + SO_2 + O_2 + H_2O = CNO^- + H_2SO_4$$

Based on the stoichiometry of this reaction, the SO_2 requirement is 2.47 g SO_2/g CN oxidized.

The SO_2/air oxidation process destroys the metal cyanide complexes typically present in metal finishing and gold mining effluents. Based on sequential sampling data from batch experiments, the preferential order of metal cyanide complex removal is:[13]

$$Zn > Fe > Ni > Cu$$

The SO_2/air oxidation system has successfully removed iron cyanide complexes from solution.[10,11] During SO_2/air treatment, iron remains in the reduced ferrous state and is not converted to the ferric state as occurs in stronger oxidizing environments. The iron cyanide complexes are removed from the solution by precipitation of metal ferrocyanide compounds of the form $Me_2Fe(CN)_6$ (where Me = Cu, Zn and Ni).[10,11] Metals liberated from the cyanide complexes of copper, zinc, and nickel are removed by precipitation of metal hydroxides at the reaction pH.

The cyanide oxidation reaction is catalyzed by the presence of copper in solution. Copper for catalysis or for precipitation is conveniently added as a $CuSO_4$ solution. Any free CN present is quickly complexed as a Cu(I) cyanide complex, which apparently is involved as a catalyst in the oxidation of CN to CNO^- by SO_2 and O_2. The effect of copper concentration in batch treatment of a synthetic cyanide effluent containing 250 mg/L CN_T is shown in Figure 14.4.1. The effect of treatment pH is shown in Figure 14.4.2. The optimum copper concentration is 50 mg/L and the optimum operating pH is in the range of 9 to 10 which can be achieved by the addition of lime.[14]

14.4.1.3 Ferrous Sulfate Treatment--

The formation of less toxic cyanide complexes such as ferro and ferric-cyanides also has been used as a method for detoxifying of cyanide wastewaters. This process involves the use of iron salts to form complex compounds with the free cyanide in the wastes. Eventually these cyanide complexes are precipitated and removed as a sludge.[15]

The major advantage of this treatment method is that it is relatively inexpensive in locations where waste ferrous sulfate is available. However, considerable quantities of sludge may be formed and the treated solutions are strongly colored. There also is evidence that ferrocyanides may be decomposed to free cyanide by sunlight. The regeneration of the cyanide under these conditions would contaminate the receiving stream.

This method has received very little acceptance by industry in this country, but appears to be used in Europe. The complexing process apparently does not completely destroy cyanide under practical operating conditions. Cyanide levels in treated solutions may be as great as 5 to 10 ppm. Thus, the sludges formed would appear to be toxic and will require substantial post-treatment prior to final disposal.

14.4.2 Pretreatment and Post-Treatment Requirements

Very little information exists in the literature concerning pretreatment and post-treatment requirements for these processes and their feed streams. Since the processes are aqueous in nature, filtration or some other solids removal process may be desirable. Adverse effects such as chemical

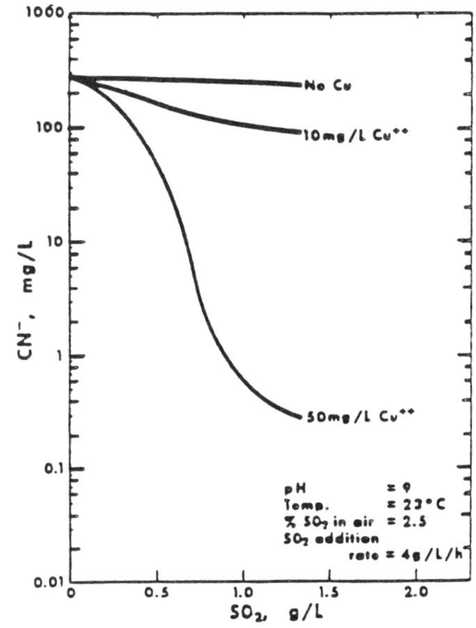

Figure 14.4.1. Effect of copper.

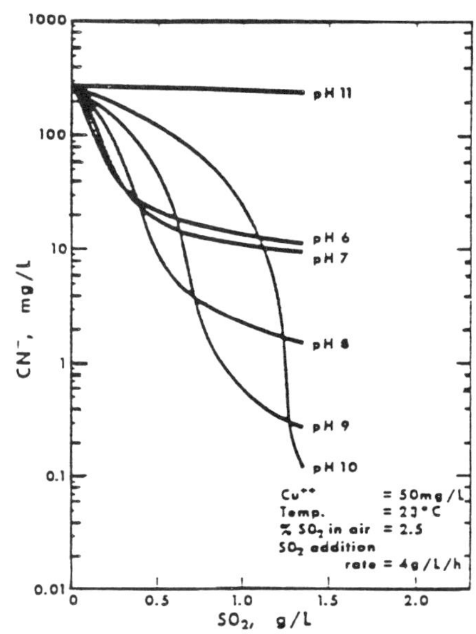

Figure 14.4.2. Effect of pH.

Source: Reference 14.

interaction, interference with pump operations, abrasion of internal parts, and fouling of internal surfaces resulting from existing or formed solids are possible problem areas, but have not been considered in the literature.

Similarly, post-treatment requirements for sulfur-based cyanide treatment technologies as reported in literature have been cursory in nature. Residuals from polysulfide treatment include thiocyanates and other oxidized sulfur compounds.[4,5] Investigators have found that at elevated temperatures thiocyanates are corrosive.[4] Therefore, additional steps may be required facilities are required in process trains to prevent water containing thiocyanate from reaching downstream equipment where heat is applied. Residuals from the INCO process have not been completely determined as of this time. Further fundamental investigations are required to define the chemical reaction mechanisms and kinetics, to determine the stability of the precipitated solids, and to assess the toxicity of treated effluents. Additional process optimization studies are also recommended.

The ferrous sulfate-cyanide treatment process suffers from the most serious residual problems. Since cyanides are merely precipitated from solution without appreciable oxidation in a voluminous sludge product, the result is a highly toxic sludge.[15] Some type of cyanide destruction and/or encapsulation process will be necessary prior to final disposal.

14.4.3 Process Performance

14.4.3.1 Polysulfide Treatment--

Currently, process performance data for the polysulfide oxidation of cyanide complexes have been limited to bench- and pilot-scale studies. Reaction rates and products depend on solution pH the SO_2 to O_2 ratio, and the catalytic and inhibitory effects of metal ions and organic compounds. Laboratory tests performed by Luthy, et al. to determine reaction pathways showed that no reaction occurred between cyanide and sulfide, however sulfur in the form of polysulfide reacts relatively quickly with the cyanide. The reaction order was determined to be 1.54 ± 0.25 with a rate constant of approximately 0.24.[16] Complex cyanides were evaluated in the presence of polysulfide at room temperature. It was observed in a survey test that $Fe(CN)_6^{-3}$ produced no or little thiocyanate in the presence of polysulfide.[16]

Subsequent investigations by Trofe, Page and Luthy, et al. sought to determine the effects of temperature and catalytic/inhibitory compounds.[5,6] The rate constant was found to double for every 12°C increase in temperature. Certain metals also had an effect on reaction rate. For example, metals ions such as CA^{+2}, Mg^{+2}, Ni^{+2}, and Zn^{+2} had a catalytic effect at low concentration.

In 1985, Ganczarczyk, et al. investigated the reaction between calcium polysulfide and concentrated cyanide solutions from electroplating operations.[17] Previously, Ganczarczyk, et al. had conducted a series of experiments to investigate the cyanide-polysulfide reaction in a 2 percent solution (20,000 mg/L CN^-) of sodium cyanide.[18,19] The reaction proceeded very rapidly, both at room temperature and at 3°C. It was 95 percent complete within 1-hour and cyanide concentrations were nondetectable within 2 weeks at a cyanide-to-polysulfide ratio of 1:2 by weight.

In the later studies, two different wastewater streams from an electroplating operation were studied. One wastewater was dragout from a rinse tank in a copper and cadmium plating process. The second wastewater was a stripper solution for removal of metal plate (Cu/Ni plate stripping liquor). Tables 14.4.1 and 14.4.2 show the liquid phase pollutant concentrations following the cyanide-polysulfide treatment of the electroplating wastewaters. Upon completion of the experiments, the following conclusions were made:[17]

- The reaction effectively converted CN^- to SCN^- within 2 to 3 days at 3°C, broke down metal-cyanide complexes, and precipitated metals generally to the levels required by municipal treatment systems.
- The polysulfide dosages necessary to achieve these goals was only about 20 percent higher than stoichiometric requirements, but at very high CN^- concentrations somewhat higher dosages might be needed.
- It seems that the cyanide-polysulfide reaction was catalyzed by the presence of metal-cyanide complexes in wastewater and was only moderately exothermic.
- SCN^- produced by CN^- conversion was partially lost during the process.

TABLE 14.4.1. LIQUID PHASE POLLUTANT CONCENTRATIONS (mg/L) IN THE TREATMENT OF THE STRIPPER WASTEWATER FROM COPPER/NICKEL PLATING, SERIES 2 (INITIAL CYANIDE-TO-POLYSULFIDE RATIO 1:1.5 BY WEIGHT)

	Duration of the experiments							
Pollutants	Initial	1 day	2 days	3 days	4 days	5 days	6 days	7 days
CN^-	56,200	87.5	25.0	31.3	50.0	31.3	28.1	31.3
SCN^-	0	20,330	16,260	20,910	19,460	14,080	12,490	10,750
Fe	110	8.9	7.4	11.5	3.4	5.5	3.8	2.6
Zn	7.1	3.2	1.8	1.3	0.9	1.5	2.2	1.1
Cu	29,020	6.5	9.3	4.1	4.0	3.6	6.2	2.2
Cd	10.5	3.0	0.5	0.8	0.5	1.0	1.1	0.9
Cr	0.6	ND[a]	ND	ND	ND	ND	ND	ND
Ni	5,130	94.1	105.0	59.8	92.0	83.5	69.4	66.9

[a]ND = Nondetectable.

Source: Reference 17.

TABLE 14.4.2. LIQUID PHASE POLLUTANT CONCENTRATIONS (mg/L) IN THE TREATMENT OF THE STRIPPER WASTEWATER FROM COPPER/CADMIUM PLATING, SERIES 3 (INITIAL CYANIDE-TO-POLYSULFIDE RATIO 1:2.0 BY WEIGHT)

	Duration of the experiments							
Pollutants	Initial	1 day	2 days	3 days	4 days	5 days	6 days	7 days
CN^-	56,200	30	ND[a]	ND	ND	ND	ND	ND
SCN^-	0	9,000	15,220	7,260	2,030	870	4,940	3,490
Fe	117.8	5.7	5.8	3.0	3.2	3.1	2.7	3.7
Zn	7.5	2.2	1.2	1.6	1.5	1.2	1.0	1.3
Cu	29,920	9.3	1.5	2.0	3.3	2.1	2.2	2.0
Cd	10.4	2.4	0.9	1.0	1.0	0.7	0.6	0.7
Cr	0.5	ND	ND	ND	ND	ND	ND	ND
Ni	5,020	25.2	28.4	20.0	17.9	7.6	22.1	20.3

[a]ND = Nondetectable.

Source: Reference 17.

Despite the large quantity of experimental data available, industrial applications of polysulfide treatment of cyanide bearing wastewaters has been limited to fluid catalytic cracking and coal gasification effluents. The only other industrial application reported in the literature is a large commercial waste treatment facility in California.[7] The cyanide treatment process at this facility is batch in nature and consists of two 18,000 gallon storage/treatment tanks into which cyanide wastes (greater than 100 ppm CN_T) are pumped. The treatment reagent used is calcium polysulfide. This reagent is stored in an adjacent fiberglass tank. The amount of reagent required to complete the cyanide to thiocyanate reaction is predetermined by onsite laboratory analysis of incoming waste for reactive CN. The process typically handles approximately 40,000 gallons of waste per month at 50 percent of its capacity.

14.4.3.2 INCO Process--

Performance data for the INCO process relates to industrial applications rather than laboratory studies of kinetic properties of the reactants. Tables 14.4.3 and 14.4.4 present typical INCO process results for selected gold mill barren and tailing liquors, and plating rinse waters.

Table 14.4.3, which contains gold mill effluent data, show that CN was consistently removed from feeds containing 40 to 2,000 mg/L down to less than 1 mg/L. Reagent requirements for these waste streams varied with the type of feed, but were generally in the range of 3 to 5 g SO_2/g CN_T for barren solutions and 4 to 7 g SO_2/g CN_T for tailing slurries.[14] Table 14.4.4 shows greater than 99 percent of the CN_T was removed from feeds containing up to 62,000 mg/L of CN_T.

14.4.3.3 Ferrous Sulfate Treatment--

No data were found for the ferrous sulfate treatment process other than that reported in Reference 15.

14.4.4 Process Costs

Table 14.4.5 contains cost data developed for an SO_2/air oxidation system (single-stage reactor) sized to treat approximately 34,500 gal/day of

TABLE 14.4.3. SELECTED GOLD MILL BARREN AND TAILING LIQUOR RESULTS

Stream	Retention time (min)	Reagent	pH	Assays (mg/L) or (wt. %)						Reagents added (g/gCN_T)		
				CN_T	SCN	Cu	Ni	Fe	Zn	SO_2	Lime	Cu^{++}
FEED A	-	-	-	1,680	820	210	0.6	2.0	758	-	-	-
STAGE 1	97	Na_2SO_3	9.3	0.13	767	0.54	0.1	0.1	3.2	3.20	0	0
FEED B	-		12.5	420	1,584	137	1.6	19	71	-	-	-
STAGE 1	26	SO_2	9.0	-	-	13	0.2	5.2	0.4	5.44	8.16	0
STAGE 2	26	SO_2	9.0	0.11	1,408	1.4	0.2	0.2	0.2	1.36	2.04	0
FEED C	-		9.5	500	270	55	53	66	53	-	-	-
STAGE 1	22	SO_2	9.0	3.0	220	13	3.2	0.2	-	3.40	4.39	0
STAGE 2	22	SO_2	9.5	1.2	216	0.4	0.8	0.2	0.4	0.85	2.19	0
FEED D	-	-	11.8	2,180	1,820	235	2.0	325	210	-	-	-
STAGE 1	60	$Na_2S_2O_5$	9.0	0.43	-	4.4	0.2	0.1	3.8	5.00	4.50	0.25
*FEED E (30% SOLIDS)	-	-	-	1,480	1,380	138	1.7	252	214	-	-	-
STAGE 1	180	$Na_2S_2O_5$	0.65	1,300	3.0	0.1	0.1	0.7	4.4	4.4	0.46	
*FEED F (24% SOLIDS)	-	-	-	40	87	1.3	1.6	12.5	.1	-	-	-
STAGE 1	15	$Na_2S_2O_5$	8.0	0.07	81	0.1	.1	0.4	.1	4.8	7.3	0.91
*FEED G (35% SOLIDS)	-	-	10.7	200	129	47	-	7.0	55	-	-	-
STAGE 1	17	H_2SO_3	8.0	6.6	91	16	-	1	.1	6.7	11.8	0.40
STAGE 2	17	-	8.5	0.2	92	0.3	-	1	.1	0	0	0

*Tailing slurries.

Source: Reference 14.

TABLE 14.4.4. TYPICAL PLATING RINSE WATER RESULTS

Stream	Retention time (min)	SO_2 Reagent	Caustic	pH	Assays (mg/L)							Reagents added (g/gCN_T)		
					CN_T	Cu	Fe	Ni	Zn	Cd	Sn	SO_2	Caustic	Cu^{++}
FEED H	-	-	-	13.1	62,400	3,600	26.4	5,400	-	-	-	-	-	-
STAGE 1	1,200	$Na_2S_2O_5$	NaOH	9.0	12.7	2.6	3.0	0.9	-	-	-	3.5	3.3	0
FEED I	-	-	-	11.4	1,280	760	6.0	2.2	-	-	1,500	-	-	-
STAGE 1	500	$Na_2S_2O_5$	NaOH	9.0	3.1	5.8	1.2	0.2	-	-	100	4.4	1.2	0
STAGE 2	500	-	-	9.1	2.8	4.6	1.0	.2	-	-	30	0	0	0
FEED J	-	-	-	11.3	540	-	0.2	-	90	-	-	-	-	-
STAGE 1	80	$Na_2S_2O_5$	NaOH	9.0	1.2	2.2	.2	-	8.4	-	-	4.7	3.5	0.09
FEED K	-	-	-	-	142	47.3	18.0	-	14.3	10.0	-	-	-	-
STAGE 1	7.2	SO_2	LIME	9.0	.4	6.0	.1	-	.1	.1	-	4.7	6.4	0
STAGE 2	7.2	SO_2	LIME	9.0	.4	2.6	0.2	-	.1	.1	-	1.4	3.2	0

Source: Reference 14.

TABLE 14.4.5. TOTAL ANNUAL COSTS FOR SO_2/AIR OXIDATION OF GOLD MILL BARREN BLEED SOLUTION

Cost item	Cost ($)[a] Reaction system exclusive of solid-liquid separation facilities	Reaction system inclusive of solid-liquid separation facilities
1. Capital cost		
(a) Major equipment	38,500	68,400
(b) Installation at 60% of major capital	23,100	41,000
(c) Total installed cost	61,600	109,400
2. Operating cost		
(a) Lime @ $46/ton	12,900	12,900
(b) SO_2 @ $230/ton	66,300	66,300
(c) Polymer @ $1.6/lb	--	900
(d) Electrical @ $0.05	1,700	1,700
(e) Labor @ $20/hr	33,700	33,700
3. Total annual cost		
(a) Amortization 10 yr @ 12%	10,900	19,400
(b) Operating cost	114,600	115,500
(c) Maintenance @ 6% of installed capital	3,700	6,600
(d) Total ($/yr)	129,200	141,500
4. Unit treatment cost		
(a) $ per 1,000 gallons	12.5	14.0

Source: Adapted from Reference 13.

[a]1987 dollars.

gold mill barren bleed solution (cyanide stream from cyanide leaching process).[13] Hydraulic retention time in the reactor was 50 minutes to allow for flow equalization. Operating costs, including reagent consumption, were based on data generated by INCO in continuous flow laboratory experiments.

Costs of the solid-liquid separation system were estimated with and without the installation of a Lamella flocculator-clarifier. Reactor design incorporates a sparger for gas transfer and a turbine-type mixer for gas-liquid contact. Automatic pH control equipment was provided. Redox potential in the reactor would be continuously recorded, but not utilized as an automatic process control variable. No provision was made for a dedicated lime feed system, dewatering equipment, or sludge disposal. Operator time associated with the treatment process was based on 2 hours/8 hour shift. Preliminary data from pilot-scale operation indicate the ORP control may be feasible, thus reducing labor requirements.

Annual unit treatment costs for this system were approximately $12.5/1,000 gallons for the reaction system, exclusive of solid-liquid separation facilities, and $14/1,000 gallons for the system incorporating the Lamella flocculator/clarifier. While the unit costs for this system are much higher than those of the alkaline chlorination system shown in Section 14.1, the greater influent cyanide concentration (1,300 mg/L) in the gold mill barren bleed contributes substantially to the higher operating costs. If equal influent cyanide concentrations were present in each cost model, the SO_2/air process would be much more competitive with alkaline chlorination. In addition, the SO_2/air process provides an added performance benefit since it is capable of removing any ferro or ferric-cyanides present in the feed stream.[13,14]

Process costs for the polysulfide and ferrous sulfate treatments of cyanide-bearing waste streams are not included in this section due to the lack of reliable cost information.

14.4.5 Status of Technology

14.4.5.1 Availability/Applications--

Sulfur-based cyanide treatment technologies, while not fully developed, have demonstrated potential for the treatment of cyanide wastes. Both reagent and equipment requirements are straightforward and simple. Application to industrial wastes is presently limited, but both polysulfide and INCO Process technologies have demonstrated high efficiencies in treating dilute and concentrated aqueous cyanide waste streams.

Licensing of the INCO Process is handled through INCO Tech, a Division of INCO, Ltd. Licensing fees are claimed to be a modest fraction of the operating costs.

14.4.5.2 Environmental Impacts--

The environmental impact of the processes discussed here relate to the unreacted contaminants and byproducts (thiocyanates) remaining in the waste stream. Additional treatment to prevent corrosion and minimize thiocyanate concentrations probably will be required. Air emissions associated with the use of these technologies will be minimal, although some care must always be observed in pH adjustments to prevent hydrogen cyanide evolution.

14.4.5.3 Advantages and Limitations--

The advantages of sulfur-based processes discussed here result from ease and simplicity of operation. Capital investments are low, relative to other cyanide oxidation processes, and reagent consumption is also low (due to nonoxidation of SCN^- to carbon dioxide and nitrogen dioxide). Disadvantages are the result of incomplete destruction and the need for subsequent treatment of the partially oxidized waste stream.

REFERENCES

1. U.S. EPA. Treatability Manual, Volume III. EPA-600/8-80-042c. July 1980.

2. Cushnie, G. C. Centec Corporation, Navy Electroplating Pollution Control Technology Assessment Manual. CR 84-019. February 1984.

3. Freeman, H. Innovative Thermal Hazardous Treatment Processes. U.S. EPA, Hazardous Waste Engineering Research Laboratory, Cincinnati, OH. 1985.

4. Ehmke, E. F. Tosco Corporation. Polysulfide stops FCCU corrosion. Hydrocarbon Processing. July 1981.

5. Luthy, R. G., and S. G. Bruce. Carnegie-Mellon University, Kinetics of Reaction of Cyanide and Reduced Sulfur Species in Aqueous Solution. Environmental Science and Technology Vol. 13, No. 12. December 1979.

6. Trode, T. W., G. C. Radian Corporation. Cyanide Removal in Coal Gasification Wastewater Using Polysulfide. 40th Industrial Waste Conference, Purdue University. 1985.

7. Roeck, D., and M. Gollands. Alliance Technologies Corporation. Hazardous Waste Treatment Facility Site Visit Report. September 1985.

8. Mitre Corporation. Manual of Practice for Wastewater Neutralization and Precipitation. EPA-600/2-81-148. August 1981.

9. Cushnie, G. C. Removal of Metals from Wastewater: Neutralization and Precipitation. Pollution Technology Review. No. 107. Noyes Publications, Park Ridge, NJ. 1984.

10. Devuyst, E. A., V. A. Ettel, and G. J. Borbely. New Method of Cyanide Destruction in Gold Mill Effluents and Tailing Slurries. Presented at the 14th Annual Operators Conference of the Canadian Mineral Processors Division of the CIM, Ottawa, Ontario. January 1982.

11. Devuyst, E. A., V. A. Ettel, and G. J. Borbely. New Process for Treatment of Wastewaters Containing Cyanide and Related Species. Presented at the 1982 AIME Annual Meeting, Dallas, Texas. February 1982.

12. Devuyst, E. A., B. R. Conard, and V. A. Ettel. Pilot Plant Operation of the Inco SO_2/Air Cyanide Removal Process", presented at the 29th Ontario Industrial Waste Conference, Toronto, Ontario. June 1982.

13. Nutt, S. G., and S. A. Zaidi. Treatment of Cyanide-Containing Wastewaters by the Copper Catalyzed SO_2/Air Oxidation Process. 38th Industrial Waste Conference, Purdue University. 1983.

14. Devuyst, E. A., B. R. Conard, and W. Hudson. Commercial Operation of INCO's SO_2/Air Cyanide Removal Process. Conference on Cyanide and the Environment Tucson, Arizona. December 1984.

15. Schiller, J. E. Removal of Cyanide and Metals from Mineral Processing Wastewaters. Bureau of Mines Report of Investigations No. 8836. 1983.

16. Nuthy, R. G., et al. Carnegie-Mellon University, Cyanide and Thiocyanate in Coal Gasification Wastewater. Journal of the Water Pollution Control Federation, Vol. 51, No. 9. September 1979.

17. Ganczarczyk, J. J., P. T. Takoaka, and D. A. Ohashi. Application of polysulfide for pretreatment of spent cyanide liquors. Journal of the Water Pollution Control Federation, Vol. 57, No. 11. November 1985.

18. Ganczarczyk, J. J. Biological Decomposition of Thiocyanate. Paper presented at the Symposium Canadian de la Recherche sur la Pollution de L'eau, St. Foy, Quebec. November 1977.

19. Ganczarczyk, J. J. Fate of Basic Pollutants in Treatment of Coke-Plant Effluents. Proc. 35th Purdue University. Ind. Waste Conference 325. 1980.

14.5 MISCELLANEOUS CYANIDE DESTRUCTION PROCESS

A variety of nonconventional or experimental processes are being studied for the treatment of cyanide-bearing wastes. The cyanide treatment technologies examined are: the Modar Process, the use of chemical oxidizing agents with and without catalysts, and catalytic oxidation.

14.5.1 Process Description

14.5.1.1 The Modar Process--

Supercritical fluid oxidation (the Modar Process) is a technology that has been proposed for the destruction of organic contaminants in wastewaters. It is basically an oxidation process conducted in a water medium at temperatures and pressures that are supercritical for water; i.e., above 374°C (705°F) and 218 atmospheres. In the supercritical region, water exhibits properties that are far different from liquid water under normal conditions; oxygen an organic compounds become totally miscible with the supercritical water (SCW) and inorganic compounds, such as salts, become very sparingly soluble. When these materials are combined in the SCW process, organics are oxidized and inorganic salts present in the feed or formed during the oxidation are precipitated from the SCW.

The oxidation reactions proceed rapidly and completely. Reaction times are less than 1 minute, as compared to reaction times of about 60 minutes used in the subcritical wet air oxidation (WAO) process. Moreover, the reaction is essentially complete. Carbon, nitrogen and hydrogen atoms within the organic contaminants are reacted to form CO_2, NO_2 and H_2O (residuals such as the low molecular weight organic acids and alcohols found in the treated WAO effluent are not found in the SCW process effluent). Heteroatoms (e.g., chlorine and sulfur) are oxidized to their corresponding acidic anion groupings. These anions, and those occurring naturally in the feed, can be neutralized by cation addition to the feed, and the total inorganic content of the waste, save that soluble in the SCW, can be precipitated and recovered by mechanical separators operating at SCW conditions.

Presently, the Modar process has not been dedicated to cyanide destruction. However, the high oxidation efficiency and rapid reaction rate of the SCW process in treating other organic compounds warrants further investigation. For a further discussion of SCW technology see References 1-7.

14.5.1.2 Other Chemical Oxidizing Agents--

As shown previously in Table 14.1.1, hydrogen peroxide, H_2O_2, and potassium permanganate, $KMnO_4$, are both relatively strong oxidizing agents. Hydrogen peroxide has been used to treat phenols, cyanides, sulfur compounds, and metal ions in dilute waste streams. Potassium permanganate is primarily used for the treatment of phenols. The choice of these and other oxidants is dependent upon such factors as toxicity, reaction rate, ease of removal of secondary products, simplicity and cost.

Oxidation with H_2O_2 is generally performed in the presence of a metal catalyst. Typical catalysts include ferrous sulfate, nickel salts, and aluminum salts. The waste is heated and then treated with H_2O_2 while being agitated. The H_2O_2 oxidation tends to proceed quickly under basic conditions.[10] The feasibility of ultraviolet catalyzed H_2O_2 oxidation has been studied, but it does not appear to be used on an industrial scale.[11] Potassium permanganate oxidation is favored under basic conditions. Raising the pH to the optimum level is accomplished by the addition of lime, soda ash, or caustic soda.

The equipment required for chemical oxidation is very simple. This includes storage vessels for the oxidizing agents and perhaps for the waste, metering equipment for both streams, and vessels with agitators to provide contact between the oxidant and the waste. Some instrumentation is required to determine the concentrations of pollutants, pH, and the degree of completion of the oxidation reaction. The process is usually monitored by an oxidation reduction (ORP) potential electrode.[12].

For the treatment of sodium, potassium, zinc, and cadmium cyanide, a hydrogen peroxide solution with formalin may be used to reduce the cyanide (Kastone Process).[13] This process is usually operated at ambient temperature and a pH between 10 and 11.5. The effluent from this process has a high biochemical oxygen demand and requires biological treatment before direct discharge to sewers.

The treatment of cyanide waste streams with alternate oxidizing agents has been limited to with batch processes or low effluent flows. The treatment of large effluent flows is generally not practicable because of a lack of suitable means of determining the correct dosage quickly and accurately enough to allow efficient use of the reagent.[14] Other limitations include chemical interference from other oxidizable species, limited shelf life (H_2O_2); inability to effectively oxidize cyanide beyond the cyanate level, and the need for catalysts. Therefore, the use of this technology is restricted to process situations where alkaline chlorination would not be feasible, i.e., waste streams containing phenols or aliphatic hydrocarbons.

14.5.1.3 Catalytic Oxidation--

One or the earliest investigations of catalytic oxidation was conducted by Battelle Laboratories in 1971 to study the adsorption of free and complexed cyanide onto activated carbon in the presence of copper.[15] Subsequent efforts were undertaken by the Calgon Corporation to also develop a cyanide detoxification method utilizing catalytic oxidation on granular activated carbon.[16] Cupric ions are added to the wastewater along with oxygen prior to passing the cyanide-bearing waste through a granular activated carbon columns. According to Calgon, "cupric ions are added to the water to accelerate and increase the efficiency of the catalytic oxidation of cyanide by granular activated carbon." In addition to improving the catalytic oxidation of the cyanide, "the presence of cupric ions results in the formation of copper cyanides, which have a greater adsorption capacity than copper of cyanide alone."[16]

14.5.2 Process Performance

According to a 1979 survey of 216 metal finishing plants practicing cyanide oxidation, three plants were found to be using hydrogen peroxide as an oxidizing agent. The process used by these plants is a proprietary treatment called the Kastone process.

The Kastone hydrogen peroxide oxidation treatment process treats both the cyanide and metals in cyanide wastewaters containing zinc or cadmium. In this process, cyanide rinse waters are heated to 49-54°C (120-130°) to break the cyanide complex, and the pH is adjusted to 10.5-11.8. Formalin (37 percent formaldehyde) is added, while the tank is vigorously agitated. After 2 to 5 minutes, a proprietary formulation (41 percent hydrogen peroxide with a catalyst and additives) is likewise added. After an hour of mixing, the reaction is complete. The cyanide is converted to cyanate and the metals are precipitated as oxides or hydroxides. The metals are then removed from solution by either settling or filtration.[13]

In terms of waste reduction performance, the Kastone process was found to be capable of reducing the cyanide level to less than 0.1 mg/L and the zinc or cadmium to less than 1.0 mg/L. Table 14.5.1 presents performance data for a treatment process using hydrogen peroxide to treat gold mill tailings. The process uses an excess of H_2O_2 to achieve rapid oxidation of cyanide ions to cyanate. A slight reduction in pH (0.2 to 0.3 units) was found to take place during the reaction. The concentration of available cyanide was reduced to less than 0.5 mg/L, but could be reduced to 0.1 mg/L at the cost of increased H_2O_2 consumption (present consumption is 0.4 to 1.0 L H_2O_2 70 percent/cubic meter of tailings). Catalytic oxidation effectiveness is shown in Table 14.5.2. While the results demonstrated that cyanide could be effectively adsorbed (80 to 99 percent) by activated carbon, regeneration efficiencies were poor (12 to 28 percent), and residuals remained in the toxic cyanide state.

Tables 14.5.3 through 14.5.5 shows the catalytic oxidation adsorption data for the treatment of copper, zinc, and cadmium wastes determined by Calgon. While Calgon has not pursued the implementations of this technology on a commercial scale, research into possible applications has continued.[7]

14.5.3 Process Costs

Due to the current level of development of these technologies, limited cost data are not available. Major costs would be associated with process equipment and the cost of chemical reagents. Pretreatment, operating, and post-treatment costs are unknown, but are expected to be similar to alkaline chlorination (Section 14.1) in the case of hydrogen peroxide and carbon adsorption (Section 8.1) for catalyzed oxidation.

TABLE 14.5.1. HYDROGEN PEROXIDE TREATMENT OF GOLD MILL TAILINGS

	Before H_2O_2 treatment	After H_2O_2 treatment
Tailings flow m^3/h (nominal)	1,100	1,100
Solids content % (nominal)	45	45
pH	10.5 - 11.0	10.2 - 10.8
Free cyanide mg/L	50 - 100	undetectable
Easily-liberatable cyanide mg/L	90 - 200	< 0.5
Total cyanide mg/L	110 - 300	1 - 10
Dissolved Cu mg/L	50 - 100	< 0.5
Zn mg/L	10 - 30	< 0.1
Fe mg/L	1 - 3	1 - 3

Source: Reference 14.

TABLE 14.5.2. RESULTS OF MULTIPLE CYCLE ADSORPTION AND REGENERATION RUNS ON CONCENTRATED ZINC CYANIDE WATERS

		Typical concentrations, ppm			Weight, pounds		Efficiency, percent	
	Component	In feed	Effluent average	Effluent at breakthrough	adsorbed	stripped	adsorption	regeneration
Cycle 1, 2,460 gal treated	Cyanide, CN	220-340	1	61	5.88	1.67	99	28
	Copper, Cu	240-317	1	30	5.47	5.06	99	92
	Iron, Fe	6-17	2	12	0.33	0.03		10
Cycle 2, 1,770 gal treated	Cyanide, CN	239-364	30	94	(4.93)	1.05	90	21
	Copper, Cu	145-252	1	43	4.10	3.09	99	75
	Iron, Fe	0.1-16.7	2	0.2	--	0.01	--	--
Cycle 3, 1,800 gal treated	Cyanide, CN	203-270	50	60	3.73	0.97	80	26
	Copper, Cu	164-214	1	32	3.06	2.87	99	94
	Iron, Fe	1.3-8.5	--	0.2	--	0.18	--	--
Cycle 4, 3,540 gal treated	Cyanide, CN	234-436	2	31	7.36	0.89	98-99	12
	Copper, Cu	170-365	1	5.8	6.72	5.59	99	83
	Iron, Fe	0.3-33	0.3	--	0.26	0.13	--	50
Cycle 5, 4,100 gal treated	Cyanide, CN	333-468	2-3	--	10.15	--	98-99	--
	Copper, Cu	224-330	1	--	9.57	--	99	--
	Iron, Fe	0.3-21	0.3	--	0.33	--	--	--

Source: Reference 15.

TABLE 14.5.3. COPPER CYANIDE WASTE TREATMENT USING CATALYTIC OXIDATION

Days on stream	Influent CN^- (mg/L)	Effluent (mg/L) CN^-	Cu	Fe
1	32	0.01	0.05	0.05
2	32	0.01	0.05	0.05
3	32	0.01	0.05	0.05
4	32	0.01	0.05	0.01
5	32	0.01	0.05	0.01
6	32	0.02	0.05	0.05
7	28	0.04	0.05	0.05
8	28	0.08	0.05	0.05
9	28	0.02	0.05	0.05
10	28	0.77	0.05	0.45
11	28	0.80	0.05	0.35
12	30	0.32	0.05	0.15
13	30	0.25	0.05	0.10
14	30	0.10	0.05	0.05
15	30	0.10	0.05	0.10

TABLE 14.5.4. ZINC CYANIDE WASTE TREATMENT USING CATALYTIC OXIDATION

Days on stream	Influent CN^- (mg/L)	Effluent (mg/L)			
		CN^-	Cu	Zn	Fe
1	22.0	0.01	0.05	0.05	0.05
2	22.0	0.02	0.05	0.05	0.05
3	22.0	0.03	0.05	0.05	0.05
4	22.0	0.01	0.05	0.05	0.05
5	22.0	0.04	0.05	0.05	0.05
6	22.0	0.04	0.05	0.05	0.05
7	30.0		0.05	0.05	0.05
8	30.0	0.08	0.05	0.05	0.05
9	30.0		0.05	0.05	0.05
10	30.0	0.05	0.05	0.05	0.05
11	25.6	0.14	0.05	0.05	0.05
12	25.6	0.04	0.05	0.05	0.05
13	25.6	0.04	0.05	0.05	0.05
14	25.6	0.07	0.05	0.05	0.05
15	25.6	0.03	0.05	0.05	0.05

TABLE 14.5.5. CADMIUM CYANIDE WASTE TREATMENT USING CATALYTIC OXIDATION

Days on stream	Influent CN^- (mg/L)	Effluent (mg/L)			
		CN^-	Cu	Zn	Fe
1	30	0.01	0.05	0.05	0.05
2	30	0.04	0.05	0.10	0.05
3	30	0.01	0.05	0.20	0.05
4	30		0.05		0.05
5	30	0.01	0.05	0.05	0.05
6	30	0.01	0.05	0.20	0.05
7	29	0.08	0.05	0.05	0.05
8	29	0.04	0.05	0.20	0.05
9	29	0.06	0.05	0.20	0.05
10	29	0.03	0.05	0.10	0.05
11	22	0.06	0.05	0.05	0.05
12	22	0.11	0.05	0.10	0.05
13	22	0.14	0.05	0.10	0.05
14	22	0.06	0.05	0.05	0.05
15	22	0.03	0.05	0.05	0.05

14.5.4 Overall Process Status

Other then hydrogen peroxide oxidation, the commercial application of these processes to free and total cyanides has yet to be established. Some level of destruction can be expected, but economic considerations have limited application. Hydrogen peroxide oxidation has been commercially applied, but typically to waste streams in which chlorine or hypochlorite oxidation would not be feasible. Hydrogen peroxide oxidation has limited application to slurries, tars, and sludges. This is due to the presence of other oxidizable components in the sludge which may be attacked indiscriminately by the oxidizing agent, thus increasing reagent consumption.

The environmental impact of the processes discussed here relate to the unreacted contaminants and byproducts remaining in the waste stream. Additional treatment usually will be required. Air emissions associated with the use of hydrogen peroxide and permanganate oxidant will be minimal, although some care must always be observed when the contaminants are high vapor pressure solvents and ignitables.

The advantages of the oxidation processes discussed here result from ease and simplicity of operation. Disadvantages are the result of incomplete destruction and the need for subsequent treatment of the oxidized waste stream.

REFERENCES

1. Josephson, J. Supercritical Fluids. Environmental Science and Technology. Volume 16, No. 10. October 1982.

2. Thomason, T. B. and M. Modell. Supercritical Water Destruction of Aqueous Wastes. Hazardous Waste. Volume 1, No. 4. 1984.

3. Franck, E. U. Properties of Water in High Temperate, High Pressure Electrochemistry in Aqueous Solutions (NACE-4). p. 109. 1976.

4. Sieber, F. MODAR Inc. Review of Draft Section, Supercritical Water Oxidation. May 16, 1986.

5. National Academy of Sciences, Medical and Biological Effects of Environmental Pollution: Nitrogen Oxides. 1977.

6. Modell, M., G. Gaudet, M. Simson, G. T. Hong, and K. Biemann. Supercritical Water Testing Reveals New Process Holds Promise. Solid Wastes Management. August 1982.

7. Freeman, H. Innovative Thermal Hazardous Waste Treatment Processes. U.S. EPA, HWERL Cincinnati, OH. 1985.

8. Prengle, H. W., Jr. Evolution of the Ozone/UV Process for Wastewater Treatment. Paper presented at Seminar on Wastewater Treatment and Disinfection with Ozone., Cincinnati, OH. September 15, 1977. International Ozone Association, Vienna, VA.

9. Harris, J. C. Ozonation. In: Unit Operations for Treatment of Hazardous Industrial Wastes. Noyes Data Corporation, Park Ridge, N.J. 1978.

10. Hackman, E. Ellsworth. Toxic Organic Chemicals-Destruction and Waste Treatment. Park Ridge, NJ, Noyes Data Corporation. 1978.

11. Sundstrom, D. W., et al. Destruction of Halogenated Aliphatics by Ultraviolet Catalyzed Oxidation with Hydrogen Peroxide. Department of Chemical Engineering, The University of Connecticut. Hazardous Waste and Hazardous Materials, 3(1). 1986.

12. Chillingworth, M. A., et al. Industrial Waste Management Alternatives for the State of Illinois, Volume IV - Industrial Waste Management Alternatives and their Associated Technologies/Processes, prepared by Alliance Technologies Corporation. February 1981.

13. Radimsky, J., et al. California Department of Health Services. Recycling and/or Treatment Capacity for Hazardous Waste Containing Cyanides. Staff Report. March 1983.

14. Knorre, M., and A. Griffiths. Cyanide Detoxification with Hydrogen Peroxide Using the Degussa Process. Cyanide and the Environment Conference Proceedings, Tucson, Arizona. December 1984.

15. Battelle Laboratories. An Investigation of Techniques for Removal of Cyanide from Electroplating Wastes. Water Pollution Control Research Series. 12010 EIE. November 1971.

16. Bernardin, F. E. Cyanide Detoxification Using Adsorption and Catalytic Oxidation on Granular Activated Carbon. Journal of the Water Pollution Control Federation. 45(2). 1973.

17. Huff, J. E., and J. Bigger. Cyanide Removal from Refinery Wastewater Using Powdered Activated Carbon. U.S. EPA Draft Report on Grant No. R804029-01. 1976.

15. Miscellaneous Cyanide Destruction Processes

The main miscellaneous cyanide destruction processes are biodegradation and thermal treatment. Biodegradation as a process for treating wastes containing cyanide is still in the developmental stage. Certain types of microorganisms have shown the ability to completely degrade low concentrations of simple cyanides. The major obstacle to implementation has been the inability of most conventional biosystems even when acclimated, to degrade fixed cyanides or simple cyanides in high concentrations. However, since the end products of complete biodegradation are nontoxic, continued research is advisable. In addition, many of the new bioaugmentation processes which can degrade fixed and/or concentrated cyanide wastes, may render biological treatment as a feasible alternative to conventional chemical or thermal destruction technologies.

Thermal treatment technologies which may be applied to cyanide-bearing hazardous wastes include incineration, evaporation, and crystallization. The processing systems involved in each of these technologies are similar to those described for management of metal-bearing hazardous wastes. Test studies have indicated high potential levels of waste destruction (i.e., in excess of 99.99 percent) for the incineration of cyanide wastes. Incineration is most typically used to destroy cyanide wastes generated in organic chemical manufacturing; e.g., acrylonitrile production. Other cyanide waste candidates for incineration are waste organic cyanide compounds such as cyanogen. Cyanide waste component recovery by evaporation or crystallization has been demonstrated to achieve yields in excess of 90 percent in certain cases.

15.1 BIOLOGICAL DESTRUCTION OF CYANIDES

15.1.1 Process Description

Microbiological degradation of cyanide is a developing technology, capable of oxidizing low concentrations of simple cyanides into carbon dioxide and ammonia.[1,2,3] The process offers several advantages over other methods of degrading or detoxifying cyanide-bearing waste streams. For example, hazardous chemicals such as caustic soda, gaseous chlorine, and hypochlorite salts are not required, thereby reducing exposure risks.[4] In addition, toxic by-products and/or sludges are not generated during processing, eliminating secondary treatment for cyanides; however, removal of the non-toxic by-products and/or sludge is still required.

However, most conventional cyanide biodegradation systems are only able to treat total cyanide concentrations of approximately 10 mg/L or less without noticeable performance impairment.[5] Furthermore, only free cyanide is biodegradable, with waste streams containing fixed cyanides experiencing the lowest removal rates.[5] These drawbacks have limited the application of biological treatment for cyanide-containing wastes.

The principal factors which control the microbial degradation process are moisture level, organic content, oxygen level, temperature, pH, and nutrient source. Typical design factors include BOD and toxic constituent removal rate, detention time, reactor surface area and type, nutrients required to sustain biological activity, and sludge production. Operating parameters, pretreatment and post-treatment requirements, and process costs have been presented in Section 11.0. Therefore, the remainder of this section will focus on performance and technological status of biological treatment of cyanides.

15.1.2 Process Performance

In 1982, wastewaters from a benzol plant were biologically treated in an upflow biotower (UBT).[5] The results were compared to 12 previous studies performed with similar wastewater in activated sludge and other types of fixed-filter reactors. Figure 15.1.1 summarizes percent cyanide removal and

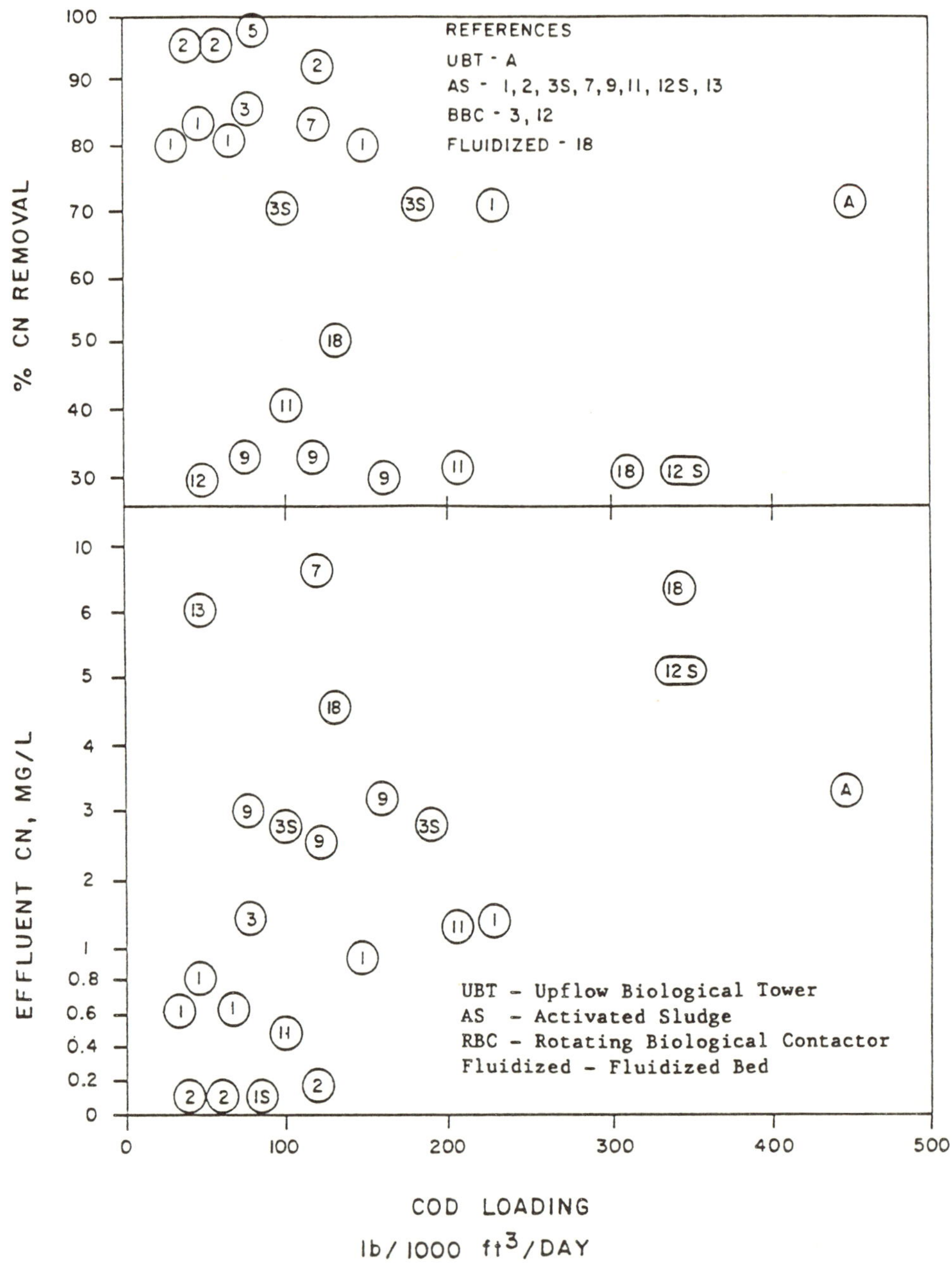

Figure 15.1.1. Cyanide removal vs. organic load for various biological reactors.

Source: Reference 5.

effluent concentrations as a function of organic loading. The data indicate that cyanide removal was only partial for the UBT, and show wide variation between the 13 studies. The inconsistency is thought to be due, in part, to the fact that investigators sometimes recorded total cyanide concentrations rather than free cyanide concentrations.[5]

However, in the overwhelming majority of cases studied, the influent cyanide concentration was less than 10 mg/L and less than 90 percent cyanide removal was achieved. These results suggest that conventional biological systems are not capable of reducing cyanide concentrations to acceptable discharge levels.

To remedy this problem, some research has been conducted to develop organisms specifically designed to degrade cyanide at levels which would normally be toxic to conventional systems. For example, Imperial Chemical Industries has recently tested and marketed the enzyme formamide hydrolase for this purpose.[6] Commercial trials were initiated in the summer of 1985 on a continuous system that provides a 2-hour residence time, 95 percent enzyme recycle, and an enzyme concentration of 10 g/L. Temperatures were approximately 30 to 35°C and the waste was maintained at a pH of 8 to 8.5.

The trial results showed that organic cyanides (nitriles) could be effectively treated after pretreatment with alkali. Some metal cyanide complexes are treatable, but the stronger ones (e.g., iron and copper) proved to be more resistant. The maximum concentration which could be handled was 10,000 ppm. The optimal feed concentration was found to be 5,000 ppm with reduction to 10 ppm achieved in 6 hours. Further reduction was reportedly possible (e.g., 1 ppm) but no data were presented to document this claim.[6] Cyanide is degraded to formamide and eventually to ammonia and carbon dioxide.

Researchers at Homestake Mining Company in South Dakota have developed a strain of _Pseudomonas paucimobilis_ that oxidizes the free and complexed cyanides and thiocyanates from the mine's wastewaters.[6] After 7 years of bench and pilot-scale evaluations, the process was commercialized as a 5.5 million gallon per day plant in the summer of 1984. The final design uses the strain and 48 rotating biological contactors. Among the alternative processes investigated by Homestake prior to commercialization were acidification/volatilization, ozonation, ion exchange, Prussian blue oxidation/precipitation, carbon adsorption, alkaline chlorination, and copper-catalyzed hydrogen peroxide (DuPont Kastone process). Biological processes investigated included activated sludge, suspended growth, and several attached growth processes, including trickling filters, biological towers, aerated biological filters, and rotating biological contactors.

The only chemical requirements for the process are soda ash and phosphoric acid. Products of the biological degradation are relatively harmless anions such as sulfates, nitrates, and carbonates. Reportedly, ammonia is not released as a by-product. Kinetics are first order until low levels of the pollutants are reached.[6]

15.1.3 Process Status

Currently, biodegradation of wastes containing cyanide is still in the developmental stage. Certain types of microorganisms have shown the ability to completely degrade low concentrations of simple cyanides.[7] The major obstacle to implementation has been the inability of most conventional biosystems, even when acclimated, to degrade fixed cyanides or simple cyanides in high concentrations.[5] However, since the end products of complete biodegradation are nontoxic, continued research is advisable. In addition, new bioaugmentation processes which degrade fixed and/or concentrated cyanide wastes appear to have substantial potential as an alternative to conventional chemical or thermal destruction technologies.

REFERENCES

1. Howe, R.H.L., "Bio-Destruction of Cyanide Wastes-Advantages and Disadvantages," Jour. Air Water Pollution, Pergamon Press, 9(1965).

2. Key, Arthur, "Gas Works Effluents and Ammonia," Institute of Gas Engineers, Britain (1938).

3. Painter, H.A., and Ware, G.C., Nature, London, 175,900 (1955).

4. Alliance Technologies Corporation. Treatment Technologies for Dioxin Containing Wastes. Contract No. 68-03-3243. August 1986.

5. Olthof, M., Oleszkiewicz, S. Benzol Plant Wastewater Treatment in a Packed-Bed Reactor. 37th Industrial Waste Conference. Purdue University. 1982.

6. Technical Insights. New Methods for Degrading/Detoxifying Chemical Wastes. Englewood, N.J. 1986.

7. Shahalom, A.M. Mixed Culture Biological Activity in Water Containing Variable Low Concentrations of Cyanide, Phenol, and BOD. 38th Industrial Waste Conference, Purdue University. 1983.

15.2 THERMAL PROCESSING OF CYANIDE-BEARING WASTES

Several of the thermal processes outlined in Section 12 for treatment of metal-bearing hazardous wastes may also be considered as alternatives for treatment of hazardous wastes containing cyanides. As discussed in Section 12, many of the cyanide-bearing hazardous wastes are generated by essentially four different industries. Wastes from electroplating and metal finishing operations comprise by far the greatest percentage of the overall volume of cyanide-bearing waste.[1] Much, if not all of those wastes also contain heavy metals such as chromium, nickel, and lead. The other important sources of cyanide-bearing waste include the metallurgical coke industry (whose wastes also contain metal constituents), acrylonitrile or cyanide-related organic compound manufacturing, and the manufacturing of cyanide salts such as sodium cyanide or potassium cyanide. The characteristics of cyanide-containing wastes from the organic chemical industry would appear to be most suitable for treatment by thermal processes such as incineration.

Economic and environmental factors constitute the most significant barriers to selection of incineration for treatment of cyanide-bearing wastes. Further, many commercial incinerators surveyed do not handle cyanide wastes at all, citing emissions of deadly cyanide gases (e.g., HCN) as a major concern. Such gases would require extensive environmental control and safety precautions, including secondary incineration. Given the expense of such procedures relative to available chemical treatment processes, most of the commercial waste processors surveyed could not recommend incineration as an option for cyanide-bearing wastes. Finally, such systems may also generate high levels of NO_x emissions, and solid and liquid waste streams requiring additional control.

15.2.1 Process Descriptions

The incineration technologies which may be employed for the disposal of cyanide-bearing hazardous wastes are similar to those identified in Section 12 of this report, and described in detail in Reference 2. Incineration and pyrometallurgical processes would lead to the destruction of the cyanide group; evaporation would result in increased concentration levels. No application of crystallization for recovery of cyanides was identified in the literature.

15.2.2 Performance Data

As detailed in Section 12, the relative ease with which hazardous wastes containing cyanides may be incinerated has been studied within the context of a general study of hazardous waste incinerability conducted by EPA.[3] A summary of the "incinerability" ratings developed by EPA for such wastes is presented in Table 15.2.1. As shown, many of the organic cyanide-bearing hazardous wastes are considered to be at least "low" potential candidates for incineration. In addition, it may be noted that most of the "incinerable" wastes may be burned in either two or three of the most widely used incineration systems. The incinerability ratings are somewhat consistent with data complied in a 1981 study of incineration risk prepared for EPA by IE Inc.,[4] a similar study conducted in 1984 by ICF,[5] and a 1982 study prepared for EPA by MITRE, Inc.,[6] in which the quantities of waste currently incinerated were estimated. These data indicate that high volumes of waste-containing cyanides from certain industries are incinerated, including acrylonitrile manufacturing, and paint production. A summary of these data is shown in Table 15.2.2. Clearly, many cyanide-bearing hazardous wastes are incinerable and are currently being incinerated. Relative to the overall volume of cyanide-bearing waste, however, the amount incinerated is very small.

The "incinerability" of cyanide-bearing wastes may be evaluated through assessment of a variety of key waste characteristics. These include:

- Physical form;
- Heat content/heat of combustion;
- Autoignition temperature/thermal stability;
- Moisture content;
- Organic content;

TABLE 15.2.1. RANKING OF INCINERABILITY OF CYANIDE-BEARING WASTES

Waste code	description[a]	Ranking[b]	Applicable technology[c]		
			LI	RK	FB
D003	Reactive wastes	Not Listed			
F006		Poor			
F007	Electroplating	Poor			
F008	sludges and	Poor			
F009	spent solutions	Poor			
F010		Poor			
F011	Heat treating	Poor			
F012	operations	Poor			
F014	Tailing pond sediment	Poor			
F019	Wastewater sludge	Not Listed			
K011		Low			
K012	Acrylonitrile	High	X	X	X
K013	production	Low			
K014		Low			
K027	Diisocyanate product	Low			
K060	Coking	Poor			
K087		Not Listed			
P013	Barium cyanide	Poor			
P021	Calcium cyanide	Poor			
P027	3 Chloropropionitrile	Low	X	X	
P029	Copper cyanide	Poor			
P030	Cyanide salts	Poor			
P031	Cyanogen	High		X	X
P032	--	Low		X	X
P033	Cyanogen chloride	Low		X	X
P052	--	High	X	X	X
P055	--	Poor			
P063	Hydrogen cyanide	High	X	X	X
P064	Isocyanic ester	High	X	X	X
P069	2 Methylacrylonitrile	Low	X	X	X
P074	Nickel cyanide	Poor			
P098	Potassium cyanide	Poor			
P099	Pot./silver cyanide	Poor			

(Continued)

TABLE 15.2.1 (continued)

Waste code description[a]	Ranking[b]	Applicable technology[c] LI	RK	FB
P104 Silver cyanide	Poor			
P106 Sodium cyanide	Poor			
P121 Zinc cyanide	Poor			
U003 Acetonitrile	Low	X	X	X
U009 Acrylonitrile	High	X	X	X
U152 Methacrylonitrile	Low	X	X	X
U223 Toluene Diisocyanate	Low	X	X	X
U246 --	Not Listed			
SIC codes:				
2869 - Solid waste from ion exchange column, acrylonitrile production	Low		X	X
2869 - Bottom stream from quench column, acrylonitrile production	High	X	X	X
2869 - Still bottoms, aniline production	High		X	X

[a]EPA and SIC waste codes determined to contain cyanides (see Section 2).

[b]General rationale for ranking as "poor", "low", or "high" is based upon heat of combustion, moisture content, solids content, and several other key waste characteristics. For a detailed explanation reference should be made to Reference 3.

[c]LI = liquid injection incinerators; RK = rotary kiln incinerators; FB = fluidized bed incinerators.

Source: Adapted from Reference 3.

TABLE 15.2.2. SUMMARY OF WASTES CONTAINING CYANIDES CURRENTLY INCINERATED OR POTENTIALLY INCINERABLE (1981)

Waste code or SIC code	Description of waste stream	Quantity incinerated or incinerable (metric ton/yr)	Data source
Currently incinerated:			
D003	Non-listed reactive wastes	12,973.7	MITRE
P063	Hydrogen cyanide	230.6	MITRE
P074	Nickel cyanide	874.1	MITRE
P106	Sodium cyanide	3.7	MITRE
U003	Acetonitrile	9,068.2	MITRE
U223	Toluene Diisocyanate	178.5	MITRE
Potentially incinerable:			
K011	Acrylonitrile Stripper Bottoms	2,700,000	IEI
K012	Acrylonitrile Bottoms	5,900	IEI
K013	Crude Acrylonitrile	6,615	IEI
K014	Acrylonitrile Purification Wastes	47,628	IEI
K027	TDI Sludge	6,284	IEI
K048	Petroleum Refining, Dissolved Air Flotation Wastes	41,600	IEI
K049	Slot Oil Solids	46,800	IEI
K050	Petroleum Industry, Heat Exchanger Sludge	1,700	IEI
K051	API Separator Sludge	312,500	IEI
K052	Leaded Tank Bottoms	1,200	IEI
K086	Printing Ink Sludges	25,200	IEI
2851	Paint Prod. Trim Sludge	33,500	IEI
2851	Paint Prod. Paint Waste	11,898	IEI
34XX	Fab. Metal Prod. Paint Waste	399,237	IEI
35XX	Machine Man. Paint Waste	62,962	IEI
36XX	Electric Eq. Man. Paint Waste	117,264	IEI
37XX	Transportation Eq. Paint Waste	248,029	IEI
2911	Crude Tank Bottoms	1,000	IEI
2834	Photochemical Wastes	661	IEI
2869	Acrylonitrile Sludges	1,852	IEI
2869	Acrylonitrile Acid Wastes	33,075	IEI
F007, F009[a]	Spent Cleaning and Electroplating Solutions	1,990	ICF
3471	Spent Electroless Nickel Plating Solutions	30.7	ICF
3471	Electroplating Rinse Water	17,300	ICF
K060	Ammonia Still Lime Sludge from Coking	72.0	ICF
F010	Heat Treatment Wastes	6.0	ICF
K011,K013,	Acrylonitrile Bottoms	3,181	ICF
K014[a],K027	TDI Residues	107.9	ICF

[a]ICF lists as combined waste stream.

Sources: Reference 4, 5, and 6.

- Solids content;
- Chlorine content;
- Viscosity; and
- Corrosivity.

A detailed discussion of each of these parameters is presented in Reference 2.

Relative to the characteristics of cyanide-bearing wastes, it is clear that, in general, several key factors are regarded as highly restrictive to incineration. First, as indicated by several officials at commercial incineration facilities contacted by Alliance in an industry survey,[7-11] the potential for formation of highly toxic cyanide gases, such as HCN, present a significant restriction on the application of incineration by these commercial facilities. Such gases would require stringent control and safety precautions, which would in turn significantly affect overall treatment costs. Second, many cyanide-bearing wastes, particularly the inorganic wastes, are aqueous waste streams. Such wastes would require blending with auxiliary fuel prior to incineration, which could constitute a significant cost increase. Third, such wastes may often exhibit highly corrosive properties, which would necessitate the usage of thermal systems which are resistant to corrosion. While such systems are common, they involve higher costs due to increased air pollution requirements and solid and liquid waste effluent handling requirements.

15.2.3 Process Costs

Process costs constitute the primary constraint to the usage of incineration (and other thermal treatment processes) for management of cyanide-bearing wastes. Overall, both capital costs and operating costs are high, due to the size of such systems and their requirements. Costs for incineration systems are detailed more fully in Reference 2.

15.2.4 Process Status

Incineration of cyanide-bearing wastes is a common method of disposal of several by-product streams in the chemical manufacturing industry, most notably, acrylonitrile manufacturing. Cyanide-bearing wastes are not commonly accepted, however, by operators of commercial incinerators, who cite economic and environmental constraints - in particular those related to the potential for generation of highly toxic cyanide gases in air emissions - as the principal deterrent.

REFERENCES

1. Versar, Inc. Technical Assessment of Treatment Alternatives for Wastes Containing Metals and/or Cyanides, Draft Final Report. Versar, Inc., Springfield, VA. Prepared for U.S. Environmental Protection Agency, Office of Solid Waste, Washington, D.C. EPA Contract No. 68-03-3149. October 31, 1984.

2. Breton, M. et al. Technical Resource Document: Treatment Technologies, for Solvent-Containing Wastes, Final Report. GCA Technology Division, Inc., Bedford, MA. Prepared for U.S. Environmental Protection Agency, Hazardous Waste Engineering Research Laboratory, Cinncinati, OH. EPA Contract No. 68-03-3243. August, 1986

3. Advanced Environmental Control Technology Research Center. Research Planning Task Group Study - Thermal Destruction. EPA-600-2-84-025. Prepared for U.S. Environmental Protection Agency, Industrial Environmental Research Laboratory, Cincinnati, OH. January 1984.

4. Industrial Economics, Inc. Interim Report on Hazardous Waste Incineration Risk Analysis (Draft). U.S. EPA-WH565. Industrial Economics, Inc., Cambridge, MA. Prepared for U.S Environmental Protection Agency, Office of Solid Waste, Washington, D.C. August 1982.

5. ICF Incorporated. RCRA Risk Cost Policy Model - Phase III Report. Prepared for U.S. Environmental Protection Agency, Office of Solid Waste, Washington, D.C. 1984.

6. MITRE Corporation. A Profile of Existing Hazardous Waste Incineration Facilities and Manufacturers in the United States. PB-84-157072 MITRE Corporation, McLean, VA. Prepared for U.S. Environmental Protection Agency, Office of Solid Waste, Washington, D.C. 1984.

7. Herrel, A. GSX Services Corp. Telephone Conversation with M. Kravett, Alliance Technologies Corp. March 1987.

8. Young, C. Waste-Tech Services, Inc. Telephone Conversation with M. Kravett, Alliance Technologies Corp. March 1987.

9. Mullen, D. SCA Chemical Services. Telephone Conversation with M. Kravett, Alliance Technologies Corp. March 1987.

10. Garcia, G. TWI, Inc. Telephone Conversation with M. Kravett, Alliance Technologies Corp. March 1987.

11. Frost, D. Rollins Environmental Services, Inc. Telephone Conversation with M. Kravett, Alliance Technologies Corp. March 1987.

12. Powers, P.W. How to Dispose of Toxic Substances and Industrial Wastes. Noyes Data Corporation, Park Ridge, N.J. 1978.

13. Anguin, M.T., and S. Anderson. Acrylonitrile Plant Air Pollution Control. Acurex Corporation, Mountain View, CA. Prepared for U.S. Environmental Protection Agency, Industrial Environmental Research Laboratory, Research Triangle Park, N.C. EPA-600/2-79-048. February 1979.

14. Ottinger, R.S., et al. Recommended Methods of Reduction, Neutralization, Recovery, or Disposal of Hazardous Waste. Volumes I-16. TRW Systems Group, Redondo Beach, CA. Prepared for U.S. Environmental Protection Agency, Office of Research and Development, Washington, D.C. EPA-670/2-73-053. August 1973.

16. Considerations for System Selection

16.1 GENERAL CONSIDERATIONS

Waste management options consist of four basic alternatives: source reduction, waste exchange, recycling/reuse, use of a treatment (e.g., precipitation) or disposal processing system or some combination of these waste handling practices. Recovery, treatment, and disposal may be performed onsite in new or existing processes or through contract with a licensed offsite firm which is responsible for the final disposition of the waste. Selection of the optimal waste management alternative will ultimately be a function of regulatory compliance and economics, with additional consideration given to factors such as safety, public and employee acceptance, liability, and uncertainties in meeting cost and treatment objectives.

Many of the technologies discussed in previous sections can be utilized to achieve waste reduction or to meet land disposal ban requirements. However, practicality will limit applications to waste streams possessing specific characteristics. Since many processes yield large economies of scale, waste volume will be a primary determinant in system selection. The physical and chemical nature of the waste stream and pertinent properties of its constituents will also determine the applicability of waste treatment processes. Economical treatment will often involve waste segregation followed by chemical reduction (e.g., chromium), precipitation (e.g., other metals), and/or oxidation (e.g., cyanides) and the use of other technologies in a system designed to progressively recover/destroy hazardous constituents. Incremental costs of contaminant removal will increase rapidly as lower concentrations are attained.[1]

16.2 WASTE MANAGEMENT PROCESS SELECTION

All generators of hazardous metal/cyanide wastes will be required to undertake certain basic steps to characterize regulated waste streams and to identify potential treatment options. Treatment process selection should involve the following fundamental steps:

1. Characterize the source, flow, and physical/chemical properties of the waste.

2. Evaluate the potential for source reduction.

3. Evaluate the potential for waste exchange.

4. Evaluate the potential for reuse or sale of recycled streams and valuable waste stream constituents; e.g., recovered metals.

5. Identify potential treatment and disposal options based on technical feasibility of meeting the land disposal restrictions. Give consideration to waste stream residuals and fugitive emissions to air.

6. Determine the availability of potential options. This includes the use of offsite services, access to markets for recovered products or waste exchange, and availability of commercial equipment and existing onsite systems.

7. Estimate total system cost for various options, including costs of residual treatment and/or disposal and value of recovered products. Cost will be a function of Items 1 through 5.

8. Screen candidate management options based on preliminary cost estimates.

9. Use mathematical process modeling techniques and laboratory/pilot-scale testing as needed to determine detailed waste management system design characteristics and process performance capabilities. The latter will define product and residual properties and identify need for subsequent processing.

10. Perform process trials of recovered products and wastes available for exchange in their anticipated end use applications. Alternatively, determine marketability based on stream characteristics.

11. Generate detailed cost analysis based on modeling and performance results.

12. Perform final system selection based on relative cost and other considerations; e.g., safety, acceptance, liability, and risks associated with data uncertainties.

Key system selection steps are discussed in more detail in the remainder of this section.

16.2.1 Waste Characterization

The first step in identifying appropriate waste management alternatives to land disposal involves characterizing the origin, flow, and quality of generated wastes. An understanding of the processing or operational practices which result in generation of the waste forms the basis for evaluating waste minimization options. Waste flow characteristics include quantity and rate. Waste quantity has a direct impact on unit waste management costs due to economies of scale in processing costs and marketability of recovered products. Flow can be continuous, periodic, or incidental (e.g., spills) and can be at a relatively constant or variable rate. This will have a direct impact on storage requirements and waste management process design; e.g., continuous or batch flow.

Waste physical and chemical characteristics are generally the primary determinant of waste management process selection for significant volume wastes. Of particular concern is whether the waste is pumpable, inorganic or organic, and whether it contains recoverable materials or constituents which may interfere with processing equipment or process performance. Waste properties such as physical form, degree of corrosivity, reactivity, compatibility with other wastes and reagents, heating value, viscosity, concentrations of metal/cyanide chemical constituents, biological and chemical oxygen demand, and solids, oil, grease, total organic, and ash content need to be determined to evaluate applicability of certain waste management processes. Individual constituent properties such as solubility (affected by the presence of chelating compounds), vapor pressure, partition coefficients, reactivity, reaction products generated with various biological and chemical (e.g., neutralizing, oxidizing, and reducing) reagents, and adsorption coefficients are similarly required to assess treatability.

Chelators and complexants will enhance metal solubility, requiring over neutralization to alkaline pH to effect metal precipitation. The presence or absence of buffers will affect neutralization reagent and pH control system requirements. Cyanides and chromium will require treatment through oxidation and reduction, respectively, prior to being combined with other metal containing wastes. Finally, wastes with high concentrations of organics may require subsequent treatment (e.g., thermal destruction for sludges, biological destruction for wastewaters) before wastes can be land disposed.

Finally, variability in waste stream characteristics will necessitate overly conservative process design and additional process controls, thereby increasing costs. Marketability of recovered products or materials offered for waste exchange will also be adversely affected by variability in waste characteristics.

16.2.2 Source Reduction Potential

Source reduction potential is highly site specific, reflecting the variability of industrial waste generating processes and product requirements. Source reduction alternatives which should be investigated include raw material substitution, product reformulation, process redesign and waste segregation. The latter may result in additional handling and storage requirements, while viability of other waste reduction alternatives may be more dependant on differential processing cost and impact on product quality.

Many opportunities exist for firms to achieve waste minimization through implementation of simple, low-cost methodologies currently proven in successful programs.[2] Lack of available techniques has been less of an impediment to increased implementation than perception that these methods are not available.[2] Historically, management has favored end-of-pipe treatment and has been reluctant to institute waste reduction and reuse practices. This reluctance is primarily due to potential for process upsets or adverse impacts on product quality. Other risks of installing waste reduction methods include uncertain investment returns and production downtime required for installation. However, in the wake of increasing waste disposal and liability costs, source reduction has repeatedly proven to be cost effective, while at the same time providing for minimal adverse health and environment impact.[2] Thus, source reduction should be considered a highly desireable waste management alternative.

16.2.3 Waste Exchange Potential

As discussed in Section 5.0, metal bearing wastes have significant potential for being managed through waste exchange whereas cyanides have limited potential. Metal bearing wastes will be good candidates for exchange if: (1) metal concentrations are high; (2) contaminant concentrations are low, consistent, and at levels which are compatible with user processes; (3) processing requirements are minimal; and (4) the waste is available in sufficient volumes on a regular basis.[3] Waste rinses and solutions recovered from processes with high purity requirements may be used directly in processes with lower specifications. An offsite reuse method with high exchange potential is metal sludge recovery through thermal processing (Section 12.0). Economics are particularly favorable when these individual wastes would have required separate treatment or costly post-treatment for organic removal. Finally, waste exchange may prove to be the least cost management option for firms with wastes that have high recovery potential, but lack the waste volume or capital to make onsite recovery viable.[3]

Potential for waste exchange is reduced when industries are faced with liability or confidentiality concerns, and stringent quality requirements. Additionally, transportation costs are frequently a limiting factor in the exchange of high volume, low concentration wastes.[2]

16.2.4 Recovery Potential

As part of the waste characterization step, the presence of potentially valuable metal waste constituents should be determined. Alternatively, in the case of concentrated acid or alkali solutions, the bulk of the waste may have recycle potential. Economic benefits can result from recovery of toxic metals from these materials if the purified solution can then be either reused in onsite applications or marketed as a saleable product. In the former case, economic benefits result from decreased consumption of virgin raw materials.

This must be balanced against possible adverse effects on process equipment or product quality resulting from buildup or presence of undesirable contaminants. Market potential is limited by the lower value of available quantity or demand. Market potential will be enhanced with improved product purity, availability, quantity, and consistency.

Onsite reuse has several advantages relative to marketing for offsite use including reduced liability and more favorable economics. Offsite sale is less profitable due to transportation costs and the reduced purchase price which offsite users can typically be charg as a result of uncertainties in product quality. Thus, economics and liability combine with factors such as concerns about confidentiality to encourage onsite reuse whenever possible.

In practice, recycling of metal/cyanide wastes has been limited to recovery of metals from concentrated solutions, such as plating and etching baths, thermal recovery of highly concentrated sludges and solids, and removal from rinses through use of membrane separation and electrolytic recovery techniques. Cyanide solutions (e.g., plating baths) are sometimes recycled using metal removal processes but are not more frequently recovered due to the low purchase price of cyanides. Recycling options have been summarized in Section 5.0 and discussed in detail in Sections 6.0 through 13.0. These technologies are summarized in Table 16.2.1 with information provided on current applications, residuals generated, and availability.

16.2.5 Identifying Potential Treatment and Disposal Options

Following an assessment of the potential for source reduction and recycling, the generator should evaluate treatment systems which are technically capable of meeting the necessary degree of hazardous constituent removal or destruction. Guideline considerations for the investigation of treatment technologies are summarized in Table 16.2.2 and discussed below.

Waste characterization steps outlined previously define inputs to the treatment process. Similarly, discharge and residual disposal requirements (e.g., land disposal restrictions on leachate concentrations) define the extent to which processing is required. Thus, restrictive waste characteristics (e.g., concentration range, flow, interfering compounds) and technological limitations of candidate treatment processes will define the field of potential technologies for a specific waste.

TABLE 16.2.1. SUMMARY OF RECYCLING TECHNOLOGIES FOR METAL WASTES

Process	Applicable waste streams	Stage of development	Performance	Residuals generated	Cost
Evaporation/ distillation	Metal plating rinses; acid pickling liquors	Well-established for treating plating rinses.	Plating solution recovered for reuse in plating bath. Rinse water can be reused.	Impurities will be concentrated, therefore, crystallization/filtration system may be required.	Can be cost-effective for recovering corrosive plating solutions from rinse waters.
Crystallization	H_2SO_4 pickling liquors; HNO_3/HF pickling liquors; caustic aluminum etch solutions.	20 to 25 systems currently in operation (fewer applications for caustic recovery).	97-98% recovery for H_2SO_4 (80-85% metal removal).	Ferrous sulfate heptahydeate crystals (can be traded or sold).	Cost-effective if treating large quantities of waste.
			99% HNO_3 and 50% HF recovered.	Metal fluoride crystals (can recover additional HF by thermal decomposition).	
			80% recovery of NaOH.	Aluminum hydroxide crystals (can be traded or sold).	
Ion exchange	Plating rinses; acid pickling baths; aluminum etching solutions; H_2SO_4 anodizing solutions; rack-stripping solutions (HF/HNO_3).	Several RFIE units in operation for treatment of corrosives.	Cocurrent systems not technically feasible for direct treatment of corrosives; can be used in conjunction with neutralization technologies to lower overall costs.	Cocurrent process generates spent regenerant, which is also corrosive.	RFIE and APU are cost-effective.
		Units for direct treatment of acid bath only available from ECO-TEC, Ltd.	RFIE units show good results. Conventional RFIE performs best with dilute solutions. APU performs best with high metal concentration (30 to 100 g/L).	Recovered metals which can be reused, treated, disposed, or marketed.	
Electrodialysis	Recovery of chromic/ sulfuric acid etching solutions.	Units currently being sold, but limited area of application. 5 in operation.	85% recovery of etching solution. 45% copper removal; 30% zinc removal.	Metals which can be treated, disposed, or regenerated for reuse.	Cost-effective for specific applications (chromic/sulfate acid etchants).
	Recovery of plating rinses (particularly chromic acid rinse water).	Several in operation.	Works best when copper concentrations are in the 2 to 4 oz/gal usage.	Chromic acid can be returned to plating bath; rinse water can be reused.	Low capital investment; cost-effective for specific application (chromic acid rinses).
	Recovery of HNO_3/HF pickling liquors.	Marketed, none in operation to date.	3 M HF/HNO_3 recorded.	2 M KOH Soln which can be recycled back to the pretreatment step for this ED application.	Cost-effective for large quantity generator.

(continued)

TABLE 16.2.1 (continued)

Process	Applicable waste streams	Stage of development	Performance	Residuals generated	Cost
Reverse osmosis	Plating rinses.	Corrosive waste membranes marketed by four companies. RD module systems applicable to corrosives available from two companies.	90% conversion achieved with cyanide plating rinses.	Recovered plating solution returned to plating bath (after being concentrated by an evaporator). Rinsewater reused.	Cost-effective for limited applications. Development of a more chemically resistant membrane would make it very cost-effective for a wider area of application.
Donnan dialysis/ coupled transport	Plating rinses; potentially applicable to acid baths.	Donnan dialysis only lab-scale tested.	Data not available for Donnan dialysis (further testing required).	Data not available for Donnan dialysis.	No cost data available for Donnan dialysis.
		Coupled transport lab and field tested. Coupled transport system is currently being marketed.	Coupled transport has demonstrated 99% recovery of chromate from plating rinses. Other plating rinses should be applicable, but not fully tested.	For chromate plating rinse applications, sodium chromate is generated; can be used elsewhere in plant or subjected to ion exchange to recover chromic acid for recycle to plating solution.	Average capital cost for plating shop is $20,000. Can be cost-effective for specific applications.
Solvent extraction	HNO_3/HF pickling liquors.	Commercial-scale systems installed for development purposes in Europe and Japan. No commercial-scale installations in U.S.	95% recovery of HNO_3; 70% recovery of HF.	Metal sludge (95% iron can be recovered by thermal decomposition).	Not available.
Thermal decomposition	Acid wastes.	Well-established for recovering spent pickle liquors generated by steel industry. Pilot-scale stage for organic wastes.	99% regeneration efficiency for pickling liquors.	98-99% purity iron oxide which can be reused, traded, or marketed.	Expensive capital investment. Only cost-effective for large quantity waste acid generators.

TABLE 16.2.2. GUIDELINE CONSIDERATIONS FOR THE INVESTIGATION OF WASTE TREATMENT, RECOVERY, AND DISPOSAL TECHNOLOGIES

A. Objectives of treatment:

- Primary function (pretreatment, treatment, mutual neutralization, residuals treatment)
- Primary mechanisms (destruction, removal, conversion, separation)
- Recover waste for reuse
- Recovery of specific chemicals, group of chemicals (acids, alkalis, metals, solvents, other organics)
- Polishing for effluent discharge (NPDES, POTW)
- Immobilization or encapsulation to reduce migration (inorganic sludge)
- Overall volume reduction of waste
- Selective concentration of constituents (acids, alkalis, metals, solvents, other organics)
- Detoxification of hazardous constituents

B. Waste applicability and restrictive waste characteristics:

- Acceptable concentration range of primary and restrictive waste constituents
- Acceptable range in flow parameters
- Chemical and physical interferences (compatibility with reagents)

C. Process operation and design:

- Batch versus continuous process design
- Fixed versus mobile process design
- Equipment design and process control complexity (pH, flow, Redox potential, conductivity, temperature, pressure, level indicatiors)
- Variability in system designs and applicability
- Spatial requirements or restrictions

(continued)

TABLE 16.2.2 (continued)

- Estimated operation time (equipment down-time)
- Feed mechanisms (wastes and reagents; solids, liquids, sludges, slurries)
- Specific operating temperature, flow, and pressure
- Sensitivity to fluctuations in feed characteristics
- Residuals removal mechanisms
- Reagent selection and requirements
- Ancillary equipment requirements (tanks, pumps, piping, heat transfer equipment)
- Utility requirements (electricity, fuel and cooling, process and make-up water)

D. Reactions and theoretical considerations:

- Waste/reagent reaction (neutralization, destruction, conversion, oxidation, reduction)
- Competition or suppressive reactions (complexants, chelators, buffers)
- Enhancing conditions (specify chemicals)
- Fluid mechanics limitations (mass and heat transfer)
- Reaction kinetics (temperature and concentration effects)
- Reactions thermodynamics (endothermic/exothermic/catalytic)

E. Process efficiency:

- Anticipated overall process efficiency
- Sensitivity of process efficiency to:
  - o feed concentration fluctuations
  - o reagent concentration fluctuations
  - o process temperature fluctuations
  - o toxic constituent concentrations (biosystems)
  - o physical form of the waste
  - o other waste characteristics

(continued)

TABLE 16.2.2 (continued)

F. Emissions and residuals management:

- Extent of fugitive and process emissions and potential sources (processing equipment, storage, handling)
- Ability (and frequency) of equipment to be "enclosed"
- Availability of emissions and residuals data/risk calculations
- Products of incomplete reaction
- Relationship of process efficiency to emissions or residuals generation
- Air pollution control device requirements
- Process residuals (fugitive/residual reagents, recovered products, filter cakes, sludges, incinerator scrubber water and ash)
- Residual constituent concentrations and leachability
- Delisting potential

G. Safety considerations:

- Safety of storing and handling reactive or corrosive wastes, reagents, products and residuals
- Special materials of construction for storage and process equipment
- Frequency and need for use of personnel protection equipment
- Requirements for extensive operator training
- Hazardous emissions (e.g., HCN) of wastes or reagents
- Minimization of operator contact with wastes or reagents
- Frequency of maintenance of equipment containing hazardous materials
- High operating temperatures and pressures
- Difficult to control temperatures
- Resistance to flows or residuals buildup
- Dangerously reactive wastes/reagents
- Dangerously volatile wastes/reagents

Estimating the appropriateness of waste treatment options requires an in-depth understanding of theoretical considerations. All unit operations have inherent limitations based on technical constraints (e.g., mass transfer limitations, reaction kinetics) and economic feasibility (e.g., restricted range of temperature, pressure, and other operating conditions; limits on materials of construction). Estimation of system performance capabilities will involve a systematic analysis of several interdependant considerations: (1) expected equilibrium products for chemical, biological, thermal, or physical processes; (2) reaction kinetics; (3) heat transfer and mass transport phenomena; and (4) process control requirements.

A key consideration in the choice of chemical treatment systems for metal/cyanide wastewaters is reagent selection (Section 10.0 for metals and Section 14.0 for cyanides). Reagents may require special handling characteristics or form hazardous or difficult to manage reaction products. Potential reagents and their associated advantages and disadvantage with respect to costs, handling, processing and sludge generation, are summarized in Table 16.2.3.

Residual characteristics will have a significant impact on ultimate reagent selection since treatment or disposal of these materials constitutes a large percentage of total waste management costs. Depending on the reagent selected and original waste characteristics, sludges will have different settling, dewatering, and compactability characteristics, as well as varying tendencies for their heavy metals to resolubilize. For wastewaters, the presence of toxic organics will also significantly add to post-treatment costs. Costs will increase with organic concentration and required removal efficiency and decrease with reactivity, volatility, adsorbability and biodegradability.[4]

Ultimately, the selection of a specific treatment system from a list of technologically feasible alternatives will depend on cost, availability, and site specific factors. These considerations are discussed below.

16.2.6 Availability of Potential Management Options

The availability of each component of a waste management system may restrict its overall applicability. Existing available capacity of onsite treatment processes (e.g., wastewater treatment systems), ancillary equipment,

TABLE 16.2.3. METALS/CYANIDE TREATMENT AND CHARACTERIZATION

Reagent	Chemical name	Common form and commercial strengh	Approximate cost/ton ($)	Handling properties	Processing properties	Sludge generation
Metals Precipitation						
High Calcium Hydrated Lime	Calcium hydroxide	Powder 72-74% CaO	46	Fair-suitable for dry feeding or for slurry.	Slow-requires 30 to 45 minutes for complete reaction.	High-reaction products are insoluble. Sludge is easily dewatered.
Caustic Soda	Sodium hydroxide	Liquid 73% NaOH	205	Good-suitable for bulk feeding or from drum corrosive.	Fast-requires 15 to 30 mintues for complete reaction.	Low-most reaction products are soluble but sludge has poor handling characteristic.
Magnesia	Magnesium oxide	Slurry 55-60 Mg $(OH)_2$	365	Fair-requires slurry feed apparatus.	Slow-reaction times are longer than hydrated lime.	Low-higher solids content and density than comparable technologies.
Sodium Sulfide	Sodium sulfuret	Flakes 60-62% Na_2S	410	Good-requires rapid mix tank corrosive.	Fast-but evolves hydrogen sulfide gas.	Low-but forms poor-settling floc.
Sodium Hydrosulfide	Sodium bisulfide	Flakes 70-72% NaHS	500	Good-requires rapid mix tank corrosive.	Fast-but evolves hydrogen sulfide gas.	Low-but forms poor-settling floc.
Ferrous Sulfide	Ferrous sulfide	Generated onsite from sodium sulfide and ferrous sulfate.	610	Fair-requires slurry apparatus.	Fair-requires 30 minutes and large excess.	High-can result in four-fold increase in sludge.
High Calcium Limestone	Calcium carbonate	Powder 95% CaO	10	Poor-must be ground and slurried.	Slow-reaction times can be 60 minutes or longer.	High-insoluble end products and high concentration of inerts.
Soda Ash	Sodium carbonate	Powder 58% Na_2O	83	Good-suitable for dry or liquid feeds.	Fair-requires 45 minutes for complete reaction.	Low-less sludge generation than calcium based technologies.

(continued)

TABLE 16.2.3 (continued)

Reagent	Chemical name	Common form and commercial strengh	Approximate cost/ton ($)	Handling properties	Processing properties	Sludge generation
Metals Reduction						
Sulfur Dioxide	Sulfur dioxide	Gas 99.9% SO_2	230	Good-can be dispensed from cylinders potential explosion hazard.	Fast-requires 15 to 30 minutes for complete reaction.	Low-all endproducts are soluble.
Sodium Metabisulfite	Sodium pyrosulfite	Flake 70-72% $Na_2S_2O_5$	64	Good-suitable for slurry or bulk liquid applications.	Fast-requires 10 to 60 minutes for complete chrome reduction.	Low-all endproducts are soluble.
Ferrous Sulfate	Ferrous sulfate heptahydrate		145	Good-available in flake or solution form.	Fast-similar to other sulfur based technologies.	High-will result in sludge product 31 times higher than busulfite reduction.
Sodium Borohydride	Sodium borohydride	Powder 97% $NaBH_4$	4,000	Good-suitable for dry or liquid feeds.	Fast-requires 5 to 60 minutes depending on metals concentration.	Low-high density high metal content sludge.
Cyanide Oxidation						
Chlorine	Chlorine	Gas 99.5% Cl_2	195	Good-can be dispensed from cylinders or bulk potential explosion hazard.	Fast-requires 30 to 60 minutes but cannot process stable complexes.	Low-mainly chlorine and caustic reaction byproducts.
Sodium Hypochlorite	Sodium hypochlorite pentahydrate	Solution 29% NaOCl	304	Good-available in flake or solution form.	Fast-similar to chlorine.	Low-will not readily react with caustic.

labor, physical space, and utilities will have a significant impact on the economic viability of a treatment system. Purchased equipment must be available in sizes and processing capabilities which meet the specific needs of the facility. Offsite disposal, recovery, and treatment facilities, and companies using exchanged materials or purchasing saleable products, must be located within a reasonable distance of the waste generator. In addition, they must have available capacity for the waste type and volume generated. Finally, time constraints may eliminate certain treatment processes from consideration as a result of anticipated delays in procurement, permitting, installation, or start-up.

In general, precipitation and chemical oxidation systems are widely applied and readily available. However, several recovery systems (e.g., Devoe-Holbein extraction for metals, INCO SO_2/air process for cyanide oxidation) and post-treatment systems for organic wastes (e.g., chemical stabilization) have only recently been applied in metal/cyanide waste treatment. Availability and uncertainty in expected cost-effectiveness will play a significant role in the decision to implement these technologies.

6.2.7 Management System Cost Estimation

The relative economic viability of waste management systems will be the primary determinant of system selection for processes which are capable of achieving comparable performance. Economic viability must be evaluated on the basis of total system costs. This includes operating and capital purchase costs as well as the availability of onsite equipment, labor and utilities, net value of recovered products, and residuals disposal costs. High capital equipment expenditures and financing constraints are frequently a limiting factor in system selection, particularly for firms with marginal profitability or high debt/equity ratios, and for processes which have higher uncertainties of success.

Costs for a given management system will be highly dependent on waste physical, chemical, and flow characteristics. Thus, real costs are very site-specific and limit the usefulness of generalizations. The reader is referred to the sections on specific technologies (Sections 6.0 through 15.0) for data on costs and their variability with respect to flow and waste characteristics. Costing methodologies have also been described in the literature[5-7] and are available in software packages[8] for select processes. Major cost centers which should be considered are summarized in Table 16.2.4.

16.2.8 Modeling System Performance and Pilot-Scale Testing

Following a preliminary cost evaluation, which will enable the generator to narrow the field of candidate waste management options, steps must be taken to finalize the selection process. Initially, these could involve the application of theoretical models to predict design and operating requirements. However, models generally sacrifice accuracy for convenience and are often not sufficiently accurate to describe complex waste streams. Laboratory or pilot-scale data are often needed as model inputs and, in most cases, are ultimately required to confirm predicted performance prior to final system selection.

Nevertheless, in many cases, modeling can minimize costly laboratory testing. Models are particularly useful in assessing relative cost-effectiveness with respect to changes in process variables and the incremental costs of achieving increasingly stringent treatment concentration levels. Thus, process evaluations should begin with the formulation of a model which incorporates the conceptual process train and the primary variables which affect process performance and design. These variables can then be assigned a range of values to reflect the previously defined source conditions. The results of computer simulations or paper studies can then be used to project anticipated full-scale results and define areas for bench-scale testing.

Bench scale studies must be designed to provide maximum accuracy and to facilitate subsequent scale-up. Equipment design parameters and operating conditions must preserve geometric, kinematic, dynamic, and thermal similarity. When possible, input parameters should therefore be arranged in the form of dimensionless variables (e.g., Reynolds number). Chemical similarity should also be maintained by using representative samples from the waste generating process. Factorial experiment design and response surface methodology techniques can be applied in bench and process trials to ensure

TABLE 16.2.4. MAJOR COST CENTERS FOR WASTE MANAGEMENT ALTERNATIVES

A. Credits:

- Material/energy recovery resulting in decreased consumption of purchased raw materials
- Sales of waste products

B. Capital costs:[a]

- Processing equipment (reagent addition, reaction vessel, recovery apparatus, sludge and other residual handling equipment)
- Ancillary equipment (storage tanks, pumps, piping)
- Pollution control equipment
- Vehicles
- Buildings, land
- Site preparation, installation, start-up

C. Operating and maintenance costs:

- Overhead, operating, and maintenance labor
- Maintenance materials
- Utilities (electricity, fuel, water)
- Reagent materials
- Disposal, offsite recovery, and waste brokering fees
- Transportation
- Taxes, insurance, regulatory compliance, and administration

D. Indirect costs and benefits:

- Impacts on other facility operations; e.g., changes in product quality as a result of source reduction or use of recycled materials
- Use of processing equipment for management of other wastes

[a]Annual costs derived by using a capital recovery factor:

$$CRF = \frac{i(1+i)^n}{(1+i)^n - 1}$$

Where: i = interest rate and n = life of the investment. A CRF of 0.177 was used to prepare treatment cost estimates in this document. This corresponds to an annual interest rate of 12 percent and an equipment life of 10 years.

that optimal performance results are obtained in the most cost-effective manner.[9] Quality control procedures should be implemented to ensure consistency and accuracy of results. Finally, precautions should be taken to ensure that measurement and control equipment employed in the process evaluation is sufficiently sensitive and versatile to assess the effect of process and feed variations on overall treatment effectiveness.

The final step in the technical approach may involve design, installation, and testing of treatment systems which have been identified as the most promising candidates for specific applications. Standard chemical engineering techniques should be utilized to scale-up control equipment from bench scale results.[5,10] Integration of a treatment technology into an industrial process will require development of energy and material balances and a detailed economic analysis. Potential process variations and upsets, impact on existing operations, ease of operation and control, safety factors, and other considerations will be incorporated into the final design. These factors will be evaluated on a case by case basis taking into consideration input data uncertainties, institutional and regulatory constraints, and the probability and consequences of failure to meet control objectives.

Many suppliers of treatment and recovery equipment use models to optimize design and operations parameters and to scale-up processes. Some equipment manufacturers are also able to provide experimental equipment and models to establish process parameters and cost.

REFERENCES

1. Allen, C.C., and B. L. Blaney. Research Triangle Institute. Techniques for Treating Hazardous Waste to Remove Volatile Organics Constituents. Performed for U.S. EPA HWERL. EPA-600/2-85- 127. March 1985.

2. Committee on Institutional Considerations in Reducing the Generation of Hazardous Industrial Wastes. Environmental Studies Board, National Research Council. Reducing Hazardous Waste Generation: An Evaluation and a Call for Action. National Academy Press, Washington, D.C. 1985.

3. GCA Technology Division, Inc. Industrial Waste Management Alternatives Assessment for the State of Illinois. Volume IV: Industrial Waste Management Alternatives and Their Associated Technologies/Processes. Final Report prepared for the Illinois Environmental Protection Agency, Division of Land Pollution Control. GCA-TR-80-80-G. February 1981.

4. Breton, M. et al. Alliance Technologies Corporation. Technical Resource Document: Treatment Technologies for Solvent Containing Wastes. Prepared for U.S. EPA HWERL under Contract No. 68-03-3243. August 1986.

5. Peters, M. S., and K. D. Timmerhaus. Plant Design and Economics for Chemical Engineers. 3rd Edition. McGraw Hill Book Company, New York, NY. 1980.

6. U.S. EPA Design Manual: Dewatering Municipal Wastewater Sludges. U.S. EPA Municipal Environmental Research Laboratory, Cincinnati, OH. EPA-625/1-82-014. October 1982.

7. MITRE Corp. Manual of Practice for Wastewater Neutralization and Precipitation. EPA-600/2-81-148. August 1981.

8. Cunningham, V. L. et al. Smith, Kline & French Laboratories. Environmental Cost Analysis System. 1986.

9. Box, G.E.P., W. G. Hunter, and J. S. Hunter. Statistics for Experimenters - An Introduction to Design Data Analysis and Model Building. John Wiley and Sons, N.Y., N.Y. 1978.

10. Johnson and Thring. Pilot Plants, Models, and Scale-Up Methods in Chemical Engineering. McGraw Hill Book Company. N.Y., N.Y. 1980.